Mars

The NASA Mission Reports
Volume 2

Incorporating files from the Beagle 2

Compiled from the archives & edited
by Robert Godwin

Rock Dusting Leaves "Mickey Mouse" Mark

This image taken by the navigation camera on the Mars Exploration Rover Spirit shows the rock dubbed "Humphrey" and the circular areas on the rock that were wiped off by the rover. The rover used a brush on its rock abrasion tool to clean these spots before examining them with its miniature thermal emission spectrometer. Later, the rover drilled into the rock with its rock abrasion tool, exposing fresh rock underneath.
Image credit: NASA/JPL/Cornell/USGS (March 5, 2004)

We acknowledge the financial support of the Government of Canada through the
Book Publishing Industry Development Program for our publishing activities.
Published by Apogee Books an imprint of Collector's Guide Publishing Inc., Box 62034, Burlington, Ontario, Canada, L7R 4K2
Printed and bound in Canada
Mars - The NASA Mission Reports Volume 2
by Robert Godwin
ISBN 1-894959-05-1

Mars

The NASA Mission Reports
Volume 2

(from the archives of the National Aeronautics and Space Administration and the European Space Agency)

A Hole in Humphrey

This image taken by the navigation camera onboard the Mars Exploration Rover Spirit shows a hole drilled by the rover in the rock dubbed "Humphrey." Spirit ground into the rock with the rock abrasion tool located on its robotic arm on the 60th martian day, or sol, of its mission. Scientists are investigating the freshly exposed rock with the rover's suite of scientific instruments, also located on the rover's arm. Spirit is on its way to a large crater nicknamed "Bonneville."

Image credit: NASA/JPL (March 5, 2004)

Special thanks to:
D.C. Agle
Colin and Judith Pillinger
Steve Squyres
Rob Manning
Jim Erickson
Dave Christensen
Philip Christensen
John Hargenrader
Dan Maas (www.maasdigital.com)
Space TV (Khaleel,Natasha, Mark)
Ruth Ann Chicoine
Dr. Marc Garneau

TABLE OF CONTENTS

PLANETARY MISSIONS STUDY TASK T2A600

MARS GLOBAL SURVEYOR

MARS EXPRESS & BEAGLE 2

MARS ODYSSEY

MARS EXPLORATION ROVERS

MARS RECONNAISSANCE ORBITER

INTRODUCTION

What is it about Mars?

For generations humanity has been drawn to that pale orange light like the proverbial moth. Our ancestors revered it as a God. Using archaic tools they tracked its every move across the heavens, taking note of its apparent indecisive journey as it backtracked against the curtain of stars. Surely this celestial vacillation indicated some form of sentience at work? The majestic parade of stars was disturbed by the roving red light and it introduced the possibility of a consciousness other than our own. Directing the heavens for who knew what end.

As our species has evolved we've never really let go of our pursuit of alien intellect. Our tools have improved, as has our understanding of the universe, we look ever further afield for signs of intellectual company but we've never really given up on Martian vitality. It still summons us.

In the late 18th century Sir William Herschel studied our red neighbour. He took note of Jean-Dominique Maraldi's observations—specifically the waxing and waning of large white spots on Mars' antipodes. He concluded that they must be polar ice caps. This may seem to be an insignificant observation but it is extraordinary when you consider that no one had ever explored the poles of the Earth at this time and the implications were crucial, if there was ice, there must also be water.

It took almost a hundred years before anyone was able to piece together the true significance of Martian ice. The eminent British astronomer Richard Proctor thought that Mars was the best place to look for extraterrestrial life. In 1870 he dismissed Mercury, Venus, Jupiter and Saturn as being too hostile to support life, but Mars and (perhaps more intriguingly) the moons of Jupiter were likely candidates for vitality.

What followed was one of the great forgotten moments in Mars' storied history. Proctor began to speculate about the implications of a Martian civilisation, but for the first time, he had facts to use as a foundation. It was May 1871.

"The force of gravity is so small at the surface of Mars that a mass which on the Earth weighs a pound, would weigh on Mars but about six and a quarter ounces ... A being shaped as men are, but fourteen feet high, would be as active as a man six feet high, and many times more powerful. On such a scale then, might the Martial navvies be built. But that is not all. The soil in which they work would weigh very much less, mass for mass, than that in which our terrestrial spadesmen labour. So that, between the far greater powers of Martial beings, and the far greater lightness of the materials they would have to deal with in constructing roads, ***canals****, bridges, or the like, we may very reasonably conclude that the progress of such labours must be very much more rapid, and their scale very much more important, than in the case of our own Earth."*

Take note of the word in bold ... we'll come back to it.

Some respected scientists today are suggesting that life here on Earth may have come from Mars. It is certain that there are indeed Martian rocks already on our planet. This is not a new idea. William Thomson, President of the British Association, made a similar claim in 1872. Here is what he said, "*When two great masses come into collision in space, it is certain that a large part of each is melted; but it also seems quite certain that, in many cases, a large quantity of debris must be shot forth in all directions ... many great and small fragments, carrying seed and living plants and animals would, undoubtedly, be scattered throughout space.*"

He went on to speculate in great detail about the implications of how these errant living molecules might have come to Earth. The theory now has the odd name of *panspermia.*

Thomson was certainly no slacker when it came to science, he was later knighted by Queen Victoria, changed his name to Lord Kelvin, invented the Kelvin temperature scale (which is still used in scientific circles today) conceived and supervised the first transatlantic cable and in his spare time invented some of the laws of thermodynamics.

The late 19th century was a fertile breeding ground for remarkable new ideas about Mars; but it was the celebrated Italian astronomer Giovanni Schiaparelli who unwittingly sparked a cacophony of theories about Martian life. In 1877 the famed Italian turned his new 8.6 inch telescope toward Mars. It was fortuitous timing. The red planet was at one of its closest approaches to Earth and Schiaparelli had a very powerful new instrument and a keen eye. What he saw didn't match any of the current maps of Mars. This was in some ways a good thing since it compelled him to draw his own map. He gave everything new Latin names, including what he perceived to be a fine latticework of channels criss-crossing the Martian surface.

That's what he called them—channels. Unfortunately for Schiaparelli the Italian word for channels is *canali.*

Now combining Schiaparelli's canali with Proctor's speculations of enormously strong Martian laborers you can see where this was heading. It was only a short step and we were suddenly introduced to advanced Martian civilisations. (I suppose we should be thankful Schiaparelli didn't use a different Italian word for channels—*scanalatura*—which translates literally to *rabbets.* We can only wonder where that might have led...apparently one wrong letter makes a difference, I'm sure Energizer batteries would have been happy.)

In the 1870's such discussions were purely speculative. None of these great thinkers could have envisioned a world where robot carriages with stereo optics and microscopes would be roaming around Mars and sending stunning images back to their creators on Earth, and yet it is only 130 years later.

It is Mars' potential for life that lures us. The significance of such life is more than most of today's pundits can comprehend. Oddly enough there is already life on Mars—we took it there—hitch-hiking on the backs of our robot emissaries; there are almost certainly a few spores that survived the journey. Similar spores survived on the moon for years when they were carried there by the Surveyor probes of the 60's and the moon is a much harsher mistress than Mars will ever be. It is the possibility of indigenous life which is truly meaningful. If we can find home-grown life on Mars (and it doesn't have to be a 14 foot tall spadesman) it will change much about how we perceive ourselves. Especially if the architecture of that life is markedly different to our own. What if we find evidence of life and it has no DNA? The implications are profound. It will eliminate the naysayers arguments. It could not be written off as spores brought from Earth (either by our probes or by Kelvin's panspermia collisions) it would have to be *alien.* It would cause a similar shift in thinking as that which accompanied Copernicus and Darwin's revelations.

However, we live in a world of budgets and electoral terms. Our most brilliant minds perform miracles on an almost daily basis and yet there are still dark pockets of ignorance who can't see the facts with their own eyes. They see the astounding pictures from the Mars Rovers and Orbiters and they conclude that they must be fake. Meanwhile there are those who still wear their ignorance like a badge of honor, telling people that if we can only stop wasting our time and money on these missions it will inevitably lead to a paradise on Earth. They ignore the fact that there are no WalMarts on Mars. They neglect to mention that the money is spent right here on Earth, employing the best minds of our society to conjure up new and wonderful manifestations of their intellect. Technologies in which we are now submerged. Yes, we could do without them. Yes, we would survive without them. But if we stopped inventing them what rational person could conclude that this would suddenly make things better for those less fortunate who can't afford them? If we didn't have CCD's how might that benefit the starving children of the world? It wouldn't. We use them every day to help with crop yields. If we didn't have robotic arms how would that guarantee a job for the unemployed? It wouldn't. Some might say that robots are replacing people but they neglect to notice that someone has to build the robots. We are gradually becoming a more skilled society. Instead of working as drones doing repetitive labor we now have robots that do this while in theory we should be retraining ourselves to build better robots. Developing these technological marvels is one of the tools in our arsenal to improve life.

They help us to improve ourselves. They are also how we improve our conditions. In the same way that the invention of soap increased our longevity. Our intellect is what has allowed us to prevail against all the odds and to look out at the universe from which we sprang.

It is these moments of unadulterated genius which give many of us hope. The human minds that have taken us from Proctor's speculations to Spirit's startling images are the best amongst us. It is up to the rest of us to pierce the veil of ignorance and appreciate what they do. Take time to marvel in the pages ahead at what our kin have accomplished.

Rob Godwin
(Editor)

NATIONAL AERONAUTICS AND SPACE ADMINISTRATION

PLANETARY MISSIONS STUDY

TASK T2A600

MARS/VENUS MANNED MISSIONS (PHASE II)

Information Book

February 24, 1967

Planetary Missions Office

ADVANCED SPACECRAFT TECHNOLOGY DIVISION

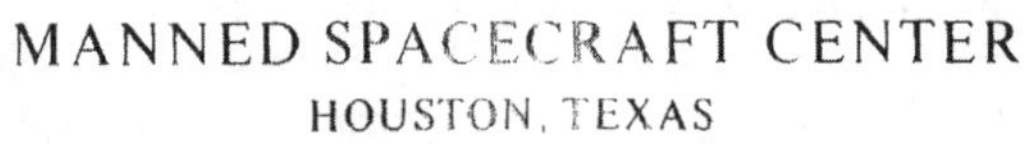

MANNED SPACECRAFT CENTER
HOUSTON, TEXAS

1.0 INTRODUCTION

1.1 Background

The Planetary Missions Office of the Advanced Spacecraft Technology Division has been assigned the responsibility for overall management and coordination of the MSC effort required to support NASA Headquarters in the 1967 Planetary Missions Study. It is planned for this study to be an extension of the study conducted in 1966, the results of which appeared in the Joint Action Group Report, "Planetary Exploration Utilizing a Manned Flight System" dated October 3, 1966. The present schedule anticipates that the detail data resulting from this study will be completed by early April, with each center furnishing NASA Headquarters supporting data for the center's part of the study by April 12, 1967. This information will be the basis for a presentation to NASA management in late April for new program consideration. It is to be noted that only the study results will be presented to Headquarters in April; however, it is intended that the MSC portion of this study be documented in a Preliminary Technical Data for Planetary Missions Report to be completed by early June.

The overall direction for the study is conducted through a Planetary Missions Joint Action Group (JAG) which is chaired by NASA Headquarters personnel and has members from KSC, MSFC, and MSC. To date, two JAG meetings have been held, one in January and one in February, and there are two remaining meetings scheduled, one at MSFC on March 22 and the last one at Headquarters on April 12, 1967.

The study is currently directed toward a 1975 & a 1979 MARS flyby and a 1977 Triple Planet Flyby and a 1978 Dual Planet Flyby. Alternate missions for a Venus flyby, MARS and Venus orbiters are being considered. It is intended that, insofar as possible, the spacecraft be designed as a Planetary Spacecraft and not just a one mission design only, so that with only relatively minor adjustments various opportunities may be considered. However, in order to provide a starting point a typical mission profile is presented in section 6 to be used as a guideline for the study.

1.2 Purpose

The purpose of this document is to serve as the instrument for the official transmission of information relative to the Planetary Missions Study between the various MSC elements involved in this project (Task T2A600).

It is planned that the overall flow of information and data generated will be generally through the Planetary Missions Office. This office will serve as the focal point for the study with the responsibility of overall management of the effort. This is not intended to hinder, in any way, the exchange of information between various elements involved in this study where this is required to accomplish an assigned task; however, it is requested that the P.M.O. be provided a copy of the information and also be appraised immediately of any resulting change in task direction that should be made.

2.0 PHILOSOPHY AND GUIDELINES

PMO Info Book Rev. B, 3-22-67 Section 2.0, page 2

RELIABILITY / MAINTAINABILITY PHILOSOPHY

1. Insofar as possible, all of the hardware subsystems shall be designed for reliable operation for the entire mission duration without maintenance and repair. It shall be the goal that actions of the crewmen will be limited to switching to redundant and/or backup elements.

2. In those cases where the above proves impossible the following will apply:

    a. Repair operations involving disassembly, complex joining operations, or use of potentially dangerous tools should not be contemplated.
    b. Detailed trouble-shooting and fault isolation techniques and equipment shall be available.
    c. All subsystems, for which in flight maintenance and repair is necessary (tube replacements, etc.), shall be accessible to crew. Subsystems and components having the highest probability of failure will be given easiest accessibility. Extravehicular inspection and maintenance shall not be contemplated. Extravehicular repair may be considered as an emergency mode.
    d. Maximum advantage should be taken of ground based assistance to crew in all phases of the maintenance and repair.

3. In-flight checkout equipment shall be included in the design. Where appropriate, maximum assistance to the crew should be provided by an integrated and semiautomatic approach to the design of this equipment.
4. For experiments which are launched to the planets, where sterilization is required, the Philosophy stated in Item 1, above, shall apply. The onboard experiments may be considered under the Philosophy covered by Items 1 and. 2, above.

MSC PMO 2-16-67

BASELINE MISSION MARS TWILIGHT FLYBY

MM (DRY)	35,000	(incl. 50% growth)
EEM	18,900	(incl. 25% growth)
MCPM	11,100	(incl. 25% growth)
EM	24,700	
EXPERIMENTS	44,800	
MM EXPENDABLES	16,800	(incl. 20% growth)
MID COURSE PROPELLANTS	13,500	
METEOROID PROTECTION	13,100	
RADIATION PROTECTION	2,100	
TOTAL S/C WT	180,000	LB.

3.0 TASK ASSIGNMENTS & ASSOCIATED INFORMATION

ENGINEERING AND DEVELOPMENT PROJECT PLAN

1. RESP ORGN	2. PROJECT CODE NO.	3. PROJECT TITLE	4. PROJECT ENGINEER (ORGN CODE, INITIALS & LAST NAME)
ET	2A600	MARS/VENUS MANNED MISSIONS(PHASE II)	ET23/C. Covington

5. PRIMARY PROJECT OBJECTIVE:

To continue the MSC contribution to the NASA in-house study of a planetary reconnaissance and surface sample retrieval mission.

6. PROJECT DESCRIPTION:	6A. TOTAL MAN DAYS	7. PLANNED START	8. PLANNED COMPLETE
	CIV.SERV. 2342 S/CONT.	2/ 4/67	6 /9 /67

This study will define a planetary program which includes manned reconnaissance and retrieval missions to Mars and Venus. MSC emphasis is upon spacecraft configuration and subsystem design, alternate missions, hyperbolic injection techniques, and flight operations, scheduling, and costing.

This Phase II culminates in a May 1967 report to Dr. Mueller and is a follow-on to the Phase I effort which was reported on in a NASA Planetary joint action Group document dated October 3, 1966.

9. PROJECT JUSTIFICATION

Definition of NASA future plans.

10. PROGRAM OFFICE (IF APPLICABLE)

11. [illegible]

FUTURE MISSIONS - PLANETARY

12. PROJECT PLAN APPROVAL

OFFICE CODE	PRINTED NAME	SIGNATURE	DATE
ET	W. E. Stoney, Jr.	[signature]	2 /17/67

MSC FORM 1171A (REV MAR 66)

13. PROJECT COSTS

ITEM	COST	ITEM	COST
(*Do not include ED in Items A-F.*) A. E&D LABOR	$105,390	G. TECHNICAL SERVICES DIVISION	
B. E&D TRAVEL	6,000	H. PHOTOGRAPHIC TECHNOLOGY LAB	200
C. SERVICE CONTRACTOR LABOR		I. COMPUTATION & ANALYSIS DIVISION	
D. SERVICE CONTRACTOR TRAVEL		J. ENGINEERING DIVISION	
E. SERVICE CONTRACTOR MAT'L		K. OTHER MSC COSTS (LIST)	
F. MATERIAL, EQUIPMENT, & SERVICES REQUIRED (LIST & ESTIMATE)		Reproduction	3000
Administrative Services	1,000		
(Graphics)			
TOTAL 13F		TOTAL (ALL COST FOR PLANNED PERIOD)	115,590

14. FUNDS AVAILABILITY CERTIFICATION

OFFICE CODE	PRINTED NAME	SIGNATURE	DATE
			/ /

15. DIVISION CONCURRENCES

DIVISION CODE	PRINTED NAME	SIGNATURE	DATE
			/ /
			/ /
			/ /
			/ /
			/ /

16. SECONDARY PROJECT OBJECTIVES (IF APPLICABLE):

16A.

E&D PROJECT PLAN

1. BRANCH SYMBOL	2. RESP ORGN		3. CODE NUMBER				4. TITLE
ET24	E	T	2	A	6	0 0	FLYBY SPACECRAFT DESIGN

5. JOB SCOPE

The results of the 1966 NASA Planetary Joint Action Group effort included an MSC configuration design for the flyby spacecraft. Since the end of that initial JAG study, studies in various areas have produced results or indications that necessitate a redesign or update of certain parts of the configuration. In view of this, areas which should be studied as having an effect upon spacecraft configuration are the following:

1. Habitability requirements - (for crew sizes of 4, 6, and 8)
2. Spacecraft orientation - (antenna, solar cells, thermal control, meteoroid protection, etc.) The magnitude of this particular study will probably be such that it will constitute a separate section of the report rather than being just mentioned with the various subsystems.
3. Experiments
4. Meteoroid shielding
5. Radiation protection
6. Abort requirements
7. Maintenance/spares/redundance/reliability - The reliability philosophy of the systems design of a long duration mission such as the Planetary Program will entail essentially that systems can be designed for long life by over-design, built in redundance and the use of major component replacement spares and a modular design. That philosophy implies that the crew will not be required to repair systems except by replacing modules or large components if it is to be advantageous relative to built-in redundance. The scope of this task is to evaluate this approach and to collect the experience of other agencies and industry towards verification of that philosophy. The obvious examples to contact for data are the Air Force and the Sandia Corporation which has faced up to similar long duration systems, both active and quiescent. The magnitude of this particular study will probably be such that it will constitute a separate section of the report.
8. Required systems/subsystems volume
9. Commonality with space station design
10. Weight

North American Aviation is currently under contract to MSFC for a study of a Mars/Venus Flyby (NAS8-18025). The results of the NAA study should be input as much as is possible in this study.

Page 1 of 2

E&D PROJECT PLAN

1. BRANCH SYMBOL	2. RESP ORGN		3. CODE NUMBER					4. TITLE
ET24	E	T	2	A	6	0	0	FLYBY SPACECRAFT DESIGN

5. JOB SCOPE Continued

It is requested that data generated by this task be available by 1 April 1967 to support the Headquarters JAG report scheduled for completion in April 1967. In addition, interim reports may be required to support the JAG meetings which are currently scheduled for 16 February, 15 March and 12 April. Further it is requested that the results of this task be documented in a final draft form by 1 May 1967 for inclusion in the Preliminary Technical Data Report for Planetary Missions.

Page 2 of 2

6. TOTAL MAN-DAYS	7. OVERTIME MAN-HOURS	8. REQUESTING DIVISION CHIEF CONCURRENCE, IF REQUIRED
300		Glenn C. Miller, Chief, Planetary Missions Off

9. CONCURRENCE	BRANCH CODE	PRINTED NAME	SIGNATURE	DATE
	ET24	Richard F. Smith		/ /

MSC FORM 1171C (MAR 66)

E&D PROJECT PLAN

1. BRANCH SYMBOL	2. RESP ORGN		3. CODE NUMBER					4. TITLE
FM8	E	T	2	A	6	0	0	ALTERNATE MISSIONS

5. JOB SCOPE

The Planetary JAG has defined four candidate missions, namely the 75 and 79 Mars Twilight Flyby and a Triple Planet Flyby in 1977 and a 1978 Dual Planet Flyby. However, it is desirable to define for consideration all possible missions in the 1970's that could be performed. At this time it is accepted that 4 Saturn V launches can be used within the KSC facility modification limits established. A fifth launch may be acceptable.

This task is to provide comparison data to allow decisions to be made as to replacement of the current JAG missions. Some of the parameters which are most important for comparison are listed below:

Number of Launches
KSC Facility Impacts
Launch Window
Earth Orbital Operations
Trip Time, Planet Passage Velocity
Earth Entry Velocity
Payload

A meeting will be held at KSC on 14 February at which a presentation of candidate replacement missions and the existing missions will be compared. These missions and S/C weights are listed on the attachment.

It is requested that the results of this task be documented in a rough draft form by 1 April 1967 for inclusion in the Preliminary Technical Data Report for Planetary Missions. In addition to reports required by February 14, interim reports may be required for future JAG meetings. The current schedule calls for JAG meetings on 15 March and 12 April.

6. TOTAL MAN-DAYS	7. OVERTIME MAN-HOURS	8. REQUESTING DIVISION CHIEF CONCURRENCE, IF REQUIRED
150		Glenn C. Miller Glenn C. Miller, Chief, Planetary Missions Office

9. CONCURRENCE	BRANCH CODE	PRINTED NAME	SIGNATURE	DATE
	FM8	Jack Funk	Jack Funk	Feb 13 67

MSC FORM 1171C (MAR 66)

MISSION	S/C WGT (LBS)	REMARKS	SPACECRAFT COMMENTS
1975 Mars Twilight Flyby	178K	P/L 44.8K Exp'mts + 24.7K Exp. Module Midcourse ΔV 1,000 ft/sec Midcourse ISP - 315 sec. 683 Day Mission	Lower Earth Entry Speed
1979 Mars Twilight Flyby	180K	P/L 44.8K Exp'mts + 24.7K Exp. Module Midcourse ΔV 1000 ft/sec Midcourse ISP - 315 sec. 692 Day Mission	BASELINE
1975 Mars Flyby with Perihelion Accel.	163K	P/L 44.8K Exp'mts + 24.7K Exp. Module Midcourse ΔV 1000 ft/sec. Midcourse ISP 315 sec. Perihelion Kick Stage-235K(Weight) 396 Day Mission (385 ISP)	Shorter Mission Avoids Asteroid Belt Higher Earth Entry Speed
1975 Mars Lightside Flyby	185K	P/L 44.8K Exp'mts + 24.7K Exp. Module Midcourse ΔV 1000 ft/sec. Midcourse ISP - 315 sec. 1,161 Day Mission	Longer Mission Avoids Asteroid Belt
1975 Mars Powered Turn	186K	P/L 44.8K Exp'mts + 24.7K Exp.Module Midcourse ΔV 1000 ft/sec. Midcourse ΔV - 315 sec. Powered Turn Stage - 209K(385 ISP) 450 Day Mission	Shorter Mission Avoids Asteroid Belt Higher Earth Entry Speed
1979 Mars Powered Turn	182K	P/L 44.8K Exp'mts + 24.7K Exp.Module Midcourse ΔV 1000 ft/sec Midcourse ISP - 315 Sec. Powered Turn Stage - 157K (385 ISP) 450 Day Mission	Shorter Mission Avoids Asteroid Belt Higher Earth Entry Speed
1972-73 Venus Flyby	80K	P/L 5.5K Exp'mts + 0 Exp'mts Module Apollo Hardware 1 Saturn-V Launch Midcourse ΔV 1000 ft/sec Midcourse ISP - 315 Sec.	Less Exp. Payload Shorter Mission Avoids Asteroid Belt Lower Earth Entry Speed 3 Men Crew
1975 Venus Lightside Flyby	160K	P/L 44.8K Exp'mts + 24.7K Exp. Module Midcourse ΔV 1000 ft/sec Midcourse ISP - 315 Sec. 367 Day Mission	Shorter Mission Avoids Asteroid Belt Lower Earth Entry Speed
1977 Triple Planet Flyby	196K	P/L 44.8K Exp'mts + 24.7K Exp. Module Midcourse ΔV - 1,500 ft/sec Powered Turn @Venus ΔV-500 ft/sec Midcourse & Turn ISP-315 689 Day Mission	Avoids Asteroid Belt Greater Midcourse ΔV Lower Earth Entry Speed
1978 Dual Planet Flyby (Free Return)	182K	P/L 44.8K Exp'mts + 24.7K Exp. Module Midcourse ΔV 1,500 ft/sec Midcourse ISP - 315 Sec 654 Day Mission	Avoids Asteroid Belt Greater Midcourse ΔV Higher Earth Entry Speed
1978 Dual Planet Flyby (Powered Turn)	200K	P/L 44.8K Exp'mts + 24.7K Exp. Module Midcourse ΔV 1,500 ft/sec Powered Turn ΔV 2,000 ft/Sec Midcourse & Turn ISP - 315 485 Day Mission	Avoids Asteroid Belt Shorter Mission Powered Turn ΔV-2000 ft/sec Higher Midcourse ΔV Higher Earth Entry Speed
1975 Venus Orbiter	168K	P/L 44.8K Exp'mts + 24.7K Exp. Module Midcourse ΔV 1,000 ft/sec Venus Retro Stage - 84K (385 ISP) Venus Injection Stage-36K (385 ISP) Midcourse ISP-315 Sec. 400 Day Mission Venus Retro & Injection 4,000 ft/sec	Avoids Asteroid Belt Shorter Mission

E&D PROJECT PLAN

1. BRANCH SYMBOL	2. RESP ORGN		3. CODE NUMBER					4. TITLE
FM8	E	T	2	A	6	0	0	ELLIPTICAL EARTH ORBIT MODE

5. JOB SCOPE

To define the various elliptical orbit modes which can be used to assemble the planetary S/C and injection stage in earth orbit. This mode is primarily effective in eliminating the large boiloff losses of liquid hydrogen in the proposed S-IVB injection stage.

The parameters which should be defined are:

1. Launch windows - both from earth and at injection for the various planetary missions.
2. Orbit parameters - ΔV, inclination altitude, period.
3. Detail mission profile from launch to injection.
4. Launch vehicle weight and performance breakdown.

It is requested that the results of this task be documented in a rough draft form by 1 April 1967 for inclusion in the Preliminary Technical Data Report for Planetary Missions. In addition, interim reports may be required to support the various JAG meetings with the first such report probably being required for the February 14, 1967 meeting. The current schedule also calls for JAG meetings on 15 March and 12 April.

6. TOTAL MAN-DAYS	7. OVERTIME MAN-HOURS	8. REQUESTING DIVISION CHIEF CONCURRENCE, IF REQUIRED
504		Glenn C. Miller (signature) Glenn C. Miller, Chief, Planetary Missions Office

9. CONCURRENCE	BRANCH CODE	PRINTED NAME	SIGNATURE	DATE
	FM8	Jack Funk	Jack Funk	2/2/67

MSC FORM 1171C (MAR 66)

E&D PROJECT PLAN

1. BRANCH SYMBOL	2. RESP ORGN		3. CODE NUMBER					4. TITLE
EC	E	T	2	A	6	0	0	PLANETARY MISSIONS CREW SIZE

5. JOB SCOPE

Conduct a study relative to establishing a rationale for crew number and general type for Mars and/or Venus missions. The broad guidelines for conducting this task were discussed with Crew Systems Division personnel on January 30, 1967, during which time it was explained that the problem should be approached from an overall human factors viewpoint and not just a work versus time standpoint. The study should include but not be limited to the following:

1. Psychological interrelationships
2. Volume/man
3. "G" requirements
4. Orientation - psychological effects?
5. Work/sleep cycle
6. Total time effects - Δ's between 450, 700, 900, 1161 days
7. Medical/environment aspects of closed ecology

It is requested that the results of this task be documented in a rough draft form by 1 April 1967 for inclusion in the Preliminary Technical Data Report for Planetary Missions.

In addition interim reports may be required to support the various JAG meetings with the first such report probably being required for the February 14, 1967, meeting. The current schedule also calls for JAG meetings on 15 March and 12 April.

6. TOTAL MAN-DAYS	7. OVERTIME MAN-HOURS	8. REQUESTING DIVISION CHIEF CONCURRENCE, IF REQUIRED
22	0	Glenn C. Miller (signature) Glenn C. Miller, Chief, Planetary Missions Offic

9. CONCURRENCE	BRANCH CODE	PRINTED NAME	SIGNATURE	DATE
	~~EC2~~ EC2	W. E. Fedderson	[signature]	/ /

MSC FORM 1171C (MAR 66)

E&D PROJECT PLAN

1. BRANCH SYMBOL	2. RESP ORGN		3. CODE NUMBER					4. TITLE
TG	E	T	2	A	6	0	0	METEOROID AND RADIATION ENVIRONMENT

5. JOB SCOPE

To develop a rationale and a procedure to be utilized in the design of meteoroid protection for interplanetary spacecraft and to present the rationale, the procedure, and the results of application of that procedure at the next Planetary JAG meeting at KSC on February 14, 1967.

Develop from the orbital parameters of the known asteroids the properties of directionality and heliocentric speed of the asteroidal flux independent of application.

Develop the mass and/or size properties of the asteroidal flux and demonstrate the validity of that development.

Develop the properties of directionality and speed relative to an arbitrarily configure spacecraft following certain interplanetary trajectories.

Develop the properties of the meteoroid protection for certain spacecraft configuration following the above mentioned interplanetary trajectories.

Develop REM vs Protection required curves for Venus missions of 380 days for both the high and low radiation cycles using both aluminum and polyethelene. Include effects of RBE.

Develop REM vs protection required curves for a Triple planet flyby(V - M - V) of 680 days duration launched in 1977 and Dual Planet Flyby launched in 78.

It is requested that the results of this task be documented in a rough draft form by 1 April 1967 for inclusion in the Preliminary Technical Data Report for Planetary Missions. In addition, interim reports may be required to support the various JAG meetings other than the February 14, 1967 meeting. The current schedule also calls for JAG meetings on 15 March and 12 April.

6. TOTAL MAN-DAYS	7. OVERTIME MAN-HOURS	8. REQUESTING DIVISION CHIEF CONCURRENCE, IF REQUIRED
60		Glenn C. Miller Glenn C. Miller, Chief, Planetary Missions Off.

9. CONCURRENCE	BRANCH CODE	PRINTED NAME	SIGNATURE	DATE
	TG	J. Modisette	[signature]	/ /

MSC FORM 1171C (MAR 66)

E&D PROJECT PLAN

1. BRANCH SYMBOL	2. RESP ORGN		3. CODE NUMBER					4. TITLE
ET23	E	T	2	A	6	0	0	ARTIFICIAL GRAVITY

5. JOB SCOPE

Determine the penalties for artificial gravity on the flyby spacecraft by use of NAA (MSFC) Flyby Study results.

It is requested that the results of this task be documented in a rough draft form by 1 April 1967 for inclusion in the Preliminary Technical Data Report for Planetary Missions. In addition, interim reports may be required to support the various JAG meetings with the first such report probably being required for the February 14, 1967, meeting. The current schedule also calls for JAG meetings on 15 March and 12 April.

6. TOTAL MAN-DAYS	7. OVERTIME MAN-HOURS	8. REQUESTING DIVISION CHIEF CONCURRENCE, IF REQUIRED
5		Glenn C Miller Glenn C. Miller, Chief, Planetary Missions Office

9. CONCURRENCE	BRANCH CODE	PRINTED NAME	SIGNATURE	DATE
	ET23	Clarke Covington	[signature]	2/12/67

MSC FORM 1171C (MAR 66)

E&D PROJECT PLAN

1. BRANCH SYMBOL	2. RESP ORGN		3. CODE NUMBER					4. TITLE
ET4	E	T	2	A	6	0	0	EARTH ENTRY ANALYSIS

5. JOB SCOPE

Conduct an investigation of the Advanced Logistics Vehicle for possible use in Earth Entry for Planetary Missions.

The following earth entry velocity extremes should be used in this study:

48,000 ft/sec.

55,000 ft/sec.

73,000 ft/sec.

The study is to include a listing of the tradeoffs and/or penalties.

It is requested that the results of this task be documented in a rough draft form by 1 April 1967 for inclusion in the Preliminary Technical Data Report for Planetary Missions. In addition, interim reports may be required to support the various JAG meetings with the first such report probably being required for the February 14, 1967, meeting. The current schedule also calls for JAG meetings on 15 March and 12 April.

6. TOTAL MAN-DAYS	7. OVERTIME MAN-HOURS	8. REQUESTING DIVISION CHIEF CONCURRENCE, IF REQUIRED
5		Glenn C. Miller, Chief, Planetary Missions Office

9. CONCURRENCE	BRANCH CODE	PRINTED NAME	SIGNATURE	DATE
	ET4	E. H. Olling		2/13/67

MSC FORM 1171C (MAR 66)

E&D PROJECT PLAN

1. BRANCH SYMBOL	2. RESP ORGN		3. CODE NUMBER					4. TITLE
FA4	E	T	2	A	6	0	0	MISSION CONTROL

5. JOB SCOPE

Define the operational philosophy for the planetary flyby mission. Consider the AAP mission control facility and determine its capability to handle the total flight schedule of AAP, Lunar and Planetary Programs. Define the mission sequence of each proposed Planetary flight and determine the MCC-S/C interface requirements during the mission. Describe the information flow between the MCC, S/C and experiments.

It is requested that the results of this task be documented in a rough draft form by 1 April 1967 for inclusion in the Preliminary Technical Data Report for Planetary Missions. In addition, interim reports may be required to support the various JAG meetings with the first such report probably being required for the February 14, 1967 meeting. The current schedule also calls for JAG meetings on 15 March and 12 April.

6. TOTAL MAN-DAYS	7. OVERTIME MAN-HOURS	8. REQUESTING DIVISION CHIEF CONCURRENCE, IF REQUIRED
180	0	Glenn C. Miller, Chief, Planetary Missions Office

9. CONCURRENCE	BRANCH CODE	PRINTED NAME	SIGNATURE	DATE
	FA4	Dennis A. Fielder		2/12/67

MSC FORM 1171C (MAR 66)

E&D PROJECT PLAN

1. BRANCH SYMBOL	2. RESP ORGN	3. CODE NUMBER	4. TITLE
EP2	E T	2 A 6 0 0	SPACECRAFT PROPULSION SYSTEM

5. JOB SCOPE

In the course of the 1966 Planetary JAG study, the S/C propulsion system for the injection from Earth orbit, Midcourse corrections, or powered turn maneuvers were defined primarily from a performance standpoint to determine S/C weights. An ISP of 310 and 385 was used for tradeoff studies and different mass fractions were used depending on the propellant loading required.

At this time the Planetary study is being reopened and it is desirable to obtain some depth to these numbers. Specifically, it is requested that the following design data be defined by February 12, 1967:

1. Recommend propellant combinations for an injection stage, a midcourse stage and a powered turn stage. The injection stage will be required to provide a ΔV of approximately 8,000 ft/sec and will be injecting a 180,000 pound S/C. The midcourse stage will provide from 1,000 to 3,500 ft/sec ΔV and will push against a 180,000 pound S/C. The powered turn stage will provide between 1500 and 12,000'/sec ΔV and again will be applied to a 180,000 pound S/C. The range of ΔV's quoted above are a result of the different requirements of the missions being analyzed.

2. Recommend an ISP for performance calculations for the above functions.

3. Recommend mass fractions of the above.

4. Recommend engine thrust levels for each.

In addition to the above design numbers it is necessary to provide an analysis and design of the engine/propellant system for the three functions outlined above. It is requested that the results of this task be documented in a rough draft form by 1 April 1967 for inclusion in the Preliminary Technical Data Report for Planetary Missions. In addition, interim reports may be required to support the various JAG meetings with the first such report probably being required for the February 14, 1967 meeting. The current schedule also calls for JAG meetings on 15 March and 12 April.

6. TOTAL MAN-DAYS	7. OVERTIME MAN-HOURS	8. REQUESTING DIVISION CHIEF CONCURRENCE, IF REQUIRED
210		Glenn C. Miller, Chief, Planetary Missions Office

9. CONCURRENCE	BRANCH CODE	PRINTED NAME	SIGNATURE	DATE
	EP2	C. W. Yodzis	Charles W. Yodzis	2/3/67

MSC FORM 1171C (MAR 66)

E&D PROJECT PLAN

1. BRANCH SYMBOL	2. RESP ORGN		3. CODE NUMBER					4. TITLE
ET35	E	T	2	A	6	0	0	PLANETARY FLYBY INJECTION ABORT REQUIREMENTS STUDY

5. JOB SCOPE

Within the design of the planetary flyby mission there is a requirement for a ΔV budget used for mid-course corrections and special thrust maneuvers unique to particular classes of trajectories. This ΔV budget requires propellant supply whose size is a function of specific impulse, for a given ΔV requirement.

This task has two major parts which are:

1. To determine what abort capability is inherent within the spacecraft using only the propellant loaded for the mid-course corrections. Complete
2. To determine what weight would have to be added to the spacecraft, over and above that required for mid-course maneuvers and power maneuvers (if applicable) in order to extend abort capability to 5 minutes after final injection stage burnout.

Both parts of this task should be done for specific impulses of 315 and 385 sec for each mission of interest.

The particular missions of interest for this task are:

1. 1975 - Mars-twilight flyby
2. 1975 - Venus lightside flyby
3. 1977 - Triple planet flyby
4. 1975 - Venus Orbiter
5. 1975 - Mars flyby with perihelion burn

It is requested that the results of this task be documented in a rough draft form by 1 April 1967 for inclusion in the Preliminary Technical Data Report for Planetary Missions. In addition, interim reports may be required to support the various JAG meetings with the first such report probably being required for the February 14, 1967 meeting. The current schedule also calls for JAG meetings on 15 March and 12 April.

6. TOTAL MAN-DAYS	7. OVERTIME MAN-HOURS	8. REQUESTING DIVISION CHIEF CONCURRENCE, IF REQUIRED
300		Glenn C. Miller, Chief, Planetary Missions Office

9. CONCURRENCE	BRANCH CODE	PRINTED NAME	SIGNATURE	DATE
	ET25	Milton A. Silveira	Milton A. Silveira	2/13/67

MSC FORM 1171C (MAR 66)

E&D PROJECT PLAN

1. BRANCH SYMBOL	2. RESP ORGN		3. CODE NUMBER					4. TITLE
EE	E	T	2	A	6	0	0	MARS/VENUS PLANETARY MISSIONS COMMUNICATIONS

5. JOB SCOPE

Conduct a study to provide technical data in the area of communications in support of the in-house Planetary Missions Study. This study is to be directed toward better defining spacecraft communications requirements based on information flow requirements. The information flow requirements are being investigated by FA4 and ET23 and will be provided as soon as available as per discussions between FA4, EE and ET23 personnel held on February 3, 1967.

The 1975 Mars Flyby Mission should be used as a baseline for communications considerations. Also, results of the North American Aviation Manned Mars and/or Venus Flyby Vehicle Systems Study (Contract NAS9-3499) should be used where feasible during the conduct of the study.

It is requested that the results of this task be documented in a rough draft form by 1 April 1967 for inclusion in the Preliminary Technical Data Report for Planetary Missions and to support the 12 April JAG meeting. In addition, an interim report may be required to support the JAG meeting currently scheduled for 15 March 1967.

6. TOTAL MAN-DAYS	7. OVERTIME MAN-HOURS	8. REQUESTING DIVISION CHIEF CONCURRENCE, IF REQUIRED
135		Glenn C. Miller, Chief, Planetary Missions Office

9. CONCURRENCE	BRANCH CODE	PRINTED NAME	SIGNATURE	DATE
	EE	R. E. Kosinski	Robert E. Kosinski	2-14-67

MSC FORM 1171C (MAR 66)

PMO Info Book-Rev.A, 5-15-67

E&D PROJECT PLAN

1. BRANCH SYMBOL	2. RESP ORGN		3. CODE NUMBER					4. TITLE
ES	E	T	2	A	6	0	0	MARS/VENUS FLYBY SPACECRAFT STRUCTURE

5. JOB SCOPE

Assess designs and provide technical consultation in the area of structural loads, stress analysis, structural materials and techniques in support of the in-house planetary mission study.

This task represents the inclusion of structural engineering in the planetary mission study and will include the following types of activities:

1. Evaluation of relative merits of various structural arrangements of the spacecraft modules (qualitative evaluation).

2. Definition of candidate structural materials and techniques and a gross evaluation of these.

3. Assisting with the structures section of the final report to be published at the end of the study. This section will include the pertinent study results; and, in particular, will be a discussion of the structural considerations necessary for the planning and design of planetary mission systems

It is requested that the results of this task be documented in a rough draft form by 1 April 1967 for inclusion in the Preliminary Technical Data Report for Planetary Missions and to support the 12 April JAG meeting.

6. TOTAL MAN-DAYS	7. OVERTIME MAN-HOURS	8. REQUESTING DIVISION CHIEF CONCURRENCE, IF REQUIRED
5		Glenn C. Miller, Chief, Planetary Missions Office

9. CONCURRENCE	BRANCH CODE	PRINTED NAME	SIGNATURE	DATE
	ES	J.N. Kotanchik	[signature]	3/10/67

MSC FORM 1171C (MAR 66)

E&D PROJECT PLAN

1. BRANCH SYMBOL	2. RESP ORGN		3. CODE NUMBER					4. TITLE
ET5	E	T	2	A	6	0	0	PLANETARY PROGRAM COST

5. JOB SCOPE

A study is currently underway at MSC relative to Mars/Venus Planetary Missions in support of the NASA Planetary Joint Action Group (JAG). This task covers the effort required to establish estimated spacecraft costs for the complete Planetary Program as covered by this study.

It is requested that data generated by this task be available by 1 April 1967 to support the Headquarters JAG report to be completed in April. Further, it is requested that the task results be documented in a final draft form by 1 May 1967 for inclusion in the Preliminary Technical Data Report for Planetary Missions.

6. TOTAL MAN-DAYS	7. OVERTIME MAN-HOURS	8. REQUESTING DIVISION CHIEF CONCURRENCE, IF REQUIRED
60		Glenn C. Miller, Chief, Planetary Missions Office

9. CONCURRENCE	BRANCH CODE	PRINTED NAME	SIGNATURE	DATE
	ET5	T. W. Briggs	[signature]	2/19/67

MSC FORM 1171C (MAR 66)

E&D PROJECT PLAN

1. BRANCH SYMBOL	2. RESP ORGN	3. CODE NUMBER	4. TITLE
EB5	E T	2 A 6 0 0	PLANETARY MISSIONS SYSTEMS CHECKOUT

5. JOB SCOPE

Conduct a study to provide technical data in the area of systems checkout in support of the in-house Planetary Missions Study. The study is to be directed toward establishing the spacecraft systems checkout requirements that should be used in designing a spacecraft for planetary missions, using the 1975 Mars Flyby as the baseline mission. The study should include considerations such as weight, volume, number and access of checkout points, crew task requirements, etc., as discussed in the meeting between Checkout Systems Branch and Planetary Missions Office personnel held on February 28, 1967.

It is requested that the results of this task be documented in rough draft form by 1 April 1967 to support the 12 April Planetary Joint Action Group Meeting. Further, it is requested that the material be documented in a form suitable for inclusion in a Preliminary Technical Data Report for Planetary Missions by the end of April.

6. TOTAL MAN-DAYS	7. OVERTIME MAN-HOURS	8. REQUESTING DIVISION CHIEF CONCURRENCE, IF REQUIRED
100		Glenn C. Miller, Chief, Planetary Missions Office

9. CONCURRENCE	BRANCH CODE	PRINTED NAME	SIGNATURE	DATE
	EB5	W. C. Bradford		3/14/67

MSC FORM 1171C (MAR 66)

PMO Info Book-Rev.A, 3-15-67

E&D PROJECT PLAN

1. BRANCH SYMBOL	2. RESP ORGN	3. CODE NUMBER	4. TITLE
ES	E T	2 A 6 0 0	MARS/VENUS FLYBY SPACECRAFT STRUCTURE

5. JOB SCOPE

Assess designs and provide technical consultation in the area of structural loads, stress analysis, structural materials and techniques in support of the in-house planetary mission study.

This task represents the inclusion of structural engineering in the planetary mission study and will include the following types of activities:

1. Evaluation of relative merits of various structural arrangements of the spacecraft modules (qualitative evaluation).

2. Definition of candidate structural materials and techniques and a gross evaluation of these.

3. Assisting with the structures section of the final report to be published at the end of the study. This section will include the pertinent study results; and, in particular, will be a discussion of the structural considerations necessary for the planning and design of planetary mission systems.

It is requested that the results of this task be documented in a rough draft form by 1 April 1967 for inclusion in the Preliminary Technical Data Report for Planetary Missions and to support the 12 April JAG meeting.

6. TOTAL MAN-DAYS	7. OVERTIME MAN-HOURS	8. REQUESTING DIVISION CHIEF CONCURRENCE, IF REQUIRED
5		Glenn C. Miller, Chief, Planetary Missions Office

9. CONCURRENCE	BRANCH CODE	PRINTED NAME	SIGNATURE	DATE
	ES	for J.N. Kotanchik		3/10/67

MSC FORM 1171C (MAR 66)

PMO Info Book-Rev.A, 3-15-67

E&D PROJECT PLAN

1. BRANCH SYMBOL	2. RESP ORGN		3. CODE NUMBER					4. TITLE
ES5	E	T	2	A	6	0	0	THERMAL ANALYSIS

5. JOB SCOPE

An in-house planetary S/C design effort has been performed by ASTD for the Planetary Joint Action Group. This group is composed of representatives from Headquarters, MSC, KSC and MSFC. It is requested that a thermal assessment be made of this design to provide guidelines on pertinent thermal problems. A sketch of the S/C design configuration is attached. The subsystems are defined in the attached report. The missions (environment) are best characterized by the following:

a. 1975 Venus Orbiter

b. 1975 Mars Twilight Flyby

c. 1979 Mars Orbiter

Spacecraft orientation effects have not been examined as yet and therefore a specific orientation has not been defined. For most purposes it is desirable to fix the S/C orientation relative to either the sun or to the earth.

It is requested that the results of this task be documented in a rough draft form by 1 April 1967 for inclusion in the Preliminary Technical Data Report for Planetary Missions and to support the 12 April JAG meeting.

6. TOTAL MAN-DAYS	7. OVERTIME MAN-HOURS	8. REQUESTING DIVISION CHIEF CONCURRENCE, IF REQUIRED
5		Glenn C. Miller, Chief, Planetary Missions Office

9. CONCURRENCE	BRANCH CODE	PRINTED NAME	SIGNATURE	DATE
	ES5	J. N. Kotanchik		/ /

MSC FORM 1171C (MAR 66)

4.0 STUDY SCHEDULES

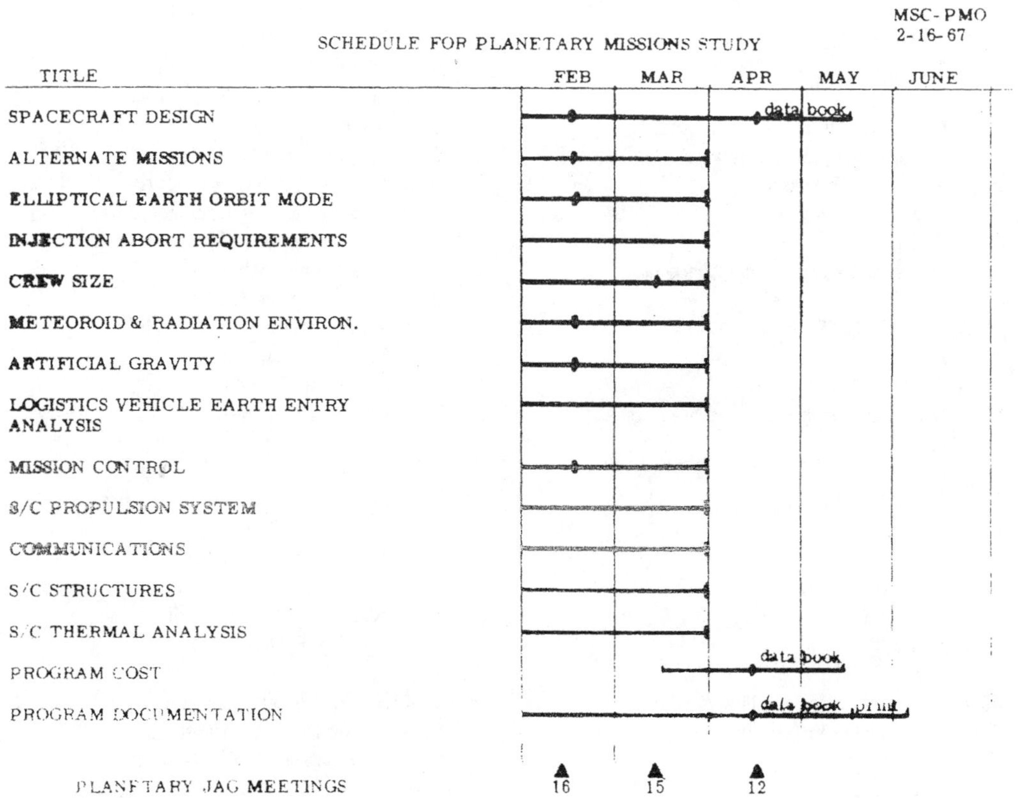

MSC-PMO
2-16-67

SCHEDULE FOR PLANETARY MISSIONS STUDY

TITLE — FEB — MAR — APR — MAY — JUNE

SPACECRAFT DESIGN (data book)
ALTERNATE MISSIONS
ELLIPTICAL EARTH ORBIT MODE
INJECTION ABORT REQUIREMENTS
CREW SIZE
METEOROID & RADIATION ENVIRON.
ARTIFICIAL GRAVITY
LOGISTICS VEHICLE EARTH ENTRY ANALYSIS
MISSION CONTROL
S/C PROPULSION SYSTEM
COMMUNICATIONS
S/C STRUCTURES
S/C THERMAL ANALYSIS
PROGRAM COST (data book)
PROGRAM DOCUMENTATION (data book print)

PLANETARY JAG MEETINGS 16 15 12

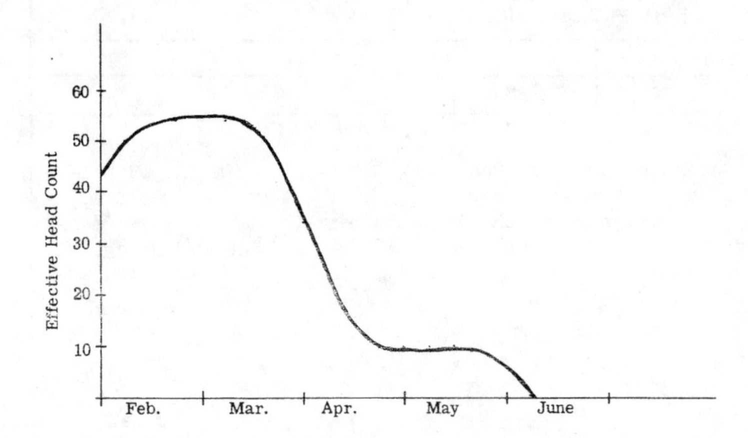

ESTIMATED MSC MANPOWER LEVEL

5.0 PRELIMINARY TECHNICAL DATA BOOK OUTLINE

PMO Info Book - Rev. A, 3-15-67

SUMMARY, VOLUME I

A. Jag Guidelines & Constraints
   1. Design
   2. Missions
   3. Experiments
B. Baseline Design
   1. Mission
   2. Experiments
   3. Configuration
   4. Weight Breakdown
   5. Subsystems
C. Design Alternatives
   1. Experiments
   2. Weights
   3. Subsystem Expendables

ENVIRONMENT & CRITERIA, VOLUME II

A. Ground Rules & Constraints
B. Meteoroid Environment
C. Radiation Environment
D. Thermal Environment
E. Spacecraft Environment Criteria.

SUBSYSTEMS, VOLUME III

A. Power
B. ECS
C. Guidance, Navigation, and Stabilization
D. Electronics
E. Crew Systems
F. Propulsion Systems
G. Checkout System
H. Data Management
I. Earth Entry Vehicle
J. Display arid Controls
K. Reliability

CONFIGURATION & WEIGHTS, VOLUME IV

A. Complete Spacecraft Configuration
   1. Two 15' cans
   2. Two 22' cans
   3. One 22' can
B. Mission Module Layouts
C. Mid-course Propulsion Module Layouts
D. Experiment Module Layout
E. Probe Configurations
F. Earth Entry Module
G. Launch Configuration
H. Launch Vehicle Description
I. Weight Analysis

MISSIONS AND MISSION OPERATIONS, VOLUME V

A. Planetary Operations Philosophy
   1. Flight Control Concept
   2. Recovery Philosophy
   3. Mission Design Procedures
   4. Real time Data Management
   5. Abort Criteria
   6. Mission Rules Criteria
B. Data Management Requirements
   1. Data Processing Requirements
   2. Ground Support System Requirements
C. Integrated Launch Schedule
   1. Total Schedule Analysis
   2. Impact on MCC-H / MSFN
D. Flight Sequence Planning
   1. Ascent to Orbit
   2. Orbital Operations
   3. Orbital Launch
   4. Transplanetary / Transearth
   5. Planetary Encounter / Capture
   6. Entry / Landing
E. Mission Performance Analysis
   1. 1975 Manned Mars Flyby
   2. Alternate Mission Studies
   3. Elliptical Parking Orbit

EXPERIMENTS, VOLUME VI

A. On-Board Experiments
B. Planetary Probes

6.0 MISSIONS AND CONFIGURATION

Rev. D-PMO, 5/19/67 Section 6.0

*1977 Triple Planet Flyby (Revised)
1979 Mars Twilight Flyby (New)

* This supersedes configuration already in Info. Book.

MISSION	Req'd S/C Weight, Lb/1000	Injection Capability From Elliptical Orbit, Lb/1000	Number Launched Req'd (10% Uprated Saturn)	Maximum Earth Entry Velocity, FPS/1000	Mars Periapsis Velocity, FPS/1000	Trip Time - Days	Maximum A.U.	Minimum A.U.
1973 Venus Flyby	160	160	2	45	38	380	1.21	0.72
*1975 Venus Flyby	160	175	2	44	37	385	1.21	0.72
*1975 Venus Orbit	160	163	5	44	37	400	1.21	0.72
1975 Mars Twilight	178	179	3	49	33	690	2.25	1.00
*1975 Mars Minimum Energy	185	199	3	45	21	1180	1.71	1.00
1975 Mars Perihelion Acc.	163	163	6	52	26	396	1.72	0.64
1975 Mars Power Turn	186							
1975 Mars Orbit								
*1977 Venus Orbit	163	163	5	44	37	400	1.21	0.72
*1977 Triple Planet Flyby	196	205	4	45	29	680	1.72	0.72
1977 Mars Orbit								
*1978 Dual Planet Flyby	182	210	4	48	25	650	1.72	0.54
1979 Venus Orbit	163	163	5	44	37	400	1.21	0.72
1979 Mars Twilight	180	185	3	48	44	680	2.20	1.00
1979 Mars Power Turn	182	182	5	69	24	475	1.72	0.60
*1979 Mars Orbit								

TYPICAL MARS / VENUS FLYBY FLIGHT OPPORTUNITIES

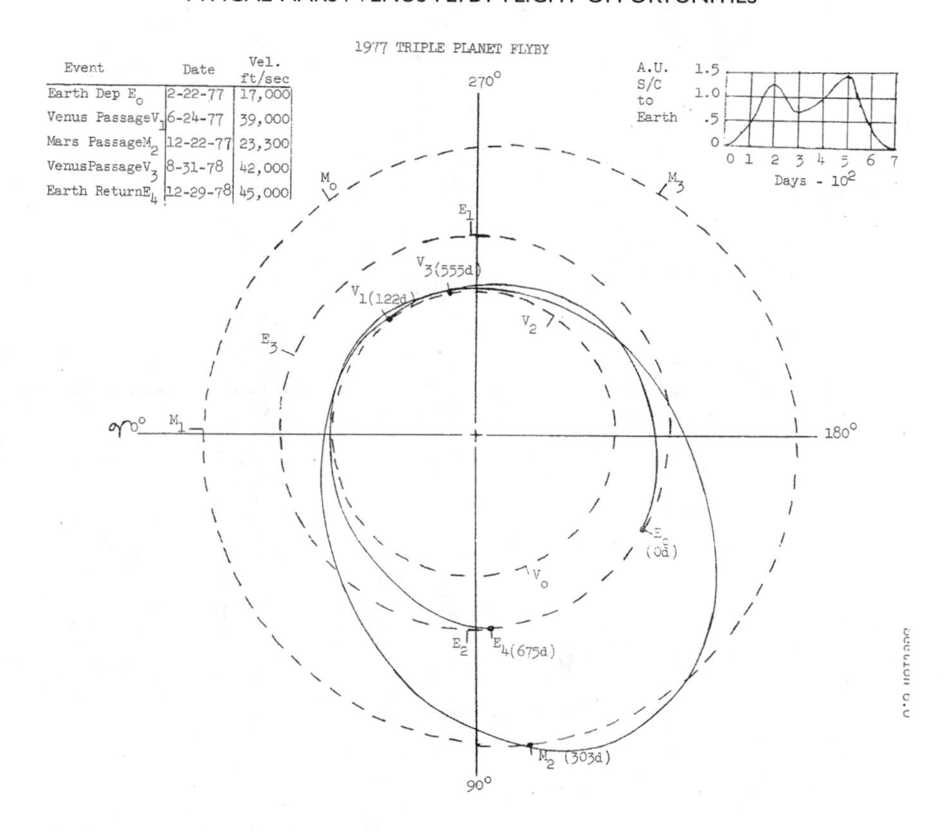

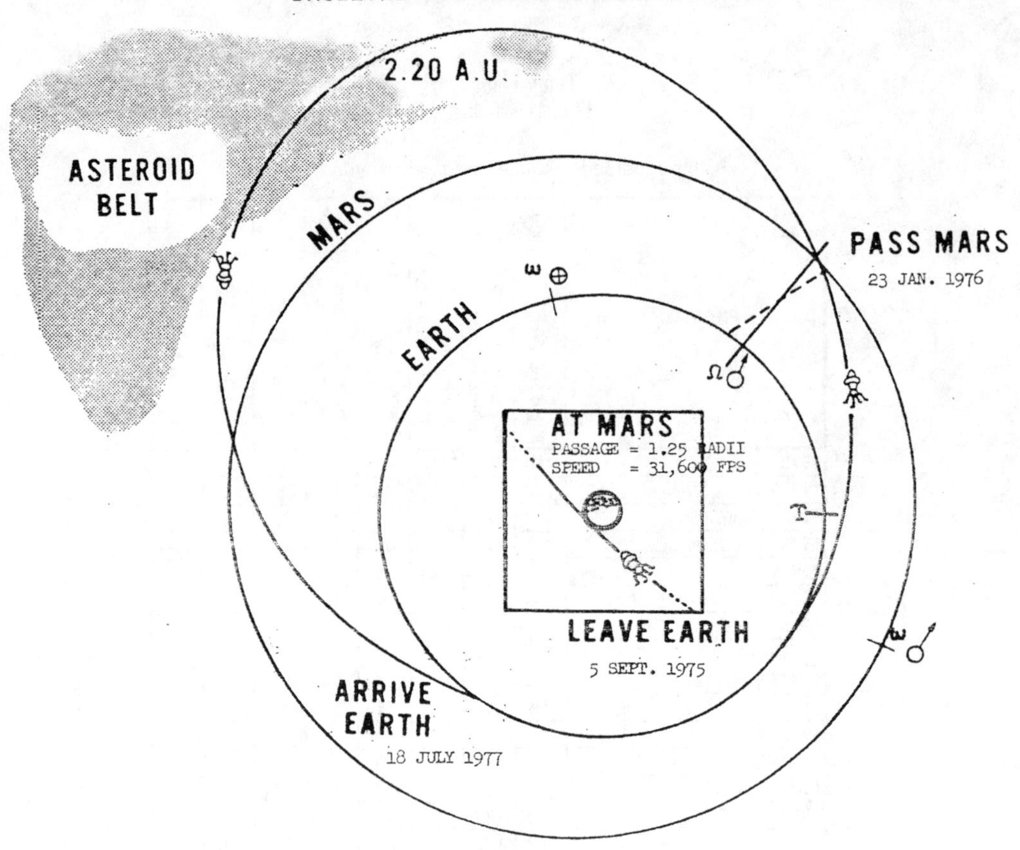

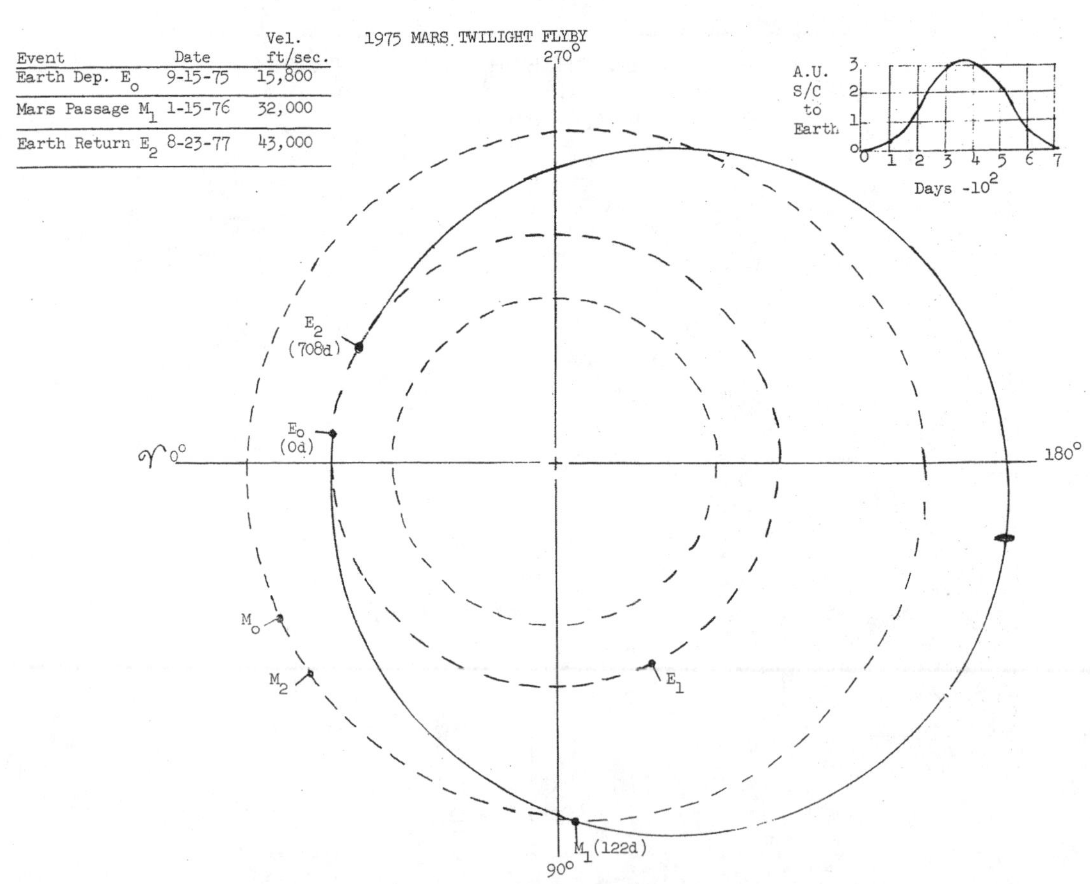

Event	Date	Vel. ft/sec.
Earth Dep. E_o	9-15-75	15,800
Mars Passage M_1	1-15-76	32,000
Earth Return E_2	8-23-77	43,000

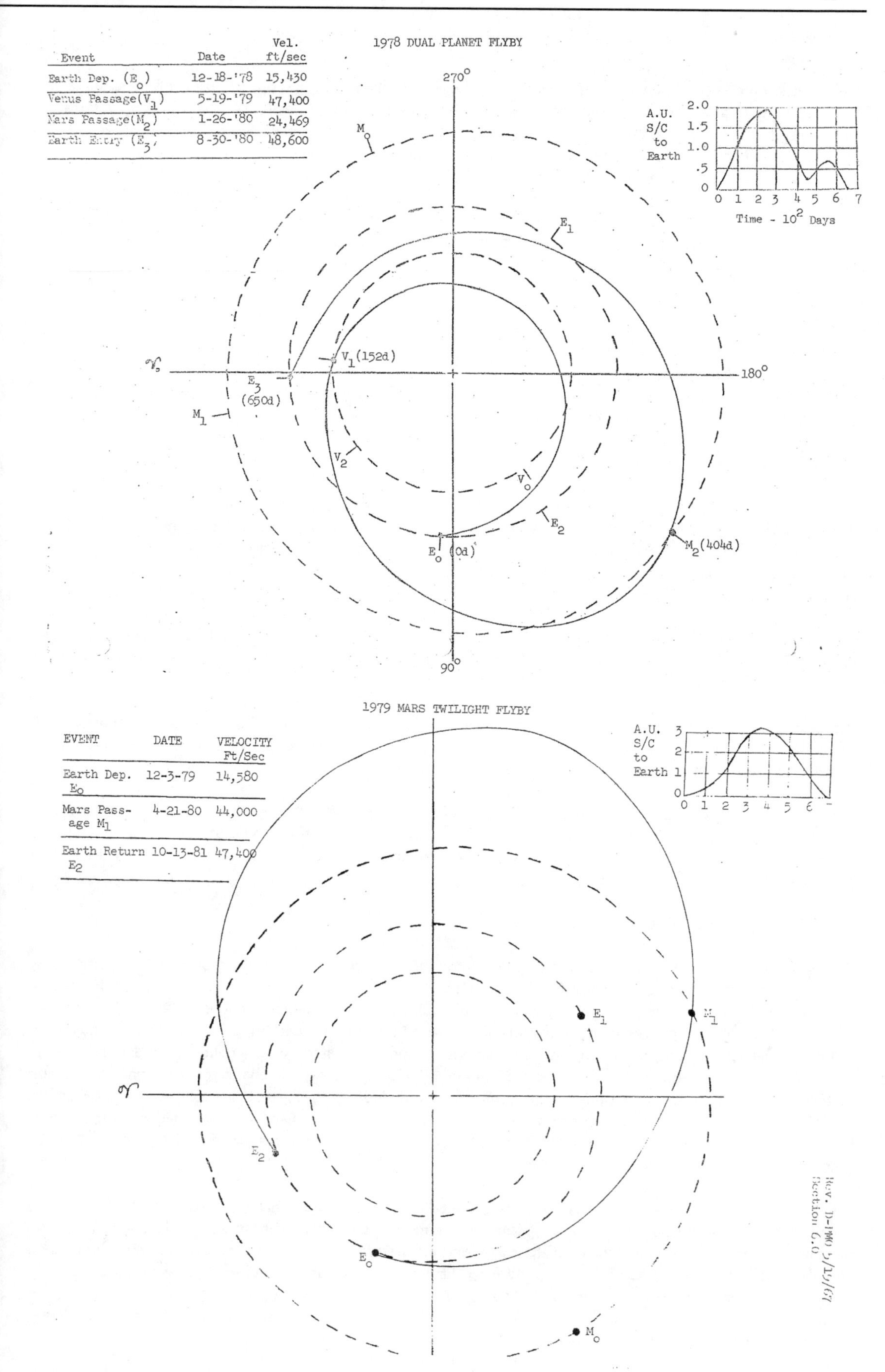
1978 DUAL PLANET FLYBY
Event | Date | Vel. ft/sec
Earth Dep. (E_o) 12-18-'78 15,430
Venus Passage(V_1) 5-19-'79 47,400
Mars Passage(M_2) 1-26-'80 24,469
Earth Entry (E_3) 8-30-'80 48,600
270°
180°
90°
V_1 (152d)
E_3 (650d)
M_2 (404d)
E_o (0d)
M_o
E_1
E_2
M_1
V_2
V_o
A.U. S/C to Earth
Time - 10^2 Days
1979 MARS TWILIGHT FLYBY
EVENT | DATE | VELOCITY Ft/Sec
Earth Dep. E_o 12-3-79 14,580
Mars Passage M_1 4-21-80 44,000
Earth Return E_2 10-13-81 47,400
A.U. S/C to Earth
E_1
M_1
E_2
E_o
M_o

6.1 ORBITAL OPERATIONS

For the first time in the manned space program, a complicated series of orbital operations involving multi-vehicle assembly, crew transfer, and integrated checkout will be required for the planetary mission. These operations are largely dictated by the limitations and capabilities of the launch vehicles used and their propulsive payloads, which become the injection stages for the planetary injection maneuver. Injection window performance penalties and lifetime limitations of the injection stages make the timing of orbital operations especially critical, both in terms of onboard crew activities and ground control support. The orbital assembly operations are further complicated by the requirement for remote docking of multiple propulsion stages, with their associated hazards. Ground-based data acquisition, tracking, and flight control monitoring will, therefore, assume new dimensions with regard to enhancement of mission success, whether or not a back-up launch concept is adopted.

6.1.1 Flight Control Support

The manned vehicle launch operation flight control will be similar to that of Apollo. However, the MOCR flight controllers will have only one or two 15-second launch opportunities per day for each of their three unmanned vehicles. All three should be launched within 48 hours or less. Holds, scrubs and rescheduling of a newly developed state will be severely constrained by this demanding schedule.

6.1.1.1 Orbital Control Operations

Since the initial task of the flight will be to check out the orbiting planetary spacecraft (MM, EM, and EEM), the SOCR flight controllers will manage the checkout for the 15 days prior to the first unmanned flight. After a "Go" decision is made based on the readiness of the spacecraft, the MOCR flight controllers will launch, insert, and rendezvous the unmanned vehicles sequentially. A 100-nautical-mile altitude parking orbit and a final 262-nautical-mile-altitude circular assembly orbit will be used.

The unmanned vehicles will be placed within 10 miles of the orbiting spacecraft by the MOCR flight controllers. After receiving a "stage is safe" Go decision from the MOCR Flight, Director, the flight crew will complete the rendezvous and docking maneuvers.

After docking, the crew and the SOCR flight controllers will check out the MM / fueled stage configuration, with the SOCR Flight Director making the Go / No Go decision for the next injection stage launch. The MOCR flight controllers will continue to support S-IVC independent data links.

When all three S-IVC stages are docked, the flight crew and the ground will cooperate in a complete integrated vehicle checkout. A Go decision is required from the SOCR Flight Director prior to preparation for transplanetary injection (TPI). Any problems encountered must be resolved prior to the expiration of the lifetime of the initially orbited injection stage.

6.1.1.2 Information Flow Analysis

During the first 15 days after the insertion of' the Mission Module into Earth orbit, the ground will direct the crew in verifying the readiness of the spacecraft and scientific payloads for the mission. A data flow summary for a typical day during the orbital assembly period is given in Table 1. Duplex voice communications to the MM is required during the Earth orbital period for coordination of EEM and MM systems checkout and repair. An updata system will be required during this period for spacecraft computer updates, limited systems control, and data relay to an onboard teleprinter. Telemetry data from the EEM and MM (including the experiment compartment and propulsion module), is required for ground support of the systems checkout. Although a single carrier uplink and a single carrier downlink could be used to convey all necessary information by using the unified carrier technique, a requirement for simultaneous telemetry from the EEM and MM would necessitate an additional downlink, which could be satisfied through the USB system on the CSM.

Following the 15 day launch pad turnaround period, the S-IVC's will next be launched in sequence and will impose additional burdens on the crew and on the ground flight controllers. Communications and tracking for each S-IVC must be maintained for ground control of the trajectory and systems to effect a safe rendezvous with the Mission Module. Updata to each S-IVC will be required for guidance computer update and may be required for systems control. Following rendezvous, systems status data will continue to be transmitted to the ground on a nearly continuous basis from each of the three S-IV's. Separate communications systems using different transmission frequencies will be required on each S-IVC.

As indicated in Table 1, a daily coverage of seven hours of contact time will be necessary for normal ground support of orbital checkout. In addition, allowances must be made to support a non-perfect orbital checkout and countdown which will require information flow related to a malfunction analysis and solution activity.

When combined with the additional coverage required for orbital assembly, a nearly continuous coverage requirement becomes evident for several days during the Earth orbital period.

The transplanetary injection support period begins two hours prior to the start of the injection burn and continues through to several minutes past cutoff. Systems data from the spacecraft and S-IVC's must be transmitted to the ground continuously during this period for Go / No Go decisions regarding the crew and systems prior to, during, and immediately after injection. Tracking data is required during this period for verification of spacecraft guidance system operation and injection trajectory.

A large percentage of the daily information transmitted to the MCC-H will be scientific data resulting from interplanetary space measurements.

Ground Support Tasks & Responsibilities	**Days : Hours : Minutes Before Injection**		**Data Rate** (BPS)		**Data Source / In Support of**
	FROM	**TO**	**UP**	**DOWN**	
Ground Support of EEM Systems Checkout	9:10:00	9:9:30	—	51.2 KBPS	EEM Systems
Voice Coordination in Support of Checkout	9:10:00	9:9:30	4K	4K	EEM Systems
Ground Orbit Computation on Spacecraft G&N Parameters	9:9:00	9:8:56	—	10K	G&N Systems
Spacecraft Computer Update	9:8:30	9:8:29	1,000	—	G&N Systems
Ground Support of MM Systems Checkout	9:6:00	9:3:00	—	90K	MM Systems
Voice Coordination in Support of Checkout	9:6:00	9:3:00	4K	4K	MM Systems
Tracking (PRN)	9:2:30	9:2:25	1 MEG	1 MEG	Ephemeris Determination
MM Systems Control	9:2:00	9:1:55	1,000	—	Systems Operation
Data Relay to Spacecraft Teleprinter for Checkout	9:1:55	9:1:45	1,000	—	MM Systems Checkout
Ground Support of MM Systems Checkout	8:20:00	8:17:30	—	90K	MM Systems
Voice Coordination in Support of Checkout	8:20:00	8:17:30	4K	4K	MM Systems
Data Relay to Spacecraft Teleprinter for Checkout	8:19:20	8:19:15	1,000	—	MM Systems
Crew Activities Scheduling - Voice	8:16:00	8:15:55	4K	4K	Crew
Ground Support of EEM Systems Checkout	8:15:30	8:15:00	—	51.2K	EEM Systems
Relay Repair Information to Spacecraft Teleprinter for Fix	8:15:20	8:15:00	1,000	—	EEM Systems
Voice Coordination of Repair & Checkout	8:15:20	8:15:00	4K	4K	EEM Systems
Ground Orbit Computation on Spacecraft G&N Parameters	8:12:00	8:11:55	—	10K	Ephemeris Determination
Tracking (PRN)	8:11:30	8:11:25	1 MEG	1 MEG	Ephemeris Determination

TABLE 1
DATA FLOW SUMMARY - EARTH ORBITAL CHECKOUT (TYPICAL DAY)

6.1.2 ORBITAL LAUNCH VEHICLE ASSEMBLY

For the baseline mode of operation, the orbital launch vehicle (OLV) to be assembled in a circular orbit at an altitude of 260 nautical miles. Four Saturn V launches will be required to meet the payload and mission velocity requirements using the S-IVB with minimum modification (J-2S engine plus 110-hour lifetime). The operations schedule for these four launches is shown in Figure 1.

The first launch is planned at T-24 days (where T-0 is interplanetary injection) and would consist of the planetary spacecraft and an Apollo Command and Service Module (CSM). The S/C launch at T-24 days allows for 14 days to turn the pad around, plus up to a 6-day hold and then the 108 hours for the launch of three Orbital Launch Vehicles. The CSM is used for crew transport and provides power in Earth orbit. The Saturn V would insert the spacecraft into a 100- by 260-nautical-mile elliptical orbit and coast to apogee. At apogee the spacecraft is inserted into the 260-nautical-mile circular orbit using the idle mode of the J-2S engine on the modified S-IVB. After insertion into the 260-nautical-mile circular orbit, the CSM is docked to the spacecraft for crew transport through the airlock.

The first OLV stage is launched at T-108 hours into an 80- by 250-nautical-mile phasing orbit. The flight plan for rendezvous is shown in Figure 2. Several orbital periods before the coelliptic maneuver is to be performed, a plane change maneuver is conducted so that the OLV is coplanar with the spacecraft. The perigee altitude would also be adjusted at apogee to phase properly the coelliptic maneuver. Up to 26 hours of coast may be required for the rendezvous if the spacecraft arrives with random phasing. In order to reduce the time required for rendezvous, a midcourse correction to adjust phasing is made by the spacecraft 2 days before the first OLV stage is launched. The rendezvous maneuver for the first OLV stage will, therefore, require less than 12 hours of coast phasing in the parking orbit, It is desirable that the first docking be completed before the second OLV stage is launched; however, this feature is not shown as a requirement for the baseline.

The terminal phase of the rendezvous is initiated about 1 hour after the coelliptical maneuver by applying a 20-fps velocity impulse at a pitch attitude of 26.6 degrees, which places the OLV on an intercept with a travel angle of 140 degrees. The braking maneuver requires a ΔV of 23 fps and occurs about one half hour later. Figure 3 shows the relative positions or the OLV with respect to the spacecraft during the coelliptic and terminal phases. The braking phase of the OLV is controlled from the spacecraft using the spacecraft radar. The OLV spacecraft relative velocity is reduced to zero at a range of 500 feet. Docking begins at this range.

Each OLV stage docks to the spacecraft or OLV spacecraft stack and is controlled remotely by a digital command link from the spacecraft to the OLV. Optical equipment on the spacecraft is used to align the spacecraft with the two targets on the OLV stage, thus assuring that the two vehicles have colinear axes as in Figure 4. Forward and braking docking velocity is applied with 1750-lb. ascent propulsion system (APS) modules. An attitude hold mode on each OLV stage insures translation only along the longitudinal axis. The OLV stage may have to be stopped and realigned at a distance of 10 to 15 feet from the stack for the final docking in order to insure accurate contact. During docking of the OLV, the CSM is backed away from the spacecraft by one crewman in order to prevent stressing the CSM docking port.

The velocity requirements for each maneuver from launch injection to docking are listed below as follows:

Change in phasing orbit + coelliptic maneuver	300 fps
Plane change	150 fps
Terminal phase initiation	20 fps
Braking theoretical	24 fps
Braking additional	48 fps
Station keeping and docking	30 fps

Transplanetary injection uses a two burn maneuver with an elliptical coast orbit between the burn. This first burn uses OLV stage III and about one fourth of OLV stage II. The resulting elliptical orbit has a period of about 4 hours with an apogee altitude of about 10,000 miles. The two-burn maneuver reduces the gravity loss from a ΔV of about 1,700 fps to one of about 450 fps, which results in a significant payload gain over injecting directly from the circular orbit. A three-burn maneuver with an elliptical coast between each stage would result in an additional reduction in gravity loss. However, a lengthened 18-hour coast would be required between OLV stages II and I.

Two launch windows present primary mission constraints on the baseline mission. The first launch window is from the launch pad to rendezvous in the assembly orbit, and the second launch window is from the assembly orbit to the interplanetary trajectory. The launch window from the pad to the assembly orbit is shown in Figure 5 for three inclinations of the assembly orbit. It is apparent that for high inclination of the assembly orbit, either large ΔV penalties must be paid or relatively short launch windows used. Several short launch windows spaced over a period of days are considered better than one large launch window for a single day. A ΔV budget of 150 fps for the launch window will provide a period of 10 minutes or greater during this launch period. The inclination of the assembly orbit cannot necessarily be varied at will to increase the launch window. The inclination of the assembly orbit must be greater than the latitude (declination) of the Earth-centered velocity vector required for the interplanetary trajectory, and this declination may be as large as 33.7 degrees for some Mars missions. Referring to Figure 1 then, there are 8 launch opportunities of 10 minutes each over a period. of three days to launch 3 OLV stages if the booster lifetime is limited to 110 hours. This will allow for a hold of any one booster for a 48 hour turn around. It will also allow for a 48 hour hold of OLV-1 and the same for OLV-2 if OLV-3 is launched on one of the two opportunities at T-84 hours. The same can be said for OLV-1 and OLV-3 if OLV-2 is launched on time.

The characteristics of the second launch window from circular orbit to heliocentric are shown in Figure 6 for several of the baseline missions. This launch window consists of two parts. The dashed line is the velocity for injection out of the assembly orbit to the heliocentric trajectory if the assembly orbit is in the plane of the heliocentric orbit. The solid line shows the effect of the Earth oblateness rotation of the assembly orbit. The inplane point of tangency between the dashed and solid lines can be controlled initially when launching the first vehicle, which in the baseline mission is the spacecraft. After the first launch, the point of minimum ΔV is set. The launch window as shown is for injection directly from circular orbit. At both extremes of the launch window it may be possible to improve the velocity requirements by making a plane change at apogee of the elliptical parking orbit. Near the center of the launch window, however, a plane change maneuver at apogee may appreciably increase the velocity requirements. The option of making a plane change maneuver at apogee of the elliptical injection orbit must, therefore, be required for the baseline approach.

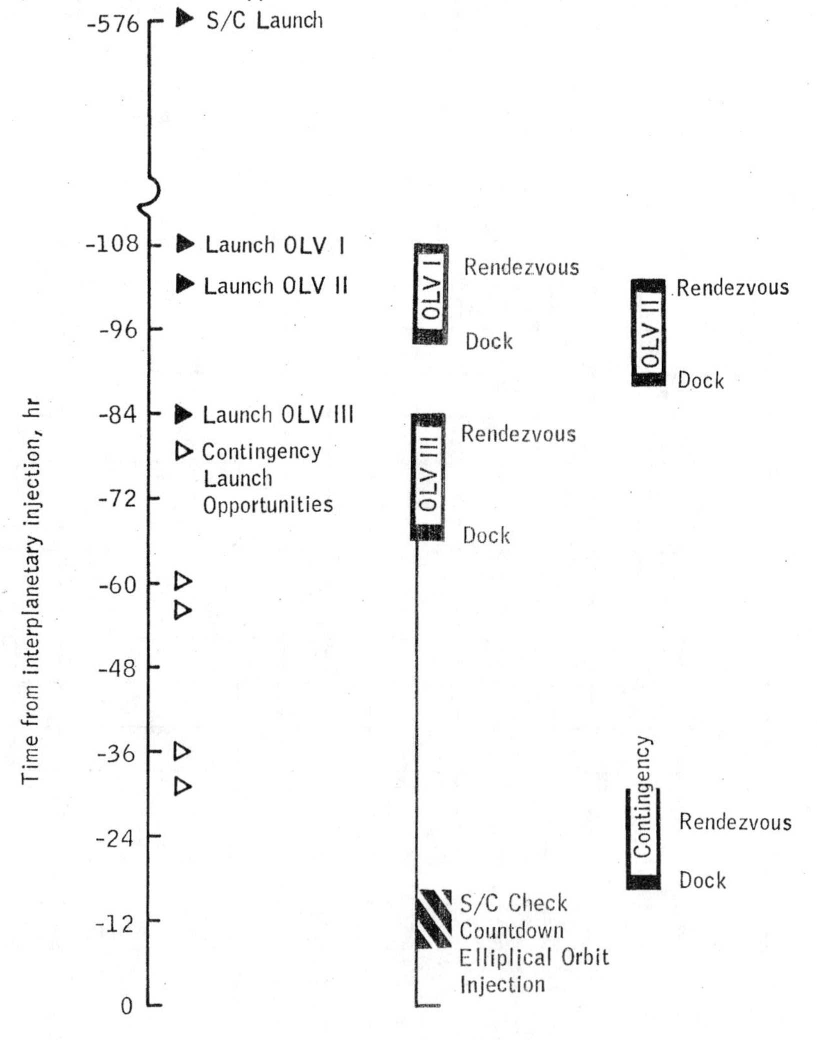

Figure 1. OLV launch sequence of events.

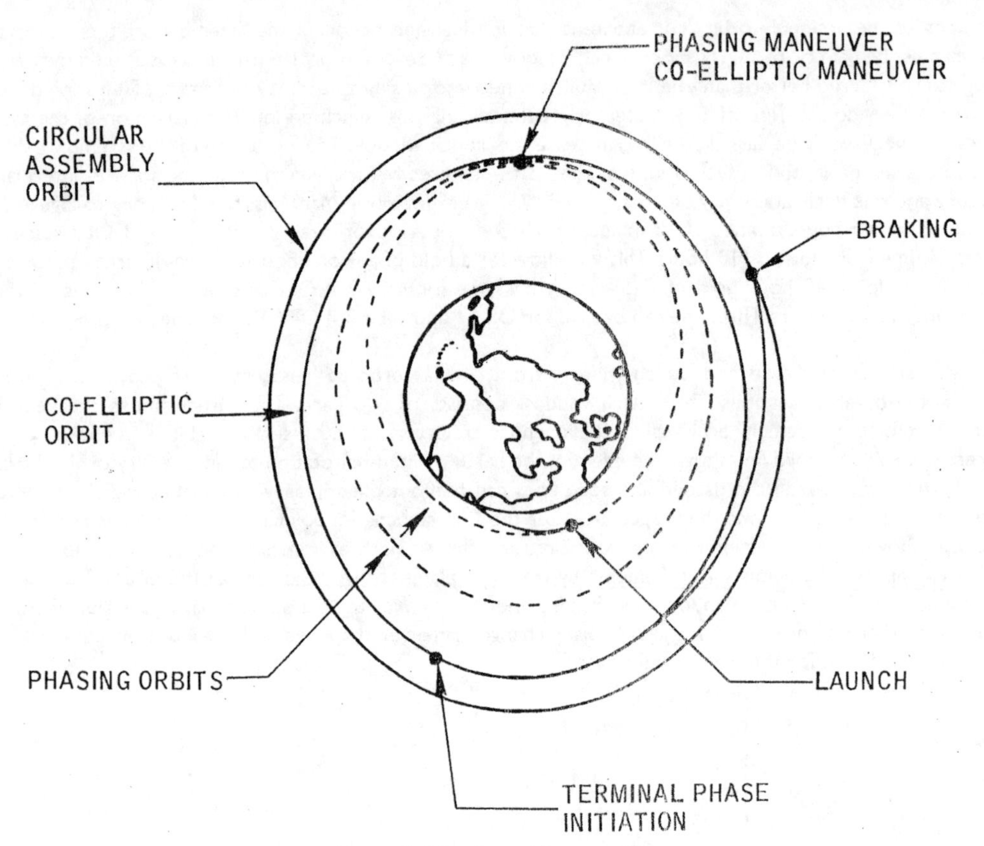

Figure 2. Circular orbit assembly.

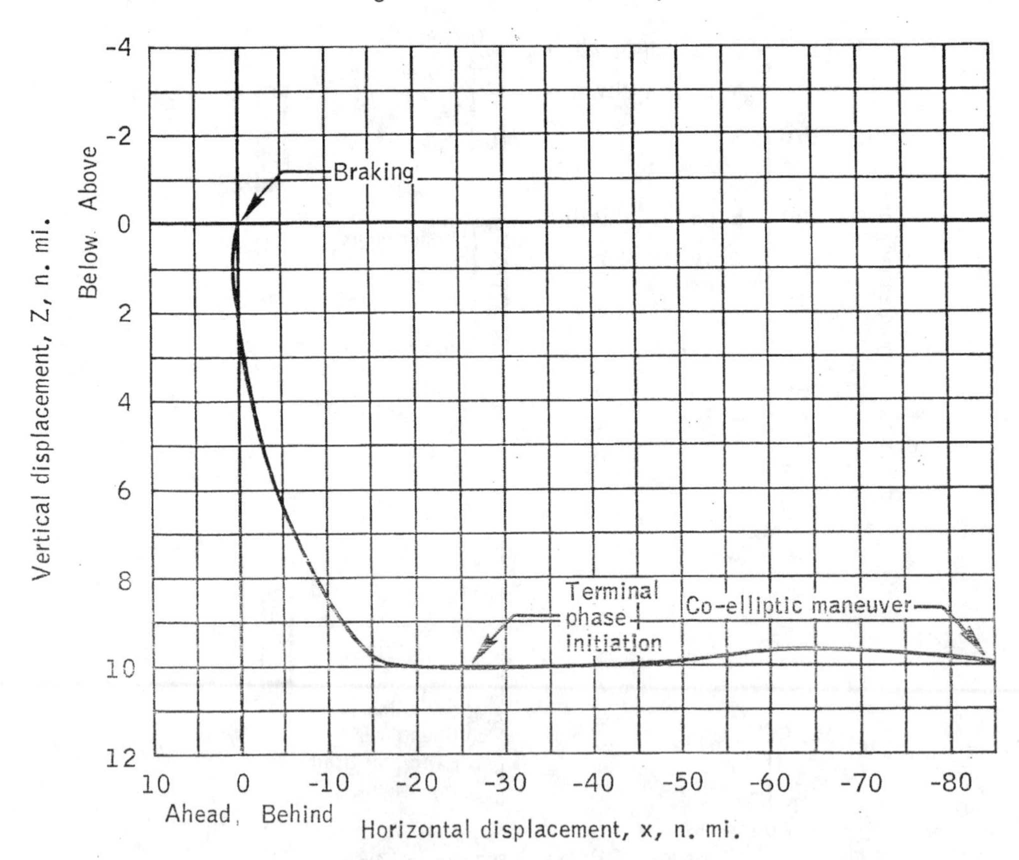

Figure 3. Relative motion of the OLV in the S/C curvilinear coordinate system from co-elliptic maneuver to braking.

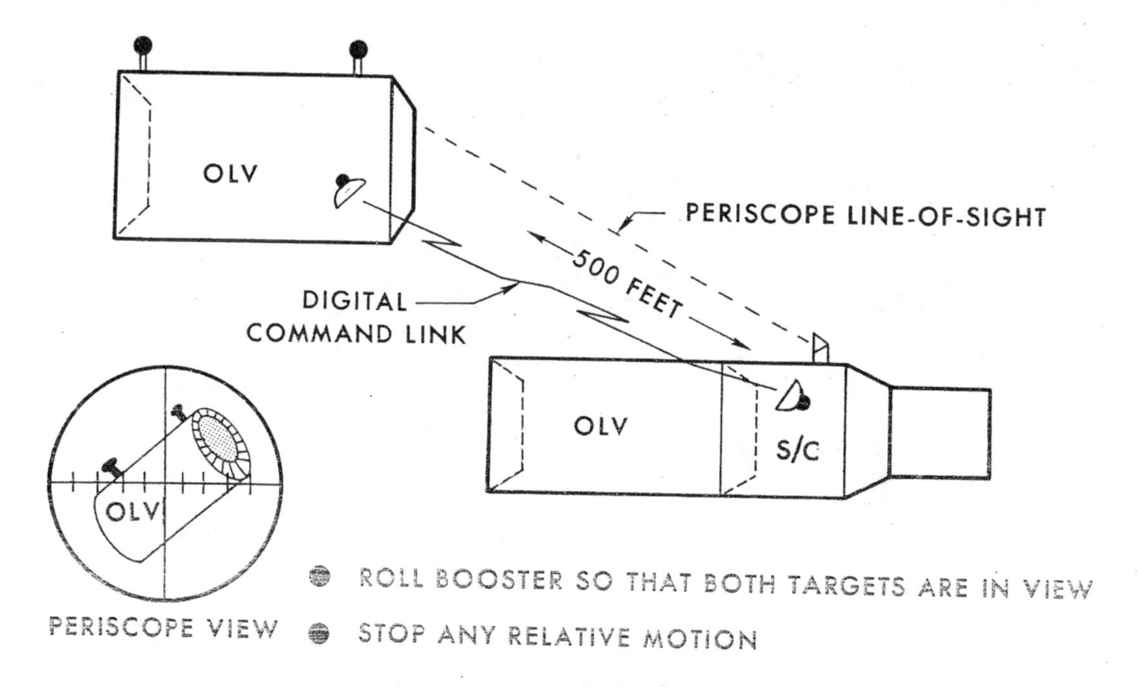

Figure 4. Orbital assembly for interplanetary missions.
(a) Acquisition

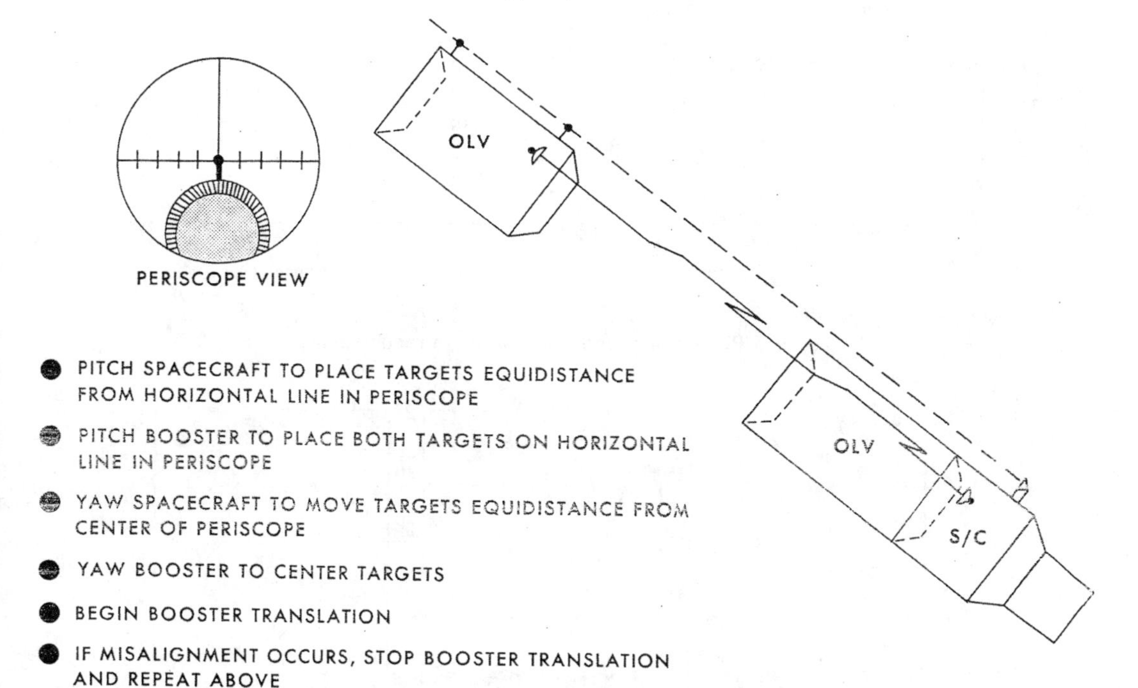

Figure 4. Orbital assembly for interplanetary missions.
(b) Alignment

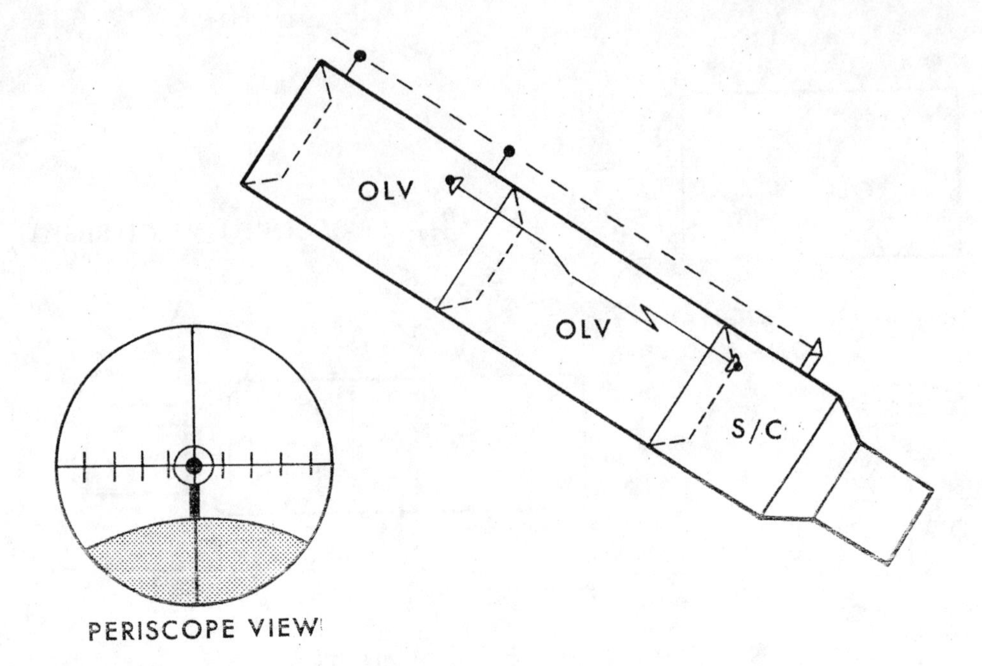

Figure 4. Orbital assembly for interplanetary missions.
(c) Docked

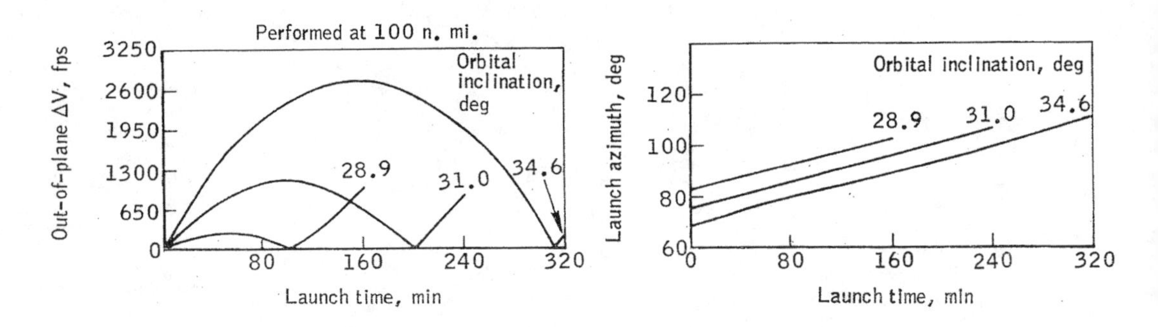

Figure 5. Plane change and launch azimuth requirements.

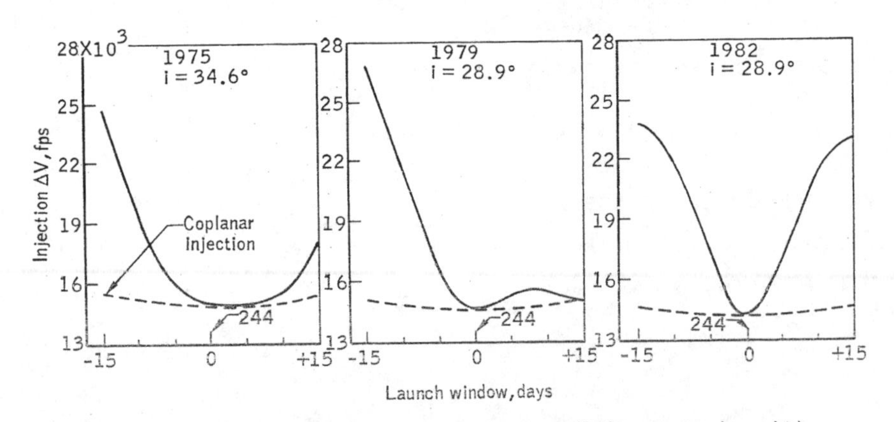

Figure 6. Mars twilight flyby injection requirements (260 n.mi. circular orbit).

6.1.3 PLANETARY ENCOUNTER OPERATIONS

6.1.3.1 Flight Control Support

The planetary encounter (PE) phase encompasses a period of from 15 days before until approximately 5 days after periapsis at the planet. It is characterized by a peak of experiments data and flight crew activity.

The four types of probes will each have an experiments flight controller assigned as responsible for monitoring the operation of that probe. The four types of probes are an orbiter, three atmospheric samplers, a lander, and a Mars Surface Sample Return (MSSR). Each of the probe flight controllers will manage the preparation and operation tasks concerned with his probe. He will insure effective probe data management under the authority of the Flight Director.

Nine days before periapsis, the flight controllers will support a third midcourse maneuver in a manner similar to that during the first two maneuvers.

Since much of the experiment operations and data analysis of this phase is time critical and limited in experiment performance opportunities, the flight control team will leave maximum manning in the ground support of systems and experiments during the encounter.

6.1.3.2 Information Flow Analysis

During the planet encounter period, several probes are to be launched toward the Mars surface and into Mars orbit. This period is a high activity period for the crew and the ground involving Mars surface photography, probe deployment, probe guidance and control, etc. Also, backup support of a spacecraft midcourse maneuver will be required, along with normal spacecraft systems monitoring by the ground.

The ground will assist the crew in probe checkout prior to deployment. The ground will also participate in the evaluation of any Mars surface photographs for real time determination of landing areas for the surface probes. Most of the data sent back to Earth during this period (up to 8×10^6 bps) will be photographs of the Mars surface selected by the crew and transmitted to Earth for ground evaluation. Navigation parameters for each probe will be sent to Earth when time permits via telemetry for ground computation of probe midcourse guidance maneuvers. This guidance data will in turn be sent to the spacecraft via the updata system for onboard control of the probes. It is anticipated that the ground will assist in the evaluation of probe data where the information may be useful for subsequent operation.

In addition to the support of the probes during the encounter period, the ground controllers will continue their normal monitoring support of the operational systems on the spacecraft for trend analysis, and will assist in crew activities scheduling.

Although it will be highly desirable to send as much data as possible during the encounter period, bandwidths and transmitter power will most assuredly be restricted. Therefore, the crew will be instrumental in selecting the more important data for transmission to Earth and recording the remainder onboard.

A detailed analysis of spacecraft and probe data flow has been conducted for the planetary encounter period, based on currently anticipated experiment and systems operations requirements. Because of the limited scope of this report, the results of this analysis are not included but are available upon request.

6.1.4 MISSION CONTROL OPERATIONS

The concept of mission control and flight operations embodies many disciplines related to the ground based support of manned space vehicles. Flight control operations begins in the prelaunch phase with comprehensive ground and flight crew simulations, continues with flight control monitoring of each launch countdown, and provides the data acquisition, flight control monitoring, and vehicle tracking throughout the flight to enhance mission success and crew safety. At the termination of the flight, recovery controllers effect deployment of search and retrieval ships and aircraft for recovery of the crew, the spacecraft, and any retrievable mission data. Each of these areas of mission control operations is governed by prescribed operating concepts, which in turn dictate recovery and flight control requirements in terms of facilities and manpower. The operating philosophies are also predicated on data management concepts, which further imply data handling systems and techniques that must be implemented with other program resource requirements.

6.1.5 Flight Control Philosophy

The Earth-based flight control operations can best be described by the tasks required of the flight controllers in support of the planetary flight sequences by a commentary on how the operations task can be implemented in a plan or concept and by pointing out the significant impact on the Earth-based flight control operations.

6.1.5.1 Flight Control Operations

The Mission Control Center in Houston (MCC-H) will have two control areas on each of the two control floors. The first control area will be the present Mission Operations Control Room (MOCR) and the second a Sustained Operations Control Room (SOCR). An SOCR will have the same basic capability as an MOCR, but less floor area will be involved. An SOCR will be configured for the control of the entire manned planetary mission. The MOCR on the same floor will monitor and control the launches of the. three S-IVC propulsion stages. One entire floor will be called up to support the mission from prelaunch through planetary injection. If mission progress remains nominal at the end of injection, the MOCR support will be terminated and the flight controllers in the SOCR will control the remainder of the mission. The Flight Director in the SOCR will be responsible for the complete planetary mission, with the MOCR Flight Director reporting to him for launch of the planetary injection stage and subsequent orbital operations.

Fifteen days before the 30-day launch window opens, the SOCR flight controllers will launch the manned vehicle. While these flight controllers support the orbital checkout of the planetary spacecraft, the MOCR flight controllers will prepare to launch three Saturn V vehicles within 48 hours or less. Precise coordination of pad support and countdowns is of extreme importance, because each of these vehicles must be properly phased in order to rendezvous with the manned vehicles. An additional constraint is imposed by the 110-hour orbital lifetime of the S-IVC injection stages.

At the beginning of transplanetary injection, the prime responsibility for crew safety shifts from the ground to the flight crew. During the injection phase, the SOCR flight controllers will monitor operation of the spacecraft and crew. The MOCR flight controllers will monitor operation of the S-IVC injection stages. The primary responsibility of these flight controllers will be to detect and respond to contingency situations and, if necessary, recommend a mission abort. If no abort is effected, MOCR mission support will end within an hour after injection. However, the expended S-IVC stages and the abandoned CSM will require further tracking and instrumentation support to preclude a non-hazardous reentry. If an abort occurs, the SOCR flight controllers will be responsible for supporting the spacecraft and flight crew. The MOCR flight controllers will be responsible for all procedures relating to the S-IVC stages.

Flight controllers will have been in a high activity mode up to this point, and their support will have been a major factor in the achievement of mission success. The spacecraft having been designed for autonomous transplanetary operation, ground control is no longer mandatory. Flight control support will be a desired function but not as necessary as before.

The mission, thus far, will have been similar to other missions from a flight control standpoint. Now, the flight control team must be revised for the remainder of the 667-day flight. The new team will consist of the Flight Director (FD) and an Assistant Flight Director (AFD). The flight dynamics personnel will only be on call, except for presence during midcourse maneuvers, planetary encounter (PE) and at other times at the request of the FD. The biomedical and flight crew activities function will be combined as a responsibility to the FD for the health, morale and welfare of the crew. A requirement for a wide range of disciplines in this area is anticipated, from psychologists and sociologists to physiologists and medical doctors. These aeromedical personnel may be on call or have a regular schedule of once- or twice-a-day visits to the SOCR. On the other hand, there may be some specialists whose duty may require continuous mission monitoring.

Probably the most significant flight control change will be in the emphasis placed on spacecraft systems support. Because of the variable but significant communications delay, real time response to contingency situations will become a function of the flight crew. However, two reasons for systems support remain, trend analysis and in-depth malfunction analysis. Trend analysis can be largely computerized. Its main value would be in consumables management and preventive analysis.

Malfunction analysis, on the other hand, requires highly specialized technicians cognizant of all the details of their assigned subsystems. Consequently, systems support will consist of a variety of specialists capable of accurate and timely work when called upon. Yet, their services may go unneeded for several days or weeks at a time. Since a morale problem may develop with these personnel, a rotation plan is recommended. Under this plan, four sets of

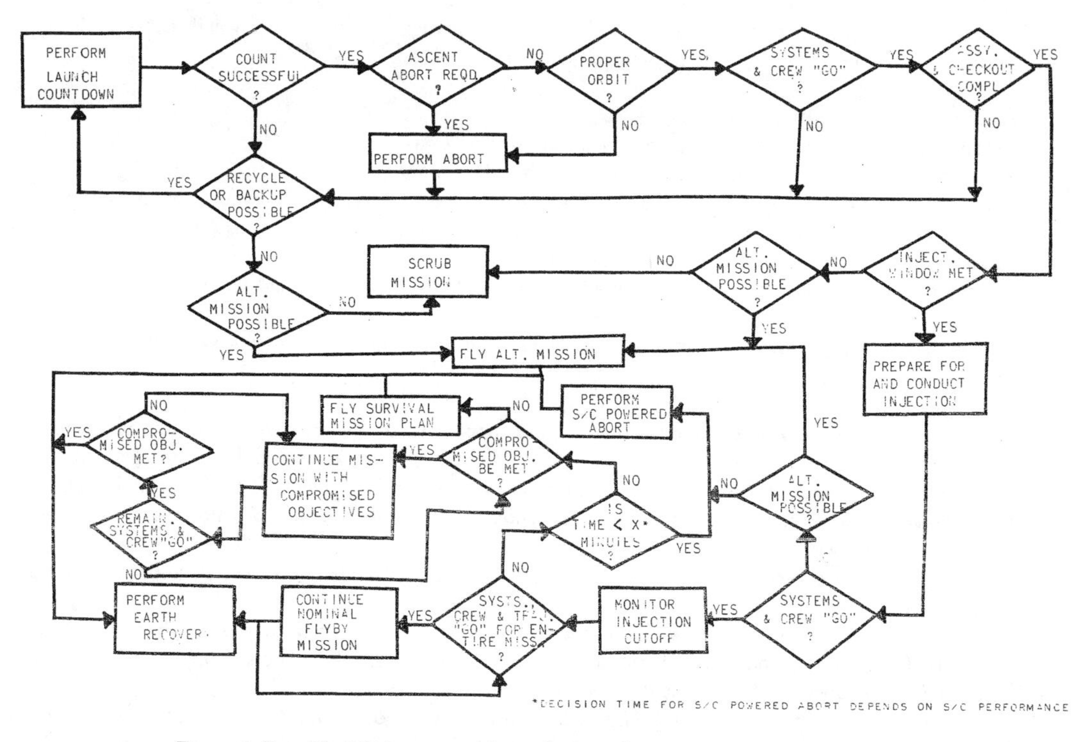

Figure 1. Simplified flight control logic diagram for planetary flyby mission.

systems flight controllers would he trained, and only one set would work during a given month. The other three sets would work on systems-related research and development projects, while retaining mission knowledge through daily flight status briefings. As a result, flight controller duty would occur only one month out of every four for each team.

As a member of the flight management function along with the FD and AFD, the Experiment Activities Officer (EAO) would manage experiment scheduling, experiment flight control procedures, and supervise the systems engineers responsible for experiment subsystems.

Finally, the flight controllers will be forced to consider three time frames simultaneously, because of the communications delay. All data and text messages received in the SOCR represent the past, the time at which they were transmitted, which can be as long as 26 minutes at spacecraft conjunction. The flight controllers work in the present time, naturally. However, any data or text message uplinked to the crew must be considered as to its effect when it is received again up to 26 minutes into the future. Since flight controllers are more comfortable in a present time frame and are accustomed to real-time data monitoring, all data and text messages received in the SOCR will be considered as real time. The SOCR clocks will be synchronized to agree with the time the data now received was transmitted. Any uplink response will have to be considered as reaching the spacecraft at a future time equal to twice the communications delay in order to compensate for the slow clocks.

Communications delay continues to increase up to aphelion, with a corresponding reduction of bit rate transmission capability because of the increasing distance between the spacecraft and Earth, Flight controllers are, therefore, less able to support the crew than was the case during TPC, and at a time when systems are in the final half of their design lifetime. It is during this phase that the communications delay approaches 26 minutes and Sun occultation of the Earth for approximately 25 days precludes effective Earth to spacecraft communications.

Prior to Sun occultation, flight controller support will prepare the crew for the communications void through a complete update of onboard information. Detailed analysis of spacecraft performance data will provide the basis for any procedural deviations from those established prior to flight. A midcourse maneuver similar to previous midcourse manuevers is required after the encounter phase. After Sun occultation and prior to the Earth entry (EE) phase, two final midcourse maneuvers will require flight control support. Otherwise, the prime flight control support emphasis will be placed on monitoring spacecraft and crew status, particularly long-term trend analyses of systems performance, and preparing the spacecraft and crew for the final entry phase.

In summary, the operations characteristics of the 1975 Mars twilight flyby mission which will be significantly different as compared to previous manned space flight experience are as follows:

a. Extended mission duration (667 days).
b. Multiple Saturn-V launch operations (three launches within 2 days or less).
c. Impossibility of early mission termination after the initial commitment to the transplanetary flight phase.
d. Communications delay (approximately 8 minutes one way at encounter and 26 minutes one way at aphelion).
e. Sun occultation of space vehicle at aphelion causing degradation of communications (approximately 25 days).
f. Extensive post-orbital-assembly launch-vehicle checkout.
g. Manned docking to fueled propulsion stage without direct flight crew vision (i.e., blind docking).

The desired mission design guidelines which are deemed required to remain consistent with stated flight-control concepts are as follows:

a. The space vehicle should be designed to use ground support to the fullest extent from the prelaunch phase through transplanetary injection. By assigning as many flight control tasks as possible to the ground for these phases, the space vehicle design need not be burdened with equipment needed only for orbital flight.
b. An automatic or semi-automatic checkout system should be designed for in-orbit checkout of the integrated Orbital Launch Vehicle (OLV).
c. Mission procedures should allow for the flight crew to be in the abort module (Earth Entry Module) during the injection maneuver.
d. The module housing the flight crew during the injection maneuver must have the capability for immediate separation from the remainder of the spacecraft to avoid the catastrophic effects of any possible explosion.
e. A post-injection abort capability should be designed into the space vehicle.
f. The space vehicle must have the capability to be autonomous in operation from injection through to Earth Entry (EE). The spacecraft's being autonomous relieves ground station constraints on trajectory design and provides onboard capability for achievement of crew safety.
g. Only if the required reliability cannot be attained in another way should an onboard systems repair capability be provided for the safe return of the crew.
h. An adequate reserve of consumables should be stored onboard to cope with unforeseen anomalies in case the crew needs to improvise either during onboard repair or to satisfy a failed onboard function.
i. A telecommunications system providing uplink and downlink text messages under the most degraded mission conditions must be designed for the spacecraft and ground support systems.
j. A duplex voice communications capability to Earth must be available in the spacecraft, despite the problems which could arise from the communications delay.

6.1.5.2 Contingency Planning and Support

A mission abort is considered to be the termination of a mission prior to attaining the mission objectives where the foremost consideration is to return the crew safely to Earth within the time constraint created by the contingency. Simplified. flight-control abort logic is shown in Figure 1. After commitment to the transplanetary phase, an abort condition reverts to an emergency requiring primary action to sustain life for the remainder of the mission with partial attainment of mission objectives a secondary consideration.

The abort capability for the launch and Earth-orbit phases will be essentially the same as for Apollo. During Earth orbit operations, the launching of the S-IVC stages, rendezvous, and hard docking, information regarding the status of the spacecraft, S-IVC systems, and crew will dictate one of the following actions:

a. Effect transplanetary injection
b. Abort the planetary mission
   (1) Proceed with alternate mission
   (2) Return crew to Earth

Alternate missions must be carefully planned prior to flight in order to take full advantage of the various combinations of hardware capability which may be available. For any alternate missions, abort and contingency planning must still be conducted to achieve acceptable crew safety. Abort will be possible using either the CSM of the EEM until such time as the CSM is jettisoned. Aborts will then be made using the EEM.

Initiation of an abort during the transplanetary injection maneuver will depend upon the status of the burn trajectory, the S-IVC and spacecraft systems, and the crew. Abort of the planetary mission will fall into two categories: an alternate Earth orbit mission or immediate return to Earth.

An alternate Earth orbit mission can be attempted if the combination of the trajectory, S-IVC, spacecraft and crew still will allow for it. Specifically, some considerations affecting the decision are: (1) the point during the injection burn when the trajectory passes from an ellipse to a hyperbola; (2) the S-IVC propulsion capability remaining; and (3) whether or not the onboard systems and crew are still able to complete a long lifetime alternate mission with a reliability commensurate to the nominal mission. These specific capabilities can only be determined when detailed spacecraft design is made.

During the injection burn, the flight crew should be in the EEM so that an abort can be executed as fast as possible. Due to the criticality of this burn, especially on flyby missions when virtually all ΔV is expended, and since a Go / No Go decision must be made very soon after injection cutoff, continuous tracking and telemetry should be available. This mission will represent the first time a vehicle assembled in space will be used for a maneuver of this type (escape) and magnitude (approximately 16,000 fps). Because of the time delay expected in ground-based data acquisition, processing, and decision execution, the crew should have an onboard systems analysis and abort implementation capability, with the ground as backup. In addition, there should be a capability to separate immediately to a safe distance from any potential hazards involving the S-IVC stages prior to an abort maneuver.

The length of time after injection burn cutoff that abort is possible can be quite short, depending on the amount of abort ΔV available. The ΔV required increases very rapidly with time, making a quick reaction time for status evaluation of the trajectory, spacecraft systems, and crew necessary. There should be enough ΔV capability for a period of time, yet to be defined, for ground tracking and computation to verify a Go decision for the planetary mission. The maximum amount of time available is a matter of a few minutes. After this time, an abort maneuver would require an enormous, amount of fuel and is, therefore, impractical.

The purpose of mission rules is to provide preplanned procedures to assist in making real-time decisions during critical prelaunch and flight phases. This brief discussion will cover three main points:

a. The need for mission rules exists for the entire mission.
b. The significant communications delay at planetary distances precludes timely mission rule implementation by the flight controllers.
c. Documented systems analysis procedures are needed as an adjunct to each mission rule.

Mission rules would be implemented by the flight controllers, except during periods of communications loss or delay when the command pilot would assume this responsibility. Ground-based implementation of mission rules would effectively be limited to a period from the prelaunch phase through a brief time beyond transplanetary injection.

After injection, an increasing communications delay essentially precludes a real time response to contingencies from the flight controllers. However, the need continues for preplanned rules to assist in making real-time decisions, even though there is no longer an abort capability onboard the space vehicle. Consequently, the command pilot would assume the mission rule implementation authority just as he would have done during a communications loss in Earth orbit.

Mission rules for phases from prelaunch through injection must be carefully analyzed for their impact on the entire mission. This analysis is especially important from a crew safety standpoint, because an "early mission termination" decision is no longer possible shortly after injection. Essentially, early mission termination rulings possible before injection are replaced by crew survival rulings after injection.

In the portion of flight when the communications delay causes the command pilot to have the mission rule implementation authority, the flight controllers will monitor space vehicle conditions and consequent mission rule applications for consistency.

Since a comprehensive onboard malfunction analysis capability is a vital requirement for this type of mission, the mission rules should also serve as an index into the proper systems analysis procedure. The flight controllers will then be able to deduce from the telemetered malfunction information which mission rule was applied and which procedure is being followed without waiting for the delayed data to confirm it. As a result, the need for updata can be determined immediately by the flight controllers and the necessary action can be taken.

6.1.6 EARTH RECOVERY PHILOSOPHY

Earth recovery of the planetary Earth entry module encompasses two basic areas of operations, nominal Earth return and abort or contingency recovery. The basic philosophy of providing recovery operations support at all times throughout the mission, commensurate with flight control procedures and probability of abort, will be followed. In general, ship and aircraft support will be reduced with respect to present Apollo support in keeping with the assumption of increased reliability for the planetary spacecraft. The major departure in type of support afforded, as compared to present flight programs, will be the inclusion of a land recovery capability as a nominal procedure.

The basic parametric recovery limits assumed for advanced planning purposes are as follows:

Entry inclinations – up to 40°
Footprint dimensions – an ellipse of about 5,000 by 600 nautical miles
High probability landing area – an ellipse about 20 nautical miles long
Entry velocities – 40,000 to 60,000 feet per second
Entry ranges – 7,000 to 10,000 nautical miles
Postlanding survival capability – 72 hours minimum

A launch abort posture of ships and aircraft will provide coverage during powered flight of the Saturn V out to 3,200 nautical miles range in a manner similar to Apollo. For support of the Earth-orbital portion of the mission, there will be two secondary recovery areas, one each in the Atlantic and Pacific Oceans, either of which could be used later in the mission as the primary water recovery area. Recovery ships and aircraft will be staged in the proximity of these areas and will remain in a standby condition until either deployed or no longer required.

Once the spacecraft has been inserted into an elliptical injection orbit, if that is the mode of planetary departure chosen, other discrete abort landing areas which coincide approximately with a return from apogee of the injection ellipse will be specified. It is expected that either or both of the orbital recovery areas will suffice for this case.

During the planetary injection phase and immediately thereafter, a series of suitable water landing sites, similar to those used in ascent to orbit, will be employed. The major recovery support, however, will be provided in the landing area resulting from an abort at the planetary Go / No Go decision point following shutdown of the injection stage engine, since this represents the high probability contingency situation for this phase. Once this decision period is passed for the planetary flyby mission, the spacecraft will not contain sufficient abort performance capability to return to Earth before the nominal mission return time, and recovery operations will be discontinued until that point.

For the nominal end of mission recovery, either land or water landing may be employed, based on conditions derived in real-time prior to the entry commit point. If the status of the spacecraft, crew, weather in the recovery area, and ground support systems is appropriate, recovery in a preselected area within the continental United States may be effected. For a water landing at the termination of the mission, a ship will have been deployed in the selected ocean area for spacecraft retrieval, and the spacecraft will be located with properly instrumented aircraft.

To allow for a land landing capability, the descent system must have the capability for maneuver when deployed in the lower atmosphere, or approximately below 20,000 feet. To provide terminal guidance data for the crew, a local ground station will be required for control at times of poor weather, nighttime or when navigational error places the parachute opening point at the outer edge of the landing zone. Included in the ground system support equipment would be a precision tracking radar, displays for approach control of the spacecraft, a general purpose computer, weather station facilities, and specialized ground support hardware and handling equipment.

Post-retrieval procedures for the spacecraft and crew will be similar to those instituted for the Apollo lunar landing mission but with hazard and contamination practices dictated by the nature of the planetary objectives. This factor is particularly relevant if the mission includes landing at the planet or when a planetary surface sample has been returned for analysis. Appropriate quarantine periods with proper support facilities and recovery crew conditioning will be planned. Postlanding medical support and analysis will be provided in the terminal landing area and during the period in which the spacecraft and crew are returned to the mission evaluation laboratory.

6.17 Data Management Concepts

6.1.7.1 Data Management Philosophy

The nature of the 1975 Mars Flyby Mission requires that a maximum of usable mission data be returned to Earth to satisfy the mission objectives. Three methods will be used for returning the mission information to Earth: (1) by transmitting the data over the communications links between the space vehicle and Earth, (2) returning the data stored on the spacecraft at the end of the mission through normal recovery techniques, and (3) by transmitting experiment data from Mars probes directly to Earth. Because of the mission importance and large expense involved, as much useful data as possible should be transmitted over the communications links to Earth during the mission to reduce the chances of a future data loss.

The importance of the crew's role in the onboard data management cannot be over emphasized. Since the communications data bandwidth is likely to be limited for a mission of this nature, the availability of the crew to interpret and evaluate the information on the spacecraft can reduce the data rate to manageable values. A data processing and monitoring system onboard designed to make effective use of man's capabilities in sorting and screening the data would be advantageous. In addition, careful design of the experiment equipment should be a goal for simplifying the experiment operations and to give acceptable results with a minimum of data flow. Some compromise in the experiment results may be required if such a compromise will significantly reduce the experiment operations complexity while still providing acceptable results.

6.1.7.2 Transmission Priority

For purposes of transmission, the data to be returned to Earth can be divided into three categories, in their order of importance to the flight control function:

a. Crew Safety – the crew safety data is that information required by the ground to monitor and make decisions on the welfare of the crew and the integrity of the spacecraft. The crew safety information would include medical data, environmental control system parameters, guidance and navigation parameters, and any further information required by the Flight Director during a spacecraft contingency.
b. Experiment – all scientific payload data is defined as experimental data. The experimental data sources will include the Mars surface and orbiter probes and supporting systems, plus all spacecraft experiment subsystems.
c. Housekeeping – housekeeping data is the routine data required periodically for monitoring the overall status of the crew and spacecraft and is the source for trend analysis by the ground-based flight controllers.

Normally all data to be transmitted to Earth will follow a predetermined data flow plan; however, during an emergency the crew safety data will have priority over all other data. If bandwidth and transmitter power restrictions occur, the experiment data may be sent in place of other data if the change in the data flow plan is approved by the Flight Director. As currently envisioned, the housekeeping data will be transmitted as required during critical operations (midcourse corrections, etc.) or during mission contingencies.

6.1.7.3 Data Transmission Formats

To optimize use of available spacecraft power, variable data rates will be used throughout the mission. The following functional data formats are assumed for transfer of the systems telemetry data to the ground and for Earth-based tracking.

FORMAT	DATA RATE, kbps		
	High	Medium	Low
EEM SYSTEMS TLM	51.2	—	1.6
MM SYSTEMS TLM	90	51.2	3.2
EEM G&N `IIM	10		
MM G&N TLM	10		
S-IVC TLM	72		
I.U. TLM	72		
PRN TRACKING (MM, EEM, S-IVC's)	1,000 (Standard Apollo)		
INFLIGHT EXPERIMENT SUBSYSTEM	Variable (up to 8 x 10^6)		
TELESCOPE SUBSYSTEM	Variable (up to 8 x 10^6)		
UPDATA	1		

SPACECRAFT TO EARTH DATA FLOW GUIDELINES

The data transmission formats in the table above are assumed for planning purposes only. Exact formats are dependent upon an accurate identification of the experiment data flow requirements. Further, the formats selected will also depend on the modulation technique required. Further reductions in spacecraft power may be possible if a unified carrier modulation technique similar to the Apollo S-band system is selected.

6.1.7.4 Data Monitoring Guidelines

In the data flow analysis previously described, daily monitoring of flight data for discrete periods were assumed for each phase of the mission, and these periods are listed below according to data source in the table below.

DATA SOURCE	AVERAGE DAILY MONITORING TIME					
	Earth Orbital	Injection	Transplan-etary Coast	Planet Encounter	Transearth Coast	Reentry
EEM SYSTEMS	1 Hr	2 Hrs 30 Min	—	—	—	4 Hrs
MM SYSTEMS	6 Hrs	2 Hrs 30 Min	1 Hr	1 Hr	1 Hr	2 Hrs
EEM OR MM G&N SYSTEM	5 Min	2 Hrs 30 Min	5 Min	5 Min	5 Min	15 Min
TRACKING	5 Min	2 Hrs 30 Min	5 Min	5 Min	5 Min	20 Min
S-IVC SYSTEMS & I.U. (During Assembly)	12 Hrs	2 Hrs 30 Min	—	—	—	—
VOICE	Variable					
UPDATA	Variable					
EXPERIMENTS	Variable					

NOTES: (1) Reentry support period lasts for 2 weeks
(2) Crew biomedical data contained in systems data
(3) Monitoring guidelines cannot be met during Sun occultation

6.1.8 Flight Control Requirements

Several communications considerations which directly influence the mission support systems required are listed as follows:

a. The radio frequency selected for interplanetary communications must, in general, remain in the range of low sky noise (1 to 10 GHz) and will be assumed to specifically remain in the two to three GHz band, compatible with the projected frequencies of the deep space network.
b. Ground to spacecraft communications modes will likely be similar to those for Apollo plus the following:
   (1) An uplink teleprinter compatible with the network updata system. The teleprinter will be used to receive flight plan changes, experiment instructions, and other hard-copy data.
   (2) A downlink teleprinter from the crew for transmitting precut messages.
   (3) Uplink television primarily for visual-aid instruction to the crew, with program material and pseudo-personal transaction a secondary function for crew morale purposes.

Manned Space Flight Network (MSFN) requirements will be based, in part, upon the development of space qualified communications hardware (antenna size, solid state transmitters, etc.) for the planetary spacecraft. Preliminary estimates show a basic network functional augmentation as follows:

a. Earth Launch
   The presently existing launch area stations in the Eastern Test Range will be adequate for launch and Earth orbital insertion support of the Mars spacecraft and launch vehicles. Since insertion cutoff may occur out of view of land stations, ship tracking and communications support will be required. The Apollo instrumentation ships can provide the required support for orbital insertion. Several links from the S-IVC's, the launch-vehicle instrumentation units, the EEM, the CSM and the MM will be required to the ground. Each link must operate on a different frequency, since all links may be active simultaneously during orbital assembly. Some augmentation may, therefore, be required for the launch area stations to accept the new frequencies.

b. Earth Orbit
   As stated earlier, the coverage requirement for support of the nominal three days of orbital assembly is essentially continuous. In addition, continuous coverage is required during the 2 hour injection support period, particularly during the critical 20-minute injection burn of the S-IVC's. In summary, the present MSFN does not provide the type of communications coverage and data bandwidth needed for Earth orbital assembly and injection. It is, therefore, assumed that an advanced communications satellite system will be made available to provide continuous communications between the Earth orbiting vehicles and the MCC-H. The communications satellite system would be required to provide tracking, voice, telemetry, and updates for two to four vehicles simultaneously. All of the spacecraft and launch vehicle data could be relayed through the satellite system directly to the MCC-H.

c. Deep Space
A continuous communications capability must be maintained throughout deep space operations, thereby imposing a requirement for three stations equipped with 210-foot antennas. Processing equipment similar to present Apollo MSFN 85-foot antenna stations will be required. Each station must have capability to support a minimum of two downlinks capable of being processed simultaneously.
It is possible that probe requirements may be satisfied with the current 85-foot stations. However, regardless of the antenna size, an arrangement similar to the present Apollo MSFN / JPL facilities co-located for lunar mission support is envisioned.

The MCC-H configuration, as defined in the program requirements document for control-center augmentation, is assured basically adequate for support of the 1975 Mars mission. That is, the mission will be supported from one floor containing a MOCR for S-IVC launch, Earth orbital and injection operations support, and a SOCR for support of the manned spacecraft during launch and Earth orbit, and sustained support of the mission thereafter. The MCC-H prelaunch support will require from two to four times the present launch data flow capability. The complexity of the systems checkout during orbital assembly points up the need for a system similar to the automatic checkout equipment installed on the Apollo spacecraft. Ground support compatible with an onboard checkout system is anticipated to have an additional impact on the computer facilities at the MCC-H, but these requirements cannot be specified until the spacecraft and its checkout procedures are better defined.

The integrated schedule used to determine impact of mission control systems includes the Planetary Joint Action Group's baseline missions, the planned lunar missions, and projected space station missions from 1968 through 1980. The schedule also includes missions required to satisfy currently assumed flight test ground rules for planetary spacecraft development.

A set of flight test ground rules have been developed to support the integrated flight schedule analysis. The ground rules will enable planners to conduct a more meaningful cost analysis of the planetary program. The following ground rules are considered mandatory for effective development of any new planetary vehicle and integration of two or more vehicles never mated before.

a. "All-up" flights planned rather than an incremental flight test approach.
b. A reentry flight test under operational limit conditions for any new Apollo-shaped entry vehicle.
c. A propulsion and aerodynamic flight test (unmanned) for any new configuration of an existing launch vehicle basic design.
d. An unmanned aerodynamic structural integrity test to flight qualify the launch vehicle with any new manned spacecraft as the payload.
e. A manned qualification test for any new spacecraft to demonstrate integration of the crew and systems before commitment to a total mission.
f. Two unmanned space tests of new propulsion modules for verification of operational running conditions and for restart qualification.
g. An unmanned transplanetary injection test to verify structural integrity of the launch vehicle and manned spacecraft.
h. A manned practice mission in near-Earth space using operational timeline events and hardware for Earth escape missions only.

More than one of the above test objectives may be accomplished on a single flight, such as those in rules 2, 3 and 4.

In addition to the baseline manned planetary flights, advanced lunar landing and Earth orbital missions are proposed in the same time period. The integrated schedule from 1972 through 1989 shows at the most three missions occurring at any given time. The support for a planetary mission requires two control areas as discussed in a previous section. The MCC-H as planned for the Apollo Applications Program is expected to be capable of controlling all missions through 1979. One problem does occur in 1980 specifically related to the support of four simultaneous missions:

a. Mars Flyby
b. Dual Planet Flyby
c. Space Station
d. Lunar (3 Saturn-V launches)

The lunar mission consists of 3 Saturn-V launches, two being unmanned and the other manned. This mission will require two control areas for launching all flights; therefore, five control areas would be required under current ground rules to support the 1980 integrated schedule. One solution would be to control both planetary mission modules from the same control area during the prelaunch through landing phase or the two unmanned lunar flights. After the unmanned flights land on the lunar surface, one control area could assume responsibility for all three lunar vehicles.

The MSFN requirements defined earlier will support the integrated schedule. No more than three Saturn-V or Saturn-IB Launches will be required in any given 48 hour period. The ETR is capable of supporting this launch rate with the possibility that one insertion ship may not be able to physically move to a designated area in time to support all insertions. The only augmentation seen at this time is the capability to receive all required orbiting vehicle frequencies. High activity Earth-orbit operations for all programs in the stated time period indicates a definite requirement for an Orbiting Data Relay Network (ODRN). Each program will make use of the ODRN and the cost of implementing such a system can be prorated among all users. The ODRN is the only new development requirement foreseen to be implemented for the MSFN to support the integrated flight schedule. The MSFN large antennas (85- and 210-foot antennas) will be reserved for the lunar and planetary missions. An additional complement of 210-foot receiving antennas is foreseen for support of both manned and unmanned planetary exploration. There may be scheduling problems; however, it is assumed JPL's Deep Space Network will be made available. If the JPL tracking facilities are not available, NASA must make other provisions for the required coverage.

6.1.9 RECOVERY CONTROL REQUIREMENTS

Three types of Earth-based facilities will be required to provide recovery support for the planetary mission: (1) Recovery Control Center (RCC), (2) Terminal Landing Facility (TLF), and (3) recovery ships and aircraft. The only change from current recovery operations is the addition of a Terminal Landing Facility, which is required to accommodate the land landing capability.

The RCC will be configured essentially as it has for all manned flights to date, i.e., real-time computer support for landing dispersion calculations, weather trends and status information, recovery network communications links, and appropriate console and display systems for recovery controllers. The RCR is located on the third floor of the Mission Control Center and would be activated prior to flight in support of mission simulations. During flight, the RCR would provide recovery control during the orbital and planetary injection phases, as well as the 4- to 7-day period prior to and immediately following end-of-mission landing.

The Terminal Landing Facility (TLT) would be located in the continental United States to support a nominal end-of-mission landing. The types of equipment required in the TLF are discussed briefly in the section Earth Recovery Philosophy, but these systems generally include landing approach control, a weather station, specialized ground handling equipment, and basic medical support.

Whether or not the ships and aircraft employed in the 1975 to 1985 time period are leased from the Department of Defense is unimportant, since both of these types of recovery vehicles will be required so long as water landing is an accepted operations concept. Ships will generally be used for long-range transportation of the spacecraft, because of their reliability and the availability of spacecraft and crew support equipment. However, large transport aircraft may be used for point to point movement of the spacecraft in special situations. Recovery aircraft will still be used for the postlanding search of the spacecraft following a water landing.

6.1.10 GROUND SUPPORT SYSTEM CONSIDERATIONS

This section treats the impact of interplanetary mission support upon existing and projected mission control systems. The following generalized statements can be made:

a. The impact of the planetary program upon the ground support systems is directly related to the data management requirements of the program. Careful design of spacecraft instrumentation and onboard data management system can greatly relieve from the ground support system impact.
b. The major impact of the planetary program will be in the area of ground receiving systems, specifically in the provision of adequate ground antenna aperture.
c. If use can be made of currently planned programs like Orbiting Data Relay Network (ODRN), the impact on the near-Earth tracking and data acquisition stations will be minimal. Otherwise, considerable effort will be required in data compression and transmission techniques.
d. The impact on the Mission Control Center will be in the area of data handling systems and the NASA Communications (NASCOM) System interface.
e. While definite costs cannot be developed, it is estimated that the Mission Control Center and Launch Data Systems will involve changes costing from $30 to $50 million. The estimated. network costs for three large-aperture ground stations would be approximately $350 million.

Mission Control Center

The data rates presently envisioned for a planetary mission far exceed the capability of the control center systems and would require extensive modification to CCATS and RTCC systems. If the ODRN concept is adopted, a data input of from 400 kbps to 1000 kbps could be received, contrasted to about 12-15 kbps presently handled. If data compression techniques are used at existing tracking and data acquisition stations, reformatting and regeneration systems will be required at the MCC-H. An expansion of the present data management and display system is likely, since rapid checkout of the spacecraft and injection stages in Earth orbit is required. Picture reconstruction equipment will be required to handle the slow-scan video data. The NASCOM interface will be changed completely to handle ODRN inputs during Earth-orbital operations and possibly COMSAT inputs during interplanetary flight. The launch data system should be expanded from two to four times the present capability to handle the multiple launches and should be redesigned to bring back the unprocessed data. It is expected that a general purpose data multiplexing and processing system similar to KSC Data Core would be used in the Control Center input.

Network

For the Earth-orbit operations, which would involve assembly and checkout of various modules, it would be useful to use an ODRN concept for data collection and relay directly to the Control Center. This would significantly reduce any expansion or changes to the present network ground stations. If the ODRN is not available, then some data compression techniques will be required, as well as considerable augmentation of the individual stations. For the interplanetary flight, the DSIF concept presently in use for Apollo lunar operations would be needed, but with much larger ground apertures to provide communications over Mars distances. The problem of providing a large enough ground aperture for the stated data rate requirement is likely the most difficult in the entire system.

A duplication of the present JPL 210-foot receiving antenna would suffice for the manned planetary mission. Another possibility is the use of phased-array techniques, which are just in the proposal stage. Large arrays of 85-foot-diameter dishes could be made to provide the needed aperture, but much research and development will be required before such a system could be operational. Assuming the availability of such a system, a wide-band communication satellite system could provide the link to the Mission Control Center.

6.2.1 ALTERNATE ORBITAL ASSEMBLY MODE

The circular orbit rendezvous mode used in this document can be improved in terms of' payload injected to Mars by utilizing an elliptical orbit rendezvous. This mode can inject slightly more than the 200,000 pound spacecraft onto the most demanding mission of this study (the triple Planet Flyby) using only 3 of the Saturn V launch vehicles considered for this study if the S-IVC propellant loading is allowed to increase to approximately 450,000 pounds. Figure 1 is the launch configuration of the three launch vehicles for this mode. Briefly, the launch sequence is as follows. The first launch with the planetary S/C is boosted to a parking orbit with a suborbital start of a standard S-IVB as in an Apollo launch. Again as in Apollo, after achieving obit the S-IVB is restarted and boosts the S/C into a 3.9 hour elliptical orbit which contains the planetary injection vector. This entire sequence is accomplished within the 4.5 hour lifetime of the S-IVB.

Launches two and three are then performed as shown in Figure 2. The S-IVC must be started suborbitally (increases payload to orbit by about 18,000 pounds). This will require that the S-IVC propellant loading be increased. The payloads of launches two and three, namely, the two S-IVC's plus the manned CSM, perform the rendezvous and docking function. One of the two S-IVC's is then used to place the other S-1VC and itself into an orbit compatible with spacecraft rendezvous. Prior to docking, the CSM is separated and docked to the spacecraft and the crew is transferred. The two injection stages are then docked to the spacecraft and the entire configuration (Figure 3) thoroughly checked out. Following checkout, the configuration is sequentially ignited and staged to perform the transplanetary injection.

In addition to the performance advantages, this mode flattens the injection window considerably due to a much smaller perturbation of the elliptical orbit compared to the circular orbit. The payload increases using this alternate mode are primarily a result of being able to load each booster to its payload capability and to the suborbital start of Saturn V third stage.

VEHICLE CONFIGURATIONS FOR THREE LAUNCHES WITH ONE CIRCULAR AND ONE ELLIPTIC ORBIT RENDEZVOUS

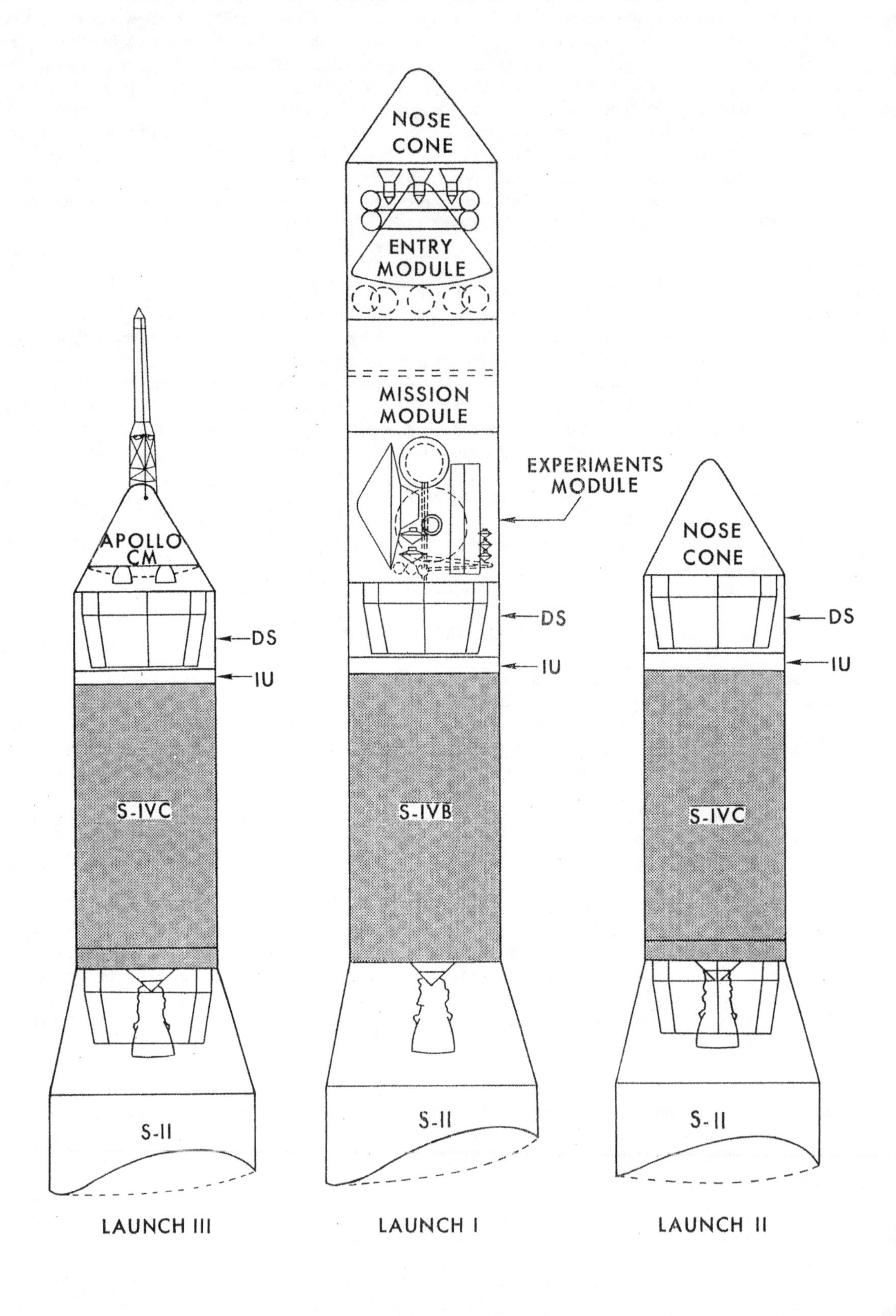

ORBIT GEOMETRY FOR LAUNCH AND RENDEZVOUS

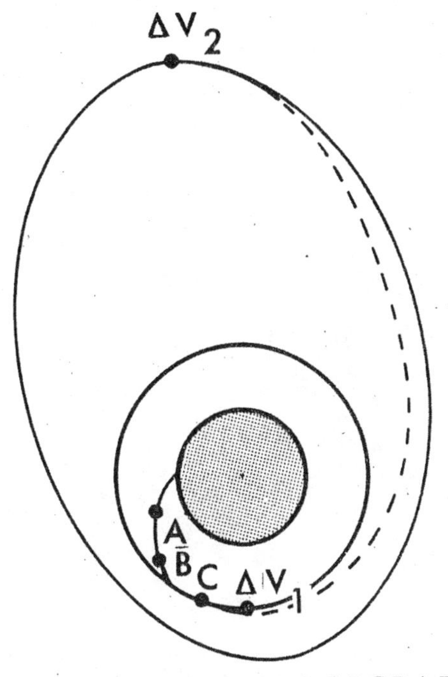

A = SI CUTOFF, SII IGNITION

B = SII CUTOFF, S-IVB IGNITION

C = S-IVB CUTOFF

ΔV_1 S-IVB THRUST

ΔV_2 S/C THRUST

LAUNCH I (SPACECRAFT)

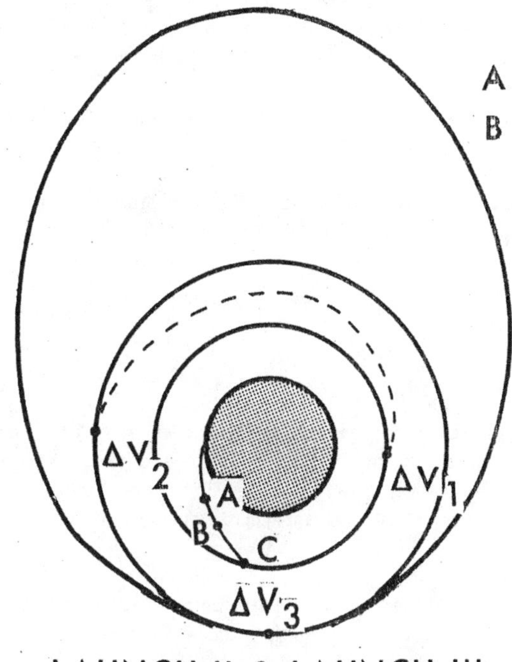

A = SI CUTOFF, SII IGNITION

B = SII CUTOFF, S-IVC IGNITION

C = S-IVC CUTOFF

ΔV_1 HOHMAN TRANSFER

ΔV_2 HOHMAN TRANSFER

ΔV_3 ELLIPTICAL ORBIT RENDEZVOUS

LAUNCH II & LAUNCH III

ASSEMBLY CONFIGURATION AFTER RENDEZVOUS IN CIRCULAR ORBIT (LOWER FIGURE) AND ELLIPTIC ORBIT (UPPER FIGURE)

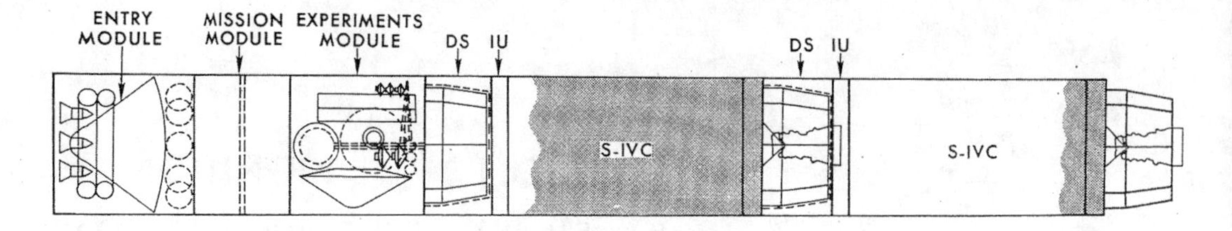

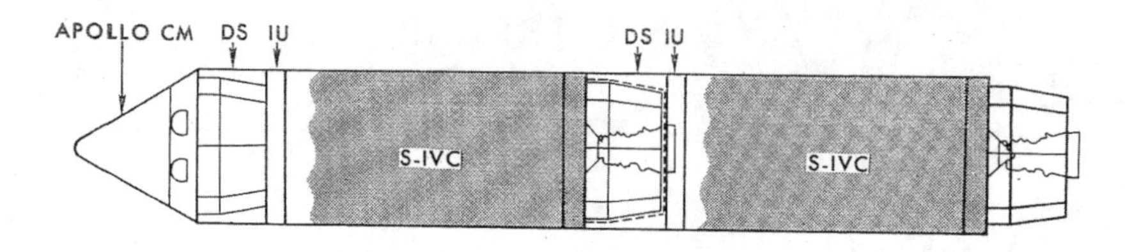

7.0 SYSTEMS

7.0 SPACECRAFT DESIGN

7.1.1 General Spacecraft Description

The spacecraft is the vehicle which (1) provides physical protection and the command and control functions to the crew necessary for experiment accomplishment, (2) protects and transports an experimental payload to planetary encounter, (3) provides the mechanism for probe deployment and recovery, (4) provides on board analysis and data return to Earth facilities, and (5) provides a safe Earth return capability for the crew and experiment data.

The preliminary design of the spacecraft is based upon maximum utilization of Apollo and Apollo Applications Program Systems and subsystems with new developments when necessary to meet the mission requirements.

The spacecraft was designed in modular fashion to provide the capability to evolve to orbital capture missions and eventually manned landing missions with a minimum of major modification. The four basic modules are:

1. Mission Module (MM)
2. Earth Entry Module (EEM)
3. Mid-Course Propulsion Module (MCPM)
4. Experiment Module (EM)

Figure 1 shows these four modules integrated in a total spacecraft general configuration. The Mid-Course Propulsion Module is located at the forward end of the spacecraft to permit the MCPM engines to be operationally ready for use by merely opening the protective enclosure. This location allows the propulsion system and the EEM, which is located in the rear of the MCPM, to be separated from the rest of the spacecraft during and immediately after planetary injection in the event an abort is necessary.

In the rear of the MCPM, with a hatch for access from the Mission Module, is the Earth Entry Module. The EEM is designed for the Earth entry function, and it is also utilized for abort and for a solar radiation shelter. The EEM is periodically checked out so that it will be ready for operation at Earth entry.

The Mission Module (MM) is located directly behind the EEM. Spacecraft controls, subsystems, and crew quarters are located in the MM. Windows and airlocks are provided for external viewing, crew ingress, and emergency extravehicular activity. All the experiments are controlled or monitored from the MM. Scientific facilities and equipment are available within the MM to conduct enroute experimentation and analysis. The astronomical telescope is housed outside the MM and is controlled from a station inside the MM.

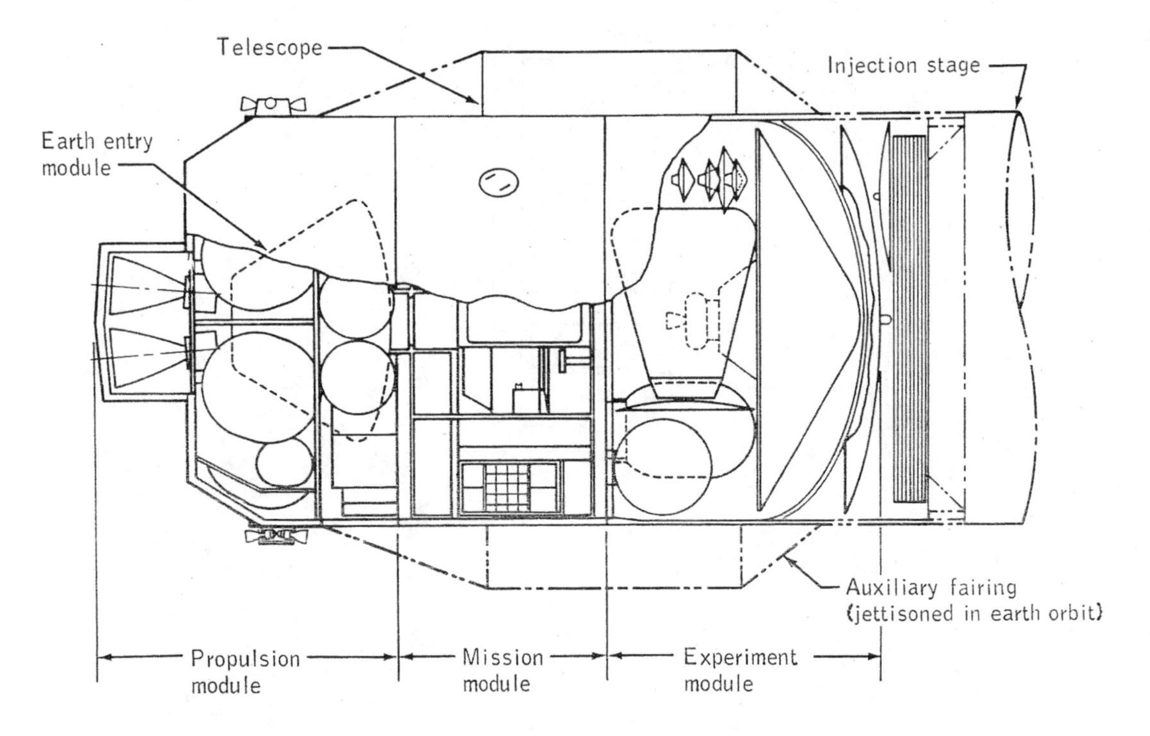

Figure 1 Spacecraft General Configuration For Mars Twilight Encounter Mission

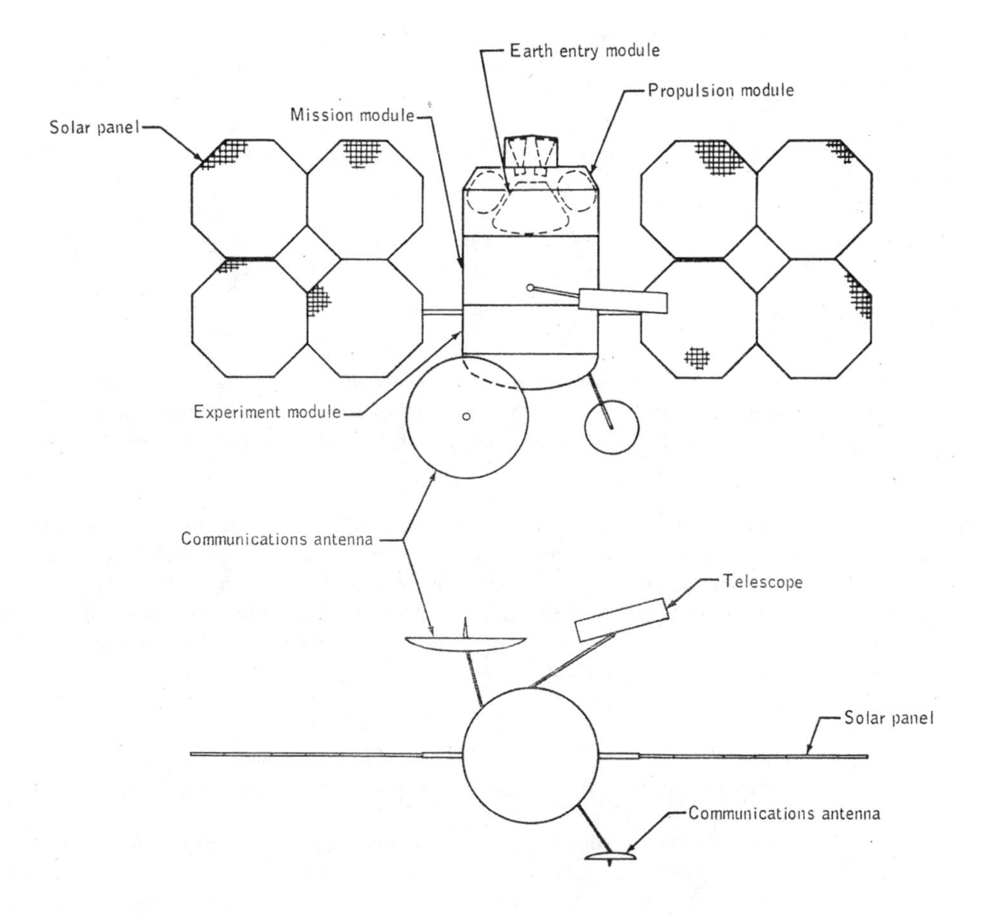

Figure 2 Spacecraft with Deployed Solar Array

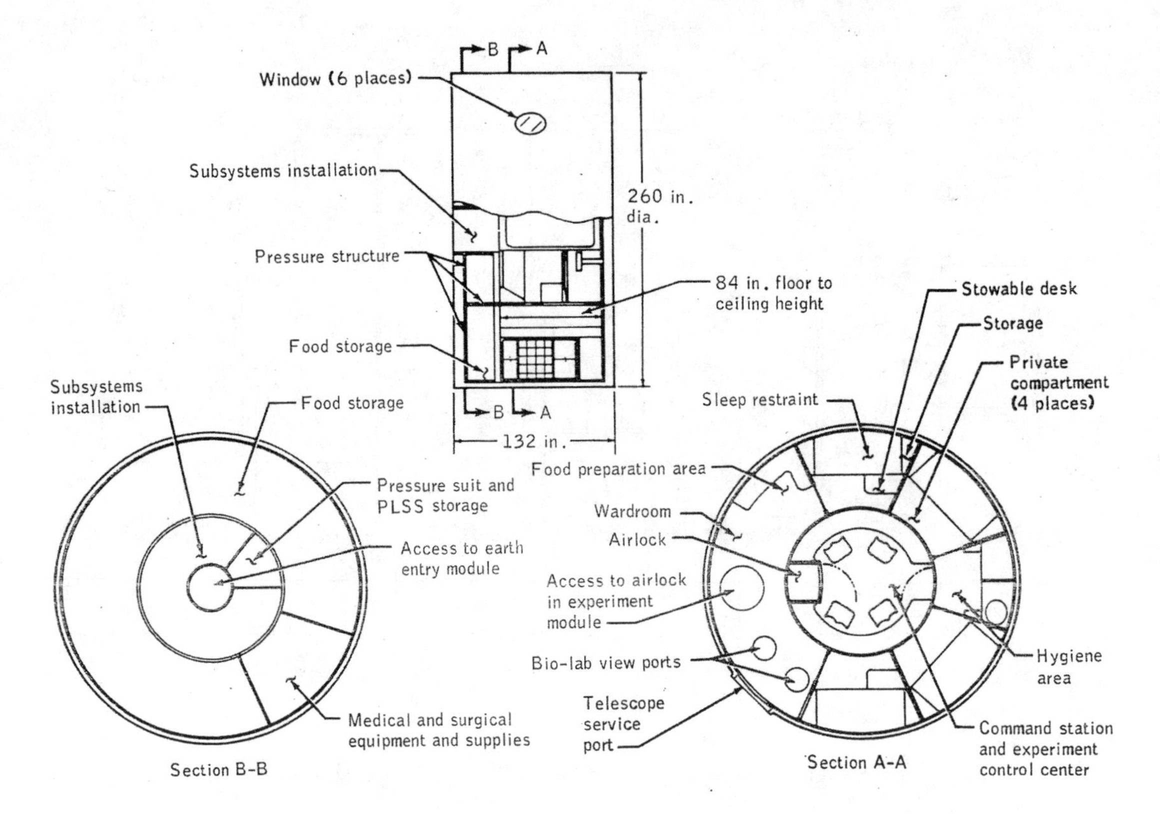

Figure 3 Four-Man Mission Module

The Experiment Module (EM) houses the Mars Surface Sample Return probe (MSSR), other probes, and a sample analysis laboratory. The MSSR probe and other probes are deployed for launch through doors in the aft bulkhead of the EM. A remotely operated system releases the MSSR and other probes from their support structure and positions them for launch.

The solar panels and the communications system antenna are stowed around and behind the EM during Earth orbit launch and planetary injection. They are deployed after launch as shown in Figure 2.

7.1.2 Mission Module (MM)

The Mission Module contains the equipment necessary for command and control operations of the spacecraft and experiments, and it contains the living accommodations for the crew. Figure 3 shows a design layout of the four man Mission Module. The MM is basically the same for each encounter mission and is also used, with modifications, as an Earth Orbital Space Station.

The MM structure is designed with two separately pressurized volumes with 7-foot ceilings. The internal volume contains the spacecraft and experiment control center and is separated from the outer crew accommodations area by a pressure-tight door and an airlock. The spacecraft outer structure is designed with a bumper and shield configuration to a thickness necessary for a 0.99 probability that the spacecraft will not be penetrated by a meteoroid. However, in the event that a penetration of the outer structure does occur, the crew can stay in or move to the control center, don pressure suits, and egress through the airlock to patch the hole.

The total pressurized area of the spacecraft is 3181 cubic feet. Internal structure, subsystems, crew systems, and food storage require 941 cubic feet, which leaves 2240 cubic feet of free, habitable volume. This 560 cubic feet of free volume per crewman exceeds the minimum volume recommended for extended duration missions.

In addition to providing adequate habitable volume, the crew accommodations are designed to provide maximum utility and privacy. The crew accommodations area contains four private compartments, a wardroom, hygiene area, food preparation area, and access to experiments. Each private compartment contains a sleep restraint system, stowable desk, storage area, and window to the outside. The wardroom is used for recreation, exercise, and eating. The wardroom wall has a window and a telescope docking port through which the telescope system may be serviced before remotely moving it from the spacecraft and decoupling it for remote operation.

The control center contains the spacecraft and experiment monitoring and control equipment and consoles, with four separate stations. Pressure suits and Portable Life Support Subsystems (PLSS) are stored and serviced in this area. Minimum hygiene facilities are provided for emergencies. The control center has access by overhead hatch to the EEM for planetary injection, EM checkout, normal Earth entry procedures, or when a solar proton event is sensed. Most of the subsystems are located in and are accessible from the control center.

The Mission Module contains five major subsystems which are:

1. Environmental Control and Life Support Subsystems (EC/LSS)
2. Electrical Power Subsystem (EPS)
3. Communications Subsystems (CS)
4. Guidance and Navigation, and Stabilization and Control (G&N and S&C)
5. Experiments Control and Data Handling

The Environmental Control and Life Support Systems provides atmospheric supply and control of carbon dioxide and trace contaminants, water supply and reclamation, thermal and humidity control of the atmosphere and equipment, water management, and waste management. The system is designed for shirt sleeve operation, but capability for long-term suited operation has been provided. The EC/LSS operations is shown functionally in the diagram in figure 4. A 7 psia oxygen-nitrogen atmosphere is provided in which the partial pressures of the two gases are approximately equal. Sub-critical cryogenic oxygen and nitrogen is stored in supply tanks in the MCPM. The atmospheric regeneration loop provides debris traps, particulate filters, charcoal filters, and a catalytic burner for contaminant control with atmosphere circulation provided by compressors in the atmospheric regeneration loop and, when necessary, by compressors in the suit loop. The atmospheric regeneration loop heat exchanger and water separators provide dehumidification and thermal conditioning of the atmosphere. Carbon dioxide is removed by regenerative molecular sieves and dumped overboard. Trace contaminant control is provided by a catalytic burner and a post-chemisorbent bed. Provisions for purging by dumping the atmosphere to space and repressurizing as many as six times during the mission allows flexibility. All water except fecal water is reclaimed and reused. Thermal control is provided by a coolant circuit to the atmospheric regeneration circuit, cabin cooling circuit, carbon dioxide collection system, and the electronic equipment cold plates. Heat is rejected through the MM structure and skin which acts as a radiator in a manner similar to the Gemini system. The EC/LSS components and circuitry are designed for redundancy. The atmospheric supply system is backed up by extra storage tanks.

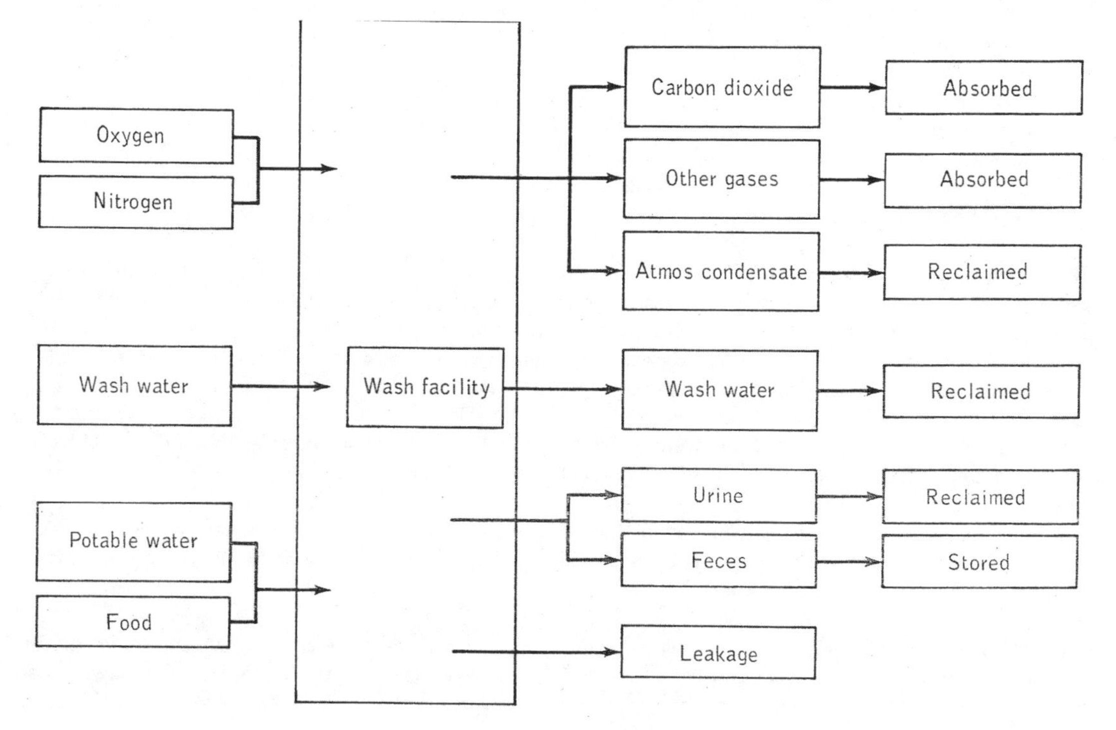

Figure 4 Environmental Control / Life Support Subsystem Diagram

The Electrical Power Subsystem (EPS) consists of 1690 square feet of deployable solar array, 1200 lb of auxiliary peaking batteries, and the accompanying regulators, controls, cable, etc. The solar cell array is folded and stored behind the EM at launch, and is deployed after injection in two independent sections fixed to two sides of the MM. The array is sized to provide a 4.71 KW conditioned power output at 2.2 A.U. from the Sun for the Mars Twilight Encounter Missions. This provides 8 KW at Mars encounter and 17 KW at Earth. Either half of the solar array will provide emergency power for safe return. Batteries provide a secondary source of power during peak power requirement periods, injection, and Mars dark side passage. A preliminary electrical load requirement is shown in Table 1. Spacecraft electrical power is supplied by 1200-hour AAP fuel cells onboard the CSM for Earth orbit operations and checkout.

NOMINAL POWER LOAD REQUIREMENT AT 2.2 A.U.

	Nominal Power Requirement, Watts
COMMUNICATIONS, INSTRUMENTATION, DATA MANAGEMENT, AND LIGHTING	2000
GUIDANCE AND NAVIGATION, STABILIZATION AND CONTROL	650
EXPERIMENTS	200
EC/LSS	940
CREW SUPPORT	305
DISTRIBUTION, CONDITIONING AND BATTERY CHARGING	615
TOTAL	4710

Table 1

The Communications Subsystem consists of two variable power, variable bandwidth FM 675-watt output transmitters which can be used together, a 50-watt output solid state PM emergency transmitter, a steerable 20.5-foot diameter parabolic antenna, and 8-foot diameter parabolic antenna, data recording and processing equipment, teletype, television and intercom. With a 210-foot antenna network at Earth and an electrical power budget of 3.8 KW at Mars encounter, this equipment will provide the following capability on the Mars twilight encounter, i.e., the most stringent mission for communications design:

1. Lunar Orbiter Quality Photographs – 6½ minutes per photo at Mars encounter
2. Commercial Quality Down Television – 35 million N.M from Earth
3. Live Apollo-quality Down Television – 100 million N.M.
4. 1 Million bits/sec data return rate – 1.12 million N.M.
5. 192,000 bits/sec – maximum range

The large antenna will time-share the data return from probes to spacecraft at encounter and until about two days past encounter. During this time, the large transmitters using the 8 foot antenna can communicate with Earth with a data rate of greater than 24,000 bits/sec. Then, the 20.5 foot antenna will be aimed at the Earth, and all the data received at encounter can be telemetered back to Earth before the electrical power output of the solar cells falls below that required for maximum transmitter power output.

The communications subsystem will be located next to the MM outer wall with virtually none of the cable from the transmitter to the antenna inside the Mission Module. This will minimize circuit losses.

The Guidance and Navigation (G&N) Subsystem consists of an inertial reference unit, a sextant, a computer, and a Control Moment Gyro (CMG) package. The Apollo G&N and Stabilization and Control Subsystem can perform basically all the necessary functions with a zero-gravity configuration.

With periodic onboard navigation the Apollo sextant accuracies are adequate, and the current computer can process the required guidance equations. Modifications to the sextant and computer may be desirable, however, for the inclusion of stadiametric data. Primary navigation will be provided by the onboard equipment and backed up by Earth based DSN tracking.

If low level artificial gravity is needed, additional equipment will be required. Spinning stability would be accomplished with a wobble damper, i.e., an electrically driven inertia wheel in a two-axis gimbal. Dampers on the gimbals absorb the wobble energy and stabilize the spin.

The Experiment Control and Data Handling Subsystem consists of computers, recorders, visual display, and digital readout equipment to monitor, store, and analyze data during the preparation for and conduct of experiments. The systems for operation are also included. The components within the system are either currently available or will evolve from Apollo and AAP activities. Improved systems with micromodules are very desirable due to size reduction and increased reliability.

7.1.3 Earth Entry Module (EEM)

The Earth Entry Module shown in Figure 5 is a 154-inch diameter entry spacecraft based on the aerodynamic shape of the Apollo Command Module (CM). It is designed for the single function of direct aerodynamic Earth entry and landing. Because of the radiation protection inherent in its design and location, it is also used as a storm shelter during solar proton events. The EEM is occupied by the crew during injection from Earth orbit in the event that abort is necessary.

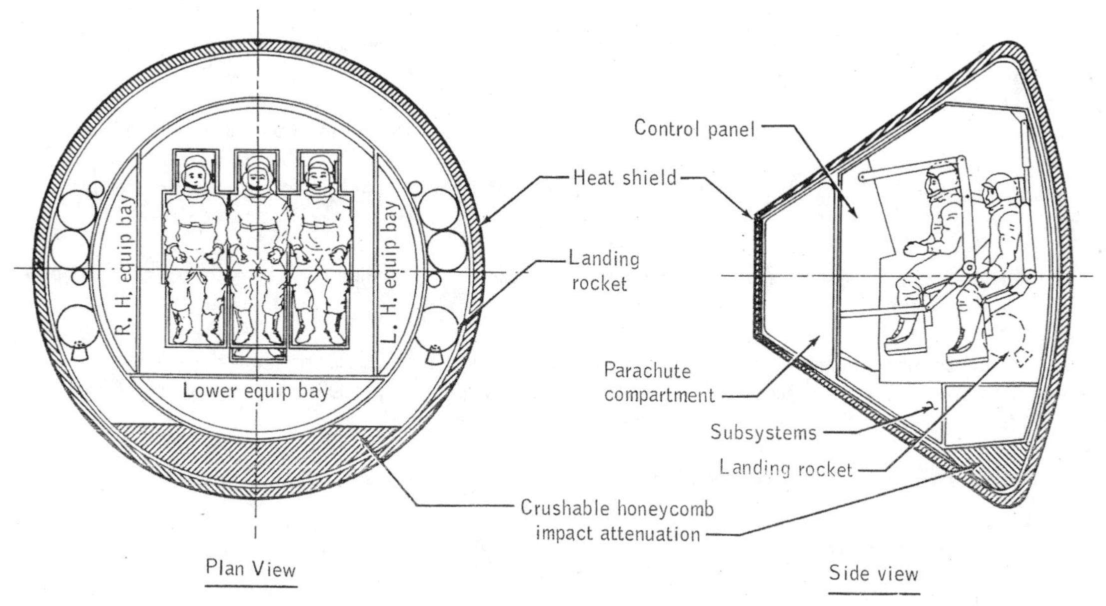

Figure 5 Four-Man Earth Entry Module

The two main parameters affecting the extent of modifications necessary to the Apollo CM are crew size and entry velocity. Four men can be accommodated in the Apollo CM space during reentry with minor CM modifications, but a crew size of six would require more extensive redesign.

The four planetary encounter missions in the flight schedule for this study have Earth entry velocities of from 45,000 to 51,000 fps, or up to 15,000 fps greater than the present Apollo design requirement of 36,000 fps. The increase in thermal loads will require an increase in heat shield thickness and weight, with a corresponding modification to the structure holding the heat shield, but major design changes to the EEM are not necessary for these entry velocities. The design incorporates solid propellant landing rockets in addition to gliding parachutes to provide a land landing capability. The Apollo CM docking structure and tunnel are deleted, and the parachutes are stored in this area.

As discussed in more detail elsewhere, the crew requires radiation protection from the highly damaging, infrequent solar proton events. The EEM itself provides a high degree of shielding inherent in its design as an entry vehicle because of the mass of its structure, closely packed dense subsystems, and its location in the spacecraft. Consequently, the crew will use the EEM as a radiation "storm shelter" when large solar events are sensed. Any radiation "holes" or "leaks" will be plugged with additional shielding material added around the EEM in the Mid-Course Propulsion Module.

As shown in Figure 5, secondary bulkheads isolate the parachute compartment, the solid propellant landing rockets, and the liquid bipropellant Reaction Control Subsystem from the pressurized interior volume. Other subsystems are inside the pressurized volume. The EEM subsystems are the:

1. Electrical Power Subsystem (EPS)
2. Environmental Control Subsystem (ECS)
3. Communication Subsystem (CS)
4. Guidance and Control Subsystem (G&C)
5. Reaction Control Subsystem (RCS)
6. Earth Recovery Subsystem (ERS)

During the mission, electrical power is supplied to the EEM by the spacecraft's primary electrical power subsystem. The EEM electrical power is supplied by batteries which are used during reentry.

The EEM Environmental Control Subsystem performs the same functions as the Mission Module ECS, but it uses a less sophisticated open cycle. A single gas atmosphere is used with high pressure gaseous storage. Lithium hydroxide canisters are used for carbon dioxide removal, and water is stored for crew consumption and heat rejection augmentation.

The Communications Subsystem provides voice communications with ground stations from the time that the crew enters the EEM in preparation for entry until recovery.

The Guidance and Control Subsystem consists of an inertial platform, a computer, an optical system and associated equipment. It provides steering and rate damping commands to the EEM RCS during Earth entry and thrusting commands to the Mid-Course Propulsion System in case of abort during and up to five minutes after planetary injection from Earth orbit.

The Earth Recovery Subsystem includes drogue and main parachutes, landing rockets, recovery aids, and other associated equipment.

The Reaction Control Subsystem consists of ablatively cooled rocket thrusters, propellant and pressurant tanks, and associated valves, regulators, controls and lines. Earth-storable, hypergolic propellants are used. The RCS provides control for reentry attitude and for roll and rate damping during entry.

7.1.4 Mid-Course Propulsion Module (MCPM)

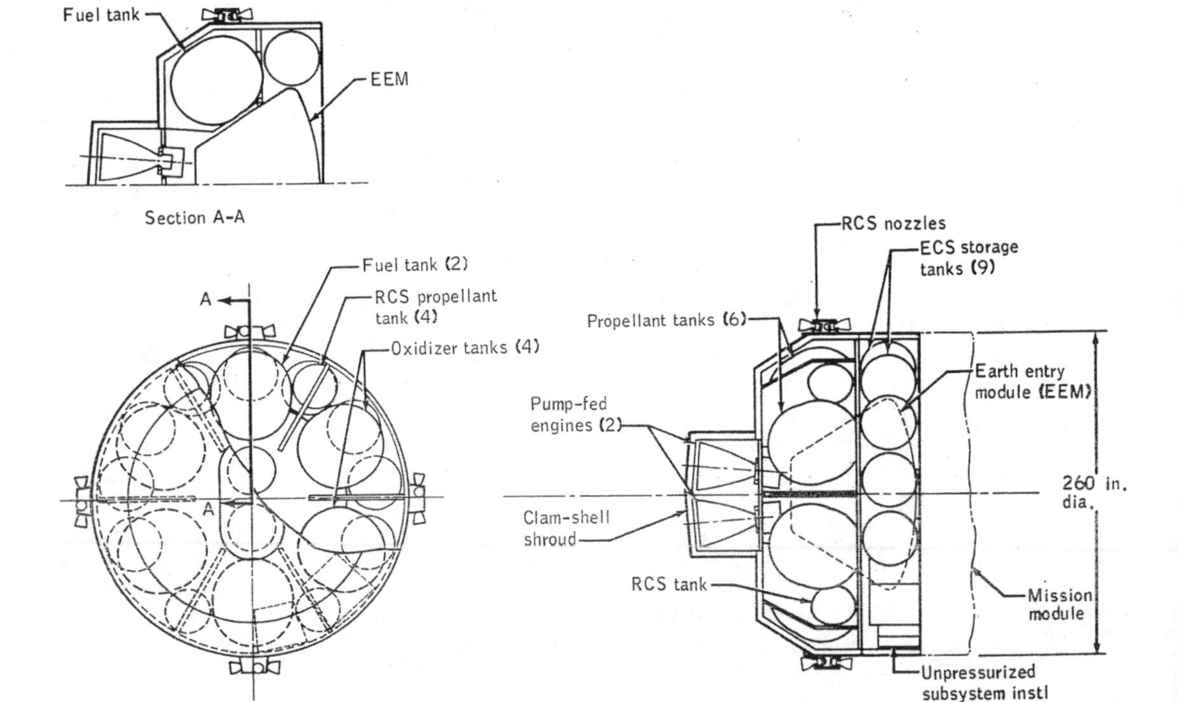

Figure 6 Mid-Course Propulsion Module

The Mid-Course Propulsion Module, shown in Figure 6, provides the propulsive thrust required for navigational mid-course maneuvers between planets. It also provides the propulsion necessary for abort from the beginning of planetary transfer injection up to five minutes after injection cutoff.

The subsystems contained within the MCPM are the:

1. Primary Propulsion Subsystem
2. Reaction Control Subsystem (RCS)

The Primary Propulsion Subsystem consists of propellant and pressurant tanks, a propellant distribution system, and two regeneratively cooled, liquid bi-propellant, pump-fed engines. The propellants and Flox and Methane, which are considered space-storable. The engines are modified RL-10's, with a thrust level of 15,000 lb each and a specific impulse of 385 lb-sec/lb. Both engines are used in case of abort, while one is used for mid-course corrections. The development schedule for this propulsion system is shown in Section 10.0.

Sixteen RCS engines are grouped in fours around the MCPM in four places. These engines are used for attitude orientation and for unloading the Control Moment Gyros.

The MCPM also houses the EEM and the Life Support Subsystem cryogenic storage tanks. The module is configured with the propellant tanks, engines, and other large masses clustered as closely as possible around the EEM to provide the maximum radiation protection to the crew when they are using the EEM as a radiation storm shelter. The propellant tanks are sized for the most stringent mission, i.e., the Triple Planet Encounter, and are off-loaded for other missions. The entire MCPM is designed for an equilibrium temperature of 70° F.

7.1.5 Experiment Module (EM)

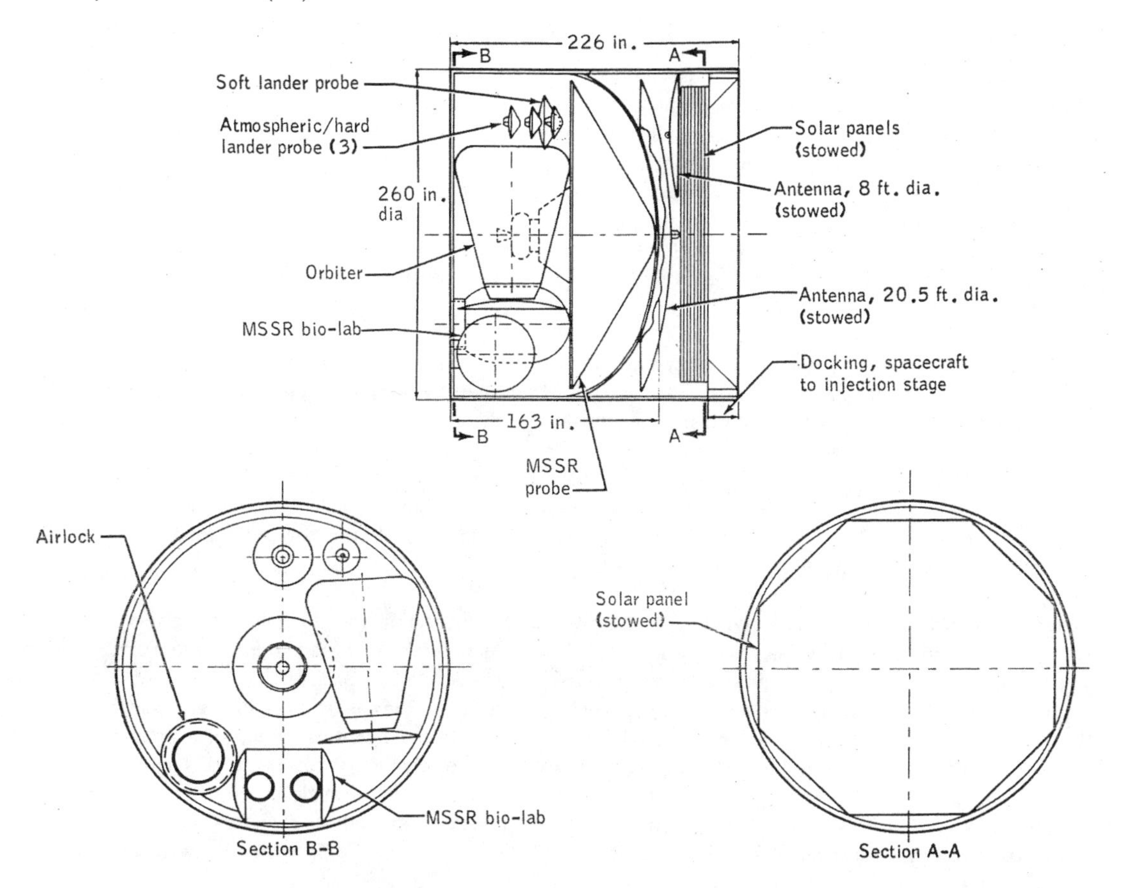

Figure 7 Experiment Module For 1975 Mars Twilight Encounter Mission

The Experiment Module is configured to structurally support, environmentally and biologically protect, and operationally deploy the experimental payload. Figure 7 shows the layout of the EM for the 1975 Twilight Encounter mission which includes:

1. Mars Surface Sample Return Probe (MSSR)
2. Soft Lander Probe
3. Orbiter Probe
4. Three Atmosphere or Aerodrag Impacter Probes
5. Remotely Operated Biological Laboratory
6. Probe Deployment Subsystem

Before installation of the EM onto the spacecraft, the probes are installed within the EM, checked out, and welded in. The entire EM is then subjected to a high-temperature heat soak for sterilization. The EM is not normally opened during the mission until probe deployment. Remote probe in-flight checkout is performed from the MM. Limited probe in-flight maintenance is possible through an airlock and biologically-shielded suit system which allows a crewman to enter the EM without contaminating the area. This is the same method in which the surface sample returned by the MSSR is manipulated in the biological laboratory and sealed in a container before being brought into the Mission Module.

The EM for the 1977 Triple Planet Encounter and 1978 Dual Planet Encounter is much larger than the one shown because of the much larger experiment payload for multiple planet encounters. This difference is reflected in the weight statement in Section 7.1.7. The three separate payloads for the 1977 mission are actually in three separate biologically-shielded volumes, and each compartment is jettisoned as its probes are deployed.

The airlock located within the EM is also used for crew ingress and egress to the spacecraft. The solar array and the two communications antennas are folded into the end of the EM and stowed there during launch and injection.

7.1.6 Environmental Protection

Meteoroid Protection

There is considerable uncertainty in the development of a meteoroid flux model for use with the Mars Twilight Encounter mission because of the unknowns of the asteroid belt beginning just beyond the Mars mean orbit distance of 1.5 A.U. Although meteoroid models have been constructed with reasonable confidence for the near-Earth cometary material, meteoroid models including asteroidal particles must be based on extrapolations between measurements made near Earth and optical and radar observations of the larger asteroids. The large uncertainties in the model meteoroid environments are largely due to these indirect methods necessary to construct such a model. However, an upper limit to asteroidal meteoroid flux between 1.9 and 4.3 A.U. is possible by incorporating observations of gegenschein light into the model, which is the basis of the meteoroid flux model used to derive the weights shown in Table 3 in Section 7.1.7.

Projections of penetration mechanics theory beyond experimental information is another source of uncertainty. Critical particle size larger than those now in use experimentally are predicted for the twilight missions. Consequently, some extrapolations of present penetration mechanics are required.

Testing and evaluation programs presently exist which should increase confidence in the experimental extrapolations. Reduction of uncertainties in the meteoroid environment itself, which account for the major uncertainty in protection requirements, could be accomplished through the use of Mariner-class or larger meteoroid measurement experiments out to 2.5 A.U.

Meteoroid shielding required is a function of the area of the spacecraft, the time-distance relationship of the trajectory, and a probability of sustaining a given number of penetrations. The additional meteoroid protection weights shown in Table 3 are for a protection system which is based on a 0.99 probability that the meteoroid shield will not be penetrated during the mission. The shielding weights for the twilight missions are also based on a directional shielding concept, in which the general unidirectionality of the asteroids in their orbits about the Sun is used to advantage to minimize shielding requirements by always orienting the spacecraft toward the oncoming asteroidal particles. Shielding requirements for the spacecraft regions are shown in Figure 8 for the twilight missions, which gives a meteoroid shielding requirement of 7720 lb over that required for structure alone. If the Twilight Encounter missions were shielded for an omnidirectional flux, the meteoroid shielding requirement would increase by 1100 pounds.

The 1977 and 1978 missions have aphelions of about 1.6 A.U. at which asteroidal flux is very small, so consequently, the spacecraft for these missions is shielded primarily for omnidirectional cometary particles which yields a lower shield weight per unit area. The larger Experiment Modules of these missions, however, still causes shielding weight requirements of 6100 and 7940 lb for the 1977 and 1978 missions, respectively.

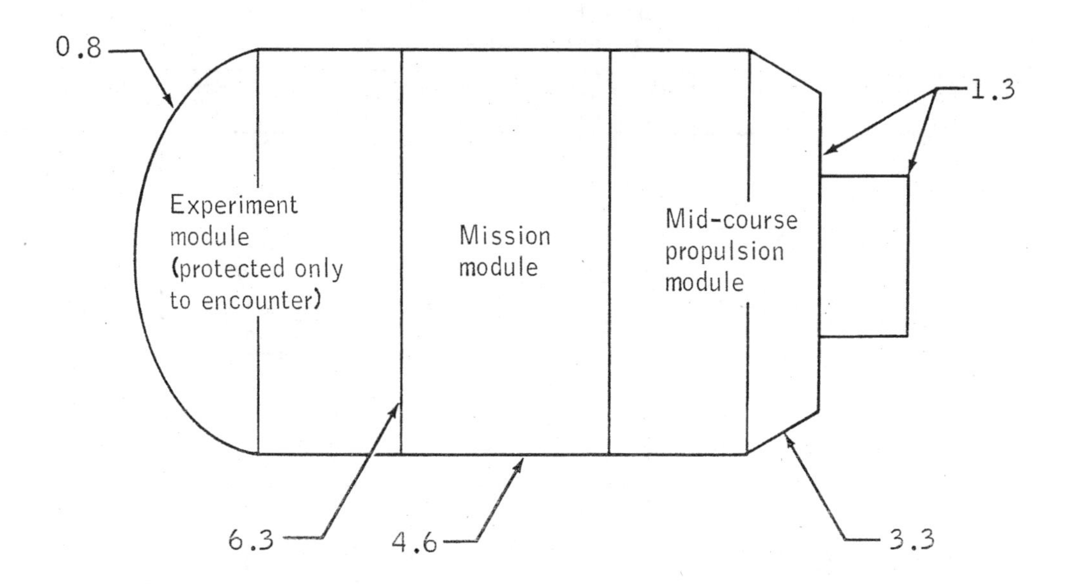

Figure 8 Total Meteoroid Protection Requirements In lbs/ft^2 For Mars Twilight Encounter Mission

Because of the thickness required for meteoroid. shielding, the shield serves as the primary load carrying structure. The structural shell consists of an outer aluminum "bumper" skin and an inner "shield" spaced several inches apart. The annular cavity is filled with a low-density, open-cell, foamed plastic. The meteoroid shielding weights shown in Table 3 are for those required over the structural weights which would be required if there were no meteoroids.

Radiation Protection

Solar proton events cannot yet be determined in advance, although present data indicates that they are more frequent during the half of the 11 year solar activity cycle when sunspots and surface disturbances are more numerous. Since solar proton events, as such, were unrecorded 15 years ago, data is available only on the last solar cycle (1954-1965). Statistical analyses of the doses produced from particle events during this (the 19th) solar cycle yields a requirement of 22 grams per square centimeter shielding to protect the crewman to a probability of 0.99 of not receiving a radiation dose greater than 100 Rems per year. Rem dose includes the effects of Relative Biological Effectiveness (RBE). This shielding requirement includes secondary radiation effects.

For these missions the Earth Entry Module is used as a radiation "storm shelter" because its structure, closely clustered subsystems, and its location among equipment of large mass in the MCPM give it a high inherent equivalent shielding effect. A preliminary solid-angle radiation shows thin areas of about 100 square feet which can be brought up to 22 gm/cm^2 with 600 lb of additional polyethelyne shielding material added to the MCPM. The 1975 mission requires no additional shielding because it occurs during the lower part of the solar cycle.

Additional data to be obtained during the current solar cycle will enable more accurate prediction of solar proton events. Current biological studies will lead to the establishment of definitive dose limits for crew members on long duration missions.

7.1.7 Preliminary Spacecraft Weight Summary

Table 2 presents a summary of the spacecraft weights for the four baseline missions. These weights do not include the experiments or the structure and deployment mechanism for the probes. Table 3 shows a weight summary for the total injected payload. The weights shown for meteoroid protection are for directional shielding (see Section 7.1.6, "Environmental Protection"), and reflect only that weight required over what would normally be needed for structure if there were no meteoroids. The radiation protection is that required to augment areas around the EEM where the shielding inherent in the design does not meet or exceed that required to protect the crew to the desired level. The 1975 mission requires no extra shielding because it comes during the lower part of the solar cycle.

SPACECRAFT WEIGHT SUMMARY
(Not including Experiments or EM)

	1975 Twilight Encounter	1977 Triple Planet Encounter	1978 Dual Planet Encounter	1979 Twilight Encounter
Mission Module	9310	9310	9310	9310
Subsystems (Dry)	15920	15250	15250	15920
Earth Entry Module	15100	15100	15100	15100
Mid-Course Propulsion Module (Dry)	20790	20790	20790	20790
Life Support Expendables	17670	17670	16970	17670
Propellant (Not Including Abort)	17000	33880	22500	17000
Total Spacecraft Weight, lbs	95790	112,000	99920	95790

Table 2

INJECTED PAYLOAD WEIGHT SUMMARY

	1975 Twilight Encounter	1977 Triple Planet	1978 Dual Planet Encounter	1979 Twilight Encounter
Spacecraft Including All Subsystems	97590	112000	99920	95790
* Additional Radiation Protection Above That Inherent in EEM	0	600	600	600
* Additional Meteoroid Protection Above That Required for Structure	7720	6100	7940	7720
Five-Minute Abort Additional Propellant Requirement	7200	0	5500	5600
Docking Adapter	2200	2200	2200	2200
Total Experiment Payload	35000	57050	56500	35000
Experiment Module	11500	21000	21000	11500
Total Injected Payload, lbs.	159,410	198,950	193,660	158,410

* See Section 7.1.6, "Environmental Protection"

Table 3

7.1.8 Reliability / Maintainability Considerations

The extended mission durations of the Mars / Venus Encounter Missions, and other planetary missions as well, dictate that spacecraft systems must be capable of operating for at least two years in space without resupply. To accomplish these missions, an approach to reliability and/or maintainability must be developed which will insure crew safety and mission success. The following guidelines and constraints have been established for spacecraft systems and subsystems design:

a. Where possible, all subsystems shall be designed for reliable operation without maintenance or repair for the entire mission. When maintenance or repair is unavoidable, the goal is that crew activity should be limited to switching to backup subsystems, "black box" modular replacement, etc., with detailed, complicated, and time consuming tasks being avoided.
b. Extravehicular activity shall be limited to emergency situations only.
c. Maintenance and repair of experiments shall be treated, in general, as are the spacecraft systems and subsystems. Probes which require sterilization should be designed for reliable operation without maintenance to avoid resterilization problems.

Even after designing for reliability, it is unavoidable that some component failures will probably occur during a two-year mission; consequently, a contingency capability must be available. An onboard checkout system of limited capability may be included. However, before an onboard checkout system can be of value, each subsystem must be evaluated independently to determine what parameters should be checked, what could be done if a difficulty were discovered, and if repair or replacement could be effected.

The development schedule for the Planetary Encounter Spacecraft shown in Section 10.0 is dependent upon the AAP and/or Earth Orbital Missions and the Planetary Spacecraft development flights to develop and prove systems and operational techniques based on this reliability / maintainability philosophy. In addition, each individual component which supports the planetary mission will have gone through development and life tests as well as selected reliability testing on the ground. As in previous manned space flight programs, these data will be utilized to support reliability estimates for the various systems.

In general, systems and subsystems in which a single failure could result in a catastrophic failure will be designed for redundancy, and the less critical systems and subsystems will use backup and/or alternate methods to assure meeting systems requirements. With this philosophy, necessary confidence should be established to proceed with Planetary Encounter Mission Spacecraft designs.

7.1.9 Alternate Design Considerations

Crew Size

In a mission the length and nature of the planetary encounter, it is expected that the mission time line will include relatively long periods of nominal crew operations interrupted by periods of peak activity associated with experiment requirements and opportunities. The nominal periods schedules may be considered as oriented around crew personal functions and spacecraft systems management, and will be decreased as other operations, especially experiments at planetary encounter, become more demanding. The crew housekeeping functions, per 24-hour day for one crew member, is as follows:

Sleep	7.5 hours
Eat	2.0 hours
Personal Hygiene	1.0 hour
Exercise and Scheduled Recreation	1.5 hours
Physiological and Performance Check	1.5 hours
Total	15.5 hours

The total of 15.5 hours leaves 8.5 hours per man 24 hours available for "work," i.e., systems management, experimentation, flight operations, and onboard crew training in critical but seldom practiced operations (reentry procedure, probe launch and management, etc.). This represents a 59.5 hour "work week" which is only moderately high by Earth standards.

Experiments, systems management, and training will utilize all of the available "work time" at encounter for a four man crew by design. However, full-time utilization becomes increasingly difficult for a five- and six-man crew. In the tightly-controlled environment of an extended mission spacecraft, it is important that most of the time be tentatively scheduled, if only in the interests of preventing boredom which may even be a problem with only 4 men at other than encounter phases of the mission.

During critical mission periods the personal time may be reduced by eliminating exercise and recreation, and by temporarily reducing sleep time by 1.5 hours. The sleep time for the mission-critical periods is that which has been accepted by NASA for the AAP mission planning. It gives the crew a basically adequate extended period of sleep (six hours) which can be supplemented when required by another shorter "nap" (1.5 hours) as fatigue builds up to the point of impairing performance.

A crew of four is fully committed for the presently defined experiment and spacecraft operational work load during the 30-day Mars approach, encounter, and departure period.

An increase in crew size from the present four up to six would require a larger Mission Module, extensive modification to the Earth Entry Module, additional mid-course propellant, additional life support EXPENDABLES, and a larger solar array. The six man spacecraft design weighs approximately 26,000 lb more than the present 4 man spacecraft deign.

Post-Injection Abort

The spacecraft is designed with the capability of aborting the mission and returning the crew to Earth up until five minutes after reaching planetary injection velocity (final injection stage cutoff). This allows time for trajectory and velocity checks with the onboard G&N equipment backed up by the ground support system, and for critical subsystems checkout before committing the spacecraft to a "Go" decision.

A five-minute abort capability requires that additional propellant be added to that already onboard for mid-course maneuvers for the 1975, 1978, and 1979 missions. The larger mid-course requirement of the 1977 Triple Planet

Encounter Mission inherently allows a post-injection abort capability of greater than five minutes. The abort velocity requirement varies with the injection velocity, so is different for each mission. The additional propellant weight required for a two-impulse burn, 1½-day return to Earth for each mission is as follows:

1975 Twilight Encounter	7200 lbs.
1977 Triple Planet Encounter	0
1978 Dual Planet Encounter	5500 lbs.
1979 Twilight Encounter	5600 lbs.

Abort is initiated by the crew, who is in the EEM during injection from Earth orbit. As shown in Figure 9, after the abort sequence has been initiated and the injection stage is cut off if still thrusting, the MCPM is separated at the forward bulkhead by pyrotechnic connectors. Solid propellant posigrade rocket motors propel the abort configuration forward and laterally at a relative velocity of approximately 100 feet per second. The abort configuration coasts away from the remaining part of the spacecraft and the injection stage for 20 seconds, during which time it is reoriented to the correct retro attitude by RCS abort thrusters. The abort configuration consists of only the EEM, propulsion system and supporting structure, and a small abort RCS. The remaining structure is jettisoned for minimum weight. Both mid-course engines are then retrofired, imparting velocity to the EEM required to put it in a 1½-day return trajectory. After cutoff, the EEM is separated from the propulsion system, and a normal entry is effected.

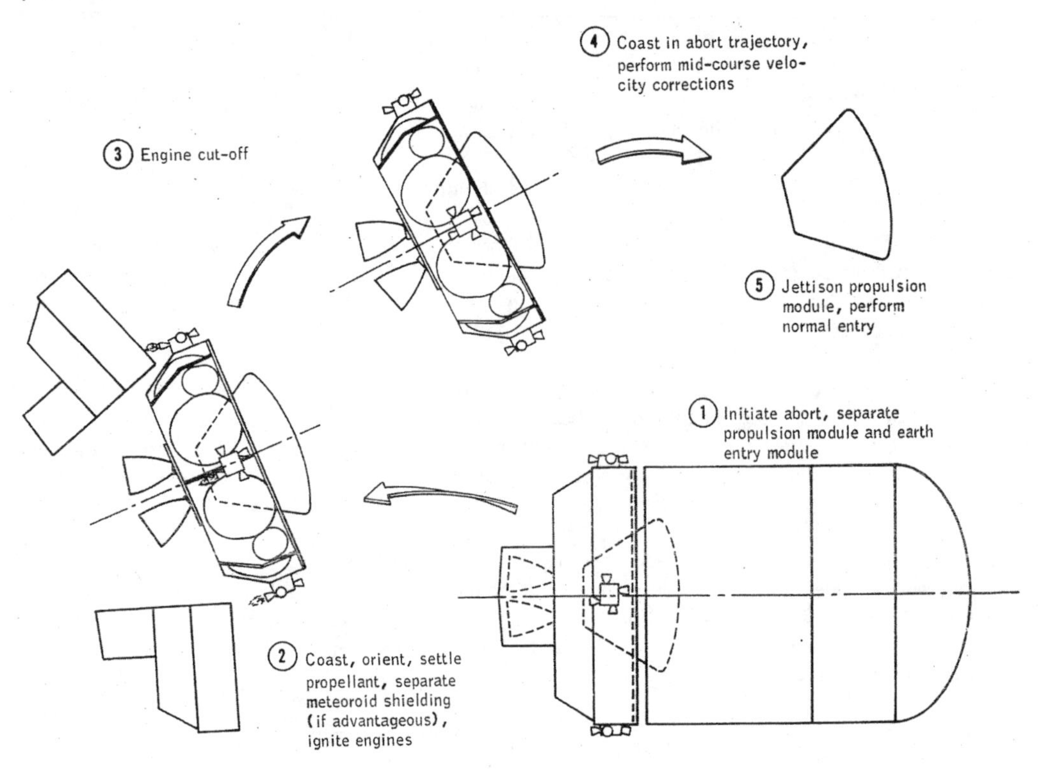

Figure 9 Abort Sequence

Artificial Gravity

Human reactions to weightlessness have been postulated as possible limiting factors in manned space flight. The three gravity influenced organ complexes most frequently implicated as potential problem areas are the cardiovascular, the musculoskeletal, and the vestibular systems.

The pilots of the last two Project Mercury flights, the crews of the Project Gemini flights (Gemini III through Gemini XII), and, reportedly, some of the Russian cosmonauts, have exhibited signs of orthostatic hypotension on assuming the upright position after landing. This condition, which if severe enough could result in fainting, is characterized by cardioacceleration (increase in pulse rate), decreased blood pressure, and narrowing of the pulse pressure when the subject assumes the erect or standing position. Postflight pulse rate and blood pressure responses to the initial standardized tilt tolerance tests have indicated progressive degradation for space flights up to 8 days duration. Marked improvement, rather than further decrement in tilt response, was observed following the 14-day Gemini VII flight. In each case the altered initial postflight cardiovascular responses have returned to control or preflight values within 50 hours, and all crewmen have been perfectly capable of performing all programmed postlanding activities. The weight losses exhibited by the Mercury and Gemini astronauts have been attributed to losses principally of body water and electrolytes (salt). Apparently the major portion of this loss occurs during the initial 24 to 48 hours of exposure to the spacecraft environment. Variations in the total weight losses are a reflection of differences in individual responses and the thermal conditions which exist during a particular mission. It would appear that the suited mode, with its attendant increased sweating (evaporative cooling) results in a greater loss of water and electrolytes than the unsuited mode. In any case, the loss appears to be inadequately compensated for by increased salt and water intake during flight, resulting in the observed postflight dehydration and decreased total body weight. The fact that the preflight weights are very nearly approached in a matter of hours during the immediate postflight period by a process of simple hydration (increased salt and water intake) lends support to the hypothesis that the weight loss is principally a matter of moderate dehydration. One might postulate that this dehydration is an adaptive response of the circulatory system to the weightless environment and that a reduced blood volume is optimal under null gravity conditions, since in these circumstances venous return is facilitated and the requirement for an expanded "1-g type" blood volume no longer exists. The specific identification of the causes of this body weight phenomenon will have to await acquisition of more precise data, including accurate water salt intake-output and periodic inflight body weights.

The utilization of more sophisticated hematological techniques relative to the longer Gemini flights has provided evidence of significant decreases in red cell mass (Gemini IV, V, VII) and plasma volume (Gemini IV and V only). The causative factors operative here are also obscure, but there is a growing body of evidence that the slightly hyperoxic atmosphere has a deleterious influence on the unsaturated lipids of the red cell membrane. Other ground-based studies have emphasized the role of prolonged inactivity or confinement followed by resumption of activity in the destruction of red blood cells. The relative importance of these factors in the etiology of the red cell loss phenomenon must await further study.

These changes are considered to be principally adaptive manifestations of the cardiovascular system to the circulatory requirements of the body while under the influence of the space flight environment. Despite these noted changes, there has been no evidence of a discrepancy between cardiovascular system function and the adequacy of circulation during any space mission.

The potential inflight preventive measures for orthostatic hypotension evaluated include vigorous regular exercise regimens, periodically inflated cuffs applied to the thighs, and negative pressure periodically applied to the lower half of the body. In addition, a modified anti-G suit may provide temporary protection against postflight orthostatic hypotension if none of the above measures prove practicable. Flight results have not thus far demonstrated a need for the incorporation of such preventive or remedial measures.

As has been observed in simulated weightlessness, the muscles of the body are expected to show some loss of mass then they are no longer working against the accustomed gravitational force. Such reduced muscular activity may also produce some demineralization of the skeletal system. Neither of these adverse effects has been observed to any significant degree on the 14-day mission. In any event, ground-based weightlessness simulation studies indicate that an adequate exercise regimen will aid in preventing both of these untoward phenomena.

The otolith apparatus, one of the organs of equilibrium, is sensitive to alterations in gravitational force fields. It has thus been postulated that during prolonged weightlessness the lack of gravity might result in changes in sensitivity of this organ leading to disorientation and mild motion sickness. Ground-based and inflight experiments performed during the Gemini VII flight have shown that the otolith organ can adapt to a weightless exposure of 14 days, then readapt to the 1-g environment without eliciting any of the untoward symptoms referred to above. With current selection and training techniques and our environmental control systems, the risk of any serious problems of performance degradation, disorientation or motion sickness in the weightless environment appears minimal and imposes no limitation on the ability of the crew members to conduct successful space missions as long as Apollo.

NASA has demonstrated that the inflight and postflight performance of man will not be hampered for flight durations up to 14 days. In order to predict accurately the ability of man to negotiate successfully missions lasting for periods of months or years, detailed medical data from experiments in a non-confining weightless environment such as will be performed in increasing increments of mission duration in the Apollo Applications Program is necessary.

By 1969, flights of up to 60 days duration will be performed in AAP. A successful demonstration of man's ability to perform for this length of time in a zero-gravity environment should be sufficient to extrapolate to a two-year zero-G mission duration with enough confidence to proceed with a planetary spacecraft design without artificial gravity.

Later AAP and space station flights will have even longer mission times which should increase confidence in the 60-day results.

If artificial gravity were required for the planetary encounter missions, the spacecraft design would be modified to a configuration similar to the one shown in Figure 10 in which the final stage of the orbital launch vehicle is used as a counterbalance. The spacecraft is configured with the propulsion module on the top of the stack, allowing injection abort as with the zero-gravity configuration. Spacecraft deployment is accomplished by the rotation of the spent stage and the mission module about the hub module and then extending the supporting arms. The spent stage, the mission module, and the hub module rotate to provide artificial gravity while the propulsion module and experiment module do not rotate. The non-rotating experiment module allows orientation of antennas, telescope and other experimental components without affecting the rotating parts of the spacecraft. Since most of the EXPENDABLES and the probes are located in the non-rotating part of the spacecraft, the center of mass of the rotating parts may easily be maintained coincident with the spin axis. A spin radius (spin axis to floor of mission module) of 75 feet is shown in Figure 10, but the dimension may be varied without greatly affecting the configuration. The solar panels shown are located for the spin axis oriented toward the Sun, but they could be mounted on the non-rotating experiment module. Crew transfer between the hub module and the mission module is via a tunnel which also houses connecting air ducts and electrical umbilicals.

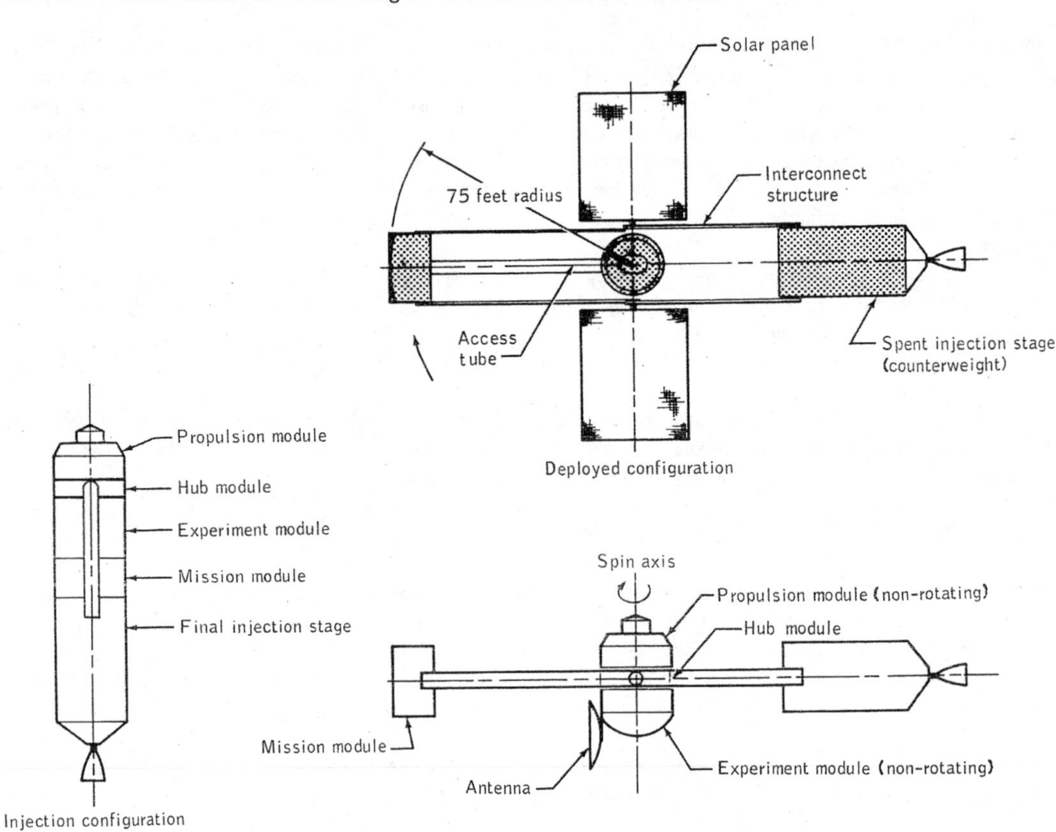

Figure 10 Alternate Configuration for Artificial Gravity

The difference in total spacecraft weight between artificial and zero-gravity configurations for each mission is the following:

1975 Twilight Encounter	40 - 44,000 lb.
1977 Triple Planet Encounter	70 - 78,000 lb.
1978 Dual Planet Encounter	60 - 66,000 lb.
1979 Twilight Encounter	40 - 44,000 lb.

Earth-Storable Propulsion Module

Figure 11 depicts an alternate Propulsion Module which contains a propulsion subsystem which is closer to present state-of-the-art than the Flox-Methane version shown in the spacecraft design. This subsystem uses Earth-storable

N_2O_4 - Aerozine 50 propellants in a modified Apollo Service Propulsion Subsystem (SPS). The expansion nozzle has been shortened to an area ratio of 40:1, which allows a smaller MCPM and clam-shell shroud around the engine for meteoroid protection. This reduction in area ratio reduces the specific impulse by 3 lb-sec/lb.

The lower specific impulse of the SPS would cause an increase in spacecraft weight over that required for the Flox-Methane engines. This increase would be 4100 lb for the Twilight Encounter mission and 6820 lb for the Triple Planet mission without an abort capability. With a post-injection abort capability of five minutes, the penalty for the Twilight mission would be 7200 pounds.

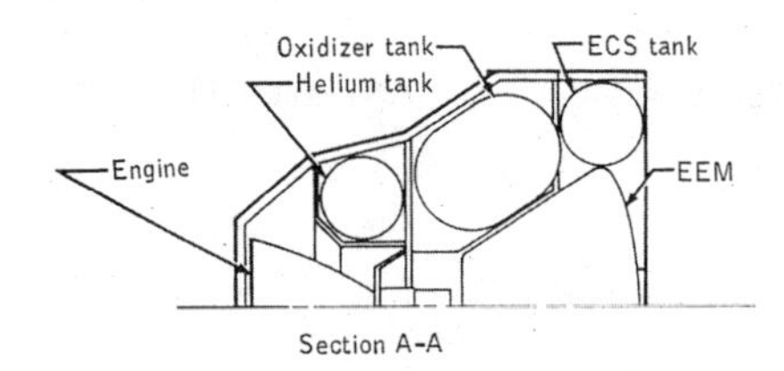

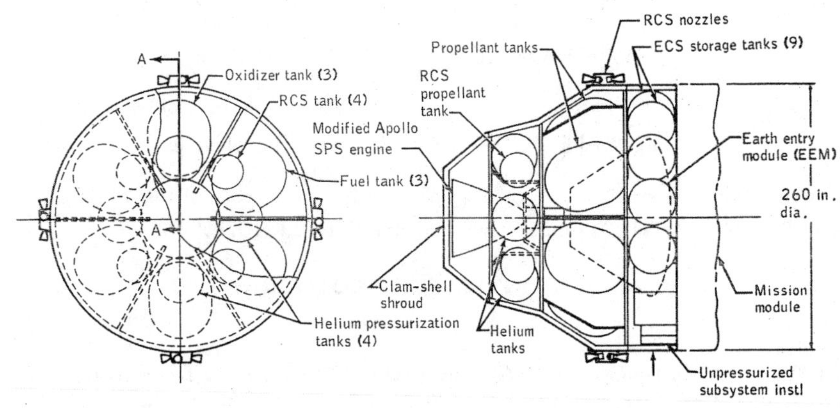

Figure 11 Alternate Mid-Course Propulsion Module

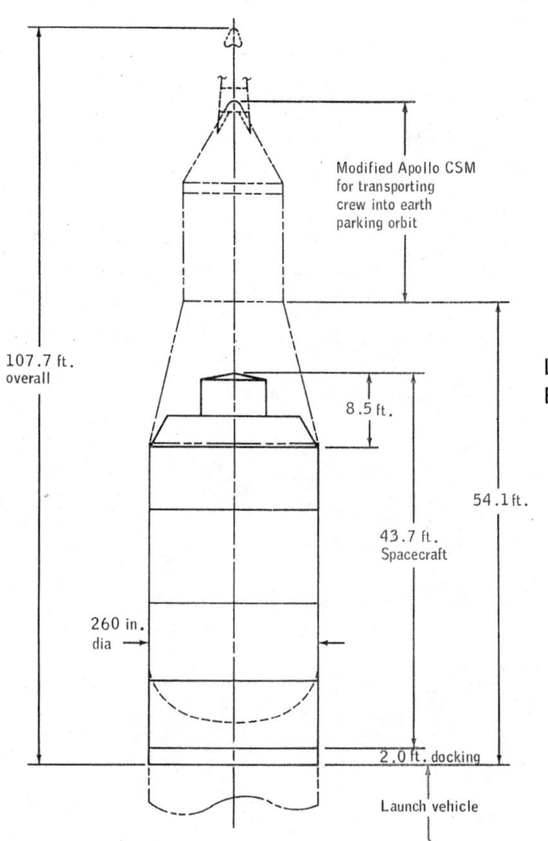

LAUNCH CONFIGURATION FOR MARS TWILIGHT ENCOUNTER MISSION

8.0 EXPERIMENTS

OBJECTIVES

INITIATE EXTENSIVE SCIENTIFIC EXPLORATION OF THE PLANETS
PROVIDE BASIS FOR FUTURE SPECIFIC SCIENTIFIC EXPLORATION

OBTAIN INFORMATION FOR DESIGN OF MANNED LANDING SYSTEMS
OBTAIN ATMOSPHERIC, SURFACE, AND BIOLOGICAL DATA FOR MANNED LANDING SYSTEM

PERFORM EN ROUTE EXPERIMENTS IN PHYSICS, ASTRONOMY, AND BIOMEDICINE

MAXIMIZE THE USE OF MAN

TARGETING AND TRACKING	COMMAND AND CONTROL	DATA DISCRIMINATION AND PROCESSING

1975 MANNED MARS FLYBY MISSION

AREAS OF INTEREST
EXOBIOLOGY, SOLID BODY AND SURFACE PROPERTIES, ATMOSPHERE

EXPERIMENTAL PAYLOADS
ON BOARD INSTRUMENTS, PROBES, IMPACTER, ORBITER, LANDER, MSSR

AREA OF INTEREST – PHOTOGRAPHY

FIELDS OF INTEREST
PLANET SHAPE, TARGETING, CLOUD COVER AND MOTION, SURFACE TOPOGRAPHY AND TEMPERATURE DISTRIBUTION, SPACECRAFT SURFACE INTERACTION, BIOLOGICAL MORPHOLOGY, ATMOSPHERIC TRANSMITTANCE

	DATA
FLYBY VEHICLE	
TELESCOPE APPROACH PHOTOGRAPHY	10^{11} BITS
HIGH RESOLUTION TELESCOPE PHOTOGRAPHY AT FLYBY	10^{12} BITS
MULTISPECTRAL APPROACH PHOTOGRAPHY	10^{12} BITS
FLYBY PANORAMIC COVERAGE	10^{13} BITS
ORBITER PROBE	
PHOTOGRAPHIC MODE	10^{11} BITS
TELEVISION MODE	10^{4} BPS TO EARTH
LANDER PROBE	
FACSIMILE CAMERA	500 BPS TO EARTH
MSSR PROBE	
FACSIMILE CAMERA SAMPLE SELECTION	10^{8} BITS
FACSIMILE CAMERA CLOUD PHOTOGRAPHY	500 BPS TO EARTH
PANORAMIC COLOR CAMERA	10^{11} BITS

AREA OF INTEREST – SOLID BODY AND SURFACE PROPERTIES

	DATA
FLYBY VEHICLE MEASUREMENTS	
I.R., VISIBLE, U.V., AND GAMMA RAY SPECTROMETERS; I.R. AND MICROWAVE RADIOMETERS	5×10^{8} BITS
ORBITER PROBE MEASUREMENTS	
MAGNETOMETER	200 BPS TO EARTH
TRACKING GRAVITY FIELD	—
GAMMA RAY SPECTROMETER	< 100 BPS

I. R., MICROWAVE RADIOMETERS	200 BPS TO EARTH
LANDER PROBE MEASUREMENTS	
SEISMOMETER	300 BPS TO EARTH
MAGNETOMETER	200 BPS TO EARTH
HEAT FLOW EXPERIMENT	LOW
SOIL MECHANICS EXPERIMENT	LOW
ALPHA BACKSCATTER EXPERIMENT	LOW
MSSR PROBE MEASUREMENTS	
ALL THE LANDER PROBE MEASUREMENTS PLUS	
SAMPLE RETURN FOR CHEMICAL ANALYSIS	4 POUND SAMPLE

AREA OF INTEREST – ATMOSPHERIC PROPERTIES

FLYBY VEHICLE MEASUREMENTS	DATA
MASS SPECTROMETER	10^4 BITS
I.R., VISIBLE AND U.V. SPECTROMETER	10^5 BITS
I.R. RADIOMETER	10^5 BITS
AERO-DRAG PROBE (IMPACTER) MEASUREMENTS	
DRAG DECELERATION / PRESSURE AND TEMPERATURE	10^4 BITS
ORBITER PROBE MEASUREMENTS	
I.R., VISIBLE AND U.V. SPECTROMETER	10^2 BPS TO EARTH
I.R. RADIOMETER	10^2 BPS TO EARTH
MICROWAVE RADIOMETER	10^2 BPS TO EARTH
TOPSIDE SOUNDER	50 BPS TO EARTH
MSSR PROBE MEASUREMENTS (SURFACE PACKAGE)	
MASS SPECTROMETER	2×10^2 BPS TO EARTH
PRESSURE AND TEMPERATURE	LOW
ANEMOMETER	30 BPS TO EARTH
LANDER PROBE MEASUREMENTS	
ALL THE MSSR PROBE MEASUREMENTS PLUS PRESSURE, TEMPERATURE AND COMPOSITION (MASS SPECTROMETER) VS. ALTITUDE FROM SOUNDING ROCKET	3×10^5 BITS

EXPERIMENT PAYLOAD

		SCIENTIFIC PAYLOAD	GROSS WEIGHT
ORBITER (1)	PHOTOGRAPHY (CIRCULAR)	200	
	ATMOSPHERE, PARTICLES AND FIELDS, SURFACE PROPERTIES	120	9,200
IMPACTER (3)	UPPER ATMOSPHERIC	45	
LANDER (1)	GEOPHYSICS AND PHOTOGRAPHY	150	
	ATMOSPHERIC SOUNDER	20	1,000
MSSR (2)	SAMPLE RETURN / FILM RETURN	10	
	GEOPHYSICS AND PHOTOGRAPHY	300	24,000
FLYBY	TELESCOPE		3,000
	REMOTE SENSORS		1,467
	BIOLOGICAL LABORATORY		1,000
	TOTAL WEIGHT		39,817

AERODYNAMIC DRAG PROBE (IMPACTER)

PURPOSE: MEASURE LOWER ATMOSPHERE DENSITY PROFILE

MISSION PROFILE: LAUNCH TO ENTER ATMOSPHERE AT M-1 MIN
PENETRATE ATMOSPHERE RECORDING DRAG DECELERATION
MEASURE PRESSURE AND TEMPERATURE AT LOW SPEED
TRANSMIT DATA TO FLYBY VEHICLE AFTER BLACKOUT

EXPERIMENTS:

AREA OF INTEREST	SCIENCE PAYLOADS WEIGHT (LBS)	DATA
ATMOSPHERIC PROPERTIES 3 AXIS ACCELEROMETER, PRESSURE & TEMPERATURE TRANSDUCER	15	10^4 BITS

GROSS LAUNCH WEIGHT: 50 LBS

ORBITER PROBE

PURPOSE: PHOTOGRAPHY OF SURFACE; PARTICLES, FIELDS, AND ATMOSPHERIC MEASUREMENTS

PROFILE:
1. LAUNCH AT M-10 DAYS WITH MIDCOURSE AT M-18 HRS AND ARRIVAL AT M-4 HRS
2. PROPULSIVE DEBOOST OVER NORTH POLE INTO 200 KM CIRCULAR POLAR ORBIT
3. SEQUENTIAL HIGH RATE PHOTOGRAPHY, STORAGE AND TRANSMISSION TO M.M. DURING FLYBY
4. LOW RATE TRANSMISSION OF T.V. PICTURES, GEOPHYSICAL AND ATMOSPHERIC DATA DIRECTLY TO EARTH FOR PROBE LIFE

GROSS LAUNCH WEIGHT = 9,200 LBS

ORBITER PROBE EXPERIMENTS

AREAS OF INTEREST	SCIENCE PAYLOAD WEIGHT (LBS)	DATA
SOLID BODY & SURFACE PROPERTIES		
PHOTOGRAPHY - CAMERA	150	10^{11} BPS AT FLYBY
T.V.	50	10^4 BPS TO EARTH
PARTICLES AND FIELDS	40	10^3 BPS TO EARTH
MAGNETOMETER, SOLAR WIND DETECTOR, MICROMETEOROID DETECTOR, COSMIC RAY TELESCOPE, GAMMA RAY SPECTROMETER		
GEODESY	—	TRACKING FROM EARTH & M.M.
THERMAL EMISSION FROM SURFACE	25	BPS TO EARTH
I.R. AND MICROWAVE RADIOMETERS		
ATMOSPHERE PROPERTIES		
REMOTE SENSING	55 (PLUS 25)	10^3 BPS TO EARTH
TOPSIDE SOUNDER, U.V. POLARIMETER, I.R. & MICROWAVE RADIOMETERS		

LANDER PROBE

PURPOSE: GEOPHYSICAL AND ATMOSPHERIC EXPLORATION
PROFILE:
1. LAUNCH AT ABOUT M-18 DAYS WITH MIDCOURSE AT M-24 HRS
2. AERODYNAMIC ENTRY AND DOPPLER CONTROLLED PROPULSIVE SOFT LANDING AT M-1R
3. LAUNCH SOUNDING ROCKET WITH DATA TRANSMITTED SIMULTANEOUSLY TO M.M. AND SURFACE PACKAGE
4. FOUR FACSIMILE PANORAMAS AND INITIAL GEOPHYSICAL DATA TRANSMITTED TO M.M. DURING FLYBY
5. ENVIRONMENTAL MONITORING CONTINUES FOR LIFE PROBE WITH DIRECT LOW RATE TRANSMISSION TO EARTH

GROSS LAUNCH WEIGHT - 1,000 LBS

LANDER PROBE EXPERIMENTS

AREAS OF INTEREST	SCIENCE PAYLOAD WEIGHT (LBS)	DATA
SOLID BODY & SURFACE PROPERTIES		
TERRAIN PHOTOGRAPHY	10	10^8 BITS AT FLYBY
SEASONAL TERRAIN PHOTOGRAPHY	—	10 BPS TO EARTH
SURFACE GEOPHYSICS	140	500 BPS TO EARTH
SEISMOMETER, MAGNETOMETER, HEAT FLOW		
ATMOSPHERE PROPERTIES		
SYNOPTIC CLOUD PHOTOGRAPHY	10 (As above)	500 BPS
SOUNDING ROCKET	20 (100 GROSS)	3×10^4 BITS AT FLYBY
PRESSURE & TEMPERATURE TRANSDUCERS		
MASS SPECTROMETER		
SURFACE WEATHER STATION		10 BPS TO EARTH

MARS SOIL SAMPLE RETURN PROBE

PURPOSE: COLLECT SOIL SAMPLE, DEPLOY EXPERIMENTS ON SURFACE

MISSION PROFILE: LAUNCH AT M-10 DAYS, MIDCOURSE AT M-12 HRS
AERODYNAMIC DEBOOST, CONTROLLED LANDING AT M-1 HR
COLLECT SAMPLES, LAUNCH RETURN ROCKET AT M-4 MIN
DEPLOY GEOPHYSICS EXPERIMENTS TO SURFACE

EXPERIMENTS:

AREAS OF INTEREST	SCIENCE PAYLOAD WEIGHT (LBS)	DATA
EXOBIOLOGY	5	SOIL SAMPLE
SOLID BODY &	20 (PHOTO CAMERA)	10^{11} BITS
SURFACE PROPERTIES	140 (SURFACE GEOPHYSICS)	500 BPS TO EARTH
ATMOSPHERIC PROPERTIES	10 (FACSIMILE CAMERA)	500 BPS TO EARTH

GROSS LAUNCH WEIGHT: 12, 000 LBS

FLYBY VEHICLE

PURPOSE: PLATFORM FOR REMOTE SURFACE AND ATMOSPHERIC MEASUREMENTS

MISSION PROFILE: HIGH ENERGY HYPERBOLIC FLYBY TRAJECTORY
PERIAPSE VELOCITY OF 10 KM/S AT 300 KM

EXPERIMENTS:

AREAS OF INTEREST	SCIENCE PAYLOAD WEIGHT (LBS)	DATA
1. SOLID BODY & SURFACE PROPERTIES		
A. REMOTE SENSING I.R., VISIBLE U.V., & GAMMA RAY SPECTROMETERS. I.R. & MICROWAVE RADIOMETERS	120	5×10^8 BITS
B. PANORAMIC SURFACE SCAN PHOTOGRAPHY	500	10^{13} BITS
C. MULTISPECTRAL PHOTOGRAPHY	800	10^{12} BITS
2. ATMOSPHERE		
A. REMOTE SENSING I.R., VISIBLE & U.V. SPECTROMETER. I.R. AND MICROWAVE RADIOMETER. U.V. POLARIMETER; TOPSIDE SOUNDER	35 (PLUS 1-A)	5×10^8 BITS
B. DIRECT SENSING TRAPPED RADIATION DETECTOR, MASS SPECTROMETER	12	10^5 BITS

C. PANORAMIC CLOUD SCAN PHOTOGRAPHY	(AS IN I-B)	(AS IN I-B)
D. MULTISPECTRAL PHOTOGRAPHY	(AS IN I-C)	(AS IN I-C)

ON-BOARD REMOTE SENSOR WEIGHT: 1467 LBS

MANNED FLYBY / VOYAGER COMPARISON

EXPERIMENT	VOYAGER	DATA TRANSMITTED	MANNED FLYBY	DATA TRANSMITTED	DATA RETURNED
LANDED ABL	X	10^9 BITS	NO	—	—
SAMPLE RETURN	NO	—	X	—	4 LBS
APPROACH PHOTOS	NO	—	X	—	10^{12} BITS
ORBITAL PHOTOS	X	10^{11} BITS	X	10^{11} BITS	10^{11} BITS
ENCOUNTER PHOTOS	NO	—	X	—	10^{13} BITS
ORBITAL GEOPHYSICS	X	10^9 BITS	X	10^9 BITS	—
LANDED GEOPHYSICS	X	10^9 BITS	X	10^9 BITS	10^8 BITS
INFLIGHT EXPERIMENTS	X	10^9 BITS	X	10^9 BITS	10^{14} BITS

EN ROUTE EXPERIMENTS

OBJECTIVES:

TAKE FULL ADVANTAGE OF SPECIAL OPPORTUNITIES OF MISSION ORBIT FOR ASTRONOMY-PHYSICS GOALS.

UTILIZE CREW AND INSTRUMENTS DURING 700 DAYS OF PRIME SPACE OBSERVING TIME.

TELESCOPIC STUDY OF MARS DURING APPROACH TO OPTIMIZE FLY-BY STRATEGY; ANALYZE SAMPLE AND FLY-BY DATA DURING RETURN.

SOLAR PHYSICS - PRIME OPPORTUNITIES

SIMULTANEOUS MANNED OBSERVATION OF SOLAR PHENOMENA FROM TWO SEPARATED STATIONS BOTH OUTSIDE EARTH'S ATMOSPHERE.

TELESCOPES ON SPACECRAFT AND IN EARTH ORBIT GIVE HIGHEST RESOLUTION PHOTOGRAPHS YET ACHIEVED WITH STEREO VIEWS AT SEVERAL WAVELENGTHS.

RADIATION MONITORING AT BOTH STATIONS GIVES CORRELATION WITH PHOTOGRAPHIC DATA.

VARYING ANGLE BETWEEN STATIONS PERMITS SEPARATION OF SPATIAL AND TIME VARIATION.

SOLAR PHYSICS EXPERIMENTS:
(ASSUMES SIMULTANEOUS EARTH ORBITAL OBSERVATIONS)

ONE METER TELESCOPE PLUS ACCESSORIES, 4000 LBS, 10^{13} BITS OF SOLAR DATA
FINE STRUCTURE OF PHOTOSPHERE, CHROMOSPHERE
UV AND IR PHOTOGRAPHS AND SPECTRA
FINE RESOLUTION OF SOLAR MAGNETIC FIELD LINES
RAPID SEQUENCE PHOTOGRAPHS OF FLARES
LIFE HISTORY OF SPOTS AND FLARES DURING SEVERAL SOLAR ROTATIONS

SOLAR WIND AND SOLAR FLARE PARTICLE AND RADIATION DETECTORS,
VARIATION WITH TIME, MAGNETIC FIELD, AND POSITION ANGLE
X-RAY AND -RAY TELESCOPES, CORRELATION WITH SPOT AND FLARE ACTIVITY OVER SOLAR DISK

RADIO OCCULTATION TO GIVE CORONAL ELECTRON DENSITY

ASTRONOMY: (USING 1 METER TELESCOPE)

PHOTOGRAPHIC AND SPECTRAL ANALYSES OF BODIES IN SOLAR SYSTEM:

	TIME IN MISSION	CLOSEST APPROACH	LEAST DISTANCE FROM EARTH
PLANETS: JUPITER	30 DAYS	4.16 A.U.	4.2 A.U.
SATURN	220 DAYS	7.64 A.U.	8.54 A.U.
PHOBOS AND DEIMOS	147 DAYS	<10,000 KM	0.5 A.U.
METEOROIDS, ASTEROIDS:			
MEDUSA	300 DAYS	0.2 A.U.	1.03 A.U.
XANTHIPPE	450 DAYS	0.14 A.U.	1.11 A.U.

DISCOVERY AND CHARTING OF UNKNOWN ASTEROIDS AND COMETS

TRIANGULATION OF SOLAR SYSTEM TO NEW ACCURACY

RADIO ASTRONOMY
LARGE DEPLOYED ANTENNA FAR FROM EARTH NOISE

PARTICLES AND FIELDS:
METEOROID FLUX, ORBITS AND COMPOSITION OVER WHOLE MISSION
PHOTOGRAPHY OF LIBRATION POINTS AND ZODIACAL LIGHT
X-RAY MAPPING OF NEW STELLAR SOURCES
TRAPPED PARTICLE SEARCH AT MARS AND VENUS

BIO-SCIENCE:
EARLY ANALYSIS OF MARTIAN SAMPLES
STUDIES OF 700-DAY ZERO G EFFECTS ON BIOLOGICAL SYSTEMS
MICROSCOPIC
MACROSCOPIC

RELATIVITY:
BENDING OF LIGHT PAST SUN
IMPROVED SHAPIRO EXPERIMENT - RADAR TRANSPONDER FROM SPACECRAFT DURING SOLAR OCCULTATION
TIME SHIFT ON SPACECRAFT USING ON-BOARD ATOMIC CLOCK:
EFFECT DUE TO DIFFERENCE IN GRAVITATIONAL POTENTIAL
EFFECT DUE TO RELATIVE VELOCITY

ADVANTAGES OF MAN AND THE MISSION MODULE

GENERAL
RELIABILITY - CHECKOUT, MAINTENANCE, SWITCHING
PROBE TARGETING AND GUIDANCE

MISSION MODULE
LARGE TELESCOPE. CAPABILITY FOR INFLIGHT EXPERIMENTS
DATA INTERPRETATION AND DISCRIMINATION
IMMEDIATE BIOANALYSIS
10^6 BPS TO EARTH

MSSR
CONTROL OF SOIL SAMPLE ACQUISITION, EXPERIMENT DEPLOYMENT, AND OPERATION BY CREW
3×10^5 BPS TO M.M. AS OPPOSED TO 10^5 BPS TO EARTH

ORBITER
DETERMINATION AND CORRECTION OF ORBITAL ELEMENTS
CONTROL OF PHOTOGRAPHIC OPERATIONS BY CREW
1 PHOTO / 5 MINUTES TO M.M. AS OPPOSED TO 1 PHOTO / DAY TO EARTH

LANDER
CONTROL OF EXPERIMENT DEPLOYMENT AND OPERATION BY CREW
3×10^5 BPS TO M.M. AS OPPOSED TO 10^3 BPS TO EARTH

9.0 WEIGHTS

WEIGHT GROWTH

All subsystem dry weight estimates should be made without a pad.

A factor of 50% should be added to the total S/C dry weights.

All expendable weights except FUEL should have a FACTOR of 20% added.

All fuel estimates should be based upon the total S/C weight.

All delta V's should have a total of 10% added for contingency.

10.0 DEVELOPMENT TESTS AND FLIGHT SCHEDULE

BIOMEDICAL STUDIES AND THE QUALIFICATION OF MAN FOR LONG SPACE MISSIONS

Within the AAP and other Earth Orbital flights, experience and scientific information must be accumulated in a timely fashion to assure the crew capability and safety for the planetary missions. The peculiar environment to which the crew will be subjected, without an abort capability, in the planetary flights are as follows:

a. Weightlessness
b. 7.0 psia O_2 and N_2 atmosphere
c. 700 day duration
d. Entry accelerations
e. Trace contaminants
f. Confinement
g. Infectious diseases
h. Monotonous diet
i. Limited volume
j. Low motivation tasks

The above environment factors of concern must be studied and experienced with rigorous attention to detail in order to establish a base from which extrapolations can be made during the development phase of the planetary spacecraft design.

The confidence in any selected course of action regarding committing man to the artificial environment represented by space vehicles must grow slowly based on actual experience. We lack sufficient understanding of the essential features of our Earth environments even to be sure that we have taken all of the critical variables into consideration in trying to devise simulations or laboratory mock-ups of space flight. The only rules we have been able to enunciate to date are first that we should follow our philosophy of incremental exposure, generally being willing to double the duration of successive manned missions as long as no unforeseen medical problems are encountered in crews returning from space flight; and second, that the more successful experience we acquire, the greater will be our ability to extend our confidence to a broader flight crew population. There is no magic number of crewman (other than one) required to have flown one increment before we can certify that a longer flight is "safe," but as we progress to longer, more sophisticated missions, we will strive to acquire all the biomedical information obtainable until we have a sufficient body of data to formulate general statements about the effects of the space flight environment on human physiology. From our standpoint, three crewmen observed over a 60 day flight are better than two. Two men, however, can provide us with the necessary data to demonstrate that a 60 day flight is tolerable and therefore a 90 or 120 day mission plan is rational.

If even one crewman encounters medical problems referable to his flight exposure, we would have to evaluate the entire body of data available as of that point in time in order to decide whether or not additional flights of duration equal to or shorter than the flight on which the problems are encountered would be required prior to attempting a longer manned mission. Thus, the faster we can build a library of biomedical information derived from manned flights of increasing duration, the greater our capability becomes to understand biomedical problems and devise appropriate corrective action when they finally occur.

By the time the launch for the 1975 Flyby would occur, many astronauts must have experienced 60 day flights in the artificial environment and some astronauts must experience 700 day flights in that environment. As stated before, there is no magic number of crewmen necessary to achieve qualification of man for long duration flights. However, there is no data in medical history which would lead us to believe that 60 days in any environment is not sufficient to establish man's predilection to succumb or survive that environment for any length of time.

The bulk of data to qualify the man can best be obtained from the rotation of crewmen every 60 days through the environment of an Earth orbital spacecraft which in itself may remain in orbit for many months to several years. In addition to this, the data from a few crewmen who have experienced the environment for 700 days would serve as the necessary culmination of the progressive steps of days in orbit which must be accomplished to create the 100 percent confidence that man can tolerate the aforementioned environmental factors.

MISSION MODULE INTEGRATED SYSTEMS TESTS

EARLY TEST REQUIREMENTS (1969-1970)

EXPERIENCE WITH MAINTENANCE REPAIR TECHNIQUES
EXPERIENCE WITH NEW SUBSYSTEMS NEEDED TO AFFECT MISSION MOD. DESIGN
2 GAS SYSTEM
MOLECULAR SIEVE FOR CO_2 REMOVAL
WATER RECLAMATION SYSTEM
WASTE MANAGEMENT SYSTEM
CONTAMINATION CONTROL SYSTEM
LONG DURATION CRYO STORAGE SYSTEM
SOLAR CELL SYSTEM
CMG STABILIZATION SYSTEM
GUIDANCE & NAVIGATION

PRECURSORY FLIGHT REQUIREMENTS

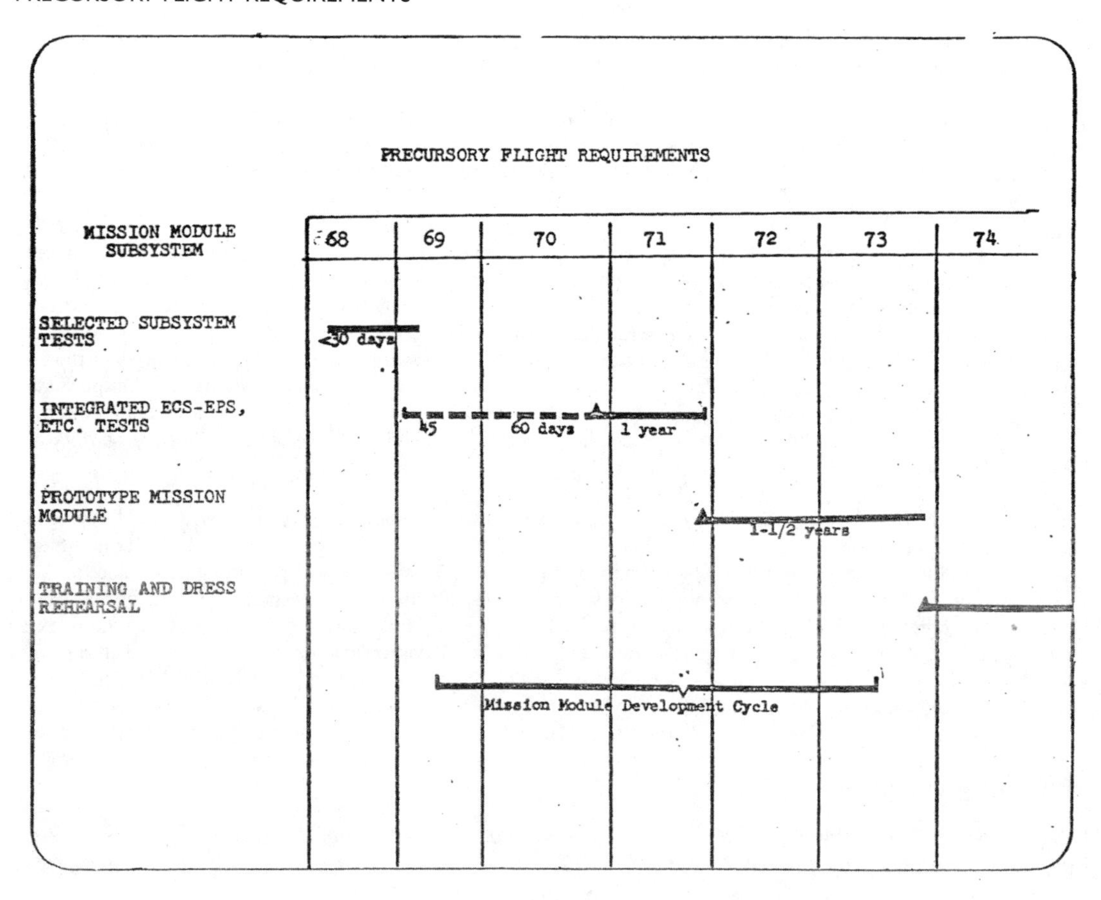

SCHEDULING

The establishment of a broad-based, flexible manned flyby capability to accomplish scientific exploration of the planets depends on the availability of unmanned probes, onboard experiments equipment, long duration systems, and necessary flight support facilities and operations capabilities. The Apollo Program, the unmanned lunar and planetary probe programs, and the Apollo Applications and Earth Orbital Flight Programs provide a sound base for the development of a manned flyby system.

Concurrent with and in conjunction with the Apollo Applications and Earth Orbital activities, a program for development, test, and evaluation of the planetary flyby systems would be conducted. Figures 1, 2 and 3 illustrate the overall activity that is necessary to make the 1975 Mars flyby mission possible. It is evident that with a hardware start in FY 1970 as shown in Figure 2, and supplemented by the precursor activities in the Apollo Applications and the Earth Orbital programs, an orderly phasing into Planetary operations starting with the 1975 flyby mission is possible.

The flight test program and operational missions are shown in Figure 3. The first test flight would take place in mid-1971 and evaluate a prototype Planetary Mission Module with a boilerplate experiment module, in Earth orbit, for one year. This test would afford this first opportunity to get actual long duration space environment test time on the Mission Module subsystems available at that time. This test could be run in conjunction with Apollo Applications or Earth Orbital missions and would permit the performance of and evaluation of experiments which would be transferred to the vehicle during this 1 year period, Four S-IB launch vehicles are available at this time (222 thru 225) for this purpose.

This test would be followed by a more complete prototype test starting in mid-1972 and extending over a two year period. The test configuration would include a boilerplate Experiment Module, a, prototype Mission Module and the first prototype of the Earth Entry and Propulsion Module. The Earth Entry Module and the Propulsion Module are of prime importance in this test as early test data is required relative to the long period of inactivity for these modules. The Earth Entry Module would be exercised, unmanned, at the conclusion of the 2 year period. The Mission Module crew would perform a multitude of Earth Orbital experiments in addition to the planetary tests. Four S-IB launch vehicles (226 thru 229) are available to transfer crew and experiments for that purpose.

Two Earth Entry Module qualification tests would be conducted in early 1973 and late 1973. The first of these would be launched by a S-IB and would be to qualify the Earth Entry Module from Earth orbit. The second would duplicate the conditions of Earth entry at 50,000 ft/sec. Failure of this mission could be backed up by the PIV Test shown in early 1974.

A test would be conducted in April 1973 to exercise the complete ground and flight operations of launch, orbit phasing, rendezvous, and final positioning and docking. The S/C would be launched by a S-V and would stay in orbit for 2 years as a further test of the long duration systems. The other two launches would be accomplished in a simulation of the 1975 Flyby launch window with the accompanying problems of phasing rendezvous which are accomplished by ground command. The final positioning and docking would be commanded by the crew in the S/C. After this test the S/C would serve as a focal point for further Earth Orbit Experiments with logistics, experiments or crew transfer supplied by the S-IB launches indicated. In addition, this S/C will perform a docking command exercise with the S-IVC (PIV launch) that is orbited in February 1974. The S-IVC would subsequently be detached from the S/C and fired in its full operational mode.

In mid-1974 an all up spacecraft plus experiments test and a planetary injection test would be performed. This would be a complete dress rehearsal and would proceed through all phases, including exercising the experiments and probes, that will be conducted the following year for the Mars flyby. This test would go through the injection burn and be directed towards one of several possible missions of scientific interest. For example, an out of the ecliptic flight could be performed with a 180 day mission time. This test would be manned throughout as it is intended to be an all up test with man playing his part in the "overall" system demonstration and evaluation. The injection velocities would be held to approximately 12,000 ft/sec which with the midcourse propellant loading sized for the Triple Planet Flyby would provide about an hour instead of 5 minutes of abort. Coupled with the above, the mission time of 180 days, the excess power available from the solar cells at 1.0 A.U., and the less severe meteoroid and solar radiation environment would make this mission a reasonably safe rehearsal of the manned flyby while performing a mission of scientific interest.

In September of 1975 the Mars Flyby Mission would be launched followed by an all up spacecraft plus experiments test in mid-1976. This latter test, configured for the 1977 Triple Planet Flyby Mission, is considered necessary to assure that all changes and special requirements necessitated by that particular mission are flight tested before the mission.

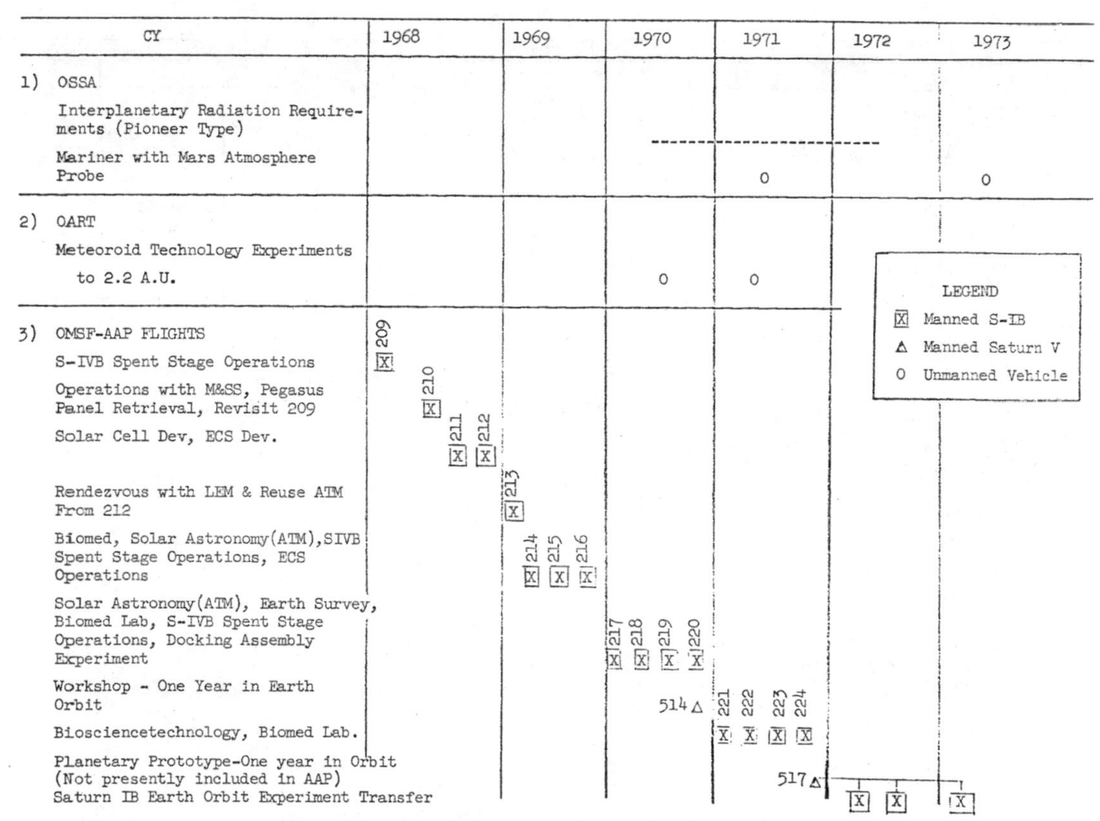

PRECURSOR FLIGHT REQUIREMENTS SCHEDULE

Figure 1

FY: 1968 | 1969 | 1970 | 1971 | 1972 | 1973 | 1974 | 1975 | 1976

CY	1968	1969	1970	1971	1972	1973	1974	1975
SPACECRAFT DEV.								
EARTH ENTRY MOD	Prog.Def.		Design, Fabrication and Test			□ □ □	□ M	□ M
MISSION MODULE	Prog.Def.		Design, Fabrication and Test	□ P	□ P	□	□ M	□ M
EXPER. MODULE	Prog.Def.		Design, Fabrication and Test	□ BP	□ BP	□ BP	□ M	□ M
PROPULSION MODULE	Prog.Def.		Design, Fabrication and Test		□ P	□	□ M	□ M
CREW TRANSFER VEHICLE	Modification, Fabrication and Test			□	□	□	□ M	□ M

LEGEND
□ R&D FLT
□ MANNED FLT
P PROTOTYPE
BP BOILERPLATE
M MISSION CONFIG

FLYBY MISSION RDT&E SCHEDULE PMO – 4/19/67

Figure 2

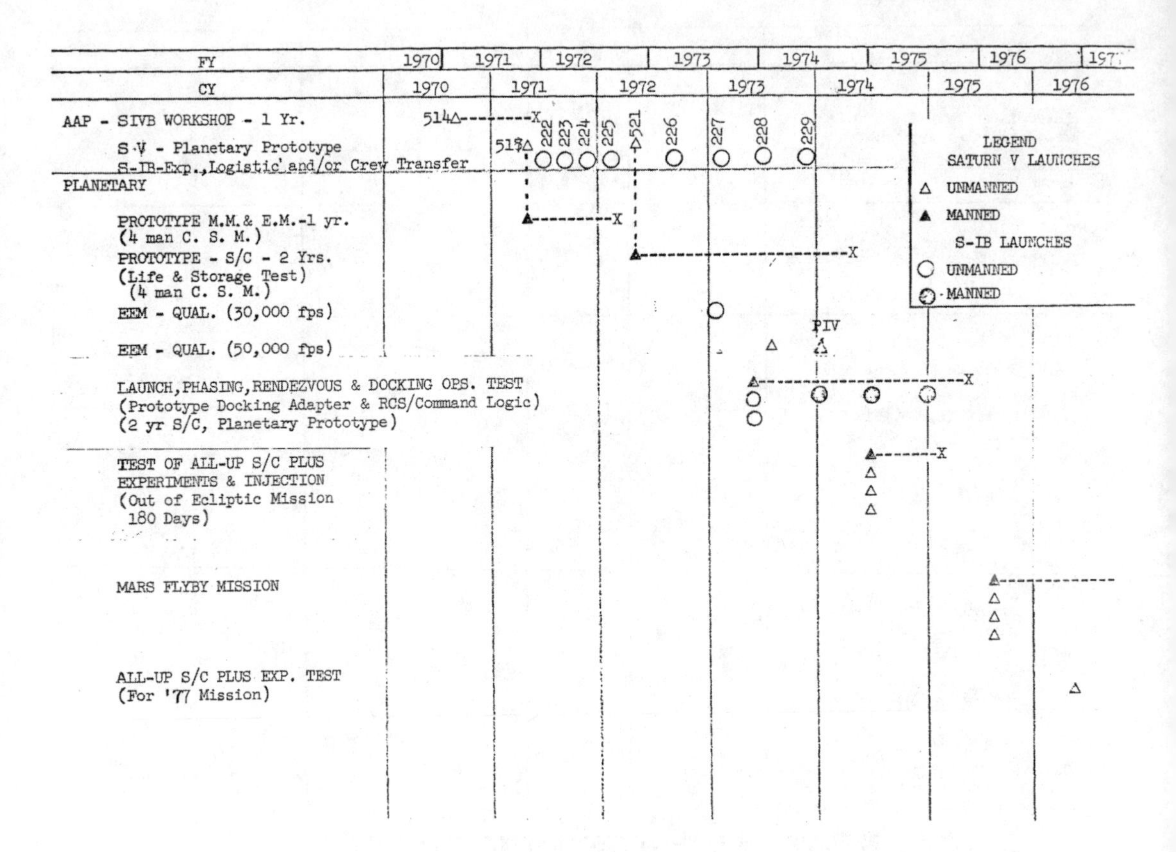

PLANETARY FLYBY FLIGHT PROGRAM SCHEDULE PMO – 4/19/67

Figure 3

Mars Global Surveyor Weekly Updates

(Jan 2000- Sept 2003)

Wednesday 5-Jan-00

(DOY 356/19:00:00 to DOY 005/19:00:00 UTC) Launch / Days since Launch = Nov 7, 1996 / 1155 days
Start of Mapping / Days since Start of Mapping = April 1, 1999 / 279 days
Total Mapping Orbits = 3707
Total Orbits = 5389
Recent Events:
The mm010 sequence continues executing nominally the daily science record and playback events. The mm010 sequence runs through January 12.

The initial MPL landing site imaging campaign has been completed, with 19 targeted scans across the predicted 1-sigma MPL landing ellipse having been executed. Unfortunately no obvious signs of the lander or its parachute has been found at this time. Note that the definition of the landing ellipse continues to change as additional atmospheric data from MGS is factored in. As a result, the current landing site imaging mosaic may not reflect the latest landing site knowledge. Additionally there were some gaps in the mosaic due to lost data at the DSN stations. As a result, a second set of imaging scans is being planned to fill in the gaps and to expand the coverage area in response to the latest atmospheric data.

In addition to completing the MPL scans, a special targeted opportunity to image the Mars Pathfinder landing site was performed Sunday December 26 at 09:18 UTC. Unfortunately due to an error in the coordinate system used to derive the targeted attitude, the wrong site was imaged. A second opportunity to image the Pathfinder landing site occurs on Sunday January 9. The mz034 mini-sequence will be built on January 6 and 7 to perform the MPF and additional MPL landing site imaging, and will execute through January 12.

The spacecraft has behaved almost flawlessly in performing all of the imaging scans.

HGA Anomaly:
The HGA inner gimbal angle continues to decrease and is currently at 53.5 degrees. The inner gimbal angle will continue decreasing, reaching the location of the gimbal obstruction at 41.5 deg in early February. STL testing begun for validation of the HGA beta supplement operations plan, which will be used to perform mapping operations for the rest of the nominal mapping mission.

Spacecraft Health:
All subsystems are reporting nominal health. As reported previously, battery 2 switched from VT curve 2 to VT curve 1 as a result of performing the MPL scans. This was expected and resulted in slightly higher battery 2 temperatures. Upon completion of the initial MPL and MPF scans, the Battery 2 VT curve was switched back to VT2. Soon after though the flight software autonomously re-commanded battery 2 back to VT1, which is where we will leave it. The trend indicates that battery 1 will shortly switch to VT1 as well.

Uplinks:
There have been 42 uplinks to the spacecraft during the last two weeks, including new star catalogs and ephemeris files and instrument command loads. Total command files radiated to the spacecraft since launch is 4300.

Upcoming Events:
1) The mm011 sequence is currently in development an is scheduled for uplink Monday January 10, with sequence start time of January 13 00:00 UTC.

2) The mz034 sequence to continue imaging the MP and MPF landing sites is scheduled for development o January 6 and 7, to begin execution on Sunday January

Wednesday 12-Jan-00

(DOY 005/19:00:00 to DOY 012/19:00:00 UTC) Launc / Days since Launch = Nov 7, 1996 / 1162 days
Start of Mapping / Days since Start of Mapping = April 1999 / 286 days
Total Mapping Orbits = 3791
Total Orbits = 5473

Recent Events:
The mm010 sequence finishes execution on Jan 13 a 01:27 UTC. The next mapping sequence, mm011, ha been uplinked to the spacecraft and will begin executio on Jan 13 at 00:00 UTC and continue executing throug Feb 4. The mm011 sequence will be the last nomina mapping sequence of the baseline mapping mission. Th next mapping sequence will be the first of the new bet supplement sequences, designed to continue mappin operations in the presence of the HGA obstruction a the 41.5 deg azimuth gimbal angle. Stay tuned for mor information about the specifics of beta supplemer mapping operations over the next few weeks

The mz034 mini- sequence was uplinked to th spacecraft on Jan 8 and finished execution on Jan 12 a 09:00 UTC. The mz034 sequence contained the secon imaging scan of the Pathfinder landing site and fou additional scans of the Mars Polar Lander landing sit The additional MPL landing site scans are intended to fi in gaps in the original imaging campaign performed prio to the holidays and to expand the coverage area to th west in response to the latest atmospheric dat Unfortunately bad luck asserted itself again on th second Pathfinder landing site image as the science dat was lost due to DSN station problems. A third attemp to image the Pathfinder site is currently being planne Additional imaging of the MPL landing site is currentl being discussed to continue expanding the coverag area.

The spacecraft continues to behave flawlessly i performing nominal mapping operations and th multitude of imaging scans.

HGA Anomaly:
The HGA inner gimbal continues to decrease and i currently at 51.7 degrees. The inner gimbal angle wi continue decreasing, reaching the location of the gimba obstruction at 41.5 deg in early February. STL testing t validate the new beta supplement operations pla continues.

Spacecraft Health:
All subsystems are reporting nominal health.

Uplinks:
There have been 18 uplinks to the spacecraft during th last two weeks, including new star catalogs an ephemeris files, instrument command loads and th mm011a and mz034a sequences. Total command file radiated to the spacecraft since launch is 4318.

Upcoming Events:
1) The mm012 sequence development kickoff meeting is scheduled for Jan 18.

2) The mz035 mini-sequence to continue imaging the MPL and MPF landing sites is scheduled for development on January 13 and 14.

3) The mz036 mini-sequence to perform additional magnetometer instrument calibrations has been approved and will begin development Jan 17.

Wednesday 19-Jan-00

(DOY 012/19:00:00 to DOY 019/19:00:00 UTC) Launch / Days since Launch = Nov 7, 1996 / 1169 days
Start of Mapping / Days since Start of Mapping = April 1, 1999 / 293 days
Total Mapping Orbits = 3876
Total Orbits = 5558

Recent Events:
The mm011 sequence began execution on Jan 13 at 00:00 UTC and runs through Feb 4. The spacecraft continues to perform flawlessly in performing the nominal mapping mission.

The mz035 mini-sequence, containing scans for a third Pathfinder landing site image and two additional MPL landing site images finished execution on Jan 19 at 13:45 UTC. Unfortunately, bad luck struck again with regard to the Pathfinder opportunity and significant portions of the Pathfinder image were lost due to more DSN problems. There is no redundant capability to return the imaging data at the high 40000 bps real-time rate and also record the data to the solid state recorders for redundant playback the next day. Consequently, return of these landing site images is a single opportunity event which requires the DSN stations to be configured and working properly during the
30 minute data return period. Hardware, weather and other factors with the stations contribute to make this far from a foolproof process. In general, the DSN has performed excellently in supporting the MGS mission.

A change request was approved by the Project to perform some HGA motion calibration tests for the Magnetometer team. These will be performed on the dark side of the planet and will have minimal effect on normal operations. The only exception is that Radio Science ingress data will be prematurely truncated on the calibration orbits. These HGA motions tests will be performed over 15 orbits, with the solar arrays being moved to different fixed positions during the HGA motions after every three orbits. Due to mini-sequence buffer size constraints, the 15 orbits will be performed in two parts. The mz036 mini-sequence was successfully transmitted to the spacecraft on Jan 19, to perform the first 9 orbits on Jan 20 and mz037 will be transmitted on Jan 21 to perform the remaining 6 orbits on Jan 21.

The Ka Band Link Experiment (KaBLE) was turned off on Jan 14, to reduce the spacecraft power loads. The KaBLE investigators are not currently utilizing the Ka Band data.

HGA Anomaly:
The HGA inner gimbal continues to decrease and is currently at 48.9 degrees. The inner gimbal angle will continue decreasing, reaching the location of the gimbal obstruction at 41.5 deg in early February. A special review was held on Jan 18 to discuss possible spacecraft testing to determine if the obstruction is still present. The recommendations from the review are expected by the end of the week. Any spacecraft testing would be performed over the next two weeks, prior to the start of beta supplement mapping operations.

Spacecraft Health:
All subsystems are reporting nominal health. A third Payload Data System (PDS) reset occurred on Jan 16. The reset data has been sent to JPL, who designed and built the PDS. The spacecraft team has been unable to correlate the data to any specific spacecraft activity.

Uplinks:
There have been 27 uplinks to the spacecraft during the last two weeks, including new star catalogs and ephemeris files, instrument command loads, KaBLE turn off and the mz036 mini-sequence load. Total command files radiated to the spacecraft since launch is 4345.

Upcoming Events:
The preliminary first beta supplement sequence will begin development Jan 20 in preparation for the start of beta supplement mapping operations on Feb 5. Additional MPL and MPF imaging opportunities will continue through Feb 4. The next imaging mini-sequence will be built on Jan 21 and will contain all the desired over-flight opportunities through Jan 26.

Wednesday 9-Feb-00

(DOY 033/19:00:00 to DOY 040/19:00:00 UTC) Launch / Days since Launch = Nov 7, 1996 / 1190 days
Start of Mapping / Days since Start of Mapping = April 1, 1999 / 314 days
Total Mapping Orbits = 4134
Total Orbits = 5816

Recent Events:
The mz041 mini- sequence, containing the "full-up"" Beta Supplement HGA motion verification test, executed on Thursday Feb 3 at 11:00 GMT. The test successfully moved the azimuth gimbal to 140°, then performed a "simulated orbit"" on the elevation gimbal, incorporating a 3-part boom interference rewind and a 3-part range-of- motion rewind, and executed the full elevation range of motion from -157° to +155°. Based upon the success of the test, the first beta supplement mapping sequence, mm012, was radiated to the spacecraft and began executing Saturday Feb 4 at 00:00 GMT. The first beta supplement orbit executed on Monday Feb 7 at 10:00 GMT and all operations on the spacecraft and with the DSN station were completely nominal. Twenty-five orbits have been successfully executed as of the writing of this report.

Spacecraft Health:
All subsystems are reporting nominal health.

Uplinks:
There have been 25 uplinks to the spacecraft during the last week, including new star catalogs and ephemeris files, instrument command loads, and the mm012 sequence. Total command files radiated to the spacecraft since launch is 4415.

Upcoming Events:
Uplink of the second beta supplement sequence, mm013, is scheduled for Thursday Feb 10.

Wednesday 16-Feb-00

(DOY 040/19:00:00 to DOY 047/19:00:00 UTC) Launch / Days since Launch = Nov 7, 1996 / 1199 days
Start of Mapping / Days since Start of Mapping = April 1, 1999 / 321 days
Total Mapping Orbits = 4219
Total Orbits = 5901

Recent Events:
Beta Supplement operations continue nominally. The mm013 sequence is currently executing on the spacecraft. The next sequence, mm014, was uplinked to the spacecraft on Feb 15 at 18:25 GMT and is scheduled to go active at 00:00 GMT on Feb 17.

Spacecraft Health:
All subsystems are reporting nominal health.

Uplinks:
There have been 30 uplinks to the spacecraft during the last week, including new star catalogs and ephemeris files, instrument command loads, and the mm013 sequence. Total command files radiated to the spacecraft since launch is 4445.

Upcoming Events:
The first fixed-HGA mapping sequence of the beta supplement era is scheduled for March 6. More details forthcoming.

Wednesday 23-Feb-00

(DOY 047/19:00:00 to DOY 0454/19:00:00 UTC) Launch / Days since Launch = Nov 7, 1996 / 1206 days
Start of Mapping / Days since Start of Mapping = April 1, 1999 / 328 days
Total Mapping Orbits = 4305
Total Orbits = 5987

Recent Events:
The pace of the new mapping Beta sequence generation problem, a build every week, and a realization of new operational constraints imposed by the new mapping scheme, has caused two sequencing errors over the last week.

On 00-048 (2/17/2000) at about 00:00 UTC, the new Beta Supplement sequence, mm014b, was scheduled to kickoff but never went active. Since that new sequence controls the tracking of the HGA to Earth, no downlink was possible. Using a 34-m HEF station, a command was successfully uplinked via LGR-2 to change the uplink rate from 125 bps to 7.8 bps. Then, at 7.8 bps, a mini-sequence was successfully radiated to auto-track the HGA for about 30 minutes per orbit for 4 consecutive orbits, allowing for a 4K real-time downlink. During the first of these orbits, spacecraft telemetry was obtained which showed that MGS was in a healthy condition. During the second of these orbits, the uplink rate was changed back to 125 bps and the new mm014c sequence was uplinked to resume normal operations. The mm014c sequence began execution nominally at 14:20 UTC and has successfully completed execution. The mm015 sequence, was uplinked and began execution on 00-051 (2/20/2000) at 00:00 UTC. The mm015 sequence is scheduled to run through 00-054 (2/23/2000). The net effect of the anomaly was the loss of most of DOY 48 recorded science telemetry.

The cause of the mm014 sequence anomaly was that it was loaded into the incorrect area of the sequence buffer, which was already executing the mm013 sequence. This error was induced by having to split the 7 day beta supplement sequence into two parts due to the size of the sequence and the much reduced uplink time available each orbit due to the complex HGA management. This splitting process introduced an ambiguity in the naming convention: mm013a was used for Part 1 of sequence 013 rather than revision level a, and mm013b was used for Part 2 of sequence 013 rather than revision level b. When a second revision of Part 1 was required, mm013c was used, since mm013b was already taken by Part 2. Due to the ambiguity in the naming convention, mm014 was run through SEQTRAN using the FINCON file from mm013c (Part 1) rather than mm013b (Part 2). The result was that mm014b was loaded into the same buffer area as Part 2 of the previous sequence. Since this previous sequence was still executing when mm014b was uplinked, the new sequence did not go active.

To confound matters, due to the current MGS to Earth range, we are not able to support two-way data (uplink and downlink data) at the high rate 40 ksps real-time downlink rate. Therefore all uplinks are currently performed during orbits in which the recorded data from the previous day is being played back. Therefore confirmation of a successful sequence uplink and activation requires waiting until the next day's data is played back. For this case there was insufficient time available to determine that the sequence load had been rejected and to rebuild a replacement sequence in time. This was due in part to a half-day loss in the schedule due to a problem with our spacecraft test lab (STL).

The second anomaly occurred with the development of the mm016a sequence, which was uplinked successfully to the spacecraft on 00-052 (2/21/200) and scheduled to begin execution on 00-055 (2/24/00) at 00:00 UTC. In this event an error in the recording and playback of the science data was discovered after the sequence had been uplinked. The sequence was successfully terminated on 00-054 (2/23/2000) at
14:00 UTC and a replacement sequence, mm016b, uplinked several orbits afterwards upon confirmation of the termination of mm016a. The root cause for this anomaly was simply a rush to uplink the sequence early enough to verify its activation in order to avoid a repeat of the mm014 sequence anomaly.

Several changes have been made to the sequence development process for the weekly sequence builds. The first change, which was already going to be made for the next sequence anyhow, was to rename the sequences. The split of the weekly sequences into two parts creates two sequences, mm016 and mm017, for example, rather than mm016a and mm016b. The second and probably the most important change made was to increase the sequence development time by three days to provide sufficient time with margin for development, review, STL validation, uplink and verification of the sequences.

One final note, the failure to get a new sequence onboard the spacecraft poses no risk to the health of the spacecraft. The spacecraft will maintain mapping nadir pointing and articulation of the solar array. Communications from the spacecraft and collection of science data will be interrupted until a new sequence can be uplinked.

Spacecraft Health:
All subsystems are reporting nominal health. A sixth PDS reset occurred on 00-019. So far no cause has been found for the recurring resets. The resets appear to have minimal if any impacts to the instruments. We continue to forward all reset data to the PDS software experts at JPL.

Uplinks:
There have been 14 uplinks to the spacecraft during the last week, including new star catalogs and ephemeris files, instrument command loads, and the mm014, mm015 and mm016 sequences. Total command files radiated to the spacecraft since launch is 4459.

Upcoming Events:
The first fixed-HGA mapping sequence of the beta supplement era is scheduled for March 6 and development of this sequence has begun. One of the limitations of the mapping beta supplement scheme is a loss of South Pole radio science atmospheric data due to HGA gimbal constraints. The project approved a three to four day period each month to perform fixed-HGA operations, in which the HGA is pointed along the spacecraft +X axis and the spacecraft is moved to point the HGA to Earth for an eight-hour period (i.e. four orbits). During the eight-hour Earth pointed period the science data recorded from the previous 24 hours of nadir pointed mapping is played back. This scheme allow radio science observations to be made at Earth occultation ingress and egress, at the expense of collecting mapping data for the eight-hour Earth pointed period. The project also approved monthly MOLA nadir off-pointed polar observations. These observations will be performed during the 24 hour nadir pointing period prior to the first eight-hour fixed-HGA Earth pointed period.

Wednesday 1-Mar-00

(DOY 054/19:00:00 to DOY 061/19:00:00 UTC) Launch / Days since Launch = Nov 7, 1996 / 1213 days
Start of Mapping / Days since Start of Mapping = April 1, 1999 / 335 days
Total Mapping Orbits = 4390
Total Orbits = 6072

Recent Events:
The spacecraft continues to operate nominally in performing the daily recording and transmission of the science data. The mm016 sequence executed successfully from 00-55 (2/24/00) through 00-58 (2/27/00). The mm017 sequence began execution on 00-59 (2/28/00) and will run through completion of 00-61 (3/1/00). The next sequence mm018, scheduled to execute from 00-62 (3/2/00) through 00-64 (3/4/00), has been successfully uplinked to the spacecraft.

The first fixed-HGA sequence of the beta supplement mapping phase, mm019, has been generated and will be uplinked Saturday March 4. The mm019 sequence will begin execution on 00-65 and continue through 00-68 (3/8/00). Included in the mm019 sequence is the first of the monthly off-nadir polar scans for the MOLA instrument, approved by the Program. The first 12 orbits of the mm019 sequence will fly with a -17.1 degree nadir off-point angle for 45 minutes centered about the South Pole and a 14.4 degree nadir off-point angle for 45 minutes centered about the North Pole. For the fixed-HGA mapping operations, the spacecraft will be nadir pointed for 24 hours, recording science data on the solid state recorders, then the spacecraft will be oriented to point the HGA bore-sight at Earth for an eight hour playback period to return the data collected on the recorders. During this eight-hour period, the spacecraft will not be nadir pointed and not collecting science data, with the exception of radio science which will perform atmospheric observations as the spacecraft enters and exits Earth occultation. As reported last week the reason for the fixed-HGA sequences is to provide radio science atmospheric observations near the South Pole as the spacecraft exits Earth occultation, a geometry not supported by the HGA gimbals in the beta supplement operations due to EMI and HGA boom interference constraints. Upon completion of the data playback at the end of the eight-hour Earth pointed period, the spacecraft will be turned back to nadir pointing to begin the next fixed-HG A cycle

Spacecraft Health:
All subsystems are reporting nominal health. Three more PDS reset occurred over the last week. The frequency has been increasing over the last two weeks. So far no cause has been found for the recurring resets. The resets appear to have minimal if any impacts to the instruments. We continue to forward all reset data to the PDS software experts at JPL.

Uplinks:
There have been 21 uplinks to the spacecraft during the last week, including new star catalogs and ephemeris files, instrument command loads, and the mm017 and mm018 sequences. Total command files radiated to the spacecraft since launch is 4480.

Upcoming Events:
The Navigation team has been looking at the ground track walk on a weekly basis to determine if an orbit trim maneuver (OTM) is required. Prior to the start of beta supplement operations, it appeared that an OTM would be needed sometime in March. However, the latest tracking data for the first few weeks of beta supplement operations shows the rate of the ground track walk decreasing and starting to knee over, which if the trend continues will not necessitate an OTM in the near term. With the likelihood of the next OTM diminishing, planning has begun for a focus calibration of the MOC to be performed in the next few weeks.

Wednesday 8-Mar-00

(DOY 061/19:00:00 to DOY 068/19:00:00 UTC) Launch / Days since Launch = Nov 7, 1996 / 1220 days
Start of Mapping / Days since Start of Mapping = April 1, 1999 / 342 days
Total Mapping Orbits = 4476
Total Orbits = 6158

Recent Events:
The spacecraft continues to operate nominally in performing the beta supplement daily recording and

transmission of science data. The mm018 sequence executed successfully from 00-62 (3/2/00) through 00-65 (3/5/00).

The mm019 sequence (i.e. first fixed-HGA sequence) began execution on 00-65 (3/5/00) and will run through completion of 00-68 (3/8/00). The mm019 sequence contains 12 orbits of MOLA off- nadir polar scans followed by three consecutive days of eight hour fixed-HGA Earth pointed periods, designed to allow radio science Earth occultation egress data, which is prohibited during beta supplement operations due to HGA boom and EMI interference constraints. Two incidents have occurred on the spacecraft during the execution of this sequence, as described below. The next sequence mm020 (another beta supplement sequence), scheduled to execute from 00-69 (3/9/00) through 00-72 (3/12/00), has been successfully uplinked to the spacecraft.

The first incident of the mm019 sequence was with star processing during the MOLA off-nadir Polar scans. The first 12 orbits of the mm019 sequence flew with a -17.1° nadir off-point angle for 45 minutes centered about the South Pole and a 14.4° nadir off-point angle for 45 minutes centered about the North Pole. This left about 30 minutes of star processing time per two-hour orbit. Anticipating that the limited star processing might be problematic, based on the previous MOLA polar scans performed last year, two things were done to ensure that the spacecraft would not lose inertial reference and enter contingency mode fault protection. First a commanded bias reset was issued two hours before the first MOLA scan to coincide with the IMU swap to high rate mode and second the lost logic limit, the number of unidentified stars allowed before an autonomous bias reset, was raised to 60 stars. Twice during the 12 orbit scan sequence, the lost logic limit was exceeded and an autonomous reset was commanded. After each reset the spacecraft successfully converged attitude. Three consecutive cumulative resets are required to actually cause entry into contingency mode. Prior to slewing the spacecraft to the first fixed- HGA Earth pointed attitude following the MOLA scans, the spacecraft attitude was nominal and star processing was properly converged. In spite of the star processing problems, the spacecraft successfully executed all of the slews and returned the data back to Earth, during the subsequent eight hour fixed-HGA Earth pointed period.

The second incident occurred when the MOLA laser temperature approached the 35° upper flight allowable limit, during the last orbit of the second eight-hour fixed-HGA Earth pointed period. The increased temperature was expected but the actual temperatures were about four degrees higher than that predicted by the thermal model. Real-time telemetry was not available for the flight team during the Earth pointed period, which is used for playback of the recorded data. The greater than expected temperature increase was noticed in the playback data on the third and last fixed-HGA Earth pointed period.

The Spacecraft Team is currently reassessing the monthly MOLA off- nadir polar orbit scans and the fixed-HGA Earth periods for radio science egress data, to determine what changes are required to ensure the safety of all the instruments and robustness of the spacecraft in performing these special events. Options will be presented to the Project at the 3/14 Mission Planning meeting. The next fixed-HGA sequence is scheduled for 00-076 (3/16/00) and will most likely be replaced with a beta supplement sequence.

Spacecraft Health:
All subsystems are reporting nominal health. As expected the battery DOD was greater during the four fixed-HGA Earth pointed orbits due to the additional transmitter on time associated with performing the radio science occultation egress event. As a result battery 2 autonomously switched to the VT 1 curve. Upon completion of the fixed-HGA sequence, battery 2 will be commanded back to VT 2.

Uplinks:
There have been 15 uplinks to the spacecraft during the last week, including new star catalogs and ephemeris files, instrument command loads, and the mm020 sequence. Total command files radiated to the spacecraft since launch is 4495.

Upcoming Events:
Planning has begun for a focus calibration of the MOC to be performed in late March.

Wednesday 15-Mar-00

(DOY 068/19:00:00 to DOY 075/19:00:00 UTC) Launch / Days since Launch = Nov 7, 1996 / 1227 days
Start of Mapping / Days since Start of Mapping = April 1, 1999 / 349 days
Total Mapping Orbits = 4562
Total Orbits = 6165

Recent Events:
The spacecraft continues to operate nominally in performing the beta supplement daily recording and transmission of science data. The mm020 sequence executed successfully from 00-63 (3/3/00) through 00-72 (3/12/00). The mm021 sequence also successfully executed from 00-63 (3/13/00) through 00-74 (3/14/00). The mm022 sequence was successfully uplinked on 00-73 (3/13/00) and began execution nominally on 00-74 (3/14/00).

Upon completion of the recent four day fixed-HGA mapping campaign performed last week, which resulted in higher than expected MOLA temperatures, the spacecraft team began looking at ways to fix the temperature problem. Additionally in assessing the first fixed-HGA campaign, it was noted that the actual transitions into and out of fixed-HGA mode of operations from the beta supplement mode of operations, were more complicated than anticipated and resulted in a larger than expected loss of science data from the other instruments. The Spacecraft Team presented at the 3/14 Project Mission Planning meeting alternatives to and options for fixing the current fixed-HGA operations. Fixed-HGA operations were to be performed 4-8 days per month for the express purpose of acquiring Radio Science Earth occultation egress measurements, which are prohibited during normal beta supplement mapping operations due to HGA gimbal constraints. From the Mission Planning meeting the favored option was to eliminate the fixed-HGA operations and perform special slews for a specified number of orbits to provide the desired longitudinal planetary coverage for the radio science occultation egress data, similar to the implementation of the MOLA off-nadir polar observations. Final details are in work and will be presented to the Project Change Board next Tuesday (3/21)

Spacecraft Health:
All subsystems are reporting nominal health. Battery 2 was successfully commanded back to VT-2 after transitioning back to beta supplement operations. The red alarm high limit for the MOLA temperature has been decreased by 3 degrees (from 35°C to 32°C) after a meeting with the MOLA engineers.

Uplinks:
There have been 20 uplinks to the spacecraft during the last week, including new star catalogs and ephemeris files, instrument command loads, and the mm021 and mm022 sequences. Two uplinks were missed after the final transition from the fixed-HGA Earth pointed attitude back to nadir pointed beta supplement operations due to a large change in the transponder temperature. The uplink sweeps for the beta supplement orbits are tailored specially for the very short uplink windows and are very sensitive to changes in the MOT temperature. The missing uplinks were successfully re-radiated. Total command files radiated to the spacecraft since launch is 4515.

Upcoming Events:
Planning has begun for a focus calibration of the MOC to be performed in late March.

Wednesday 22-Mar-00

(DOY 075/19:00:00 to DOY 082/19:00:00 UTC) Launch / Days since Launch = Nov 7, 1996 / 1234 days
Start of Mapping / Days since Start of Mapping = April 1, 1999 / 356 days
Total Mapping Orbits = 4648
Total Orbits = 6251

Recent Events:
The spacecraft continues to operate nominally in performing the beta supplement daily recording and transmission of science data. The mm022 sequence executed successfully from 00-75 (3/15/00) through 00-78 (3/18/00). The mm023 sequence also successfully executed from 00-79 (3/19/00) through 00-81 (3/21/00). The mm024 sequence was successfully uplinked on 00-81 (3/21/00) and began execution nominally on 00-82 (3/22/00). The mm025 sequence is scheduled to be uplinked on 00-85 (3/25/00) and will execute from 00-86 (3/26/00) through 00-88 (3/28/00).

Work has begun on the development of a series of MOC star scans to optimize the camera focus. Five total star scans of the Pleiades star cluster will be performed, one scan every other day, with the scan being done on the night-side of the orbit. The scans are currently scheduled on the days of 3/30, 4/01, 4/04, 4/05 and 4/07.

A series of FSW memory readout commands were performed over the last week and a half in order to determine the cause of an increased number of floating point overflow counts since the start of beta supplement operations. The results of the MROs determined that the overflows were coming from the polynomial evaluation function called by the articulation code and were a result of the HGA gimbals passing through 0 and 90 degrees, which is happening frequently during beta supplement due to the HGA management. The overflows appear to be benign and at this time, the SCT is still ascertaining whether any action is required.

Spacecraft Health:
All subsystems are reporting nominal health.

Uplinks:
There have been 18 uplinks to the spacecraft during the last week, including new star catalogs and ephemeris files, instrument command loads, and the mm023 and mm024 sequences. Total command files radiated to the spacecraft since launch is 4533.

Upcoming Events:
MOLA Off-nadir North and South Pole scans scheduled for 4/8 and 4/9

Wednesday 29-Mar-00

(DOY 082/19:00:00 to DOY 089/19:00:00 UTC) Launch / Days since Launch = Nov 7, 1996 / 1241 days
Start of Mapping / Days since Start of Mapping = April 1, 1999 / 363 days
Total Mapping Orbits = 4734
Total Orbits = 6337

Recent Events:
The spacecraft continues to operate nominally in performing the beta supplement daily recording and transmission of science data. The mm024 sequence executed successfully from 00-83(3/23/00) through 00-86 (3/26/00). The mm025 sequence also successfully executed from 00-87 (3/27/00) through 00-89 (3/29/00). The mm026 sequence was successfully uplinked on 00-88 (3/28/00) and will begin execution on 00-89 (3/29/00). The mm027 sequence is scheduled to be uplinked on 00-92 (4/1/00) and will execute from 00-94 (4/3/00) through 00-96 (4/5/00). The mz049 mini-sequence, containing four MOC star scans for the purpose of improving the camera focus, was also uplinked successfully on 00-89 (3/29/00). The first star scan will occur on 00-90 at 14:35:56 GMT on the spacecraft.

The mm028 and mm029 sequences are currently in development. A fifth and final MOC focus star scan will be included in the mm028 sequence. It is intended to repeat the MOC focus scans every six months. Also included in the mm028 sequence is the next MOLA North and South Pole off-nadir scans. The MOLA polar scans will begin at approximately 12:00 GMT on 00-99 (4/8/00). The scans will take place over 12 consecutive orbits to provide complete longitudinal coverage of the poles.

Spacecraft Health:
All subsystems are reporting nominal health.

Uplinks:
There have been 18 uplinks to the spacecraft during the last week, including new star catalogs and ephemeris files, instrument command loads, and the mm026, mm027 and mz049 sequences. Total command files radiated to the spacecraft since launch is 4551.

Upcoming Events:
Planning for the replacement of the monthly fixed-HGA

periods for the purpose of obtaining radio science atmospheric occultation egress data, prohibited in normal beta supplement operations due to HGA gimbal constraints, proceeds.

Wednesday 5-Apr-00

(DOY 089/19:00:00 to DOY 096/19:00:00 UTC) Launch / Days since Launch = Nov 7, 1996 / 1245 days
Start of Mapping / Days since Start of Mapping = April 1, 1999 / 370 days
Total Mapping Orbits = 4820
Total Orbits = 6423

Recent Events:
The spacecraft continues to operate nominally in performing the beta supplement daily recording and transmission of science data. The mm026 sequence executed successfully from 00-90(3/30/00) through 00-93 (4/3/00). The mm027 sequence also successfully executed from 00-94 (4/4/00) through 00-96 (4/6/00). The mm028 sequence was successfully uplinked on 00-95 (4/5/00) and will begin execution on 00-97 (4/8/00). Included in the mm028 sequence, is the fifth and last star scan of the current MOC focus campaign, and a 24-hour period of MOLA off-nadir scans of the North and South Poles, scheduled on 00-99 through 00-100 (4/8-4/9/00). The mm029 sequence is scheduled to be uplinked on 00-99 (4/8/00) and will execute from 00- 101 (4/10/00) through 00-103 (4/12/00).

The mz049 mini-sequence, containing the first four MOC star scans, has begun execution. The first three scans have completed to date. The MOC successfully acquired the desired stars from the first scan, but the stars were on the edge of their image. The second scan resulted in the MOC not seeing the stars. On the third scan the stars were acquired but were again on the edge of the image. The spacecraft attitude control errors were very small during each of the scans, leading us to believe there is some kind of knowledge error in the design of the scans, resulting in the stars not being centered in the image. AACS is currently analyzing the data and working with JPL and the MOC team to determine the cause of this error. Depending on the success of the fourth and fifth scans, additional scans may be performed in the next couple of weeks to ensure the MOC focus objective.

Spacecraft Health:
All subsystems are reporting nominal health.

Uplinks:
There have been 14 uplinks to the spacecraft during the last week, including new star catalogs and ephemeris files, instrument command loads, and the mm028 and mm029 sequences. Total command files radiated to the spacecraft since launch is 4565.

Upcoming Events:
Planning for the replacement of the monthly fixed-HGA periods for the purpose of obtaining radio science atmospheric occultation egress data, prohibited in normal beta supplement operations due to HGA gimbal constraints, proceeds. The first Radio Science occultation egress scans are scheduled for 5/20-5/22/00.

Wednesday 12-Apr-00

(DOY 096/19:00:00 to DOY 0103/19:00:00 UTC)
Launch / Days since Launch = Nov 7, 1996 / 1253 days
Start of Mapping / Days since Start of Mapping = April 1, 1999 / 377 days
Total Mapping Orbits = 4906
Total Orbits = 6509

Recent Events:
The spacecraft continues to operate nominally in performing the beta supplement daily recording and transmission of science data. The mm028 sequence executed successfully from 00-97 (4/6/00) through 00-100 (4/9/00). The mm029 sequence also successfully executed from 00-101 (4/10/00) through 00-103 (4/12/00). The mm030 sequence was successfully uplinked on 00-102 (4/11/00) and will begin execution on 00-104 (4/13/00).

Included in the mm028 sequence, was the fifth and last star scan of the current MOC focus campaign, and a 24-hour period of MOLA off-nadir scans of the North and South Poles, on 00-99 through 00-100 (4/8-4/9/00). The spacecraft performance during the final MOC star scan was nominal, but the stars were not seen in the image. Thus two of the five scans resulted in not seeing the stars. All indications are that the missed stars were due to a knowledge error in designing the scans. A retry of the two missing star scans will be scheduled for later in April, with an updated scan attitude.

Spacecraft performance during the MOLA scans was also nominal, with a single STAREX (star processing code) anomaly occurring proceeding the first MOLA off-nadir orbit. A STAREX autonomous reset occurred shortly after the sequence commanded the IMU to high rate mode and issued an intentional bias reset prior to the first off-nadir slew. After the autonomous reset STAREX converged properly and performed nominally for the duration of the MOLA scans.

Spacecraft Health:
All subsystems are reporting nominal health.

Uplinks:
There have been 16 uplinks to the spacecraft during the last week, including new star catalogs and ephemeris files, instrument command loads, and the mm030 and mm031 sequences. Total command files radiated to the spacecraft since launch is 4581.

Upcoming Events:
Planning for the replacement of the monthly fixed-HGA periods for the purpose of obtaining radio science atmospheric occultation egress data, prohibited in normal beta supplement operations due to HGA gimbal constraints, proceeds. The first Radio Science occultation egress scans are scheduled for 5/20-5/22/00. Work has been underway over the last month on determining the feasibility of performing a bistatic radar experiment, in which the Spacecraft signal will be reflected off the Martian surface to Earth. All design hurdles have been cleared and work is now underway to implement this experiment for a May 14 opportunity.

Wednesday 19-Apr-00

(DOY 0103/19:00:00 to DOY 0110/19:00:00 UTC)
Launch / Days since Launch = Nov 7, 1996 / 1260 days
Start of Mapping / Days since Start of Mapping = April 1, 1999 / 384 days
Total Mapping Orbits = 4992
Total Orbits = 6595

Recent Events:
The spacecraft continues to operate nominally in performing the beta supplement daily recording and transmission of science data. The mm030 sequence executed successfully from 00-104 (4/13/00) through 00-107 (4/16/00). The mm031 sequence also successfully executed from 00-108 (4/17/00) through 00-110 (4/19/00). The mm032 sequence was successfully uplinked on 00-108 (4/17/00) and will begin execution on 00-111 (4/20/00).

A retry of the two missed MOC focus star scans is scheduled for 4/28 and 4/30. The second attempts will utilize an updated scan attitude to account for the gyro bias error propagated when the IMU is switched from low rate to high rate mode for the slews.

Spacecraft Health:
All subsystems are reporting nominal health.

Uplinks:
There have been 11 uplinks to the spacecraft during the last week, including new star catalogs and ephemeris files, instrument command loads, and the mm032 and mm033 sequences. Total command files radiated to the spacecraft since launch is 4592.

Upcoming Events:
A bistatic radar experiment, in which the Spacecraft signal will be reflected off the Martian surface to Earth, is on schedule for execution on 5/14/00, upon Project approval expected next week.

The implementation of the first Radio Science occultation egress scans, replacing the previous fixed-HGA periods for obtaining occultation egress data, are also on schedule for execution on 5/20-5/22/00.

Wednesday 26-Apr-00

(DOY 110/19:00:00 to DOY 117/19:00:00 UTC) Launch / Days since Launch = Nov 7, 1996 / 1267 days
Start of Mapping / Days since Start of Mapping = April 1, 1999 / 391 days
Total Mapping Orbits = 5078
Total Orbits = 6681

Recent Events:
MGS reached the 5000 mapping orbit milestone on 00-111 (4/20/00) at 12:31:55 UTC.

The spacecraft continues to operate nominally in performing the beta supplement daily recording and transmission of science data. The mm032 sequence executed successfully from 00-111 (4/20/00) through 00-114 (4/23/00). The mm033 sequence also successfully executed from 00-115 (4/24/00) through 00-117 (4/26/00). The mm034 sequence was successfully uplinked on 00-116 (4/25/00) and will begin execution on 00-118 (4/27/00). The mz050 mini-sequence, containing the retry of the two missed MOC focus star scans on 4/28 and 4/30, is scheduled for uplink on 00-118 (4/27/00).

Spacecraft Health:
All subsystems are reporting nominal health.

Uplinks:
There have been 19 uplinks to the spacecraft during the last week, including new star catalogs and ephemeris files, instrument command loads, and the mm033 and mm034 sequences. Total command files radiated to the spacecraft since launch is 4611.

Upcoming Events:
A bistatic radar experiment, in which the Spacecraft signal will be reflected off the Martian surface to Earth, is on schedule for execution on 5/14/00.

The implementation of the first Radio Science occultation egress scans, replacing the previous fixed-HGA periods for obtaining occultation egress data, is also on schedule for execution on 5/20-5/22/00.

Wednesday 3-May-00

(DOY 117/19:00:00 to DOY 124/19:00:00 UTC) Launch / Days since Launch = Nov 7, 1996 / 1274 days
Start of Mapping / Days since Start of Mapping = April 1, 1999 / 398 days
Total Mapping Orbits = 5164
Total Orbits = 6767

Recent Events:
The spacecraft continues to operate nominally in performing the beta supplement daily recording and transmission of science data. The mm034 sequence executed successfully from 00-118 (4/27/00) through 00-121 (4/30/00). The mm035 sequence also successfully executed from 00-122 (5/1/00) through 00-124 (5/3/00). The mm036 sequence was successfully uplinked on 00-123 (5/2/00) and will begin execution on 00-125 (5/4/00).

The mz050 mini-sequence, containing the retry of the two missed MOC focus star scans, successfully executed on 4/28 and 4/30. The updated quaternion for these scans correctly accounted for the bias error due to the IMU being commanded to high rate mode and resulted in the stars being correctly centered in the image.

Spacecraft Health:
All subsystems are reporting nominal health.

Uplinks:
There have been 11 uplinks to the spacecraft during the last week, including new star catalogs and ephemeris files, instrument command loads, and the mm035 and mm036 sequences. Total command files radiated to the spacecraft since launch is 4622.

Upcoming Events:
A bistatic radar experiment, in which the Spacecraft signal will be reflected off the Martian surface to Earth, is in final development and will execute on May 14.

The next monthly MOLA Off-Nadir Polar Scans is about to begin development. The scans are scheduled for 12 consecutive orbits beginning on May 17.

The implementation of the first Radio Science occultation egress scans, replacing the previous fixed-HGA periods for obtaining occultation egress data, is also on schedule for execution on 5/20-5/22/00.

Wednesday 10-May-00

(DOY 124/19:00:00 to DOY 131/19:00:00 UTC) Launch / Days since Launch = Nov 7, 1996 / 1281 days
Start of Mapping / Days since Start of Mapping = April 1, 1999 / 405 days
Total Mapping Orbits = 5250
Total Orbits = 6853

Recent Events:
The spacecraft continues to operate nominally in performing the beta supplement daily recording and transmission of science data. The mm036 sequence executed successfully from 00-125 (5/4/00) through 00-128 (5/7/00). The mm037 sequence also successfully executed from 00-129 (5/8/00) through 00-131 (5/10/00). The mm038 sequence was successfully uplinked on 00-130 (5/9/00) and will begin execution on 00-132 (5/11/00).

Spacecraft Health:
All subsystems are reporting nominal health.

Uplinks:
There have been 16 uplinks to the spacecraft during the last week, including new star catalogs and ephemeris files, instrument command loads, and the mm037 and mm038 sequences. Total command files radiated to the spacecraft since launch is 4638.

Upcoming Events:
A bistatic radar experiment, in which the Spacecraft signal will be reflected off the Martian surface to Earth, is in final development and will execute on May 14. The commands to perform this experiment have been incorporated into the mm039 sequence, which is scheduled to be uplinked on 00-134 (5/13/00)

The next monthly MOLA off-nadir polar scan mini-sequence (mz051) is also in final development and will execute over 12 consecutive orbits beginning on May 17.

The implementation of the first Radio Science occultation egress scans, replacing the previous fixed-HGA periods for obtaining occultation egress data, is also in development at this time and is scheduled for execution on 5/20- 5/22/00.

Wednesday 24-May-00

(DOY 131/19:00:00 to DOY 145/19:00:00 UTC) Launch / Days since Launch = Nov 7, 1996 / 1295 days
Start of Mapping / Days since Start of Mapping = April 1, 1999 / 419 days
Total Mapping Orbits = 5418
Total Orbits = 7021

Recent Events:
Background Sequences — The spacecraft continues to operate nominally in performing the beta-supplement daily recording and transmission of science data. The following background sequences executed successfully: mm038 from 00-132 (5/11/00) through 00-135 (5/14/00); mm039 from 00-136 (5/15/00) through 00-138 (5/17/00); mm040 from 00- 139 (5/18/00) through 00-142 (5/21/00); and mm041 from 00-143 (5/22/00) through 00-145 (5/24/00). The mm042 sequence was successfully uplinked on 00-144 (5/23/00) and will begin executing on 00-146 (5/25/00).

Science Experiments - Three science experiment campaigns took place during the past two weeks. They were Bistatic Radar, MOLA Off-NADIR Polar Scans, and Radio Science Occultation Egress Scans. Commands for the Bistatic Radar experiment were included in the mm038 sequence as part of DOY 135 activities. Commands for the MOLA Off-NADIR Polar Scans composed the mz051 mini-sequence. They executed over 11 consecutive orbits on DOY 138 and DOY 139. Finally, MGS performed Radio Science Occultation Egress Scans on DOY 141,142, and 143 according to the mz052 mini-sequence. Overall, spacecraft performance during all experiments was nominal. The following paragraphs provide additional information regarding AACS performance.

In the MOLA experiment, we used the same CSA offset quaternions as the previous two polar opportunities. The SARBES (Starex reset) command was eliminated from this sequence and replaced with a general memory load of high rate gyro bias estimates and a separate software command to load the high rate steady state gyro biases. We also retained the divided North and South Pole slews implemented with the last MOLA polar opportunity. We are happy to report that the Starex problems observed in previous MOLA polar experiments have disappeared. Starex performance was completely, totally, and utterly nominal. Slew performance was superb, as it was for the previous polar experiments.

For the first time during a Radio Science Occultation Egress experiment, we used an Off-NADIR pointing scheme. AACS performance proved nominal, comparing well with predicts. Again, Starex performed superbly, converging in 10 stars. We successfully used the same method of gyro bias loading in the Radio Science experiment that we used in the MOLA Polar Scan experiment. Preliminary results indicate that the science teams received useful data during these experiments. Therefore, we plan to use the same spacecraft pointing and sequence development techniques for future MOLA and Radio Science experiments.

Spacecraft Health:
All subsystems report nominal health.

Uplinks:
There have been 28 uplinks to the spacecraft during the last two weeks, including new star catalogs and ephemeris files, instrument command loads, and the background sequences cited above. Total command files radiated to the spacecraft since launch is 4666.

Upcoming Events:
Normal beta-supplement sequences are planned between now and Solar Conjunction. We are in the final stages of Solar Conjunction planning. The mm050 sequence will cover the Solar Conjunction period between 6/21/00 and 7/12/00.

Wednesday 7-Jun-00

(DOY 145/19:00:00 to DOY 159/19:00:00 UTC) Launch / Days since Launch = Nov 7, 1996 / 1309 days
Start of Mapping / Days since Start of Mapping = April 1, 1999 / 433 days
Total Mapping Orbits = 5589
Total Orbits = 7192

Recent Events:
Background Sequences — The spacecraft continues to operate nominally in performing the beta-supplement daily recording and transmission of science data. The following background sequences executed successfully: mm042 from 00-146 (5/25/00) through 00-149 (5/28/00); mm043 from 00-150 (5/29/00) through 00-152 (5/31/00); mm044 from 00- 153 (6/01/00) through 00-156 (6/04/00); and mm045 from 00-157 (6/05/00) through 00-159 (6/07/00). The mm046 sequence was successfully uplinked on 00-158 (6/06/00) and will begin executing on 00-160 (6/08/00).

Spacecraft Health:
All subsystems report nominal health.

Uplinks:
There have been 19 uplinks to the spacecraft during the last two weeks, including new star catalogs and ephemeris files, instrument command loads, and the background sequences cited above. Total command files radiated to the spacecraft since launch is 4685.

Upcoming Events:
Normal beta-supplement sequences are planned between now and Solar Conjunction. We are in the initial stages of Solar Conjunction command sequence generation and testing. The mm050, and mm051 sequences will contain the lion's share of spacecraft commands for the Solar Conjunction period, 6/21/00 through 7/12/00. There will be some spacecraft configuration commands related to Solar Conjunction contained in the mm049 and mm052 sequences.

Wednesday 14-Jun-00

(DOY 159/19:00:00 to DOY 166/19:00:00 UTC) Launch / Days since Launch = Nov 7, 1996 / 1316 days
Start of Mapping / Days since Start of Mapping = April 1, 1999 / 440 days
Total Mapping Orbits = 5674
Total Orbits = 7277

Recent Events:
Background Sequences — The spacecraft continues to operate nominally in performing the beta-supplement daily recording and transmission of science data. The mm046 background sequence executed successfully from 00-160 (6/08/00) through 00-163 (6/11/00). The mm047 background sequence has been performing nominally since 00-164 (6/12/00) and will terminate at the end of day 00-166 (6/14/00). The mm048 sequence was successfully uplinked on 00-165 (6/13/00) and will begin executing on 00-167 (6/15/00).

Spacecraft Health:
All subsystems report nominal health.

Uplinks:
There have been 17 uplinks to the spacecraft during the last week, including new star catalogs and ephemeris files, instrument command loads, and the background sequences cited above. Total command files radiated to the spacecraft since launch is 4702.

Upcoming Events:
We are in the final stages of Solar Conjunction command sequence generation and testing. The mm050 and mm051 sequences will contain the lion's share of spacecraft commands for the Solar Conjunction period, 6/21/00 through 7/12/00. To prepare for Solar Conjunction, the MOC will be turned off using commands resident in the mm049 sequence. We expect to return to normal spacecraft configuration and beta-supplement command sequences starting with mm052 on 00- 195 (7/13/00).

Wednesday 28-Jun-00

(DOY 173/19:00:00 to DOY 180/19:00:00 UTC) Launch / Days since Launch = Nov 7, 1996 / 1330 days
Start of Mapping / Days since Start of Mapping = April 1, 1999 / 454 days
Total Mapping Orbits = 5846
Total Orbits = 7449

Recent Events:
Background Sequences — The mm050 and mm051 Solar Conjunction sequences were uplinked successfully and the MOC was turned off by commands in the mm049 sequence. The mm050 sequence became active on 00-175 (6/23/00). We are now in the Solar Conjuction phase of the mission.

Spacecraft Health:
All subsystems report nominal health. We are seeing more noise in the downlink signal as the Sun-Earth-Mars angle decreases. This is expected. The Sun-Earth-Mars angle will reach a minimum of 0.87 degrees on 00-183 (7/1/00). The downlink signal will gradually improve with time beyond that date.

Uplinks:
Only one file was uplinked to the spacecraft during the last week. It was the mm051 sequence. Total command files radiated to the spacecraft since launch is 4717.

Upcoming Events:
We expect to return to normal spacecraft configuration and beta-supplement command sequences starting with the mm052 sequence on 00-195 (7/13/00). Until then, we will monitor spacecraft health and the actual vs. predicted timing of orbital events. The spacecraft attempts to contact Earth during three of the twelve orbits each day during the Solar Conjunction period. When the downlink signal is good, we receive engineering telemetry and the actual Mars equator crossing time for that orbit. We compare the actual equator crossing time with the predicted equator crossing time to determine the accuracy with which we placed spacecraft maneuvers. We expect the accuracy of our orbital timing to degrade with time. Tracking the rate at which the timing degrades will help us ensure the spacecraft health during the next few weeks. We are prepared to cancel or rebuild stored sequences if timing degradation reaches unacceptable levels. Presently our predictions deviate from actual orbit timing by less than six seconds. This is well within the established parameters (Gee, I sound like Data). Once the spacecraft emerges from Solar Conjunction and we are

able to obtain two-way Doppler tracking data, our orbital timing estimates will return to normal.

Wednesday 5-Jul-00

(DOY 180/19:00:00 to DOY 187/19:00:00 UTC) Launch / Days since Launch = Nov 7, 1996 / 1337 days
Start of Mapping / Days since Start of Mapping = April 1, 1999 / 461 days
Total Mapping Orbits = 5932
Total Orbits = 7535

Recent Events:
Background Sequences — We are in the Solar Conjunction phase of the mission. The mm050 sequence is progressing nominally. The mm051 sequence will commence on 00-190, (7/8/00).

Spacecraft Health:
All subsystems report nominal health. We saw the noise in the downlink signal increase as the Sun-Earth-Mars (SEM) angle approached the minimum value of 0.87 degrees on 00-183 (7/1/00). We received no telemetry from 00-181 (6/29/00) through 00-183 (7/1/00) due to Solar Conjunction. Now, the downlink signal is gradually improving as the SEM angle increases.

Uplinks:
No commands were uplinked this week. Total command files radiated to the spacecraft since launch is 4717.

Upcoming Events:
We expect to return to normal spacecraft configuration and beta- supplement command sequences starting with the mm052 sequence on 00-195 (7/13/00). We continue to monitor spacecraft health and the actual vs. predicted timing of orbital events. The spacecraft attempts to contact Earth during three of the twelve orbits each day during the Solar Conjunction period. When the downlink signal is good, we receive engineering telemetry and the actual Mars equator crossing time for that orbit. We compare the actual equator crossing time with the predicted equator crossing time to determine the accuracy with which we placed spacecraft maneuvers. We expect the accuracy of our orbital timing to degrade with time. Tracking the rate at which the timing degrades helps us ensure the spacecraft health during the Solar Conjunction. We are prepared to cancel or rebuild stored sequences if timing degradation reaches unacceptable levels. Presently our predictions deviate from actual orbit timing by less than twenty-two seconds. This is well within the established timing envelope. Timing would have to deviate by 75 seconds before we considered canceling the stored sequence. We anticipate satisfactory timing throughout the remainder of mm050 and mm051. We expect to obtain satisfactory two-way Doppler tracking data on 00-190 (7/8/00). Orbital timing estimates will return to normal in the week following. Sequence development of mm052 and mm053 is proceeding nominally.

Wednesday 12-Jul-00

(DOY 187/19:00:00 to DOY 194/19:00:00 UTC) Launch / Days since Launch = Nov 7, 1996 / 1344 days
Start of Mapping / Days since Start of Mapping = April 1, 1999 / 468 days
Total Mapping Orbits = 6017
Total Orbits = 7620

Recent Events:
Background Sequences — We are in the last day of the Solar Conjunction phase of the mission and sequences continue to execute nominally. The mm051 sequence will terminate and the mm052 sequence will begin on 00-195 (7/13/00). mm052, the first beta- supplement sequence to execute since 00-174 (6/22/00), will start recording science data on 00-195 (7/13/00). Playbacks of science data commence on 00-196 (7/14/00).

Spacecraft Health:
All subsystems report nominal health.

Uplinks:
There have been 10 uplinks to the spacecraft during the last week, including new star catalogs and ephemeris files and the mm052 background sequence. We turned off two-way non-coherent mode to regain two-way Doppler capability and we set the Command Loss Timer to its nominal value of 104 hours. The total number of command files radiated to the spacecraft since launch is 4727.

Upcoming Events:
The MOC will be turned on tomorrow, 00-195 (7/13/00), via real-time commands. The mm053 background will be uplinked 00-196 (7/14/00).

Wednesday 19-Jul-00

(DOY 194/19:00:00 to DOY 201/19:00:00 UTC) Launch / Days since Launch = Nov. 7, 1996 / 1351 days
Start of Mapping / Days since Start of Mapping = April 1, 1999 / 475 days
Total Mapping Orbits = 6103
Total Orbits = 7706

Recent Events:
Background Sequences — The spacecraft continues to operate nominally in performing the beta-supplement daily recording and transmission of science data. The mm052 sequence executed successfully from 00-195 (7/13/00) through 00-197 (7/15/00). The mm053 sequence has performed well since it started on 00-198 (7/16/00). It terminates on 00-201 (7/19/00). The mm054 sequence, successfully uplinked on 00-200 (7/18/00), begins executing on 00-202 (7/20/00).

Spacecraft Health:
All subsystems report nominal health.

Uplinks:
There have been 12 uplinks to the spacecraft during the last week, including new star catalogs and ephemeris files, MOC turn-on commands, instrument command loads, and the background sequences cited above. There have been 4739 command files radiated to the spacecraft since launch.

Upcoming Events:
The mm055 background sequence will be uplinked 00-203 (7/21/00).

Wednesday 26-Jul-00

(DOY 201/19:00:00 to DOY 208/19:00:00 UTC) Launch / Days since Launch = Nov. 7, 1996 / 1358 days
Start of Mapping / Days since Start of Mapping = April 1, 1999 / 482 days
Total Mapping Orbits = 6189
Total Orbits = 7792

Recent Events:
Background Sequences — The spacecraft continues to operate nominally in performing the beta-supplement daily recording and transmission of science data. The mm054 sequence executed successfully from 00-202 (7/20/00) through 00-204 (7/22/00). The mm055 sequence has performed well since it started on 00-205 (7/23/00). It terminates on 00-208 (7/26/00). The mm056 sequence, successfully uplinked on 00-207 (7/25/00), begins executing on 00-209 (7/27/00).

Spacecraft Health:
All subsystems report nominal health.

Uplinks:
There have been 13 uplinks to the spacecraft during the last week, including new star catalogs and ephemeris files, instrument command loads, and the background sequences cited above. On 00-203 (7/21/00) we successfully demonstrated the capability to perform uplinks using the new DSS-14 X-band transmitter. There have been 4752 command files radiated to the spacecraft since launch.

Upcoming Events:
The mm057 background sequence will be uplinked 00-210 (7/28/00). The MOLA will be turned- on on 00-214 (8/1/00). Radio Science Occultation Egress Scans will take place 00-221 (8/8/00) and 00-222 (8/9/00).

Wednesday 2-Aug-00

(DOY 208/19:00:00 to DOY 215/19:00:00 UTC) Launch / Days since Launch = Nov. 7, 1996 / 1365 days
Start of Mapping / Days since Start of Mapping = April 1, 1999 / 489 days
Total Mapping Orbits = 6274
Total Orbits = 7877

Recent Events:
Background Sequences — The spacecraft continues to operate nominally in performing the beta-supplement daily recording and transmission of science data. The mm056 sequence executed successfully from 00-209 (7/27/00) through 00-211 (7/29/00). The mm057 sequence has performed well since it started on 00-212 (7/30/00). It terminates on 00-215 (8/02/00). The mm058 sequence, successfully uplinked on 00-214 (8/01/00), begins executing on 00-216 (8/03/00).

On 00-214 (8/01/00), the MOLA was turned on as part of the mm057 sequence. It had been off for a two-month period that included Solar Conjunction. MOLA laser transmitter life is directly related to its on-time. We plan to double its remaining useful life by operating the instrument every other month.

Spacecraft Health:
All subsystems report nominal health.

Uplinks:
There have been 13 uplinks to the spacecraft during the last week, including new star catalogs and ephemeris files, instrument command loads, and the background sequences cited above. There have been 4765 command files radiated to the spacecraft since launch.

Upcoming Events:
The mz053 mini-sequence will be uplinked 00-216 (8/03/00). It contains the Radio Science Occultation Egress Scans that will take place on 00- 221 (8/8/00) and 00-222 (8/9/00). The mm059 background sequence will be uplinked 00-217 (8/04/00).

Wednesday 9-Aug-00

(DOY 215/19:00:00 to DOY 222/19:00:00 UTC) Launch / Days since Launch = Nov. 7, 1996 / 1372 days
Start of Mapping / Days since Start of Mapping = April 1, 1999 / 496 days
Total Mapping Orbits = 6360
Total Orbits = 7963

Recent Events:
Background Sequences — The spacecraft continues to operate nominally in performing the beta-supplement daily recording and transmission of science data. The mm058 sequence executed successfully from 00-216 (8/03/00) through 00-218 (8/05/00). The mm059 sequence has performed well since it started on 00-219 (8/06/00). It terminates on 00-222 (8/09/00). The mm060 sequence, successfully uplinked on 00-222 (8/09/00), begins executing on 00-223 (8/10/00).

The spacecraft successfully completed twelve Radio Science Occultation Egress Scans on 00-221 (8/8/00) and 00-222 (8/9/00). The AACS performed well using the same off-NADIR pointing scheme that proved successful last May.

Spacecraft Health:
All subsystems report nominal health.

Uplinks:
There have been 9 uplinks to the spacecraft during the last week, including new star catalogs and ephemeris files, instrument command loads, and the background sequences cited above. There have been 4774 command files radiated to the spacecraft since launch.

Upcoming Events:
The mm061 background sequence will be uplinked 00-224 (8/11/00). MOLA Polar Scans are scheduled for 00-242 (8/29/00) through 00-246 (9/2/00).

Wednesday 16-Aug-00

(DOY 222/19:00:00 to DOY 229/19:00:00 UTC) Launch / Days since Launch = Nov. 7, 1996 / 1379 days
Start of Mapping / Days since Start of Mapping = April 1, 1999 / 503 days
Total Mapping Orbits = 6446
Total Orbits = 8049

Recent Events:
Background Sequences — The spacecraft continues to operate nominally in performing the beta-supplement daily recording and transmission of science data. The mm060 sequence executed successfully from 00-223 (8/10/00) through 00-225 (8/12/00). The mm061 sequence has performed well since it started on 00-226 (8/13/00). It terminates on 00-229 (8/16/00). The mm062 sequence, successfully uplinked on 00-228 (8/15/00), begins executing on 00-230 (8/17/00).

Spacecraft Health:
All subsystems report nominal health.

Uplinks:
There have been 14 uplinks to the spacecraft during the last week, including new star catalogs and ephemeris files, instrument command loads, and the background sequences cited above. There have been 4788 command files radiated to the spacecraft since launch.

Upcoming Events:
The mm063 background sequence will be uplinked 00-231 (8/18/00). MOLA Polar Scans are scheduled for 00-243 (8/30/00) through 00-246 (9/2/00).

Wednesday 30-Aug-00

(DOY 236/19:00:00 to DOY 243/19:00:00 UTC) Launch / Days since Launch = Nov. 7, 1996 / 1393 days
Start of Mapping / Days since Start of Mapping = April 1, 1999 / 517 days
Total Mapping Orbits = 6,617
Total Orbits = 8,220

Recent Events:
Background Sequences — The spacecraft continues to operate nominally in performing the beta- supplement daily recording and transmission of science data. The mm064 sequence executed successfully from 00-237 (8/24/00) through 00-239 (8/26/00). The mm065 sequence has performed well since it started on 00-240 (8/27/00). It terminates on 00-243 (8/30/00). The mm066 sequence, successfully uplinked on 00-242 (8/29/00), begins executing on 00-244 (8/31/00).

Other - The mz054 mini- sequence containing the MOLA Polar Scans was uplinked 00-241 (8/28/00). Between 00-243 (8/30/00) and 00-246 (9/2/00) MGS will specifically target the North and South poles with the laser altimeter during thirteen orbits covering 360 degrees of latitude.

Spacecraft Health:
All subsystems report nominal health. As MGS approaches aphelion, the average energy margin per orbit is dropping as expected. Power projections indicate that the average energy margin per orbit will drop below the 40 Watt-hour threshold between early September 2000 and mid- January 2001. Though the spacecraft is capable of operating nominally below this threshold, the project has elected to maintain the average energy margin at or above 40 Watt-hours. The on-board command scripts that control solar array motion will be modified to provide the desired margin. This change will take effect 00-251 (9/7/00).

Uplinks:
There have been 9 uplinks to the spacecraft during the last week, including new star catalogs and ephemeris files, instrument command loads, and the background sequences cited above. There have been 4,813 command files radiated to the spacecraft since launch.

Upcoming Events:
The mm067 background sequence will be uplinked 00-245 (9/1/00). Radio Science Occultation Egress Scans, scheduled for 00-252 (9/8/00) through 00-253 (9/09/00), are contained in the mz055 mini- sequence. It will be uplinked 00-250 (9/6/00). MOC Focus Calibrations are scheduled for 00-262 (9/18/00) through 00-272 (9/28/00).

Wednesday 6-Sep-00

(DOY 243/19:00:00 to DOY 250/19:00:00 UTC) Launch / Days since Launch = Nov. 7, 1996 / 1400 days
Start of Mapping / Days since Start of Mapping = April 1, 1999 / 524 days
Total Mapping Orbits = 6,703
Total Orbits = 8,306

Recent Events:
Background Sequences — The spacecraft continues to operate nominally in performing the beta- supplement daily recording and transmission of science data. The mm066 sequence executed successfully from 00-244 (8/31/00) through 00-246 (9/02/00). The mm067 sequence has performed well since it started on 00-247 (9/3/00). It terminates on 00-250 (9/6/00). The mm068 sequence, successfully uplinked on 00-249 (8/5/00), begins executing on 00-251 (9/7/00).

Other - Between 00-243 (8/30/00) and 00-246 (9/2/00) MGS specifically targeted the North and South poles with the laser altimeter during thirteen orbits covering 360 degrees of latitude. Now that the mz054 MOLA Polar Scan mini-sequence has successfully terminated, scientists are processing the resulting data set.

Spacecraft Health:
All subsystems report nominal health. As MGS approaches aphelion, the average energy margin per orbit is dropping as expected. Two new solar array motion scripts have been uplinked to maintain 40 Watt-hours of average energy margin per orbit while MGS travels near aphelion. They take effect on 00-251 (9/7/00).

Uplinks:
There have been 17 uplinks to the spacecraft during the last week, including instrument command loads, the previously cited background sequences, the two new solar array motion scripts, and the mz055 mini-sequence. The mz055 mini-sequence will command Radio Science Occultation Egress Scans between 00-252 (9/8/00) and 00-253 (9/09/00).

There have been 4,830 command files radiated to the spacecraft since launch.

Upcoming Events:
The mm069 background sequence will be uplinked 00-252 (9/8/00). MOC Focus Calibrations, scheduled for 00-262 (9/18/00) through 00-272 (9/28/00), will be contained in the mz056, mz057, and mz058 mini-sequences.

Wednesday 13-Sep-00

(DOY 250/19:00:00 to DOY 257/19:00:00 UTC) Launch / Days since Launch = Nov. 7, 1996 / 1407 days
Start of Mapping / Days since Start of Mapping = April 1, 1999 / 531 days
Total Mapping Orbits = 6,788
Total Orbits = 8,391

Recent Events:
Background Sequences — The spacecraft continues to operate nominally in performing the beta- supplement daily recording and transmission of science data. The mm068 sequence executed successfully from 00-251 (9/7/00) through 00-253 (9/09/00). The mm069

sequence has performed well since it started on 00-254 (9/10/00). It terminates on 00-257 (9/13/00). The mm070 sequence, successfully uplinked on 00-256 (9/12/00), begins executing on 00-258 (9/14/00).

Other - MGS successfully completed twelve Radio Science Occultation Egress Scans, between 00-252 (9/8/00) and 00-253 (9/09/00). They were contained in the mz055 mini-sequence. Scientists are analyzing the resulting data set.

Spacecraft Health:
All subsystems report nominal health. Because MGS is approaching aphelion, the average energy margin per orbit has been dropping as expected. Two new solar array motion scripts, designed to increase power margin, began operating on 00- 251 (9/7/00). The spacecraft is now maintaining 60 Watt-hours of average energy margin per orbit compared to 35 Watt-hours just before the new scripts took effect. We expect the spacecraft to maintain an average energy margin per orbit greater than 40 Watt-hours through aphelion.

Uplinks:
There have been 15 uplinks to the spacecraft during the last week, including new star catalogs and ephemeris files, instrument command loads, and the background sequences cited above. There have been 4,845 command files radiated to the spacecraft since launch.

Upcoming Events:
The mm071 background sequence will be uplinked 00-259 (9/15/00). MOC Focus Calibrations, scheduled for 00-262 (9/18/00) through 00-272 (9/28/00), will be contained in the mz056, mz057, and mz058 mini-sequences.

Wednesday 20-Sep-00

(DOY 257/19:00:00 to DOY 264/19:00:00 UTC) Launch / Days since Launch = Nov. 7, 1996 / 1414 days
Start of Mapping / Days since Start of Mapping = April 1, 1999 / 538 days
Total Mapping Orbits = 6,874
Total Orbits = 8,477

Recent Events:
Background Sequences — The spacecraft continues to operate nominally in performing the beta- supplement daily recording and transmission of science data. The mm070 sequence executed successfully from 00-258 (9/14/00) through 00-260 (9/16/00). The mm071 sequence has performed well since it started on 00-261 (9/17/00). It terminates on 00-264 (9/20/00). The mm072 sequence, successfully uplinked on 00-263 (9/19/00), begins executing on 00-265 (9/21/00).

Other - MGS successfully completed the first of five MOC Focus Calibrations on 00-262 (9/18/00). The MOC focus calibration image contained the three targeted stars (Alcyone, Maia and Taygeta). Downtrack position was good and crosstrack position improved from previous calibrations.

Spacecraft Health:
All subsystems report nominal health.

Uplinks:
There have been 17 uplinks to the spacecraft during the last week, including new star catalogs and ephemeris files, instrument command loads, and the background sequences cited above. There have been 4,862 command files radiated to the spacecraft since launch.

Upcoming Events:
The mm073 background sequence will be uplinked 00-266 (9/22/00). MOC Focus Calibrations #2 - #5, contained in the mz057 and mz058 mini-sequences, will be performed 00-266 (9/22/00) through 00-272 (9/28/00). MOLA Polar Scans will be performed 00-281 (10/7/00) through 00-285 (10/11/00). They will be contained in the mz059 mini-sequence.

Wednesday 27-Sep-00

(DOY 264/19:00:00 to DOY 271/19:00:00 UTC) Launch / Days since Launch = Nov. 7, 1996 / 1421 days
Start of Mapping / Days since Start of Mapping = April 1, 1999 / 545 days
Total Mapping Orbits = 6,959
Total Orbits = 8,562

Recent Events:
Background Sequences — The spacecraft continues to operate nominally in performing the beta- supplement daily recording and transmission of science data. The mm072 sequence executed successfully from 00-265 (9/21/00) through 00-267 (9/23/00). The mm073 sequence has performed well since it started on 00-268 (9/24/00). It terminates on 00-271 (9/27/00). The mm074 sequence, successfully uplinked on 00-270 (9/26/00), begins executing on 00-272 (9/28/00).

Other - This week MGS completed the second, third, and forth of five scheduled MOC Focus Calibrations. The results of calibration #2 were inconclusive due to gaps in the data. The Deep Space Network antenna had difficulty maintaining the data downlink due to poor weather at the receiver site. MOC Focus Calibrations #3 and #4 were successful in imaging most of the targeted stars, providing scientists and engineers with data that will be used to improve the precision MOC targeting capabilities of MGS.

Spacecraft Health:
All subsystems report nominal health. During the beta-supplement phase of this mission we have been contending with the issue of how to retrieve from MGS all of the science and engineering data that is recorded. We knew that eventually planetary position geometry combined with the need to maneuver the spacecraft around the HGA gimbal obstruction would result in our inability to playback all of the data within the available downlink windows. We have attempted two modifications in the Solid State Recorder (SSR) record/playback strategy over the past weeks. The modification that works the best is analogous to pressing the "pause"" button on a typical tape recorder. So, every time we playback SSR Track 7, we command the SSR to "pause"" just before MGS becomes occulted by Mars or it maneuvers around the HGA gimbal obstruction. We pick up the playback of SSR Track 7 during the next downlink opportunity. Presently, it takes three separate downlink opportunities to playback SSR Track 7 completely. With less than 45 minutes of each 2 hour orbit available for playback, it already takes 12 downlink opportunities over 6 complete orbits to playback SSR Tracks 1 through 6. Therefore, MGS

requires 15 - 16 hours of downlink each day to playback all recorded data.This situation will gradually improve as planetary geometry becomes more favorable.The beta-supplement phase of this mission ends in late June of 2001.

Uplinks:
There have been 13 uplinks to the spacecraft during the last week, including instrument command loads and sequences cited above.There have been 4,875 command files radiated to the spacecraft since launch.

Upcoming Events:
The mm075 background sequence will be uplinked 00-273 (9/29/00). MOC Focus Calibration #5 contained in the mz058 mini- sequence, will be performed on 00-272 (9/28/00). An additional MOC Focus Calibration (#6) is scheduled for 00-277 (10/3/00) to gain the data lost during MOC Focus Calibration #2. It's uplink designation will be mz060. MOLA Polar Scans, contained in the mz059 mini-sequence, will be performed 00-281 (10/7/00) through 00-285 (10/11/00). Radio Science Occultation Egress Scans are scheduled for 00-287 (10/13/00) and 00-288 (10/14/00).

Wednesday 4-Oct-00

(DOY 271/19:00:00 to DOY 278/19:00:00 UTC) Launch / Days since Launch = Nov. 7, 1996 / 1428 days
Start of Mapping / Days since Start of Mapping = April 1, 1999 / 552 days
Total Mapping Orbits = 7,045
Total Orbits = 8,648

Recent Events:
Background Sequences — The spacecraft continues to operate nominally in performing the beta- supplement daily recording and transmission of science data. The mm074 sequence executed successfully from 00-272 (9/28/00) through 00-274 (9/30/00). The mm075 sequence has performed well since it started on 00-275 (10/1/00). It terminates on 00-278 (10/4/00).The mm076 sequence, successfully uplinked on 00-277 (10/3/00), begins executing on 00-279 (10/5/00).

Other - This week MGS completed the fifth and sixth MOC Focus Calibrations. Both MOC Calibration Images hit the three targeted stars (Alcyone, Maia and Taygeta) with no image drops near the stars. Cross-track positions measured 2.1 - 2.2 milli-radians from the center on the same side of the detector as observed in previous star images. This is within the MGS pointing accuracy error budget.

Spacecraft Health:
All subsystems report nominal health.

Uplinks:
There have been 16 uplinks to the spacecraft during the last week, including new star catalogs and ephemeris files, instrument command loads and the sequences cited above. There have been 4,891 command files radiated to the spacecraft since launch.

Upcoming Events:
The mm077 background sequence will be uplinked on 00-280 (10/6/00). MOLA Polar Scans, contained in the mz059 mini- sequence, will be performed on 00-281 (10/7/00) through 00-285 (10/11/00). Radio Science Occultation Egress Scans, scheduled for 00-287 (10/13/00) and 00-288 (10/14/00), will be commanded by the mz061 mini-sequence.

Wednesday 11-Oct-00

(DOY 278/19:00:00 to DOY 285/19:00:00 UTC) Launch / Days since Launch = Nov. 7, 1996 / 1435 days
Start of Mapping / Days since Start of Mapping = April 1, 1999 / 559 days
Total Mapping Orbits = 7,131
Total Orbits = 8,734

Recent Events:
Background Sequences — The spacecraft continues to operate nominally in performing the beta- supplement daily recording and transmission of science data. The mm076 sequence executed successfully from 00-279 (10/5/00) through 00-281 (10/7/00). The mm077 sequence has performed well since it started on 00-282 (10/8/00). It terminates on 00-285 (10/11/00). The mm078 sequence, successfully uplinked on 00-285 (10/11/00), begins executing on 00-286 (10/12/00).

Other - MGS completed fifteen MOLA Polar Scans over five days from 00-281 (10/7/00) through 00-285 (10/11/00).They were part of the mz059 mini-sequence.

Spacecraft Health:
All subsystems report nominal health.

Uplinks:
There have been 13 uplinks to the spacecraft during the last week, including new star catalogs and ephemeris files, instrument command loads and the sequences cited above. There have been 4,904 command files radiated to the spacecraft since launch.

Upcoming Events:
The mm079 background sequence will be uplinked on 00-287 (10/13/00). The next set of specialized science observations occur in November. MOLA Polar Scans, contained in the mz062 mini-sequence, will be performed on 00-316 (11/11/00) through 00-319 (11/14/00). Radio Science Occultation Egress Scans, scheduled for 00-322 (11/17/00) through 00-325 (11/20/00), will be commanded by the mz062 mini-sequence.

Wednesday 18-Oct-00

(DOY 285/19:00:00 to DOY 292/19:00:00 UTC) Launch / Days since Launch = Nov. 7, 1996 / 1442 days
Start of Mapping / Days since Start of Mapping = April 1, 1999 / 566 days
Total Mapping Orbits = 7,217
Total Orbits = 8,820

Recent Events:
Background Sequences — The spacecraft continues to operate nominally in performing the beta- supplement daily recording and transmission of science data. The mm078 sequence executed successfully from 00-286 (10/12/00) through 00-288 (10/14/00). The mm079 sequence has performed well since it started on 00-289 (10/15/00). It terminates on 00-292 (10/18/00). The mm080 sequence, successfully uplinked on 00-291 (10/17/00), begins executing on 00-293 (10/19/00).

Other - MGS successfully completed twelve Radio

Science Occultation Egress Scans on 00-287 (10/13/00) and 00-288 (10/14/00). They were part of the mz061 mini- sequence.

Spacecraft Health:
All subsystems report nominal health.

Uplinks:
There have been 13 uplinks to the spacecraft during the last week, including new star catalogs and ephemeris files, instrument command loads and the sequences cited above. There have been 4,917 command files radiated to the spacecraft since launch.

Upcoming Events:
The mm081 background sequence will be uplinked on 00-294 (10/20/00). The next set of specialized science observations occur in November. MOLA Polar Scans, contained in the mz062 mini-sequence, will be performed on 00-316 (11/11/00) through 00- 319 (11/14/00). Radio Science Occultation Egress Scans, scheduled for 00-322 (11/17/00) through 00-325 (11/20/00), will be commanded by the mz062 mini-sequence.

Wednesday 25-Oct-00

(DOY 292/19:00:00 to DOY 299/19:00:00 UTC) Launch / Days since Launch = Nov. 7, 1996 / 1449 days
Start of Mapping / Days since Start of Mapping = April 1, 1999 / 573 days
Total Mapping Orbits = 7,303
Total Orbits = 8,906

Recent Events:
Background Sequences — The spacecraft continues to operate nominally in performing the beta- supplement daily recording and transmission of science data. The mm080 sequence executed successfully from 00-293 (10/19/00) through 00-295 (10/21/00). The mm081 sequence has performed well since it started on 00-296 (10/22/00). It terminates on 00-299 (10/25/00). The mm080 sequence, successfully uplinked on 00-298 (10/24/00), begins executing on 00-300 (10/26/00).

Other - No special science scans were conducted this week.

Spacecraft Health:
All subsystems report nominal health.

Uplinks:
There have been 7 uplinks to the spacecraft during the last week, including instrument command loads and the sequences cited above. There have been 4,924 command files radiated to the spacecraft since launch.

Upcoming Events:
The mm083 background sequence will be uplinked on 00-301 (10/27/00). The next set of specialized science observations occur in November. MOLA Polar Scans, contained in the mz062 mini-sequence, will be performed on 00-316 (11/11/00) through 00-319 (11/14/00). Radio Science Occultation Egress Scans, scheduled for 00-322 (11/17/00) through 00-325 (11/20/00), will be commanded by the mz063 mini-sequence.

Wednesday 8-Nov-00

(DOY 306/19:00:00 to DOY 313/19:00:00 UTC) Launch / Days since Launch = Nov. 7, 1996 / 1463 days
Start of Mapping / Days since Start of Mapping = April 1, 1999 / 587 days
Total Mapping Orbits = 7,473
Total Orbits = 9,076

Recent Events:
Background Sequences — The spacecraft continues to operate nominally in performing the beta- supplement daily recording and transmission of science data. The mm084 sequence executed successfully from 00-307 (11/2/00) through 00-309 (11/4/00). The mm085 sequence has performed well since it started on 00-310 (11/6/00). It terminates on 00-313 (11/8/00). The mm086 sequence, successfully uplinked on 00-312 (11/7/00), begins executing on 00-314 (11/9/00).

Other - No special science scans were conducted this week.

Spacecraft Health:
All subsystems report nominal health. The Safe Mode mission-phase-relays have been commanded to the Mapping state. Therefore, in the unlikely event that MGS enters Safe Mode, it will remain in a positive power margin state.

Uplinks:
There have been 8 uplinks to the spacecraft during the last week, including instrument command loads, the sequences cited above, and the command to change the mission- phase-relays to Mapping. There have been 4,947 command files radiated to the spacecraft since launch.

Upcoming Events:
The mm087 background sequence will be uplinked on 00-315 (11/10/00). MOLA Polar Scans, contained in the mz062 mini-sequence, will be performed on 00-316 (11/11/00) through 00-319 (11/14/00). Radio Science Occultation Egress Scans, scheduled for 00-322 (11/17/00) through 00-325 (11/20/00), will be commanded by the mz063 & mz064 mini-sequences.

Wednesday 15-Nov-00

(DOY 313/19:00:00 to DOY 320/19:00:00 UTC) Launch / Days since Launch = Nov. 7, 1996 / 1470 days
Start of Mapping / Days since Start of Mapping = April 1, 1999 / 594 days
Total Mapping Orbits = 7,559
Total Orbits = 9,162

Recent Events:
Background Sequences — The spacecraft continues to operate nominally in performing the beta- supplement daily recording and transmission of science data. The mm086 sequence executed successfully from 00-314 (11/9/00) through 00-316 (11/11/00). The mm087 sequence has performed well since it started on 00-317 (11/13/00). It terminates on 00-320 (11/15/00). The mm088 sequence, successfully uplinked on 00-319 (11/14/00), begins executing on 00-321 (11/16/00).

Other - MOLA Polar Scans, contained in the mz062 mini-sequence, were performed successfully on 00- 316 (11/11/00) through 00-319 (11/14/00). Data processing will begin as soon as the data are available on the central

database.
Spacecraft Health:
All subsystems report nominal health. Telecom link margin is improving as the distance between Earth and Mars decreases. This will allow us to increase the record and playback rates by a factor of two. We plan to increase the record rate to 8 ksps on 00-331 (11/26/00) and the playback rate to 42 ksps on 00-332 (11/27/00).

Uplinks:
There have been 15 uplinks to the spacecraft during the last week, including new star catalogs and ephemeris files, instrument command loads, and the sequences cited above. Unfortunately, seven uplinks attempted today (00-320) were unsuccessful due to bad spacecraft frequency predictions. The affected command files will be uplinked tomorrow morning, 00-321 (11/16/00). There have been 4,947 command files radiated to the spacecraft since launch.

Upcoming Events:
The mm089 background sequence will be uplinked on 00-322 (11/17/00). Radio Science Occultation Egress Scans, scheduled for 00-322 (11/17/00) through 00-325 (11/20/00), will be commanded by the mz063 & mz064 mini-sequences. The mz063a command file will be retransmitted to the spacecraft tomorrow, 00-321 (11/16/00), because today's attempt failed as explained above.

Wednesday 22-Nov-00

(DOY 320/19:00:00 to DOY 327/19:00:00 UTC) Launch / Days since Launch = Nov. 7, 1996 / 1477 days
Start of Mapping / Days since Start of Mapping = April 1, 1999 / 601 days
Total Mapping Orbits = 7,645
Total Orbits = 9,248

Recent Events:
Background Sequences — The spacecraft continues to operate nominally in performing the beta- supplement daily recording and transmission of science data. The mm088 sequence executed successfully from 00-321 (11/16/00) through 00-323 (11/18/00). The mm089 sequence has performed well since it started on 00-324 (11/20/00). It terminates on 00-327 (11/22/00). The mm090 sequence, successfully uplinked on 00-326 (11/21/00), begins executing on 00-328 (11/23/00).

Other - Radio Science Occultation Egress Scans, contained in the mz063 and mz064 mini- sequences, were accomplished successfully on 00-322 (11/17/00) through 00-325 (11/20/00). Twenty-four total scans were performed.

Spacecraft Health:
All subsystems report nominal health. Telecom link margin is improving as the distance between Earth and Mars decreases. This will allow us to increase the record and playback rates by a factor of two. We plan to increase the record rate to 8 ksps on 00-331 (11/26/00) and the playback rate to 42 ksps on 00-332 (11/27/00).

Uplinks:
There have been 17 uplinks to the spacecraft during the last week, including new star catalogs and ephemeris files, instrument command loads, and the sequences cited above. There have been 4,979 command files radiated to the spacecraft since launch.

Upcoming Events:
The mm091 background sequence will be uplinked on 00-329 (11/24/00). Science Campaign F commences on 00-345 (12/10/00) and lasts until 00-356 (12/21/00). Radio Science Occultation Egress Scans, scheduled for 00-353 (12/18/00) through 00-356 (12/21/00), will be commanded by the mz065 & mz066 mini-sequences.

Wednesday 29-Nov-00

(DOY 327/19:00:00 to DOY 334/19:00:00 UTC) Launch / Days since Launch = Nov. 7, 1996 / 1484 days
Start of Mapping / Days since Start of Mapping = April 1, 1999 / 608 days
Total Mapping Orbits = 7,731
Total Orbits = 9,334

Recent Events:
Background Sequences — The spacecraft continues to operate nominally in performing the beta- supplement daily recording and transmission of science data. The mm090 sequence executed successfully from 00-328 (11/23/00) through 00-330 (11/25/00). The mm091 sequence has performed well since it started on 00-331 (11/26/00). It terminates on 00-334 (11/29/00). The mm092 sequence, successfully uplinked on 00-333 (11/28/00), begins executing on 00-335 (11/30/00).

The background sequences now command SSR record and playback rates of 8 ksps and 42 ksps respectively. We were able to double the previous rates because of the improved telecom link margin available, as Earth and Mars get closer to each other. The change took effect on 00-331 (11/26/00).

Other - No special science scans were conducted this week.

Spacecraft Health:
All subsystems report nominal health.

Uplinks:
There have been 13 uplinks to the spacecraft during the last week, including new star catalogs and ephemeris files, instrument command loads, and the sequences cited above. There have been 4,992 command files radiated to the spacecraft since launch.

Upcoming Events:
The mm093 background sequence will be uplinked on 00-336 (12/1/00). Science Campaign F commences on 00-345 (12/10/00) and lasts until 00-356 (12/21/00). Radio Science Occultation Egress Scans, scheduled for 00-353 (12/18/00) through 00-356 (12/21/00), will be commanded by the mz065, mz066, & mz067 mini-sequences.

Wednesday 6-Dec-00

(DOY 334/19:00:00 to DOY 341/19:00:00 UTC) Launch / Days since Launch = Nov. 7, 1996 / 1491 days
Start of Mapping / Days since Start of Mapping = April 1, 1999 / 615 days
Total Mapping Orbits = 7,817
Total Orbits = 9,420

Recent Events:
Background Sequences — The spacecraft continues to operate nominally in performing the beta- supplement daily recording and transmission of science data. The

mm092 sequence executed successfully from 00-335 (11/30/00) through 00-337 (12/2/00). The mm093 sequence has performed well since it started on 00-338 (11/3/00). It terminates on 00-341 (11/6/00). The mm094 sequence, successfully uplinked on 00-340 (12/5/00), begins executing on 00-342 (12/7/00).

Other - No special science scans were conducted this week.

Spacecraft Health:
All subsystems report nominal health.

Uplinks: There have been 8 uplinks to the spacecraft during the last week, including instrument command loads and the sequences cited above. There have been 5,000 command files radiated to the spacecraft since launch.

Upcoming Events:
The mm095 background sequence will be uplinked on 00-343 (12/8/00). Science Campaign F commences on 00-345 (12/10/00) and lasts until 00-356 (12/21/00). Radio Science Occultation Egress Scans, scheduled for 00-353 (12/18/00) through 00-356 (12/21/00), will be commanded by the mz065,mz066, & mz067 mini-sequences.

Wednesday 13-Dec-00

(DOY 341/19:00:00 to DOY 348/19:00:00 UTC) Launch / Days since Launch = Nov. 7, 1996 / 1498 days
Start of Mapping / Days since Start of Mapping = April 1, 1999 / 622 days
Total Mapping Orbits = 7,902
Total Orbits = 9,505

Recent Events:
Background Sequences — The spacecraft continues to operate nominally in performing the beta- supplement daily recording and transmission of science data. The mm094 sequence executed successfully from 00-342 (12/7/00) through 00-344 (12/9/00). The mm095 sequence has performed well since it started on 00-345 (12/10/00). It terminates on 00-348 (12/13/00). The mm096 sequence, successfully uplinked on 00-347 (12/12/00), begins executing on 00-349 (12/14/00).

Other - No special science scans were conducted this week. Science Campaign F started on 00-345 (12/10/00) and lasts until 00-356 (12/21/00). MGS is receiving 24 hour/day coverage during the campaign, greatly increasing the amount of high resolution real-time science data that it transmits to Earth.

Spacecraft Health:
All subsystems report nominal health.

Uplinks:
There have been 7 uplinks to the spacecraft during the last week, including instrument command loads and the sequences cited above. There have been 5,007 command files radiated to the spacecraft since launch.

Upcoming Events:
The mm097 background sequence will be uplinked on 00-350 (12/15/00). Radio Science Occultation Egress Scans, scheduled for 00-353 (12/18/00) through 00-356 (12/21/00), will be commanded by the mz065,mz066, & mz067 mini-sequences.

Wednesday 20-Dec-00

(DOY 348/19:00:00 to DOY 355/19:00:00 UTC) Launch / Days since Launch = Nov. 7, 1996 / 1505 days
Start of Mapping / Days since Start of Mapping = April 1, 1999 / 629 days
Total Mapping Orbits = 7,988
Total Orbits = 9,591

Recent Events:
Background Sequences — The spacecraft continues to operate nominally in performing the beta- supplement daily recording and transmission of science data. The mm096 sequence executed successfully from 00-349 (12/14/00) through 00-351 (12/16/00). The mm097 sequence has performed well since it started on 00-352 (12/17/00). It terminates on 00-355 (12/20/00). The mm098 sequence, successfully uplinked on 00-354 (12/19/00), begins executing on 00-356 (12/21/00).

Other — Science Campaign F, which started on 00-345 (12/10/00), finishes tomorrow 00-356 (12/21/00). MGS received 24 hour/day coverage during the campaign, greatly increasing the amount of high resolution real-time science data that it transmitted to Earth. The first of two back-to-back Radio Science Occultation Egress Scan experiments was successfully completed yesterday 00-354 (12/19/00). The second experiment started today and terminates tomorrow 00-356 (12/21/00).

Spacecraft Health:
All subsystems report nominal health.

Uplinks:
There have been 19 uplinks to the spacecraft during the last week, including new star catalogs and ephemeris files, instrument command loads, and the sequences cited above. There have been 5,026 command files radiated to the spacecraft since launch.

Upcoming Events:
The mm099, mm100, mm101, and mm102 background sequences will be uplinked on 00-357 (12/22/00), 00-361 (12/26/00), 00-364 (12/29/00), and 01-002 (01/02/01) respectively. Two Delta Differential One-Way Range (DDOR) experiments will be conducted on 01-009 (1/9/01). Data from these DDOR experiments will assist the Mars Odyssey Program in calibrating the interplanetary navigation system. Radio Science Occultation Egress Scans, scheduled for 01-010 (1/10/01) through 01-011 (01/11/01), will be commanded by the,mz068, & mz069 mini-sequences.

Wednesday 10-Jan-01

(DOY 003/19:00:00 to DOY 010/19:00:00 UTC) Launch / Days since Launch = Nov. 7, 1996 / 1526 days
Start of Mapping / Days since Start of Mapping = April 1, 1999 / 650 days
Total Mapping Orbits = 8,245
Total Orbits = 9,848

Recent Events:
Background Sequences — The spacecraft continues to operate nominally in performing the beta- supplement daily recording and transmission of science data. The mm102 sequence executed successfully from 01-004 (1/4/00) through 01-006 (1/6/01). The mm103 sequence

has performed well since it started on 01-007 (1/7/01). It terminates on 01-010 (1/10/01). The mm104 sequence, successfully uplinked on 01-009 (1/9/01), begins executing on 01-011 (1/11/01).

Other — Two Delta Differential One-Way Range (DDOR) experiments were conducted on 01-009 (1/9/01). Data from these DDOR experiments are being analyzed and should assist the Mars Odyssey Program in calibrating the interplanetary navigation system.

Spacecraft Health:
All subsystems report nominal health.

Uplinks: There have been 16 uplinks to the spacecraft during the last two weeks, including new star catalogs and ephemeris files, instrument command loads, the background sequences cited above, and the mz068 and mz069 mini-sequences. There have been 5,065 command files radiated to the spacecraft since launch.

Upcoming Events:
The mm105 background sequence will be uplinked on 01-012 (1/12/01). Radio Science Occultation Egress Scans, scheduled for 01-010 (1/10/01) through 01-011 (01/11/01), will be commanded by the,mz068, & mz069 mini-sequences. Another DDOR experiment will be conducted on 01-013 (1/13/01). MOLA Polar Scans are scheduled for 01-018 (1/18/01) through 01-022 (1/22/01). On 01-025 (1/25/01), a Roll Only Targeted Observation (ROTO) demonstration will be conducted on the spacecraft. It will verify our readiness to conduct these observations during the extended mission. Two more DDOR experiments are scheduled for 01-024 (1/24/01) and 01-027 (1/27/01). 01-031 (1/31/01) will mark the end of the MGS primary mission and 01-032 (2/1/01) will mark our entry into the extended mission.

Wednesday 17-Jan-01

(DOY 010/19:00:00 to DOY 017/19:00:00 UTC) Launch / Days since Launch = Nov. 7, 1996 / 1533 days
Start of Mapping / Days since Start of Mapping = April 1, 1999 / 657 days
Total Mapping Orbits = 8,331
Total Orbits = 10,014

Recent Events:
Background Sequences — The spacecraft continues to operate nominally in performing the beta- supplement daily recording and transmission of science data. The mm104 sequence executed successfully from 01-011 (1/11/00) through 01-013 (1/13/01). The mm105 sequence has performed well since it started on 01-014 (1/14/01). It terminates on 01-017 (1/17/01). The mm106 sequence, successfully uplinked on 01-016 (1/16/01), begins executing on 01-018 (1/18/01).

Other - MGS successfully completed twelve Radio Science Occultation Egress Scans from 01-010 (1/10/01) to 01-011 (01/11/01). One Delta Differential One-Way Range (DDOR) experiment was conducted on 01-013 (1/13/01). Data from this and other DDOR experiments are being analyzed and should assist the Mars Odyssey Program in calibrating the interplanetary navigation system.

Spacecraft Health:
All subsystems report nominal health.

Uplinks:
There have been 8 uplinks to the spacecraft during th last week, including instrument command loads, th background sequences cited above, and the mz070 min sequence. There have been 5,073 command file radiated to the spacecraft since launch.

Upcoming Events:
The mm107 background sequence will be uplinked o 01-019 (1/19/01). MOLA Polar Scans, contained in th mz070 mini-sequence, are scheduled for 01-01 (1/18/01) through 01-022 (1/22/01). On 01-02 (1/25/01), a Roll Only Targeted Observation (ROTC demonstration will be conducted on the spacecraft. will verify our readiness to conduct these observatior during the extended mission. Two more DDO experiments are scheduled for 01-024 (1/24/01) and 0 027 (1/27/01). 01-031 (1/31/01) will mark the end of th MGS primary mission and 01- 032 (2/1/01) will mar our entry into the extended mission.

Wednesday 24-Jan-01

(DOY 017/19:00:00 to DOY 024/19:00:00 UTC) Launc / Days since Launch = Nov. 7, 1996 / 1540 days
Start of Mapping / Days since Start of Mapping = April 1999 / 664 days
Total Mapping Orbits = 8,416
Total Orbits = 10,099

Recent Events:
Background Sequences — The spacecraft continues t operate nominally in performing the beta- suppleme daily recording and transmission of science data. Th mm106 sequence executed successfully from 01-0 (1/18/00) through 01-020 (1/20/01). The mm1C sequence has performed well since it started on 01-02 (1/21/01). It terminates on 01-024 (1/24/01). The mm1(sequence, successfully uplinked on 01-023 (1/23/01 begins executing on 01-025 (1/25/01).

Other - Two unrelated spacecraft anomalies occurre on 01-018 (1/18/91).

The first anomaly occurred within the first two hours 01-018 (1/18/91). A Yaw Axis momentum unlo terminated prematurely, failing to completely unload Axis momentum to 1.0 Nms. Instead, the momentu unload terminated at 2.5 Nms. This required flig software to immediately command another Yaw A momentum unload to complete the task of reaching 1 Nms. Subsequent analysis revealed that the unload w prematurely terminated by the "give up"" timer used prevent unloads from persisting beyond the time sp allowed. The Yaw unload in question was induced by t mz070 MOLA Polar Scan mini-sequence. Comman within the mini- sequence lowered the unload thresho to 2.0 Nms in order to dump momentum before t MOLA Polar Scans began. This routine procedure w required because MOLA Polar Scans do not allc proper Yaw momentum unloading. Unfortunately, t timing of the induced unload was out-of-phase with t momentum profile for that particular orbit. Flig software waited appropriately until the unload could accomplished in-phase with the momentum profi However, the delay stretched the execution of t unload beyond the established "give up"" boundary a the unload terminated early. Flight software respond appropriately throughout this scenario, eventua reducing the Z-Axis momentum to 1.0 Nms witho ground intervention. The MOLA Polar scans for the subsequent three orbits executed nominally. Corrective action to increase the "give-up"" timer parameter is in work to ensure that future momentum unloads do not terminate prematurely due to worst-case phasing errors.

The second anomaly involved a hardware failure on 01-018 (1/18/01). At approximately 15:35:43 Spacecraft Event Time (SCET) Universal Time Coordinated (UTC), the X-axis Reaction Wheel Assembly (RWA-X) failed. Redundancy Management (REDMAN) code within flight software reacted appropriately by switching X-axis attitude control to the redundant "Skew"" Reaction Wheel Assembly (RWA-S). Stored-sequence command processing proceeded nominally, allowing MGS to continue mapping without interruption. A command was uplinked on 01-019 (1/19/01) to terminate the mz070 mini-sequence, thereby ensuring the MOLA Polar Scans would not occur until the anomaly was understood. On 01-020 (1/20/91), a command was uplinked to keep RWA-S reaction wheel speeds within comfotable limits by reducing the nominal Yaw Axis momentum unload threshold from 10 Nms to 6 Nms. The mz071C mini-sequence was uplinked and executed on 01-024 (1/24/91) to determine the future availability of RWA-X. RWA- X did not respond during the 20-second test, verifying that it had failed. The spacecraft team is continuing its probe into the probable root cause(s) of the failure. The leading theory is that a component failed in the RWA-X electronics. Our goal is to determine the most probable root cause(s) of the failure so that we can adequately maintain spacecraft health and capability to achieve extended mission objectives. Meanwhile, the MGS spacecraft is operating as designed on the redundant reaction wheel.

One Delta Differential One-Way Range (DDOR) experiment was conducted on 01-024 (1/24/01).

Spacecraft Health:
RWA-X has failed, forcing the Attitude and Articulation Control Subsystem (AACS) to operate on the redundant "skew"" reaction wheel for attitude control about the X- axis. The Electrical Power Subsystem (EPS) experienced a fault that probably originated in the RWA-X electronics. The fault has cleared and the EPS is operating normally. The TWTA-2 filament of the Telecommunications subsystem switched OFF in response to the EPS fault, as designed. It is back on line and operating nominally. The Payload Data Subsystem (PDS) experienced a "warm"" Power On Reset (POR) and the payload bus current momentarily dropped by 0.2 amps. The PDS RAM remained intact, allowing the PDS to recover within two telemetry frames. All other subsystems report nominal health.

Uplinks:
There have been 18 uplinks to the spacecraft during the last week, including new star catalogs and ephemeris files, instrument command loads, the background sequences, anomaly related commands, and the mz071 mini-sequence. There have been 5,091 command files radiated to the spacecraft since launch.

Upcoming Events:
The mm109 background sequence will be uplinked on 01-026 (1/26/01). The Roll Only Targeted Observation (ROTO) demonstration scheduled for this weekend has been delayed pending results from the anomaly investigation. Another DDOR experiment is scheduled for 01-027 (1/27/01). It is contained in the mm108 sequence. 01-031 (1/31/01) will mark the end of the MGS primary mission and 01- 032 (2/1/01) will mark our entry into the extended mission phase.

Wednesday 7-Feb-01

(DOY 031/19:00:00 to DOY 038/19:00:00 UTC) Launch / Days since Launch = Nov. 7, 1996 / 1554 days
Start of Mapping / Days since Start of Mapping = April 1, 1999 / 678 days
Total Mapping Orbits = 8,575
Total Orbits = 10,258

Recent Events:
Background Sequences — The spacecraft continues to operate nominally in performing the beta- supplement daily recording and transmission of science data. The mm110 sequence executed successfully from 01-032 (2/1/00) through 01-034 (2/3/01). The mm111 sequence has performed well since it started on 01-035 (2/4/01). It terminates on 01-038 (2/7/01). The mm112 sequence, successfully uplinked on 01-037 (2/6/01), begins executing on 01-039 (2/08/01).

Other - Another successful DDOR experiment was performed on 01-034 (2/3/01) as part of the mm110 background sequence.

Spacecraft Health:
All subsystems report nominal health. Investigation into the root cause of the X-axis Reaction Wheel failure continues. Although we are confident that a short circuit within the RWA-X electronics caused the failure, isolating the exact piece-part that failed seems unlikely. Work continues in identifying operational impacts associated with using RWA-S for attitude control. We are also investigating methods of protecting the spacecraft against a second RWA failure.

Uplinks:
There have been 16 uplinks to the spacecraft during the past week, including new star catalogs and ephemeris files, instrument command loads, and the background sequences cited above. We also increased the Momentum Unload Give Up Timer to 120 minutes so that manually induced momentum unloads do not terminate prematurely due to worst-case phasing errors. (See the 1/24/01 MGS Status Report) Finally, we reset the Audit Queue Counter to a value consistent with the Audit Queue activity that occurred since the counter reached its maximum value of 255. This particular counter does not rollover automatically and must be reset with a ground command when it reaches its maximum value. There have been 5,113 command files radiated to the spacecraft since launch.

Upcoming Events:
The mm113 background sequence will be uplinked on 01-040 (2/09/01). Development of the mz072 Roll Only Targeted Observation (ROTO) Demonstration mini-sequence will commence on 01-043 (2/12/01).

Wednesday 21-Feb-01

(DOY 045/19:00:00 to DOY 052/19:00:00 UTC) Launch / Days since Launch = Nov. 7, 1996 / 1568 days
Start of Mapping / Days since Start of Mapping = April 1,

1999 / 692 days
Total Mapping Orbits = 8,759
Total Orbits = 10,442

Recent Events:
Background Sequences — The spacecraft continues to operate nominally in performing the beta- supplement daily recording and transmission of science data. The mm114 sequence executed successfully from 01-046 (2/15/00) through 01-048 (2/17/01). The mm115 sequence has performed well since it started on 01-049 (2/18/01). It terminates on 01-052 (2/21/01). The mm116 sequence, successfully uplinked on 01-051 (2/20/01), begins executing on 01-053 (2/22/01).

Other - MGS performed flawlessly during the Roll Only Targeted Observation demonstration executed on orbit #8690 of 01-047 (2/16/00). Post-test analyses of the engineering data verified nominal performance by all subsystems.

Spacecraft Health:
All subsystems report nominal health.

Uplinks:
There have been 6 uplinks to the spacecraft during the past week, including instrument command loads and the background sequences cited above. There have been 5,135 command files radiated to the spacecraft since launch.

Upcoming Events:
The mm117 background sequence will be uplinked on 01-054 (2/23/01). Because the ROTO demonstration was successful, standard procedure will be to uplink two ROTO mini-sequences each week. The ROTO mini-sequences for this next week are mz073 and mz074.

Wednesday 28-Feb-01

(DOY 052/19:00:00 to DOY 059/02:00:00 UTC) Launch / Days since Launch = Nov. 7, 1996 / 1575 days
Start of Mapping / Days since Start of Mapping = April 1, 1999 / 699 days
Total Mapping Orbits = 8,836
Total Orbits = 10,519

Recent Events:
Background Sequences — The spacecraft continues to operate nominally in performing the beta- supplement daily recording and transmission of science data. The mm116 sequence executed successfully from 01-053 (2/22/00) through 01-055 (2/24/01). The mm117 sequence has performed well since it started on 01-056 (2/25/01). It terminates on 01-059 (2/28/01). The mm118 sequence, successfully uplinked on 01-058 (2/27/01), begins executing on 01-060 (3/01/01).

Other - MGS performed five Roll Only Targeted Observations since the last status report, bringing the total number of ROTOs to six.

Spacecraft Health:
All subsystems report nominal health.

Uplinks:
There have been 9 uplinks to the spacecraft during the past week, including instrument command loads, the background sequences cited above, and ROTO mini-sequences mz073 and mz074. There have been 5,144 command files radiated to the spacecraft since launch.

Upcoming Events:
A flight software patch to the Safe Mode code will be uplinked on 01-060 (3/01/01). The purpose of this patch is to prevent the HGA from entering the HGA exclusion zone during a Safe Mode entry. Presently, when the S/C enters Safe Mode, the HGA automatically slews directly to a predetermined target position. Unfortunately, this type of slew poses a risk in some cases due to the HGA gimbal obstruction. It is possible that the HGA could drive through the HGA exclusion zone on its way to the target position. The project has mitigated this risk to date by building background sequences that avoid commanding the HGA into potentially hazardous positions. The unfortunate trade-off has been reduced communications time each orbit. The HGA Safe Mode software patch causes the HGA to "freeze"" its position upon Safe Mode entry. Consequently, the project regains communications time every orbit by moving the risk mitigation from nominal sequencing to the Safe Mode fault response.

The mm119 background sequence will be uplinked on 01-061 (3/02/01). ROTO mini-sequences mz075 and mz076 will be uplinked and executed this next week.

Wednesday 7-Mar-01

(DOY 059/02:00:00 to DOY 066/19:00:00 UTC) Launch / Days since Launch = Nov. 7, 1996 / 1582 days
Start of Mapping / Days since Start of Mapping = April 1, 1999 / 706 days
Total Mapping Orbits = 8,930
Total Orbits = 10,613

Recent Events:
Background Sequences — The spacecraft continues to operate nominally in performing the beta- supplement daily recording and transmission of science data. The mm118 sequence executed successfully from 01-060 (3/01/01) through 01-062 (3/03/01). The mm119 sequence has performed well since it started on 01-063 (3/04/01). It terminates on 01-066 (3/07/01). The mm120 sequence, successfully uplinked on 01-065 (3/06/01), begins executing on 01-067 (3/08/01).

Other - MGS performed 11 Roll Only Targeted Observations since the last status report, bringing the total number of ROTOs to 17.

A flight software patch to the Safe Mode code was uplinked on 01-060 (3/01/01). The purpose of this patch is to prevent the HGA from entering the HGA exclusion zone during a Safe Mode entry. The patch reduces HGA risk and allows increased SSR playback time or increased real-time data transmission during every communications orbit.

A parameter update was uplinked on 01-064 (3/05/01) to ensure entry into the Safe Mode response if another RWA failure occurs.

Spacecraft Health:
All subsystems report nominal health.

Uplinks:
There have been 26 uplinks to the spacecraft during the

past week, including new star catalogs and ephemeris files, instrument command loads, the background sequences cited above, the Safe Mode patch and parameter update, and ROTO mini-sequences mz075 and mz076. There have been 5,170 command files radiated to the spacecraft since launch.

Upcoming Events:
The mm121 background sequence will be uplinked on 01-068 (3/09/01). ROTO mini-sequences mz077 and mz078 will be uplinked and executed this next week.

Wednesday 14-Mar-01

(DOY 066/19:00:00 to DOY 073/19:00:00 UTC) Launch / Days since Launch = Nov. 7, 1996 / 1589 days
Start of Mapping / Days since Start of Mapping = April 1, 1999 / 713 days
Total Mapping Orbits = 9,016
Total Orbits = 10,699

Recent Events:
Background Sequences — The spacecraft continues to operate nominally in performing the beta- supplement daily recording and transmission of science data. The mm120 sequence executed successfully from 01-067 (3/08/01) through 01-069 (3/10/01). The mm121 sequence has performed well since it started on 01-070 (3/11/01). It terminates on 01-073 (3/14/01). The mm122 sequence, successfully uplinked on 01-072 (3/13/01), begins executing on 01-074 (3/15/01).

Other - MGS performed 9 Roll Only Targeted Observations since the last status report, bringing the total number of ROTOs to 26.

Spacecraft Health:
All subsystems report nominal health.

Uplinks:
There have been 20 uplinks to the spacecraft during the past week, including new star catalogs and ephemeris files, instrument command loads, the background sequences cited above, and ROTO mini- sequences mz077 and mz078. There have been 5,190 command files radiated to the spacecraft since launch.

Upcoming Events:
The mm123 background sequence will be uplinked on 01-075 (3/16/01). ROTO mini-sequences mz079 and mz080 will be uplinked and executed this next week. On 01-082 (3/23/01), the mz081 mini-sequence will command MGS to a more fuel-efficient attitude for approximately four hours. The data collected during this demonstration will allow us to characterize the thermal, momentum buildup, fuel usage, and Starex processing profiles of the spacecraft in the new attitude. We will also gather high frequency body rate data to diagnose the health of the -Y solar array hinge. This knowledge will help us plan for future mission activities.

Wednesday 21-Mar-01

(DOY 073/19:00:00 to DOY 080/19:00:00 UTC) Launch / Days since Launch = Nov. 7, 1996 / 1596 days
Start of Mapping / Days since Start of Mapping = April 1, 1999 / 720 days
Total Mapping Orbits = 9,102
Total Orbits = 10,785

Recent Events:
Background Sequences — The spacecraft continues to operate nominally in performing the beta- supplement daily recording and transmission of science data. The mm122 sequence executed successfully from 01-074 (3/15/01) through 01-076 (3/17/01). The mm123 sequence has performed well since it started on 01-077 (3/18/01). It terminates on 01-080 (3/21/01). The mm124 sequence, successfully uplinked on 01-079 (3/20/01), begins executing on 01-081 (3/22/01).

Other - MGS performed 10 Roll Only Targeted Observations since the last status report, bringing the total number of ROTOs to 36.

Spacecraft Health:
All subsystems report nominal health.

Uplinks:
There have been 10 uplinks to the spacecraft during the past week, including instrument command loads, the background sequences cited above, the ROTO mini-sequence mz080, and the Relay-22 Demo mini-sequence mz081. There have been 5,200 command files radiated to the spacecraft since launch.

Upcoming Events:
The mm125 background sequence will be uplinked on 01-082 (3/23/01). ROTO mini-sequences mz082 and mz083 will be uplinked and executed this next week. On 01-082 (3/23/01), the mz081 mini- sequence will command MGS to a more fuel-efficient attitude for approximately four hours. The data collected during this demonstration will allow us to characterize the thermal, momentum buildup, fuel usage, and Starex processing profiles of the spacecraft in the new attitude. We will also gather high frequency body rate data to diagnose the health of the -Y solar array hinge. This knowledge will help us plan for future mission activities.

Wednesday 28-Mar-01

(DOY 080/19:00:00 to DOY 087/19:00:00 UTC) Launch / Days since Launch = Nov. 7, 1996 / 1603 days
Start of Mapping / Days since Start of Mapping = April 1, 1999 / 727 days
Total Mapping Orbits = 9,162
Total Orbits = 10,845

Recent Events:
Background Sequences — The spacecraft continues to operate nominally in performing the beta- supplement daily recording and transmission of science data. The mm124 sequence executed successfully from 01-081 (3/22/01) through 01-083 (3/24/01). The mm125 sequence has performed well since it started on 01-084 (3/25/01). It terminates on 01-087 (3/28/01). The mm126 sequence, successfully uplinked on 01-086 (3/27/01), begins executing on 01-088 (3/29/01).

Other - On 01-082 (3/23/01), the mz081 Relay-22 mini-sequence commanded MGS to a more fuel- efficient attitude by inducing a +22 degree Pitch offset. The spacecraft maintained this attitude for approximately eight hours. The data collected helped characterize the thermal, momentum buildup, fuel usage, and Starex processing profiles of the spacecraft in the new attitude.

During the Relay-22 Demo, Thermal subsystem performance was nominal. The predicted temperature response agreed well with the observed data. Temperature changes were generally small with the single exception of the MOC secondary mirror, which showed a 5°C temperature increase as predicted. This increase is not a concern. There are no thermal issues for the expected final pitch angle of somewhat less than 22 degrees.

Attitude Control also looked nominal but requires more investigation since the MOC team reported Roll offpoints large enough to be noticed by their instrument. The spin momentum buildup rate was - 0.5 Nms/hr, which would result in an unload event every 18 hours vs. the present 7.5 hours. Adjusting current fuel use to the new rate of spin momentum buildup, we estimate the daily fuel use for the new configuration to be approximately 15 g/day (11 g/day spin, 4 g/day yaw). This represents approximately half the fuel consumption of the nominal attitude profile. We anticipate a Pitch offset near 16 degrees to be near optimum. As expected, there was no noticeable change in the Yaw momentum buildup rate. Starex performed nominally, converging in 11 stars. No major changes in gyro biases (versus nominal) were observed.

We were unable to collect high frequency body rate data because of a subtle error in the way the Solid State Recorders were commanded. We expect to collect this data in the near future in order to diagnose the health of the -Y solar array hinge. The data will also help us determine what effect solar array motion has on TES instrument data.

The Power subsystem performed nominally during the Relay-22 Demonstration.

Three new Solar Array Autonomous Eclipse Management scripts were loaded into memory this week. They will go active on 01-088 (3/29/01), returning Solar Array management from a 4-motion strategy to a 3- motion strategy. This may extend the life of the solar array gimbals and may reduce solar array noise effects upon the TES data.

MGS performed 4 Roll Only Targeted Observations since the last status report, bringing the total number of ROTOs to 40.

Spacecraft Health:
All subsystems report nominal health.

Uplinks: There have been 13 uplinks to the spacecraft during the past week, including instrument command loads, the background sequences cited above, ROTO mini-sequences mz082 and mz083, and three modified Solar Array Autonomous Eclipse Management scripts. There have been 5,213 command files radiated to the spacecraft since launch.

Upcoming Events:
The mm127 background sequence will be uplinked on 01-089 (3/30/01). ROTO mini- sequences mz084 and mz085 will be uplinked and executed this next week. Radio Science Occultation Egress Scans, contained in the mz085 mini-sequence, will take place on 01-095 (4/5/01) and 01-096 (4/6/01).

Wednesday 4-Apr-01

(DOY 087/19:00:00 to DOY 094/19:00:00 UTC) Launch / Days since Launch = Nov. 7, 1996 / 1610 days
Start of Mapping / Days since Start of Mapping = April 1, 1999 / 734 days
Total Mapping Orbits = 9,272
Total Orbits = 10,955

Recent Events:
Background Sequences — The spacecraft continues to operate nominally in performing the beta- supplement daily recording and transmission of science data. The mm126 sequence executed successfully from 01-088 (3/29/01) through 01-090 (3/31/01). The mm127 sequence has performed well since it started on 01-091 (4/01/01). It terminates on 01-094 (4/04/01). The mm128 sequence, successfully uplinked on 01-093 (4/03/01), begins executing on 01-095 (4/05/01).

Other - MGS performed 6 Roll Only Targeted Observations since the last status report, bringing the total number of ROTOs to 46.

Spacecraft Health:
All subsystems report nominal health.

Uplinks: There have been 15 uplinks to the spacecraft during the past week, including new star catalogs and ephemeris files, instrument command loads, the background sequences cited above, ROTO mini-sequence mz084, and Radio Science Occultation Egress Scan mini-sequence mz085. There have been 5,228 command files radiated to the spacecraft since launch.

Upcoming Events:
The mm129 background sequence will be uplinked on 01-096 (4/06/01). ROTO mini-sequences mz086 and mz087 will be uplinked and executed this next week. Radio Science Occultation Egress Scans, contained in the mz085 mini-sequence, will take place on 01-095 (4/5/01) and 01- 096 (4/6/01).

Wednesday 18-Apr-01

(DOY 094/19:00:00 to DOY 108/19:00:00 UTC) This report covers two weeks.

Launch / Days since Launch = Nov. 7, 1996 / 1624 days
Start of Mapping / Days since Start of Mapping = April 1, 1999 / 748 days
Total Mapping Orbits = 9,444
Total Orbits = 11,127

Recent Events:
Background Sequences — The spacecraft continues to operate nominally in performing the beta-supplement daily recording and transmission of science data. The mm128, mm129, and mm130 sequences executed successfully from 01-095 (4/05/01) through 01-104 (4/14/01). The mm131 sequence has performed well since it started on 01-105 (4/15/01). It terminates on 01-108 (4/18/01). The mm132 sequence, successfully uplinked on 01-107 (4/17/01), begins executing on 01-109 (4/19/01).

Other - MGS successfully performed twelve Radio Science Occultation Egress Scans on 01-095 (4/5/01) and 01-096 (4/6/01). They were contained in the mz085 mini-sequence. MGS has performed 65 Roll Only

argeted Observations to date.
pacecraft Health:
All subsystems report nominal health.

Jplinks:
There have been 30 uplinks to the spacecraft during the ast two weeks, including new star catalogs and phemeris files, instrument command loads, the ackground sequences cited above, a thruster ccumulator reset command, and ROTO mini-sequences nz086 through mz090.

Commands were uplinked to modify the Reaction Vheel Assembly fault protection software and ssociated scripts. The modified response will allow a uicker recovery should another reaction wheel fail.

he first of three solar array position management cripts was modified to offset the commanded solar rray positions by 25 degrees. Our goal is to extend the fe of the Partial Shunt Assemblies by reducing the mount of excess power generated by the solar arrays. Ve will update the other two scripts after verifying the xpected spacecraft performance under the first script.

here have been 5,258 command files radiated to the pacecraft since launch.

Jpcoming Events:
he mm133 background sequence will be uplinked on 1-110 (4/20/01). On 01-115 (4/25/01) of that sequence, IGS will transmit high frequency spacecraft body rate ata to Earth. The data will provide insight into the ealth of the -Y solar array hinge and the effects of solar rray motion on TES data. MOC Defocus Calibration cans will be performed by the mm134 and mm135 equences between 01-116 (4/26/01) and 01-123 5/02/01). Radio Science Occultation Egress Scans are cheduled for 01-132 (5/12/01) and 01-133 (5/13/01).

OTO mini-sequences mz090 and mz091 will execute is next week.

Vednesday 25-Apr-01

DOY 108/19:00:00 to DOY 115/19:00:00 UTC)This eport covers two weeks.
aunch / Days since Launch = Nov. 7, 1996 / 1631 days
tart of Mapping / Days since Start of Mapping = April 1, 999 / 755 days
otal Mapping Orbits = 9,530
otal Orbits = 11,213

ecent Events:
ackground Sequences — The spacecraft continues to perate nominally in performing the beta- supplement aily recording and transmission of science data. The m132 sequence executed successfully from 01-109 4/19/01) through 01-111 (4/21/01). The mm133 equence has performed well since it started on 01-112 4/22/01). It terminates on 01-115 (4/25/01). The mm134 equence, successfully uplinked on 01-114 (4/24/01), egins executing on 01-116 (4/26/01).

Other - MGS transmitted high frequency spacecraft ody rate data to Earth on 01-115 (4/25/01). The data ill provide insight into the health of the -Y solar array inge and the effects of solar array motion on TES data. IGS has performed 75 Roll Only Targeted Observations to date.

Spacecraft Health:
All subsystems report nominal health.

Uplinks:
There have been 12 uplinks to the spacecraft during the past week, including instrument command loads, the background sequences cited above, and ROTO mini-sequences mz091 through mz092.

The last two of three solar array position management scripts were modified to offset the commanded solar array positions by 25 degrees. Our goal is to extend the life of the Partial Shunt Assemblies by reducing the amount of excess power generated by the solar arrays.

There have been 5,270 command files radiated to the spacecraft since launch.

Upcoming Events:
The mm135 background sequence will be uplinked on 01-117 (4/27/01). MOC Defocus Calibration Scans will be performed by the mm134 and mm135 sequences between 01-116 (4/26/01) and 01-123 (5/02/01). Radio Science Occultation Egress Scans are scheduled for 01-132 (5/12/01) and 01-133 (5/13/01). ROTO mini-sequences mz092 and mz093 will execute this next week.

Wednesday 2-May-01

(DOY 115/19:00:00 to DOY 122/19:00:00 UTC) Launch / Days since Launch = Nov. 7, 1996 / 1638 days
Start of Mapping / Days since Start of Mapping = April 1, 1999 / 762 days
Total Mapping Orbits = 9,615
Total Orbits = 11,298

Recent Events:
Background Sequences — The spacecraft is now in Contingency Mode (C-Mode), so background sequence processing has terminated until the Spacecraft Team intervenes. The mm134 sequence executed successfully from 01-116 (4/26/01) through 01-118 (4/28/01). The mm135 sequence performed well from the time it started on 01-119 (4/29/01) until it terminated on 01-122 (5/02/01) at 10:10:55 Spacecraft Event Time (SCET) with a C-Mode entry. The mm136 sequence, successfully uplinked on 01-121 (5/01/01), was to begin executing on 01-123 (5/03/01), but will not due to the C-Mode entry.

Other - MGS transmitted high frequency spacecraft body rate data to Earth on 01-115 (4/25/01). The data will provide insight into the health of the -Y solar array hinge and the effects of solar array motion on TES data. Four of five MOC Defocus Calibration Scans were performed by the mm134 and mm135 sequences before C-Mode was entered. The fifth scan may be made-up at some future date. All four ROTOs contained in the mz093 mini-sequence were performed. The mz094 mini-sequence, containing five ROTOs, will not be uplinked due to C- Mode recovery. MGS has performed 84 Roll Only Targeted Observations to date.

Spacecraft Health:
All subsystems report good health and safe status consistent with C-Mode entry. Downlink telemetry is consistent with the following C-Mode state: Spacecraft coning about the sun with the solar arrays automatically tracking the sun and generating adequate power; HGA "frozen" in the position it was in when the spacecraft

entered C-Mode; 7.8125 bps uplink and 10 bps downlink communications provided through the LGA; science instruments powered off.

MGS project engineers have initiated the C-Mode Recovery Plan.

We know that the Sun Monitor Ephemeris fault response sent MGS into C-Mode but we do not know why. The spacecraft continually compares the sun position calculated from on-board sun sensors to the expected sun position provided by the Spacecraft Team. When the two positions diverge greater that five degrees for greater than two seconds, the Sun Monitor Ephemeris fault response initiates C-Mode.

Uplinks:
There have been 16 uplinks to the spacecraft during the past week, including new star catalogs and ephemeris files, instrument command loads, the background sequences cited above, and ROTO mini-sequence mz093.

There have been 5,286 command files radiated to the spacecraft since launch.

Upcoming Events:
The mm137 background sequence will not be uplinked on 01-124 (5/04/01) because we are recovering from C-mode. If the recovery effort goes well, background sequence processing will start with mm138. Radio Science Occultation Egress Scans are scheduled for 01-133 (5/13/01) and 01-134 (5/14/01). An MGS Mars Relay On-orbit UHF Test, initiated by the Mars Exploration Rover project, will be conducted with Stanford University on 01-135 (5/15/01) and 01-136 (5/16/01).

Wednesday 9-May-01

(DOY 122/19:00:00 to DOY 129/19:00:00 UTC) Launch / Days since Launch = Nov. 7, 1996 / 1645 days
Start of Mapping / Days since Start of Mapping = April 1, 1999 / 769 days
Total Mapping Orbits = 9,701
Total Orbits = 11,384

Recent Events:
Background Sequences - There have been no background sequences operating since MGS entered C-Mode a week ago on 01-122 (5/02/01). Last week, the mm135 sequence terminated approximately fourteen hours earlier than expected due to C-Mode entry. The mm136 sequence, though uplinked, was prevented from starting on 01-123 (5/3/01) by C-Mode. The mm137 sequence was never uplinked. Recovery efforts have gone well the past two days and we expect to start normal background sequencing with mm138. It will start today, 01-129 (5/9/01), at 23:57:00 UTC Spacecraft Event Time. mm139 is scheduled for uplink on 01-131 (5/11/01). It will commence on 01- 133 (5/13/01).

Other - No ROTOs or other special science scans were conducted this week. The mz095 mini-sequence was used to return MGS to its normal NADIR pointing attitude. It also triggered an on-board command script to control movement of the HGA and operate the downlink transmitter during the transition from C-Mode to nominal mapping. The mz096 mini-sequence was prepared in the optimistic hope that we might recover in time to conduct a ROTO. It was never uplinked due to the intense effort required to prepare and uplink the C- Mode recovery command files. The mz098 mini-sequence was uplinked today to terminate the HGA command script shortly before the mm138 background sequence goes active.

Spacecraft Health:
All subsystems report good health and status. All science instruments have been powered-on except the TES. The TES was powered-on, then off when it did not respond correctly to the NIPC commands prepared by the TES Team. The TES team suspects that the problem may be with the NIPC file itself and not the with the TES instrument.

The Sun Monitor Ephemeris fault detection and response have been disabled to prevent another C-Mode entry while the Spacecraft Team investigates the root cause of the failure. Other fault protection routines will continue to protect MGS from losing its attitude knowledge.

Uplinks:
There have been 66 C-Mode recovery uplinks to the spacecraft during the past week. There have been 5,352 command files radiated to the spacecraft since launch.

Upcoming Events:
The Radio Science Occultation Egress Scans contained in the mz097 mini-sequence should take place as scheduled on 01-133 (5/13/01) and 01-134 (5/14/01). An MGS Mars Relay On-orbit UHF Test, initiated by the Mars Exploration Rover project, will be conducted with Stanford University on 01-135 (5/15/01) and 01-136 (5/16/01).

Wednesday 16-May-01

(DOY 129/19:00:00 to DOY 136/19:00:00 UTC) Launch / Days since Launch = Nov. 7, 1996 / 1652 days
Start of Mapping / Days since Start of Mapping = April 1, 1999 / 776 days
Total Mapping Orbits = 9,786
Total Orbits = 11,469

Recent Events:
Background Sequences - Except for the TES being off, the spacecraft is operating nominally in performing the beta-supplement daily recording and transmission of science data. The mm138 sequence executed successfully from 01-130 (5/10/01) through 01-132 (5/12/01). The mm139 sequence has performed well since it started on 01-133 (5/13/01). It terminates on 01-136 (5/16/01). The mm140 sequence, successfully uplinked on 01-135 (5/15/01), begins executing on 01-137 (5/17/01).

Other - Twelve Radio Science Occultation Egress Scans contained in the mz097 mini- sequence were successfully performed on 01-133 (5/13/01) and 01-134 (5/14/01). An MGS Mars Relay On-orbit UHF Test, initiated by the Mars Exploration Rover project, was conducted successfully with Stanford University on 01-135 (5/15/01) and 01-136 (5/16/01). Three more Roll Only Targeted Observations (ROTOs) were performed since the last report. MGS has completed a total of 87 ROTOs to date.

Spacecraft Health:
All subsystems report good health and status. All science instruments have been powered-on except the TES. During C-Mode recovery, the TES was powered-on, then off when it did not respond correctly to the NIPC commands prepared by the TES Team. The TES team determined the problem to be with the NIPC file itself and not the with the TES instrument. The TES will be powered-up on 01-138 (5/18/01).

The Sun Monitor Ephemeris fault protection continues to be disabled to prevent another C-Mode entry while the Spacecraft Team investigates the root cause of the failure. Other fault protection routines will continue to protect MGS from losing its attitude knowledge.

Uplinks:
There have been 22 uplinks to the spacecraft during the past week, including three C-Mode recovery clean-up commands, new star catalogs and ephemeris files, instrument command loads, the background sequences cited above, the mz097 Radio Science mini-sequence, and ROTO mini- sequences mz099 and mz100. There have been 5,374 command files radiated to the spacecraft since launch.

Upcoming Events:
The Thermal Emission Spectrometer will be powered-up on 01-138 (5/18/01). Within the next few weeks, we will update the three solar array position management scripts to offset the commanded solar array positions by 35 degrees instead of the present offset of 25 degrees. This update is needed to extend the life of the Partial Shunt Assemblies by reducing the amount of excess power generated by the solar arrays.

Another MGS Mars Relay On-orbit UHF Test will be conducted with Stanford University between 01-177 (6/26/01) and 01-180 (6/29/01).

Wednesday 23-May-01

(DOY 136/19:00:00 to DOY 143/19:00:00 UTC) Launch / Days since Launch = Nov. 7, 1996 / 1659 days
Start of Mapping / Days since Start of Mapping = April 1, 1999 / 783 days
Total Mapping Orbits = 9,871
Total Orbits = 11,554

Recent Events:
Background Sequences - The spacecraft is operating nominally in performing the beta-supplement daily recording and transmission of science data. The mm140 sequence executed successfully from 01-137 (5/17/01) through 01-139 (5/19/01). The mm139 sequence has performed well since it started on 01-140 (5/20/01). It terminates on 01-143 (5/23/01). The mm142 sequence, successfully uplinked on 01-142 (5/22/01), begins executing on 01-144 (5/24/01).

Other - Eleven more Roll Only Targeted Observations (ROTOs) were performed since the last report. MGS has completed a total of 98 ROTOs to date.

Spacecraft Health:
All subsystems report good health and status. All science instruments have been powered-on. The TES was powered-up on 01-138 (5/18/01).

The Sun Monitor Ephemeris fault protection continues to be disabled to prevent another C-Mode entry while the Spacecraft Team investigates the root cause of the failure. Other fault protection routines will continue to protect MGS from losing its attitude knowledge.

Uplinks:
There have been 15 uplinks to the spacecraft during the past week, including the TES power-on command file, instrument command loads, the background sequences cited above, and ROTO mini-sequences mz100b, mz101, and mz102. There have been 5,389 command files radiated to the spacecraft since launch.

Upcoming Events:
Within the next week, we will update the three solar array position management scripts to offset the commanded solar array positions by 35 degrees instead of the present offset of 25 degrees. This update is needed to extend the life of the Partial Shunt Assemblies by reducing the amount of excess power generated by the solar arrays.

Another MGS Mars Relay On-orbit UHF Test will be conducted with Stanford University between 01-177 (6/26/01) and 01-180 (6/29/01).

Wednesday 30-May-01

(DOY 143/19:00:00 to DOY 150/19:00:00 UTC) Launch / Days since Launch = Nov. 7, 1996 / 1666 days
Start of Mapping / Days since Start of Mapping = April 1, 1999 / 790 days
Total Mapping Orbits = 9,957
Total Orbits = 11,640

Recent Events:
Background Sequences - The spacecraft is operating nominally in performing the beta-supplement daily recording and transmission of science data. The mm142 sequence executed successfully from 01-144 (5/24/01) through 01-146 (5/26/01). The mm143 sequence has performed well since it started on 01-147 (5/27/01). It terminates on 01-150 (5/30/01). The mm144 sequence, successfully uplinked on 01-149 (5/29/01), begins executing on 01-151 (5/31/01).

Other - Nine more Roll Only Targeted Observations (ROTOs) were performed since the last report. MGS has completed a total of 107 ROTOs to date.

Spacecraft Health:
All subsystems report good health and status.

The Sun Monitor Ephemeris fault protection continues to be disabled to prevent another C-Mode entry while the Spacecraft Team investigates the root cause of the failure. Other fault protection routines will continue to protect MGS from losing its attitude knowledge.

Uplinks:
There have been 17 uplinks to the spacecraft during the past week, including new star catalogs and ephemeris files, instrument command loads, the background sequences cited above, and ROTO mini- sequence mz103.

Three solar array position management scripts were updated to offset the commanded solar array positions by 35 degrees instead of the previous offset of 25 degrees. The update should extend the life of the Partial Shunt Assemblies by reducing the amount of excess power generated by the solar arrays

There have been 5,406 command files radiated to the spacecraft since launch.

Upcoming Events:
DOY 01-172 (6/21/01) marks the end of the beta-supplement phase and the beginning of the nominal mapping phase of the mission. Planetary and orbital geometry will allow the HGA to auto-track the Earth without impacting the HGA boom. Nominal mapping sequences will not be as command intensive as the beta-supplement sequences. Therefore, 28-day background sequences will be the norm instead of the 3 and 4 day beta-supplement sequences.

Another MGS Mars Relay On-orbit UHF Test will be conducted with Stanford University between 01-177 (6/26/01) and 01-180 (6/29/01).

Wednesday 6-Jun-01

(DOY 150/19:00:00 to DOY 157/19:00:00 UTC) Launch / Days since Launch = Nov. 7, 1996 / 1673 days
Start of Mapping / Days since Start of Mapping = April 1, 1999 / 797 days
Total Mapping Orbits = 10,043
Total Orbits = 11,726

Recent Events:
Background Sequences - The spacecraft is operating nominally in performing the beta- supplement daily recording and transmission of science data. The mm144 sequence executed successfully from 01-151 (5/31/01) through 01-153 (6/02/01). The mm145 sequence has performed well since it started on 01-154 (6/03/01). It terminates on 01-157 (6/06/01). The mm146 sequence, successfully uplinked on 01-156 (6/05/01), begins executing on 01-158 (6/07/01).

Other - Ten more Roll Only Targeted Observations (ROTOs) were performed since the last report. MGS has completed a total of 117 ROTOs to date.

Spacecraft Health:
All subsystems report good health and status.

The Sun Monitor Ephemeris fault protection is disabled to prevent another C-Mode entry due to sun-sensor shadowing by the HGA. Sun-sensor shadowing is the most likely cause of the 5/2/01 entry into C-Mode. The shadowing was caused by a combination of planetary geometry and beta-supplement HGA positioning. The 35 degree Solar Panel offpoint caused the SAM sun-sensor instead of the SAP sun-sensor to be selected for sensing sun position. Flight software could not accurately compute sun position when the SAM sun-sensor became shadowed by the HGA. Other fault protection routines continue to protect MGS from losing its attitude knowledge.

Uplinks:
There have been 13 uplinks to the spacecraft during the past week, including instrument command loads, the background sequences cited above, and ROTO mini-sequences mz104, mz105, & mz106. There have been 5,419 command files radiated to the spacecraft since launch.

Upcoming Events:
DOY 01-172 (6/21/01) marks the end of the beta-supplement phase and the beginning of the nominal mapping phase of the mission. Planetary and orbital geometry will allow the HGA to auto-track the Earth without impacting the HGA boom. Nominal mapping sequences will not be as command intensive as the beta-supplement sequences. Therefore, 28-day background sequences will be the norm instead of the 3 and 4 day beta-supplement sequences.

Another MGS Mars Relay On-orbit UHF Test will be conducted with Stanford University between 01- 177 (6/26/01) and 01-179 (6/28/01). MOLA Polar Scans are scheduled for 01-193 (7/12/01) and 01-194 (7/13/01).

Wednesday 13-Jun-01

(DOY 157/19:00:00 to DOY 164/19:00:00 UTC) Launch / Days since Launch = Nov. 7, 1996 / 1680 days
Start of Mapping / Days since Start of Mapping = April 1, 1999 / 804 days
Total Mapping Orbits = 10,129
Total Orbits = 11,812

Recent Events:
Background Sequences - The spacecraft is operating nominally in performing the beta- supplement daily recording and transmission of science data. The mm146 sequence executed successfully from 01-158 (6/7/01) through 01-160 (6/09/01). The mm147 sequence has performed well since it started on 01-161 (6/10/01). It terminates on 01-164 (6/13/01). The mm148 sequence, successfully uplinked on 01-163 (6/12/01), begins executing on 01-165 (6/14/01).

Other - Nine more Roll Only Targeted Observations (ROTOs) were performed since the last report. MGS has completed 126 ROTOs to date.

Spacecraft Health:
All subsystems report good health and status.

Uplinks:
There have been 18 uplinks to the spacecraft during the past week, including new star catalogs and ephemeris files, instrument command loads, the background sequences cited above, and ROTO mini-sequences mz107 & mz108. There have been 5,437 command files radiated to the spacecraft since launch.

Upcoming Events:
DOY 01-172 (6/21/01) marks the end of the beta-supplement phase and the beginning of the nominal mapping phase of the mission. Planetary and orbital geometry will allow the HGA to auto-track the Earth without impacting the HGA boom. Nominal mapping sequences will not be as command intensive as the beta-supplement sequences. Therefore, 28-day background sequences will be the norm instead of the 3 and 4 day beta-supplement sequences.

Another MGS Mars Relay On-orbit UHF Test will be conducted with Stanford University between 01-177 (6/26/01) and 01-179 (6/28/01). MOLA Polar Scans are scheduled for 01-193 (7/12/01) and 01-194 (7/13/01).

Wednesday 20-Jun-01

(DOY 164/19:00:00 to DOY 171/19:00:00 UTC) Launch / Days since Launch = Nov. 7, 1996 / 1687 days
Start of Mapping / Days since Start of Mapping = April 1, 1999 / 811 days
Total Mapping Orbits = 10,214
Total Orbits = 11,897

Recent Events:
Background Sequences - The spacecraft is operating nominally in performing the beta- supplement daily recording and transmission of science data. The mm148 sequence executed successfully from 01-165 (6/14/01) through 01-167 (6/16/01). The mm149 sequence has performed well since it started on 01-168 (6/17/01). It terminates on 01-171 (6/20/01). The mm150 sequence, successfully uplinked on 01-170 (6/19/01), begins executing on 01-172 (6/21/01). The start of mm150 marks the end of the beta-supplement phase and the beginning of the nominal mapping phase of the mission. Planetary and orbital geometry will allow the HGA to auto-track the Earth without impacting the HGA boom. Therefore, background command sequences can take advantage of the on-board command scripts originally designed for the nominal mapping mission. 28-day background sequences will become the norm instead of the 3 and 4 day beta-supplement sequences.

Other - Eleven more Roll Only Targeted Observations (ROTOs) were performed since the last report. MGS has completed 137 ROTOs to date.

Spacecraft Health:
All subsystems report good health and status.

Uplinks:
There have been 12 uplinks to the spacecraft during the past week, including instrument command loads, the background sequences cited above, and ROTO mini-sequences mz109 & mz110. We also updated Solar Array AEM Script #1 timing to avoid potential electromagnetic interference (EMI) with the solar arrays during HGA transmissions to Earth. There have been 5,449 command files radiated to the spacecraft since launch.

Upcoming Events:
Another MGS Mars Relay On- orbit UHF Test will be conducted with Stanford University between 01-177 (6/26/01) and 01-179 (6/28/01). MOLA Polar Scans are scheduled for 01-193 (7/12/01) and 01-194 (7/13/01).

Wednesday 27-Jun-01

(DOY 171/19:00:00 to DOY 178/19:00:00 UTC) Launch / Days since Launch = Nov. 7, 1996 / 1694 days
Start of Mapping / Days since Start of Mapping = April 1, 1999 / 818 days
Total Mapping Orbits = 10,299
Total Orbits = 11,982

Recent Events:
Background Sequences - The spacecraft is operating nominally in performing the beta- supplement daily recording and transmission of science data. The mm150 sequence has performed well since it started on 01-172 (6/21/01). It terminates on 01-199 (7/18/01).
Other - An MGS Mars Relay On-orbit UHF experiment with Stanford University started on 01-177 (6/26/01) and finishes on 01-179 (6/28/01). So far, the experiment has progressed nominally. A more detailed report will be forthcoming following the data analyses.

Eight more Roll Only Targeted Observations (ROTOs) were performed since the last report. MGS has completed 145 ROTOs to date.

Spacecraft Health:
All subsystems report good health and status.

Uplinks:
There have been 17 uplinks to the spacecraft during the past week, including new star catalogs and ephemeris files, instrument command loads, the mz112 UHF Relay-16 experiment, and ROTO mini- sequences mz111 & mz113. There have been 5,466 command files radiated to the spacecraft since launch.

Upcoming Events:
MOLA Polar Scans are scheduled for 01-193 (7/12/01) and 01-194 (7/13/01). A Bistatic Radar Test is planned for 01-217 (8/5/01). More MOLA Polar Scans are scheduled for 01-219 (8/7/01) and 01-220 (8/8/01). Finally, the first of several Delta-DOR experiments will take place on 01-225 (8/13/01).

Wednesday 18-Jul-01

(DOY 192/19:00:00 to DOY 199/19:00:00 UTC) Launch / Days since Launch = Nov. 7, 1996 / 1715 days
Start of Mapping / Days since Start of Mapping = April 1, 1999 / 839 days
Total Mapping Orbits = 10,556
Total Orbits = 12,239

Recent Events:
Background Sequences - The spacecraft is operating nominally in performing daily recording and transmission of science data. The mm150 sequence has performed well since it started on 01-172 (6/21/01). It terminates on 01-199 (7/18/01). The mm151 sequence, successfully uplinked on 01-197 (7/16/01), begins executing on 01-200 (7/19/01).

Other - Dust storms spreading across the planet are making narrow angle photographic imaging of Mars impracticable. Therefore, ROTO operations remain suspended until the global dust storms subside. MGS has completed 155 ROTOs to date.

An in-flight MOLA diagnostic test (MZ115) was performed on 01-195 (7/14/01) to gain some insight into the MOLA anomaly reported last week. Unfortunately, the MOLA instrument did not assume normal operation, but the data indicate that the MOLA computer is working and that the instrument responds to commanding. We expect further diagnostic testing following the MOLA team's analyses of the most recent data.

Spacecraft Health:
All subsystems report good health and status except for the MOLA payload instrument that malfunctioned on 01-181 (6/30/01). The MOLA instrument team is still investigating the anomaly.

Uplinks:
There have been 13 uplinks to the spacecraft during the past week, including new star catalogs and ephemeris

files, instrument command loads, the MM151 background sequence, and the MZ115 MOLA diagnostic test. 5,500 command files have been radiated to the spacecraft since launch.

Upcoming Events:
A Bistatic Radar experiment is planned for 01-217 (8/5/01). The first of several Delta-DOR experiments will take place on 01-236 (8/24/01).

Wednesday 25-Jul-01

(DOY 199/19:00:00 to DOY 206/19:00:00 UTC) Launch / Days since Launch = Nov. 7, 1996 / 1722 days
Start of Mapping / Days since Start of Mapping = April 1, 1999 / 846 days
Total Mapping Orbits = 10,642
Total Orbits = 12,325

Recent Events:
Background Sequences - The spacecraft is operating nominally in performing daily recording and transmission of science data. The mm151 sequence has performed well since it started on 01-200 (7/19/01). It terminates on 01-227 (8/15/01).

Other - Dust storms spreading across the planet are making narrow angle photographic imaging of Mars impracticable. Therefore, ROTO operations remain suspended until the global dust storms subside. MGS has completed 155 ROTOs to date.

A second in-flight MOLA diagnostic test (MZ116) was performed on 01-206 (7/25/01) to gain more insight into the MOLA anomaly. Detailed analyses will be performed after the recorded data is transmitted to Earth on 01-207 (7/26/01).

Spacecraft Health:
All subsystems report good health and status except for the MOLA payload instrument that malfunctioned on 01-181 (6/30/01). The MOLA instrument team is still investigating the anomaly.

Uplinks:
There have been 6 uplinks to the spacecraft during the past week, including instrument command loads and the MZ116 MOLA diagnostic test. 5,506 command files have been radiated to the spacecraft since launch.

Upcoming Events:
A Bistatic Radar experiment is planned for 01-217 (8/5/01). The first of several Delta-DOR experiments will take place on 01-236 (8/24/01).

Wednesday 8-Aug-01

(DOY 213/19:00:00 to DOY 220/19:00:00 UTC) Launch / Days since Launch = Nov. 7, 1996 / 1736 days
Start of Mapping / Days since Start of Mapping = April 1, 1999 / 860 days
Total Mapping Orbits = 10,813
Total Orbits = 12,496

Recent Events:
Background Sequences - The spacecraft is operating nominally in performing daily recording and transmission of science data. The mm151 sequence has performed well since it started on 01-200 (7/19/01). It terminates on 01-227 (8/15/01).

Other - MGS performed the MZ118 Bistatic Radar experiment on 01-217 (8/5/01) and the MZ121 MOLA Diagnostic Test #3 on 01-218 (8/06/01). The spacecraft operated nominally during both events. The Radio Science and MOLA teams are in the process of analyzing the data collected from their respective experiments. The MOLA team was successful in patching the MOLA software to provide higher sampling rates into the data needed to determine what hardware component failed on 01-181 (6/30/01).

Seven more Roll Only Targeted Observations (ROTOs) were performed since the last report. MGS has completed 165 ROTOs to date.

Spacecraft Health:
All subsystems report good health and status except for the MOLA payload instrument that malfunctioned on 01-181 (6/30/01). The MOLA instrument team is still investigating the anomaly.

Uplinks:
There have been 17 uplinks to the spacecraft during the past week, including new star catalogs and ephemeris files, instrument command loads, the Bistatic Radar and MOLA mini-sequences referenced above, and the MZ122 and MZ123 ROTO mini-sequences. 5,532 command files have been radiated to the spacecraft since launch.

Upcoming Events:
On 01-228 (8/16/01) the nominal spacecraft orientation will change. The spacecraft will be commanded to pitch +16 degrees from NADIR in order to better align its attitude with the gravity gradient, thereby reducing fuel consumption and extending spacecraft life.

The first of several Delta-DOR experiments will take place on 01-236 (8/24/01).

Wednesday 15-Aug-01

(DOY 220/19:00:00 to DOY 227/19:00:00 UTC) Launch / Days since Launch = Nov. 7, 1996 / 1743 days
Start of Mapping / Days since Start of Mapping = April 1, 1999 / 867 days
Total Mapping Orbits = 10,898
Total Orbits = 12,581

Recent Events:
Background Sequences - The spacecraft is operating nominally in performing daily recording and transmission of science data. The mm151 sequence has performed well since it started on 01-200 (7/19/01). It terminates on 01-227 (8/15/01). The mm152 sequence, successfully uplinked on 01-222 (8/10/01), starts on 01-228 (8/16/01).

Other - On 01-226 (8/14/01), the MGS High Gain Antenna (HGA) signal became erratic and then disappeared at 01-226/13:16, seven minutes before the scheduled Loss of Signal (LOS). The spacecraft team realized that the spacecraft was having difficulty maintaining its attitude due to an expired mapping ephemeris file. They uplinked a new file through the Low Gain Antenna subsequently reacquiring the signal at the scheduled time of 01-226/23:54. Quick coordination by the DSN schedulers and Ops Chief with the Voyager program resulted in an extension to the MGS DSN track. The extension provided the time needed to successfully uplink the new file. Their efforts helped prevent MGS from entering C-Mode, thereby saving valuable science data.

Seven of nine scheduled Roll Only Targeted Observations (ROTOs) were successfully performed since the last report. The last two were aborted during the LOS recovery. MGS has completed 172 ROTOs to date.

Spacecraft Health:
All subsystems report good health and status except for the MOLA payload instrument that malfunctioned on 01-181 (6/30/01). The MOLA instrument team is still investigating the anomaly.

Uplinks:
There have been 50 uplinks to the spacecraft during the past week, including LOS recovery commands, new star catalogs and ephemeris files, instrument command loads, command script updates, the mm152 sequence, and the MZ124 and MZ125 ROTO mini-sequences. 5,582 command files have been radiated to the spacecraft since launch.

Upcoming Events:
On 01-228 (8/16/01) the nominal spacecraft orientation will change. The spacecraft will be commanded to pitch +16 degrees from NADIR in order to better align its attitude with the gravity gradient, thereby reducing fuel consumption and extending spacecraft life.

The first of several Delta-DOR experiments will take place on 01-236 (8/24/01).

Wednesday 29-Aug-01

(DOY 234/19:00:00 to DOY 241/19:00:00 UTC) Launch / Days since Launch = Nov. 7, 1996 / 1757 days
Start of Mapping / Days since Start of Mapping = April 1, 1999 / 881 days
Total Mapping Orbits = 11,070
Total Orbits = 12,753

Recent Events:
Background Sequences - The spacecraft is operating nominally in performing daily recording and transmission of science data. The mm152 sequence has performed well since it started on 01-228 (8/16/01). It terminates on 01-255 (9/12/01).

A Delta-DOR experiment was performed on 01-236 (8/24/01).

Other - Five Roll Only Targeted Observations (ROTOs) were successfully performed since the last report. MGS has completed 183 ROTOs to date.

Spacecraft Health:
All subsystems report good health and status except for the MOLA payload instrument that malfunctioned on 01-181 (6/30/01). The MOLA instrument team is still investigating the anomaly.

Uplinks:
There have been 13 uplinks to the spacecraft during the past week, including instrument command loads, and the MZ128 and MZ129 ROTO mini-sequences. Also a new ROTO script was uplinked. It will allow MGS to perform ROTOs during scheduled communications orbits. Previously MGS has been restricted to performing ROTOs only during scheduled non-communications orbits. 5,590 command files have been radiated to the spacecraft since launch.

Upcoming Events:
Delta-DOR experiments are scheduled for 01-247 (9/4/01), 01-253 (9/10/01), 01-267 (9/24/01), 01-269 (9/26/01).

A Solar Array (SA) and High Gain Antenna (HGA) flexible modes baseline diagnostic test will be performed on 01-154 (9/11/01). Performing this test periodically should give us insight into any additional SA hinge or HGA gimbal degradation if it occurs.

MOLA Diagnostic Test #4 will take place on 01-156 (9/13/01).

Wednesday 5-Sep-01

(DOY 241/19:00:00 to DOY 248/19:00:00 UTC) Launch / Days since Launch = Nov. 7, 1996 / 1764 days
Start of Mapping / Days since Start of Mapping = April 1, 1999 / 888 days
Total Mapping Orbits = 11,155
Total Orbits = 12,838

Recent Events:
Background Sequences - The spacecraft is operating nominally in performing daily recording and transmission of science data. The mm152 sequence has performed well since it started on 01-228 (8/16/01). It terminates on 01-255 (9/12/01).

A Delta-DOR experiment was performed on 01-247 (9/4/01).

Other - Fourteen Roll Only Targeted Observations (ROTOs) were successfully performed since the last report. MGS has completed 197 ROTOs to date.

Spacecraft Health:
All subsystems report good health and status except for the MOLA payload instrument that malfunctioned on 01-181 (6/30/01). The MOLA instrument team is still investigating the anomaly. MOLA Diagnostic Test #4 is scheduled for 01-256 (9/13/01).

Uplinks:
There have been 12 uplinks to the spacecraft during the past week, including new star catalogs and ephemeris files, instrument command loads, and the MZ130 ROTO mini-sequences. 5,615 command files have been radiated to the spacecraft since launch.

Upcoming Events:
Delta-DOR experiments are scheduled for 01-253 (9/10/01), 01-267 (9/24/01), 01-269 (9/26/01).

A Solar Array (SA) and High Gain Antenna (HGA) flexible modes baseline diagnostic test will be performed on 01-154 (9/11/01). Performing this test periodically should give us insight into any additional SA hinge or HGA gimbal degradation if it occurs.

MOLA Diagnostic Test #4 will take place on 01-156

(9/13/01).

Wednesday 12-Sep-01

(DOY 248/19:00:00 to DOY 255/19:00:00 UTC) Launch / Days since Launch = Nov. 7, 1996 / 1771 days
Start of Mapping / Days since Start of Mapping = April 1, 1999 / 895 days
Total Mapping Orbits = 11,241
Total Orbits = 12,924

Recent Events:
Background Sequences - The mm152 background sequence terminated unexpectedly on 01-249 (9/6/01) at 05:34:41 SCET UTC when MGS entered Contingency Mode (C-Mode) due to star processing difficulties. The MGS team has been successful in recovering from C-Mode and playing back the data recorded on 01-248 (9/5/01) and 01-249 (9/6/01) up to the point of C-Mode entry. MGS is presently in the Relay-16 attitude with the MOC and MAG/ER instruments powered back on. MOLA will be powered during the MOLA Diagnostic Test #4 on 01-256 (9/13/01) and TES will be powered on 01-258 (9/15/01). The spacecraft will be operating nominally once the mz153 background sequence goes active later today at 01-155 (9/12/01) 23:59:00.

Other - The Delta-DOR experiment scheduled for 01-253 (9/10/01) and the Solar Array (SA) and High Gain Antenna (HGA) flexible modes baseline diagnostic test scheduled for 01-154 (9/11/01) did not take place due to C-Mode recovery.

One Roll Only Targeted Observation (ROTO) was successfully performed before the spacecraft entered C-Mode. MGS has completed 198 ROTOs to date.

Spacecraft Health:
All subsystems report good health and status. The spacecraft team is investigating the spacecraft's difficulty in processing stars leading to the C-Mode entry. The MOLA instrument team is still investigating the MOLA anomaly of 01-181 (6/30/01). MOLA Diagnostic Test #4 is scheduled for 01-256 (9/13/01).

Uplinks:
There have been 112 uplinks to the spacecraft during the past week, including C-Mode recovery files, new star catalogs and ephemeris files, instrument command loads, and the MM153 background sequence. 5,727 command files have been radiated to the spacecraft since launch.

Upcoming Events:
Delta-DOR experiments are scheduled for 01-267 (9/24/01) and 01-269 (9/26/01). MOLA Diagnostic Test #4 will take place on 01-156 (9/13/01).

Wednesday 19-Sep-01

(DOY 255/19:00:00 to DOY 262/19:00:00 UTC) Launch / Days since Launch = Nov. 7, 1996 / 1778 days
Start of Mapping / Days since Start of Mapping = April 1, 1999 / 902 days
Total Mapping Orbits = 11,327
Total Orbits = 13,010

Recent Events:
Background Sequences - The spacecraft is operating nominally in performing daily recording and transmission of science data. The mm153 sequence ha performed well since it started 01-156 (9/13/01). I terminates on 01-290 (10/17/01).

Other - MOLA Diagnostic Test #4 was performed o 01-256 (9/13/01). The MOLA team is in the process c analyzing the resulting data.

Roll Only Targeted Observations (ROTO) wer temporarily suspended pending investigation of the C Mode entry root cause. MGS entered C-Mode o 9/6/01 because the STAREX star processing softwar could not converge on a good inertial attitude followin a ROTO. The ROTO did not cause the C-Mode entr but STAREX performance may impact how we perforr ROTOs in the future. MGS has completed 198 ROTO to date.

Spacecraft Health:
All subsystems report good health and status. Th spacecraft team continues to investigate the STARE anomaly that caused the 9/6/01 C-Mode entry. Th MOLA instrument team is still investigating the MOL anomaly of 01-181 (6/30/01). The MOC instrumen required a hardware reboot to recover from an anomal that temporarily prevented it from returning proper formatted science data.

Uplinks:
There have been 28 uplinks to the spacecraft during th past week, including new star catalogs and ephemer files, TES instrument turn- on, MOC instrument re boots, and standard instrument command loads. 5,75 command files have been radiated to the spacecra since launch.

Upcoming Events:
Delta-DOR experiments are scheduled for 01-26 (9/24/01) and 01-269 (9/26/01).

Wednesday 26-Sep-01

(DOY 262/19:00:00 to DOY 269/19:00:00 UTC) Launc / Days since Launch = Nov. 7, 1996 / 1785 days
Start of Mapping / Days since Start of Mapping = April 1999 / 909 days
Total Mapping Orbits = 11,412
Total Orbits = 13,095

Recent Events:
Background Sequences - The spacecraft is operatin nominally in performing daily recording an transmission of science data. The mm153 sequence ha performed well since it started 01-256 (9/13/01). terminates on 01-290 (10/17/01).

Other - Delta-DOR experiments were performed o 01-267 (9/24/01) and 01-269 (9/26/01).

No Roll Only Targeted Observations (ROTOs) wer performed this past week. ROTOs were temporari suspended pending investigation of the C-Mode entr root cause. MGS entered C-Mode on 9/6/01 becaus the STAREX star processing software could nc converge on a good inertial attitude following a ROTC The ROTO did not cause the C-Mode entry, bu STAREX performance following ROTOs may impac how we perform ROTOs in the future. MGS ha completed 198 ROTOs to date.

Spacecraft Health:
All subsystems report good health and status. The spacecraft team is investigating ways to improve the probability that STAREX will converge on a good inertial attitude following each ROTO. The MOLA instrument team is still investigating the MOLA anomaly of 01-181 (6/30/01) and will be requesting a fifth diagnostic test in the near future.

Uplinks:
There have been 16 uplinks to the spacecraft during the past week, including new star catalogs and ephemeris files, instrument command loads, and the MZ136A ROTO mini-sequence. 5,771 command files have been radiated to the spacecraft since launch.

Upcoming Events:
On 01-270 (9/27/01), MGS will perform a ROTO during a communications orbit for the first time. Previously, MGS had been restricted to performing ROTOs only during non- communications orbits. Two more ROTOs during communications orbits are scheduled for the MZ037 mini-sequence.

A Solar Array (SA) and High Gain Antenna (HGA) flexible modes baseline diagnostic test will be performed on 01-278 (10/05/01). It is designated MZ139. Performing this test periodically should give us insight into any additional SA hinge or HGA gimbal degradation, if it occurs.

Wednesday 3-Oct-01

(DOY 269/19:00:00 to DOY 276/19:00:00 UTC) Launch / Days since Launch = Nov. 7, 1996 / 1792 days
Start of Mapping / Days since Start of Mapping = April 1, 1999 / 916 days
Total Mapping Orbits = 11,498
Total Orbits = 13,181

Recent Events:
Background Sequences - The spacecraft is operating nominally in performing daily recording and transmission of science data. The mm153 sequence has performed well since it started 01-256 (9/13/01). It terminates on 01-290 (10/17/01).

Other - Three Roll Only Targeted Observations (ROTOs) were performed this past week. All three occurred during scheduled communications orbits. Previously, MGS had been restricted to performing ROTOs only during non-communications orbits. This new capability of performing ROTOs during communications orbits will allow MGS to perform ROTOs during the Odyssey aerobraking support phase, when MGS has 24 hours of DSN coverage each day. MGS has completed 201 ROTOs to date.

Spacecraft Health:
All subsystems report good health and status.

The MOLA instrument team has completed their investigation of the MOLA anomaly of 01-181 (6/30/01). The collective results from the instrument tests indicate the anomaly was caused by an interruption of the 100-MHz signal from the precision oscillator in the MOLA altimeter electronics. Without the oscillator signal, the 10-Hz laser trigger signal cannot occur and the laser cannot fire. All testing since July, 2001 shows that the instrument responds properly to commanding and that all other parts of the MOLA are operating normally. The diagnostic tests also demonstrated that the MOLA is fully capable of continuing to collect data in passive radiometer mode.

Uplinks:
There have been 19 uplinks to the spacecraft during the past week, including new star catalogs and ephemeris files, instrument command loads, and the MZ137 and MZ138 ROTO mini-sequences. 5,790 command files have been radiated to the spacecraft since launch.

Upcoming Events:
A Solar Array (SA) and High Gain Antenna (HGA) flexible modes baseline diagnostic test will be performed on 01-278 (10/05/01). It is designated MZ139. Performing this test periodically should give us insight into any additional SA hinge or HGA gimbal degradation, if it occurs.

Five MOC defocus calibrations will be performed between 01-281 (10/8/01) and 01-289 (10/16/01).

A fifth and final MOLA diagnostic test will be conducted on 01-283 (10/10/01) after which the MOLA will be left on to operate in passive radiometer mode.

Wednesday 7-Nov-01

(DOY 304/19:00:00 to DOY 311/19:00:00 UTC) Launch / Days since Launch = Nov. 7, 1996 / 1827 days
Start of Mapping / Days since Start of Mapping = April 1, 1999 / 951 days
Total Mapping Orbits = 11,926
Total Orbits = 13,609

Recent Events:
Background Sequences - The spacecraft is operating nominally in performing daily recording and transmission of science data. TES and MOC images are being supplied to the Mars Odyssey team to support their aerobraking campaign. The MM154B sequence has performed well since it started on 01-295 10/22/01). It terminates on 01-319 (11/15/01).

Other - Roll Only Targeted Observations (ROTOs) have been suspended to reduce the probability of entering C-Mode during the initial stages of Mars Odyssey aerobraking. The C-Mode investigation continues. Its purpose is to determine the root cause of the latest C-Mode entry and identify risk mitigation steps that will allow MGS to resume ROTO operations during Odyssey aerobraking. MGS has completed 207 ROTOs to date.

Spacecraft Health:
All subsystems report good health and status. The spacecraft team is investigating the spacecraft's difficulty in processing stars leading to the C-Mode entry.

Uplinks:
There have been 19 uplinks to the spacecraft during the past week, including new star catalog and ephemeris files, and instrument command loads. 5,950 command files have been radiated to the spacecraft since launch.

Upcoming Events:
MGS will continue to support Odyssey during the aerobraking phase of the Odyssey mission by supplying

TES and MOC images of the Martian surface. Martian dust storms can bloom into the upper atmosphere, significantly increasing the atmospheric density and posing a threat to Odyssey.The MGS images will provide the Odyssey team the opportunity to avoid the dust storms if they occur.

Next week the Sun Monitor Ephemeris fault detection will be re-enabled with updated parameters. The new parameters will prevent transient shadowing of the sun sensors from triggering C-mode entry and allow additional C-mode protection if attitude knowledge becomes corrupted.

Wednesday 14-Nov-01

(DOY 311/19:00:00 to DOY 318/19:00:00 UTC) Launch / Days since Launch = Nov. 7, 1996 / 1834 days
Start of Mapping / Days since Start of Mapping = April 1, 1999 / 958 days
Total Mapping Orbits = 12,011
Total Orbits = 13,694

Recent Events:
Background Sequences - The spacecraft is operating nominally in performing daily recording and transmission of science data. TES and MOC images are being supplied to the Mars Odyssey team to support their aerobraking campaign.The MM154B sequence has performed well since it started on 01-295 10/22/01). It terminates on 01-319 (11/15/01). MM155A starts tonight at 01-318 (11/14/01), 23:57 SCET UTC.

Other - Roll Only Targeted Observations (ROTOs) have been suspended to reduce the probability of entering C-Mode during the initial stages of Mars Odyssey aerobraking. The C-Mode investigation continues. Its purpose is to determine the root cause of the latest C-Mode entry and identify risk mitigation steps that will allow MGS to resume ROTO operations during Odyssey aerobraking. MGS has completed 207 ROTOs to date.

The Sun Monitor Ephemeris fault detection was re-enabled with updated parameters.The new parameters will prevent transient shadowing of the sun sensors from triggering C-mode entry and allow additional C-mode protection if attitude knowledge becomes corrupted.

Spacecraft Health:
All subsystems report good health and status. The spacecraft team is investigating the spacecraft's difficulty in processing stars leading to the C-Mode entry.

Uplinks:
There have been 12 uplinks to the spacecraft during the past week, including new star catalog and ephemeris files, instrument command loads, and the MM155 background sequence. 5,962 command files have been radiated to the spacecraft since launch.

Upcoming Events:
MGS will continue to support Odyssey during the aerobraking phase of the Odyssey mission by supplying TES and MOC images of the Martian surface. Martian dust storms can bloom into the upper atmosphere, significantly increasing the atmospheric density and posing a threat to Odyssey.The MGS images will provide the Odyssey team the opportunity to avoid the dust storms if they occur.

Wednesday 28-Nov-01

(DOY 325/19:00:00 to DOY 332/19:00:00 UTC) Launc / Days since Launch = Nov. 7, 1996 / 1848 days
Start of Mapping / Days since Start of Mapping = April 1999 / 972 days
Total Mapping Orbits = 12,182
Total Orbits = 13,865

Recent Events:
Background Sequences - The spacecraft is operatin nominally in performing daily recording an transmission of science data. TES and MOC images ar being supplied to the Mars Odyssey team to suppor their aerobraking campaign.The MM155A sequence ha performed well since it started on 01-318 11/14/01). terminates on 01-347 (12/13/01).

Other - Roll Only Targeted Observations (ROTOs) hav been suspended to reduce the probability of entering C Mode during the initial stages of Mars Odysse aerobraking. The C-Mode investigation continues. It purpose is to determine the root cause of the latest C Mode entry and identify risk mitigation steps that wi allow MGS to resume ROTO operations during Odysse aerobraking. MGS has completed 207 ROTOs to date.

Spacecraft Health:
All subsystems report good health and status. Th spacecraft team is investigating the spacecraft's difficult in processing stars leading to the C-Mode entry. A min sequence is being developed to collect in-flight dat while the gyros are at high-rate. It is possible that som attitude knowledge error is accumulating while th gyros are at high-rate during ROTO slews.

Uplinks:
There have been 22 uplinks to the spacecraft during th past week, including new star catalog and ephemeri files, and instrument command loads. 5,995 comman files have been radiated to the spacecraft since launch.

Upcoming Events:
MGS will continue to support Odyssey during th aerobraking phase of the Odyssey mission by supplyin TES and MOC images of the Martian surface. Martia dust storms can bloom into the upper atmospher significantly increasing the atmospheric density an posing a threat to Odyssey.The MGS images will provid the Odyssey team the opportunity to avoid the dus storms if they occur.

Wednesday 5-Dec-01

(DOY 332/19:00:00 to DOY 339/19:00:00 UTC) Launc / Days since Launch = Nov. 7, 1996 / 1855 days
Start of Mapping / Days since Start of Mapping = April 1999 / 979 days
Total Mapping Orbits = 12,268
Total Orbits = 13,951

Recent Events:
Background Sequences - The spacecraft is operatin nominally in performing daily recording an transmission of science data. TES and MOC images ar being supplied to the Mars Odyssey team to suppor their aerobraking campaign.The MM155A sequence ha performed well since it started on 01-318 11/14/01). terminates on 01-347 (12/13/01).

Other - Roll Only Targeted Observations (ROTOs) have been suspended to reduce the probability of entering C-Mode during Mars Odyssey aerobraking. The Spacecraft Team continues to investigate MGS difficulties in star processing following slew maneuvers like ROTOs. The team is designing tests and identifying steps needed to mitigate C-mode entry following ROTOs. MGS has completed 207 ROTOs to date.

Spacecraft Health:
All subsystems report good health and status. The MZ147 and MZ148 mini-sequences are being developed to collect in-flight data while the gyros are at high-rate. It is possible that some attitude knowledge error is accumulating while the gyros are at high-rate during ROTO slews.

Uplinks:
There have been 10 uplinks to the spacecraft during the past week, including new star catalog and ephemeris files, and instrument command loads. 6,005 command files have been radiated to the spacecraft since launch.

Upcoming Events:
The MZ147 and MZ148 High- rate Gyro Data Collection mini-sequences are scheduled to execute 01-345 12/11/01.

MGS will continue to support the Odyssey mission during the aerobraking phase by supplying TES and MOC images of the Martian surface. Martian dust storms can bloom into the upper atmosphere, significantly increasing the atmospheric density and posing a threat to Odyssey. The MGS images will provide the Odyssey team the opportunity to avoid the dust storms if they occur.

Wednesday 12-Dec-01

(DOY 339/19:00:00 to DOY 346/19:00:00 UTC) Launch / Days since Launch = Nov. 7, 1996 / 1862 days
Start of Mapping / Days since Start of Mapping = April 1, 1999 / 986 days
Total Mapping Orbits = 12,354
Total Orbits = 14,037

Recent Events:
Background Sequences - The spacecraft is operating nominally in performing daily recording and transmission of science data. TES and MOC images are being supplied to the Mars Odyssey team to support their aerobraking campaign. The MM155A sequence has performed well since it started on 01-318 (11/14/01). It terminates on 01-347 (12/13/01). MM156A starts tonight at 01-346 (12/12/01), 23:57 SCET UTC.

Other - Roll Only Targeted Observations (ROTOs) have been suspended to reduce the probability of entering C-Mode during Mars Odyssey aerobraking. The Spacecraft Team continues to investigate MGS difficulties in star processing following slew maneuvers like ROTOs. The team is designing tests and identifying steps needed to mitigate C-mode entry following ROTOs. MGS has completed 207 ROTOs to date.

The MZ147 High-rate Gyro Data Collection mini-sequence executed successfully on 01-345 (12/11/01). The MZ148 contingency mini-sequence was uplinked, then terminated as planned due to the success of MZ147. The data show that the high-rate gyro bias estimates used during the September and October ROTOs would not have caused the loss of attitude knowledge that resulted in C-mode entry.

Spacecraft Health:
All subsystems report good health and status.

Uplinks:
There have been 17 uplinks to the spacecraft during the past week, including new star catalog and ephemeris files, instrument command loads, and the High-rate Gyro Data Collection mini-sequences cited above. 6,022 command files have been radiated to the spacecraft since launch.

Upcoming Events:
MGS will continue to support the Odyssey mission during the aerobraking phase by supplying TES and MOC images of the Martian surface. Martian dust storms can bloom into the upper atmosphere, significantly increasing the atmospheric density and posing a threat to Odyssey. The MGS images will provide the Odyssey team the opportunity to avoid the dust storms if they occur.

Wednesday 19-Dec-01

(DOY 346/19:00:00 to DOY 353/19:00:00 UTC) Launch / Days since Launch = Nov. 7, 1996 / 1869 days
Start of Mapping / Days since Start of Mapping = April 1, 1999 / 993 days
Total Mapping Orbits = 12,439
Total Orbits = 14,122

Recent Events:
Background Sequences - The spacecraft is operating nominally in performing daily recording and transmission of science data. TES and MOC images are being supplied to the Mars Odyssey team to support their aerobraking campaign. The MM156A sequence has performed well since it started on 2001-346 (12/12/01). It terminates on 2002- 017 (1/17/01).

Other - Roll Only Targeted Observations (ROTOs) have been suspended to reduce the probability of entering C-Mode during Mars Odyssey aerobraking. The Spacecraft Team continues to investigate MGS difficulties in star processing following slew maneuvers like ROTOs. The team is designing tests and identifying steps needed to mitigate C-mode entry following ROTOs. Some of these steps will be implemented in January 2002. The effectiveness of these steps will be assessed during a few specialized ROTOs in January. MGS has completed 207 ROTOs to date.

Spacecraft Health:
All subsystems report good health and status.

Uplinks:
There have been 12 uplinks to the spacecraft during the past week, including new star catalog and ephemeris files, and instrument command loads. 6,034 command files have been radiated to the spacecraft since launch.

Upcoming Events:
MGS will continue to support the Odyssey mission during the aerobraking phase by supplying TES and MOC images of the Martian surface. Martian dust storms can bloom into the upper atmosphere, significantly increasing the atmospheric density and

posing a threat to Odyssey.The MGS images will provide the Odyssey team the opportunity to avoid potential dust storms. Odyssey should complete the aerobraking phase in mid-January 2002.

Wednesday 2-Jan-02

(DOY 353/19:00:00 to DOY 002/19:00:00 UTC)This report covers two weeks.
Launch / Days since Launch = Nov. 7, 1996 / 1883 days
Start of Mapping / Days since Start of Mapping = April 1, 1999 / 1007 days
Total Mapping Orbits = 12,611
Total Orbits = 14,294

Recent Events:
Background Sequences - The spacecraft is operating nominally in performing daily recording and transmission of science data. MGS support of Mars Odyssey should end by mid- January with the end of their aerobraking campaign.The MM156A sequence has performed well since it started on 2001-346 (12/12/01). It terminates on 2002- 017 (1/17/01).

Other - Roll Only Targeted Observations (ROTOs) have been suspended to reduce the probability of entering C-Mode during Mars Odyssey aerobraking.The Spacecraft Team continues to investigate MGS difficulties in star processing following slew maneuvers like ROTOs. The team is designing tests and identifying steps needed to mitigate C-mode entry following ROTOs. Some of these steps will be implemented this month, January 2002.The effectiveness of these steps will be assessed during a few specialized ROTOs in January. Our goal is to get back to performing ROTOs routinely with little risk of entering C- mode. MGS has completed 207 ROTOs to date.

Spacecraft Health:
All subsystems report good health and status.

Uplinks:
There have been 22 uplinks to the spacecraft during the past two weeks, including new star catalog and ephemeris files, and instrument command loads. 6,057 command files have been radiated to the spacecraft since launch.

Upcoming Events:
MGS will continue to support Odyssey for the remainder of their aerobraking phase. A specialized mini-sequence is being prepared to perform ROTOs by mid-January. It implements several changes designed to improve the MGS spacecraft's ability to maintain attitude knowledge following ROTOs.

Wednesday 9-Jan-02

(DOY 002/19:00:00 to DOY 009/19:00:00 UTC) Launch / Days since Launch = Nov. 7, 1996 / 1890 days
Start of Mapping / Days since Start of Mapping = April 1, 1999 / 1014 days
Total Mapping Orbits = 12,696
Total Orbits = 14,379

Recent Events:
Background Sequences - The spacecraft is operating nominally in performing daily recording and transmission of science data. MGS support of Mars Odyssey will end on 2002- 011 (1/11/02) with the end of their aerobraking campaign. The MM156A sequence has performed well since it started on 2001-346 (12/12/01). It terminates on 2002-017 (1/17/02).

Other - While the investigation into MGS star processing difficulties remains ongoing, the Spacecraft Team is focusing primarily on implementing steps to reduce the risk of entering C-mode following ROTOs. On 2002-011 (1/11/02) MGS will execute two ROTOs. The most significant change to the ROTO profile is commanding MGS to the nadir attitude following target imaging.While at the nadir attitude, MGS will utilize the MHSA (horizon sensor) to assist in acquiring good inertial attitude knowledge. MGS will then return to the Relay-16 attitude in order to conserve fuel. Relay-16 is an attitude that pitches the spacecraft +16 degrees from nadir to better align its attitude with the gravity gradient, thereby reducing fuel consumption and extending spacecraft life. MGS has completed 207 ROTOs to date.

Spacecraft Health:
All subsystems report good health and status.

Uplinks:
There have been 16 uplinks to the spacecraft during the past two weeks, including new star catalog and ephemeris files, and instrument command loads. 6,073 command files have been radiated to the spacecraft since launch.

Upcoming Events:
The mz149b mini-sequence will perform two ROTOs on 02-011 (1/11/02). It implements several changes designed to improve the MGS spacecraft's ability to maintain attitude knowledge following ROTOs.

Wednesday 16-Jan-02

(DOY 009/19:00:00 to DOY 016/19:00:00 UTC) Launch / Days since Launch = Nov. 7, 1996 / 1897 days
Start of Mapping / Days since Start of Mapping = April 1, 1999 / 1021 days
Total Mapping Orbits = 12,782
Total Orbits = 14,465

Recent Events:
Background Sequences - The spacecraft is operating nominally in performing daily recording and transmission of science data.The MM156A sequence has performed well since it started on 2001-346 (12/12/01). It terminates on 2002-017 (1/17/02). MM157A starts tonight at 01-016 (1/16/02), 23:57 SCET UTC.

Other - While the investigation into MGS star processing difficulties remains ongoing, the Spacecraft Team is focusing primarily on implementing steps to reduce the risk of entering C-mode following ROTOs. We now command MGS to the nadir attitude following each ROTO.While at the nadir attitude, MGS utilizes the MHSA (horizon sensor) to help update its inertial attitude knowledge. After gyro biases successfully converge, MGS returns to the Relay-16 attitude to conserve fuel. MGS successfully executed two ROTOS on 2002-011 (1/11/02) and one ROTO on 2002- 015 (1/15/02).All ROTO events occurred on schedule. Gyro biases converged between 9 and 39 minutes, well within the hour allocated. The MHSA allowed MGS to make significant attitude corrections sooner than in the past. Subsequently, the number of misidentified stars remained low following the ROTOs. MGS has completed 210 ROTOs to date.

Spacecraft Health:
All subsystems report good health and status.

Uplinks:
There have been 17 uplinks to the spacecraft during the past two weeks, including new star catalog and ephemeris files, instrument command loads, the MM157 background sequence, and the MZ149 and MZ150 ROTO mini-sequences. 6,090 command files have been radiated to the spacecraft since launch.

Upcoming Events:
The MZ150 mini-sequence will perform another ROTO later today. More ROTOs will be performed this next week in the upcoming MZ151 and MZ152 mini-sequences.

Wednesday 23-Jan-02

(DOY 016/19:00:00 to DOY 023/19:00:00 UTC) Launch / Days since Launch = Nov. 7, 1996 / 1904 days
Start of Mapping / Days since Start of Mapping = April 1, 1999 / 1028 days
Total Mapping Orbits = 12,867
Total Orbits = 14,550

Recent Events:
Background Sequences - The spacecraft is operating nominally in performing daily recording and transmission of science data. The MM157A sequence has performed nominally since it started on 2002-016 (1/16/02). It terminates on 2002-031 (1/31/02).

Other - MGS successfully executed two ROTOS on 2002-016 (1/16/02) and 2002-019 (1/19/02). All ROTO events occurred on schedule. Gyro biases converged at 12 and 39 minutes respectively, well within the hour allocated. MGS continues to make significant attitude corrections following the ROTOs based upon inputs from the MHSA (horizon sensor). The Star Identification Software (SIS) successfully identified all available stars during the gyro-bias convergence process. MGS has completed 212 ROTOs to date.

Spacecraft Health:
All subsystems report good health and status.

Uplinks:
There have been 25 uplinks to the spacecraft during the past two weeks, including new star catalog and ephemeris files, instrument command loads, and the MZ151, MZ152, and MZ153 ROTO mini-sequences. 6,115 command files have been radiated to the spacecraft since launch.

Upcoming Events:
The MZ152 mini-sequence will perform a ROTO later today and MZ153 will perform a ROTO early tomorrow, 02-024 (1/24/02). At least five more ROTOs will occur this next week in subsequent mini-sequences. Also on 02-024 (1/24/02), the MOLA will be commanded back to radiometry mode with the MZ154 mini-sequence. It was commanded to normal science mode over the weekend to check the status of its laser.

Wednesday 30-Jan-02

(DOY 023/19:00:00 to DOY 030/19:00:00 UTC) Launch / Days since Launch = Nov. 7, 1996 / 1911 days
Start of Mapping / Days since Start of Mapping = April 1, 1999 / 1035 days
Total Mapping Orbits = 12,953
Total Orbits = 14,636

Recent Events:
Background Sequences - The spacecraft is operating nominally in performing daily recording and transmission of science data. The MM157A sequence has performed nominally since it started on 2002-016 (1/16/02). It terminates on 2002-031 (1/31/02). MM158B starts tonight at 02-030 (1/30/02), 23:57 SCET UTC.

Other - The MZ154 mini-sequence commanded the MOLA back to radiometry mode on 02-024 (1/24/02). It had previously been commanded to normal science mode to check the status of its laser. Unfortunately, the laser is still inoperative.

MGS successfully executed seven Roll Only Targeting Opportunity (ROTO) imaging scans in the MZ152, MZ153, and MZ155 mini-sequences. All ROTO events occurred on schedule. Gyro biases converged between 12 and 46 minutes, within the hour allocated. MGS continues to make significant attitude corrections following ROTOs based upon inputs from the MHSA (horizon sensor). AACS engineers are investigating ways to define the gyro biases and scale factors more precisely in order to reduce the attitude errors that accumulate during ROTOs. MGS has completed 219 ROTOs to date.

Spacecraft Health:
All subsystems report good health and status.

Uplinks:
There have been 10 uplinks to the spacecraft during the past week, including instrument command loads, the mini- sequences referenced above, and the MM158 background sequence. 6,125 command files have been radiated to the spacecraft since launch.

Upcoming Events:
More ROTOs will be performed in the MZ156 and MZ157 mini- sequences.

Wednesday 6-Feb-02

(DOY 030/19:00:00 to DOY 037/19:00:00 UTC) Launch / Days since Launch = Nov. 7, 1996 / 1918 days
Start of Mapping / Days since Start of Mapping = April 1, 1999 / 1042 days
Total Mapping Orbits = 13,039
Total Orbits = 14,722

Recent Events:
Background Sequences - The spacecraft is operating nominally in performing daily recording and transmission of science data. The MM158B sequence has performed nominally since it started on 02-030 (1/30/02). It terminates on 02-058 (2/27/02).

Other - MGS successfully executed three Roll Only Targeting Opportunity (ROTO) imaging scans in the MZ156 and MZ157 mini-sequences. All ROTO events occurred on schedule. Gyro biases converged within the hour allocated. MGS continues to make significant attitude corrections following ROTOs based upon inputs from the MHSA (horizon sensor). AACS engineers are investigating ways to define the gyro biases and scale factors more precisely in order to reduce the attitude errors that accumulate during ROTOs. MGS has completed 222 ROTOs to date.

Spacecraft Health:
All subsystems report good health and status. We updated the solar array positioning command scripts this week to increase the power output of the solar arrays. As Mars travels toward aphelion causing MGS power margins to decrease, we reduce solar array offset angles to maintain an appropriate level of excess power. By effectively regulating excess power, we also extend the life of the Partial Shunt Assemblies.

Uplinks:
There have been 19 uplinks to the spacecraft during the past week, including instrument command loads, new star catalogs and ephemeris files, solar array script updates, and the ROTO mini-sequences mentioned above. 6,144 command files have been radiated to the spacecraft since launch.

Upcoming Events:
MGS will perform more ROTOs in the MZ158 and MZ159 mini-sequences.

We will update the Y-axis gyro low-rate scale factor on 02-042 (2/11/02) to reduce the gyro propagation errors we see building during ROTOs.

Wednesday 13-Feb-02

(DOY 037/19:00:00 to DOY 044/19:00:00 UTC) Launch / Days since Launch = Nov. 7, 1996 / 1925 days
Start of Mapping / Days since Start of Mapping = April 1, 1999 / 1049 days
Total Mapping Orbits = 13,124
Total Orbits = 14,807

Recent Events:
Background Sequences - The spacecraft is operating nominally in performing daily recording and transmission of science data. The MM158B sequence has performed nominally since it started on 02-030 (1/30/02). It terminates on 02-058 (2/27/02).

Other - MGS successfully executed three Roll Only Targeting Opportunity (ROTO) imaging scans in the MZ158 and MZ159 mini-sequences. All ROTO events occurred on schedule and gyro biases converged within the hour allocated. We still see significant attitude corrections following each of the ROTOs. On 02-042 (2/11/02), we updated the Y-axis gyro low-rate scale factor to reduce gyro propagation errors during ROTOs. We are pleased to see a corresponding 27% reduction in gyro bias variations. We anticipate that the scale factor update will result in smaller attitude corrections in the next series of ROTOs. MGS has completed 225 ROTOs to date.

Spacecraft Health:
All spacecraft subsystems report good health and status. The TES team corrected a couple of TES instrument malfunctions that surfaced recently. The TES is now operating nominally.

Uplinks:
There have been 28 uplinks to the spacecraft during the past week, including instrument command loads, new star catalogs and ephemeris files, a gyro scale-factor update, and the ROTO mini-sequences mentioned above. 6,172 command files have been radiated to the spacecraft since launch.

Upcoming Events:
MGS will perform more ROTOs in the MZ161 and MZ162 mini-sequences.

Wednesday 20-Feb-02

(DOY 044/19:00:00 to DOY 051/19:00:00 UTC) Launch / Days since Launch = Nov. 7, 1996 / 1932 days
Start of Mapping / Days since Start of Mapping = April 1, 1999 / 1056 days
Total Mapping Orbits = 13,210
Total Orbits = 14,893

Recent Events:
Background Sequences - The spacecraft is operating nominally in performing daily recording and transmission of science data. The MM158B sequence has performed nominally since it started on 02-030 (1/30/02). It terminates on 02-058 (2/27/02).

Other - MGS successfully executed five Roll Only Targeting Opportunity (ROTO) imaging scans in the MZ160 mini-sequence. All ROTO events occurred on schedule and gyro biases converged within the hour allocated. We still see significant attitude corrections following each of the ROTOs. On 02- 042 (2/11/02), we updated the Y-axis gyro low-rate scale factor to reduce gyro propagation errors during ROTOs. We are pleased to see a corresponding 29% reduction in the standard deviation of the bias.

With a sine fit to the bias, we observe a 40% reduction in the amplitude of the variation.

We did not get the expected reduction in the pitch correction. However, this could be due to roll and yaw errors cross coupling into pitch error. MGS has completed 230 ROTOs to date.

Spacecraft Health:
All spacecraft subsystems report good health and status.

Uplinks:
There have been 7 uplinks to the spacecraft during the past week, including instrument command loads and the ROTO mini-sequences mentioned above. 6,179 command files have been radiated to the spacecraft since launch.

Upcoming Events:
MGS will perform more ROTOs in the MZ161 and MZ162 mini-sequences. Another Solar Array and HGA Flexible Modes diagnostic test will be conducted on 02-063 (3/4/02). On 02-073 (3/14/02), MGS will return to the Beta-Supplement mode of operating in order to avoid the HGA obstruction zone. We enter this mode of operation whenever the Earth Beta Angle decreases to less than 43 degrees.

Wednesday 27-Feb-02

(DOY 051/19:00:00 to DOY 058/19:00:00 UTC) Launch / Days since Launch = Nov. 7, 1996 / 1939 days
Start of Mapping / Days since Start of Mapping = April 1, 1999 / 1063 days
Total Mapping Orbits = 13,295
Total Orbits = 14,978

Recent Events:
Background Sequences - The spacecraft is operating nominally in performing daily recording and transmission of science data. The MM158B sequence has performed nominally since it started on 02-030 (1/30/02). It terminates on 02-058 (2/27/02). MM159A starts tonight at 02-058 (2/28/02), 23:57 SCET UTC.

Other - MGS successfully executed three Roll Only Targeting Opportunity (ROTO) imaging scans in the MZ161 & MZ162 mini-sequences. All ROTO events occurred on schedule and gyro biases converged within the hour allocated. We still see significant attitude corrections following each of the ROTOs. MGS has completed 233 ROTOs to date.

Spacecraft Health:
All spacecraft subsystems report good health and status.

Uplinks:
There have been 20 uplinks to the spacecraft during the past week, including new star catalogs and ephemeris files, instrument command loads, the mini-sequences referenced above, the MM159 background sequence, and a gyro scale-factor update. 6,199 command files have been radiated to the spacecraft since launch.

Upcoming Events:
MGS will perform more ROTOs in the MZ163 and MZ164 mini-sequences. Another Solar Array and HGA Flexible Modes diagnostic test will be conducted on 02-063 (3/4/02). On 02-073 (3/14/02), MGS will return to the Beta-Supplement mode of operating in order to avoid the HGA obstruction zone. We enter this mode of operation whenever the Earth Beta Angle decreases to less than 43 degrees.

Wednesday 6-Mar-02

(DOY 058/19:00:00 to DOY 065/19:00:00 UTC) Launch / Days since Launch = Nov. 7, 1996 / 1946 days
Start of Mapping / Days since Start of Mapping = April 1, 1999 / 1070 days
Total Mapping Orbits = 13,381
Total Orbits = 15,064

Recent Events:
Background Sequences

The MM158B background sequence terminated when the spacecraft entered Contingency Mode (C-mode).

At 1854 UTC on 02-058 (2/27/02), we sent commands to update the Y-axis gyro scale factor in order to improve gyro propagation accuracy. Such an update requires cycling the star processing software (STAREX), presenting the small possibility of misidentifying stars as STAREX attempts to converge on a solution. In this case, the number of unidentified stars climbed and spacecraft attitude drifted slightly as MGS approached Earth occultation at 2039 UTC. MGS emerged from Earth occultation at 2112 UTC and the DSN stations detected no signal. The MGS team began recovery procedures.

The MGS spacecraft has been recovered. Analysis of real-time telemetry indicates that the spacecraft drifted more than 5 degrees during the occultation period, thereby triggering the Sun Monitor Ephemeris fault protection response and subsequent C-Mode entry. Future playbacks of the spacecraft data recorded on 02-058 should confirm this scenario.

Presently, MGS is nadir pointed with communications re-established through the high-gain antenna. The team is in the process of completing the C-mode recovery procedure.

Spacecraft Health:
All spacecraft subsystems report good health and status.

Uplinks:
There have been 252 uplinks to the spacecraft during the past week, all related to C-mode recovery. 6,451 command files have been radiated to the spacecraft since launch.

Upcoming Events:
The MOC and MAG/ER instruments will be powered later today. The MM159B background sequence, soon to be uplinked, will go active today at 2357 SCET UTC. The MOLA and TES instruments will be powered on 02-067 (3/08/02). The MZ066 ROTO mini-sequence executes on 02-070 (3/11/02). On 02-073 (3/14/02), MGS will return to the Beta-Supplement mode of operating in order to avoid the HGA obstruction zone. We enter this mode of operation whenever the Earth Beta Angle decreases to less than 43 degrees.

Wednesday 13-Mar-02

(DOY 065/19:00:00 to DOY 072/19:00:00 UTC) Launch / Days since Launch = Nov. 7, 1996 / 1953 days
Start of Mapping / Days since Start of Mapping = April 1, 1999 / 1077 days
Total Mapping Orbits = 13,466
Total Orbits = 15,149

Recent Events:
Background Sequences - The spacecraft is operating nominally in performing daily recording and transmission of science data. The MM159B sequence has performed nominally since it started on 02-065 (3/06/02). It terminates on 02-073 (3/14/02). MM160A starts tonight at 02-072 (3/13/02), 23:57 SCET UTC. It marks the transition from nominal mapping to beta-supplement mapping.

Other - The MGS spacecraft has completely recovered from C-mode. All science instruments are powered and operating nominally. Analysis of playback telemetry confirms that the spacecraft drifted more than 5 degrees following the STAREX reset on on 02-058 (2/27/02), thereby triggering the Sun Monitor Ephemeris fault protection response and subsequent C-Mode entry. The Spacecraft team is continuing its intensive analyses of the playback data.

MGS successfully executed a Roll Only Targeting Opportunity (ROTO) imaging scan on 02-070 (3/11/02) during a scheduled non-comm period. This is significant, as ROTOs performed during the beta- supplement mission phase must occur during non-comm orbits to prevent damaging the HGA. All ROTO events occurred on schedule and gyro biases converged within the hour allocated. MGS has completed 234 ROTOs to date.

Spacecraft Health:
All spacecraft subsystems report good health and status.

Uplinks:
There have been 37 uplinks to the spacecraft during the past week, including miscellaneous C-mode recovery commands, new star catalogs and ephemeris files, instrument power-on commands, instrument command loads, the MM159B and MM160A background sequences, and the MZ166A ROTO mini- sequence. 6,488 command files have been radiated to the spacecraft since launch.

Upcoming Events:
The MM160A beta-supplement background sequence will go active today at 2357 SCET UTC. Returning to beta-supplement operations is necessary to avoid the HGA obstruction zone. We enter this mode of operation whenever the Earth Beta Angle decreases to less than 43 degrees.

Wednesday 20-Mar-02

(DOY 072/19:00:00 to DOY 079/19:00:00 UTC) Launch / Days since Launch = Nov. 7, 1996 / 1960 days
Start of Mapping / Days since Start of Mapping = April 1, 1999 / 1084 days
Total Mapping Orbits = 13,552
Total Orbits = 15,235

Recent Events:
Background Sequences - The spacecraft is operating nominally in performing daily recording and transmission of science data. MGS transitioned to the beta-supplement mode of mapping at the beginning of the MM160A sequence. MM160A started at 02-072 (3/13/02), 23:57 SCET UTC. Since then, the MM160A and MM161C sequences have performed nominally. MM162A starts tonight at 02-079 (3/20/02), 23:57 SCET UTC.

Other - The Spacecraft team is continuing its intensive analyses of the data leading up to the latest C-mode entry. We know that the root cause of the C-mode entry was that the star processing software (STAREX) misidentified a star during the initialization cycle following a scale factor update. The project has adopted two new flight rules to preclude C-mode entry while the spacecraft team investigates ways of improving STAREX performance. The first rule states that STAREX may not be manually reset or re-enabled unless the AACS mode is "Mapping"" or "Sun-Star-Init"". It guarantees that MGS will be nadir pointed with the horizon sensor (MHSA) in the attitude control loop, able to provide independent corrections should the spacecraft attitude start drifting because of a misidentified star. The second rule prohibits resetting STAREX at points in the orbit where there are detectable stars within one degree of an on- board swath catalog star. This reduces the risk that STAREX will introduce significant attitude knowledge errors by misidentifying stars due to clustering. Swath star catalogs are created and uplinked to MGS weekly. AACS engineers will now remove stars from the catalogs that exhibit this clustering signature. STAREX only tries to identify stars in its star catalog. Therefore, even though the CSA still detects the star clusters, STAREX will ignore them during star processing because they are no longer in its star catalog.

The solar array positioning command scripts were updated this week to increase the power output of the solar arrays. MGS power margins decrease as it travels toward aphelion. We compensate by reducing the solar array offset angles to maintain an appropriate level of excess power. By effectively regulating excess power, we also extend the life of the Partial Shunt Assemblies.

Roll Only Targeting Opportunity (ROTO) imaging scans were briefly suspended during the transition to beta-supplement. They should resume next week. MGS has completed 234 ROTOs to date.

Spacecraft Health:
All spacecraft subsystems report good health and status.

Uplinks:
There have been 18 uplinks to the spacecraft during the past week, including new star catalogs and ephemeris files, instrument command loads, the MM161C and MM162A background sequences, and the new solar array scripts. 6,506 command files have been radiated to the spacecraft since launch.

Upcoming Events:
The mm163 background sequence will be uplinked on 02-081 (3/22/02). ROTO mini-sequence builds should resume next week.

Wednesday 27-Mar-02

(DOY 079/19:00:00 to DOY 086/19:00:00 UTC) Launch / Days since Launch = Nov. 7, 1996 / 1967 days
Start of Mapping / Days since Start of Mapping = April 1, 1999 / 1091 days
Total Mapping Orbits = 13,637
Total Orbits = 15,320

Recent Events:
Background Sequences - The spacecraft is operating nominally in performing daily recording and transmission of science data. The mm162 sequence executed successfully from 02-080 (3/21/02) through 02-082 (3/23/02). The mm163 sequence has performed well since it started on 02-083 (3/24/02). It terminates on 02-086 (3/27/02). The mm164 sequence, successfully uplinked on 02-085 (3/26/02), begins executing on 02-087 (3/28/02).

Other - The Spacecraft team is continuing its intensive analyses of the data leading up to the latest C-mode entry. We know that the root cause of the C-mode entry was that the star processing software (STAREX) misidentified a star during the initialization cycle following a scale factor update. However, attitude knowledge is being corrupted quicker and to a greater degree than we would expect with a misidentified star. We are continuing our efforts to recreate the C-mode entry in the spacecraft test lab.

Roll Only Targeting Opportunity (ROTO) imaging scans

were briefly suspended during the transition to beta-supplement.They will resume on 02-088 (03/29/02) with the mz167 mini-sequence. MGS has completed 234 ROTOs to date.

Spacecraft Health:
All spacecraft subsystems report good health and status.

Uplinks:
There have been 18 uplinks to the spacecraft during the past week, including new star catalogs and ephemeris files, instrument command loads, and the mm163 and mm164 background sequences. 6,524 command files have been radiated to the spacecraft since launch.

Upcoming Events:
The mm165 background sequence will be uplinked on 02-088 (3/29/02). ROTO mini-sequence mz167 will execute on 02-088 (03/29/02).

Wednesday 3-Apr-02

(DOY 086/19:00:00 to DOY 093/19:00:00 UTC) Launch / Days since Launch = Nov. 7, 1996 / 1974 days
Start of Mapping / Days since Start of Mapping = April 1, 1999 / 1098 days
Total Mapping Orbits = 13,723
Total Orbits = 15,406

Recent Events:
Background Sequences - The mm164 sequence executed successfully from 02-087 (3/28/02) through 02-089 (3/30/02).The mm165 sequence performed well from 02-090 (3/31/02) until it was terminated prematurely by Contingency mode (C-mode). The spacecraft entered C-mode at 22:22:21 SCET UTC on 02-091 (4/01/02) due to a star processing problem. Presently the spacecraft team is progressing through the C- mode recovery procedure. The spacecraft is performing nominally in the C-mode configuration. The DSN stations are able to lock onto the 10 bps downlink signal and command at the 7.8125 bit rate.

MGS should be able to achieve inertial reference later this evening, allowing us to return to a nadir attitude within the next couple of days. Playing back data, powering up science instruments, and returning to background stored sequence control will follow. The recovery process will also include modifying flight software to prevent corrupting attitude knowledge when star processing is reset while MGS is pointed off-nadir. The spacecraft team has discovered that a small subroutine within the flight software is incompatible with sustained off-nadir operations. MGS software was designed for a nadir pointed mapping mission. Off-nadir mapping operations started in August 2001 to extend mission life by conserving propellant.

Other - MGS successfully executed one ROTO imaging scan on 02- 088 (03/29/02) in the MZ167 mini-sequence. MGS has completed 235 ROTOs to date.

Spacecraft Health:
All spacecraft subsystems report good health and status.

Uplinks:
There have been 37 uplinks to the spacecraft during the past week, including new star catalogs and ephemeris files, instrument command loads, the mm165 background sequence, and C-mode recovery commands. 6,561 command files have been radiated to the spacecraft since launch.

Upcoming Events:
The project will complete the C-mode recovery, fix the flight software, power the science instruments, and return MGS to background stored sequence control.

Wednesday 10-Apr-02

(DOY 093/19:00:00 to DOY 100/19:00:00 UTC) Launch / Days since Launch = Nov. 7, 1996 / 1981 days
Start of Mapping / Days since Start of Mapping = April 1, 1999 / 1105 days
Total Mapping Orbits = 13,809
Total Orbits = 15,492

Recent Events:
Background Sequences - We have recovered MGS from C-Mode. It is nadir pointed with downlink communications through the HGA. An on-board script that is triggered every time the spacecraft emerges from solar eclipse is controlling HGA movement and TWTA cycling.The mm168 background sequence begins tonight at 23:57:00 SCET UTC on 02-101 (4/10/02). It will return the spacecraft to the relay-16 attitude and commence normal operations.

Other - All on-board recorded data have been played back. A STAREX flight software patch has been uplinked to prevent attitude knowledge from being corrupted if star processing is reset while MGS is pointed off-nadir. The MOC has been powering up and is imaging MARS.

No ROTOs were accomplished this week. MGS has completed 235 ROTOs to date.

Spacecraft Health:
All spacecraft subsystems report good health and status.

Uplinks:
There have been 60 uplinks to the spacecraft during the past week, including C-mode recovery commands, new star catalogs and ephemeris files, instrument command loads, STAREX flight software patch, the mz168 return to nadir mini-sequence, the mz169 data playback mini-sequence, the mm168 background sequence, and the mz170 science instrument turn-on minisequence. 6,561 command files have been radiated to the spacecraft since launch.

Upcoming Events:
The MOLA, MAG/ER, and TES science instruments will be powered on 02-101 (4/10/02).

Wednesday 17-Apr-02

(DOY 100/19:00:00 to DOY 107/19:00:00 UTC) Launch / Days since Launch = Nov. 7, 1996 / 1988 days
Start of Mapping / Days since Start of Mapping = April 1, 1999 / 1112 days
Total Mapping Orbits = 13,894
Total Orbits = 15,577

Recent Events:
Background Sequences - The spacecraft is operating nominally in performing the beta- supplement daily recording and transmission of science data. The mm168 sequence executed successfully from 02-101 (4/11/02)

through 02-103 (4/13/02). The mm169 sequence has performed well since it started on 02-104 (4/14/02). It terminates on 02-107 (4/17/02). The mm170 sequence, successfully uplinked on 02-106 (4/16/02), begins executing on 02-107 (4/17/02).

Other - All science instruments are powered and operating nominally. No ROTOs were accomplished this week. MGS has completed 235 ROTOs to date.

Spacecraft Health:
All spacecraft subsystems report good health and status.

Uplinks:
There have been 20 uplinks to the spacecraft during the past week, including new star catalogs and ephemeris files, instrument command loads, and the mm169 and mm170 background sequences. 6,641 command files have been radiated to the spacecraft since launch.

Upcoming Events:
ROTOs will resume next week with the mz171 mini-sequence.

Wednesday 24-Apr-02

(DOY 107/19:00:00 to DOY 114/19:00:00 UTC) Launch / Days since Launch = Nov. 7, 1996 / 1995 days
Start of Mapping / Days since Start of Mapping = April 1, 1999 / 1119 days
Total Mapping Orbits = 13,979
Total Orbits = 15,662

Recent Events:
Background Sequences - The spacecraft is operating nominally in performing the beta- supplement daily recording and transmission of science data. The mm170 sequence executed successfully from 02-108 (4/18/02) through 02-110 (4/20/02). The mm171 sequence has performed well since it started on 02-111 (4/21/02). It terminates on 02-114 (4/24/02). The mm172 sequence, successfully uplinked on 02-113 (4/23/02), begins executing on 02-114 (4/24/02).

Other - No ROTOs were performed this week. There were no MER landing site targets within the available windows of opportunity. The team is identifying, building and testing flight software modifications to improve star processing and reduce the risk of entering C-mode. The goal is to safely eliminate the need for some of the ROTO restrictions so that ROTOs can be performed more frequently. MGS has completed 235 ROTOs to date.

Spacecraft Health:
All spacecraft subsystems report good health and status.

Uplinks:
There have been 8 uplinks to the spacecraft during the past week, including instrument command loads, and the mm171 and mm172 background sequences. 6,649 command files have been radiated to the spacecraft since launch.

Upcoming Events:
MOC focus calibrations are scheduled for mid-May.

Wednesday 1-May-02

(DOY 114/19:00:00 to DOY 121/19:00:00 UTC) Launch / Days since Launch = Nov. 7, 1996 / 2002 days
Start of Mapping / Days since Start of Mapping = April 1, 1999 / 1126 days
Total Mapping Orbits = 14,066
Total Orbits = 15,749

Recent Events:
Background Sequences - The spacecraft is operating nominally in performing the beta- supplement daily recording and transmission of science data. The mm172 sequence executed successfully from 02-115 (4/25/02) through 02-117 (4/27/02). The mm173 sequence has performed well since it started on 02-118 (4/28/02). It terminates on 02-121 (5/01/02). The mm174 sequence, successfully uplinked on 02-120 (4/30/02), begins executing on 02-121 (5/01/02).

Other - No ROTOs were performed this week. The team is identifying, building and testing flight software modifications to improve star processing and reduce the risk of entering C-mode. The goal is to safely eliminate the need for some of the ROTO restrictions so that ROTOs can be performed more frequently. MGS has completed 235 ROTOs to date.

Two new 4-part motion solar array positioning command scripts were uplinked this week. MGS will transition from the 3-part motion scripts to the 4-part motion scripts when the mm174 background sequence begins executing. The new scripts will compensate for the decreasing power levels MGS experiences as it approaches aphelion.

Spacecraft Health:
All spacecraft subsystems report good health and status.

Uplinks:
There have been 28 uplinks to the spacecraft during the past week, including new star catalogs and ephemeris files, instrument command loads, and the mm173 and mm174 background sequences. 6,677 command files have been radiated to the spacecraft since launch.

Upcoming Events:
MOC focus calibrations are scheduled for mid-May.

Wednesday 8-May-02

(DOY 121/19:00:00 to DOY 128/19:00:00 UTC) Launch / Days since Launch = Nov. 7, 1996 / 2009 days
Start of Mapping / Days since Start of Mapping = April 1, 1999 / 1133 days
Total Mapping Orbits = 14,152
Total Orbits = 15,835

Recent Events:
Background Sequences - The spacecraft is operating nominally in performing the beta- supplement daily recording and transmission of science data. The mm174 sequence executed successfully from 02-122 (5/2/02) through 02-124 (5/04/02). The mm175 sequence has performed well since it started on 02-125 (5/05/02). It terminates on 02-128 (5/08/02). The mm176 sequence, successfully uplinked on 02-127 (5/07/02), begins executing on 02-128 (5/08/02).

Other - No ROTOs were performed this week. The team has uplinked two software modifications (patches) to improve star processing and reduce the risk of entering C-mode. The first patch, uplinked during the April C-mode recovery, prevents attitude knowledge from being corrupted if star processing is reset while MGS is pointed off-nadir. The second patch, uplinked this week, corrects the way STIME is calculated. STIME is the time that has passed since the last identified star. STAREX used to simply count the number of seconds since the last identified star. Because STAREX processing is briefly inhibited whenever the HGA or solar arrays stop and start moving, the counter also stopped. This introduced small timing errors into star processing that resulted in STAREX discarding stars that it normally would have identified. STAREX now calculates STIME by subtracting the time of the last identified star from the present time, so brief interruptions to star processing do not adversely affect STIME. With these flight software improvements in place, the spacecraft team recommends relaxing some of the ROTO restrictions currently in place. MGS has completed 235 ROTOs to date.

Spacecraft Health:
All spacecraft subsystems report good health and status.

Uplinks:
There have been 21 uplinks to the spacecraft during the past week, including new star catalogs and ephemeris files, instrument command loads, the STIME patch, the mm175 and mm176 background sequences, and no-op commands to test the new command system scheduled to be online this summer. 6,698 command files have been radiated to the spacecraft since launch.

Upcoming Events:
The first three of five MOC focus calibrations are scheduled for 02-132 (5/12/02), 02-133 (5/13/02), and 02-134 (5/14/02). The last two are scheduled for the following week. ROTOs will resume following the MOC focus calibrations.

Wednesday 22-May-02

(DOY 135/19:00:00 to DOY 142/19:00:00 UTC) Launch / Days since Launch = Nov. 7, 1996 / 2023 days
Start of Mapping / Days since Start of Mapping = April 1, 1999 / 1147 days
Total Mapping Orbits = 14,322
Total Orbits = 16,005

Recent Events:
Background Sequences - The spacecraft is operating nominally in performing the beta- supplement daily recording and transmission of science data. The mm178 sequence executed successfully from 02-136 (5/16/02) through 02-138 (5/18/02). The mm179 sequence has performed well since it started on 02-139 (5/19/02). It terminates on 02-142 (5/22/02). The mm180 sequence, successfully uplinked on 02-141 (5/21/02), begins executing on 02-142 (5/22/02).

Other - MGS completed the last two of five MOC focus calibrations this week. No ROTOs were performed this week. They will resume on 02-143 (5/23/02). MGS has completed 235 ROTOs to date.

Spacecraft Health:
All spacecraft subsystems report good health and status.

Uplinks:
There have been 19 uplinks to the spacecraft during the past week, including new star catalogs and ephemeris files, instrument command loads, the mz172b & mz173c MOC Focus Calibration mini- sequences, the mm179a and mm180a background sequences, and the mz174a ROTO mini- sequence. 6,737 command files have been radiated to the spacecraft since launch.

Upcoming Events:
ROTOs will resume on 02-143 (5/23/02). They should continue on a regular basis thereafter depending upon the availability of non-communications orbits in each background sequence. During the beta- supplement phase of the mission, in order to prevent the HGA from impacting the HGA boom, we restrict ROTOs to non-communications orbits when the HGA remains fixed with respect to the spacecraft body.

Wednesday 29-May-02

(DOY 142/19:00:00 to DOY 149/19:00:00 UTC) Launch / Days since Launch = Nov. 7, 1996 / 2030 days
Start of Mapping / Days since Start of Mapping = April 1, 1999 / 1154 days
Total Mapping Orbits = 14,407
Total Orbits = 16,090

Recent Events:
Background Sequences - The spacecraft is operating nominally in performing the beta- supplement daily recording and transmission of science data. The mm180 sequence executed successfully from 02-143 (5/23/02) through 02-145 (5/25/02). The mm181 sequence has performed well since it started on 02-146 (5/26/02). It terminates on 02-149 (5/29/02). The mm182 sequence, successfully uplinked on 02-148 (5/28/02), begins executing on 02-149 (5/29/02).

Other - Two Roll Only Targeted Observations (ROTOs) were performed successfully on 02-143 (5/23/02) by the mz174 minisequence. MGS has completed 237 ROTOs to date.

Spacecraft Health:
All spacecraft subsystems report good health and status.

Uplinks:
There have been 13 uplinks to the spacecraft during the past week, including new star catalogs and ephemeris files, instrument command loads, and the mm181a and mm182a background sequences. 6,750 command files have been radiated to the spacecraft since launch.

Upcoming Events:
ROTOs are planned for next week in the mz175 and mz176 mini- sequences.

Wednesday 5-Jun-02

(DOY 149/19:00:00 to DOY 156/19:00:00 UTC) Launch / Days since Launch = Nov. 7, 1996 / 2037 days
Start of Mapping / Days since Start of Mapping = April 1, 1999 / 1161 days
Total Mapping Orbits = 14,494
Total Orbits = 16,177

Recent Events:
Background Sequences - The spacecraft is operating nominally in performing the beta- supplement daily

recording and transmission of science data. The mm182 sequence executed successfully from 02-150 (5/30/02) through 02-152 (6/01/02). The mm183 sequence has performed well since it started on 02-153 (6/02/02). It terminates on 02-156 (6/05/02). The mm184 sequence, successfully uplinked on 02-155 (6/04/02), begins executing on 02-156 (6/05/02).

Other - Three Roll Only Targeted Observations (ROTOs) were performed successfully by the mz175 & mz176 minisequences. MGS has completed 240 ROTOs to date.

Spacecraft Health:
All spacecraft subsystems report good health and status. Attitude and Articulation Control Subsystem engineers (AACS) are monitoring an interesting shift in the z-axis gyro bias trend. They are investigating whether the shift correlates to any known physical phenomenon. They are also developing alternate methods of controlling the spacecraft should there be further degradation within the subsystem. Presently, the z-axis gyro bias appears to be stabilizing at a new value within specifications. Other telemetry indications appear nominal.

Uplinks:
There have been 20 uplinks to the spacecraft during the past week, including new star catalogs and ephemeris files, instrument command loads, the mz174 and mz175 ROTO mini-sequences, and the mm183a and mm184a background sequences. 6,770 command files have been radiated to the spacecraft since launch.

Upcoming Events:
The MOLA Make-Up Heater will be turned off on 02-156 (6/05/02) to maintain adequate power margin as MGS approaches aphelion. ROTOs are planned for next week in the mz177 and mz178 mini-sequences.

Wednesday 19-Jun-02

(DOY 163/19:00:00 to DOY 170/19:00:00 UTC) Launch / Days since Launch = Nov. 7, 1996 / 2051 days
Start of Mapping / Days since Start of Mapping = April 1, 1999 / 1175 days
Total Mapping Orbits = 14,665
Total Orbits = 16,348

Recent Events:
Background Sequences - The spacecraft is operating nominally in performing the beta- supplement daily recording and transmission of science data. The mm186 sequence executed successfully from 02-164 (6/13/02) through 02-166 (6/15/02). The mm187 sequence has performed well since it started on 02-167 (6/16/02). It terminates on 02-170 (6/19/02). The mm188 sequence, successfully uplinked on 02-169 (6/18/02), begins executing on 02-170 (6/19/02).

Other - Five Roll Only Targeted Observations (ROTOs) were performed successfully by the mz179 & mz180 minisequences. MGS has completed 253 ROTOs to date.

Spacecraft Health: All spacecraft subsystems report good health and status. The star identification software had some difficulty identifying stars following four of the ROTOs, but converged nicely in each case after performing an autonomous reset. This proves that the software modifications we made several weeks ago are working as designed. The characteristics of some of the stars may have changed since the original source sta catalog was developed in 1950, so we are in the proces of updating the star catalog to reflect those changes.

Uplinks:
There have been 17 uplinks to the spacecraft during th past week, including new star catalogs and ephemeri files, instrument command loads, the mz180 and mz18 ROTO mini-sequences, and the mm187 and mm18 background sequences. 6,805 command files have bee radiated to the spacecraft since launch.

Upcoming Events:
ROTOs are planned for next week in the mz181 an mz182 mini- sequences.

On 02-172 (6/21/02) and 02-177 (6/26/02), DSS-15 wi perform acquisition and tracking tests with MGS t characterize the current performance capabilities c both MGS receivers. It is suspected that the acquisitio and tracking performance of one or both spacecra receivers has degraded since launch. No measuremer of this performance has been conducted since 199 Knowledge of the acquisition and tracking performanc of each transponder helps ensure that the proper uplin sweep parameters will be used for every uplink.

Wednesday 26-Jun-02

(DOY 170/19:00:00 to DOY 177/19:00:00 UTC) Launc / Days since Launch = Nov. 7, 1996 / 2058 days
Start of Mapping / Days since Start of Mapping = April 1999 / 1183 days
Total Mapping Orbits = 14,751
Total Orbits = 16,434

Recent Events:
Background Sequences - The spacecraft is operatin nominally in performing the beta- supplement dai recording and transmission of science data. The mm18 sequence executed successfully from 02-171 (6/20/0 through 02-173 (6/22/02). The mm189 sequence ha performed well since it started on 02-174 (6/23/02). terminates on 02-177 (6/26/02). The mm190 sequenc successfully uplinked on 02-176 (6/25/02), begir executing on 02-177 (6/26/02).

Other - On 02-172 (6/21/02) and 02-177 (6/26/02 DSS-15 performed acquisition and tracking tests wit MGS to characterize the current performanc capabilities of both MGS receivers. Updating ou knowledge of the acquisition and tracking performanc of the receivers will allow us to update the uplink swee parameters for MGS.

Seven Roll Only Targeted Observations (ROTOs) we performed successfully by the mz181 & mz18 minisequences. MGS has completed 260 ROTOs to dat

Spacecraft Health: All spacecraft subsystems repo good health and status. Similar to last week, the st identification software had some difficulty identifyi stars following two of the ROTOs. The flight softwa modifications we made several weeks ago continue t work as designed and STAREX converged aft autonomously resetting. We have instituted a proce that should improve STAREX performance by modifyi the star catalog when we determine that problemat stars are about to enter the CSA field of view.

Uplinks:
There have been 23 uplinks to the spacecraft during the past week, including new star catalogs and ephemeris files, instrument command loads, the mz182 and mz183 ROTO mini-sequences, and the mm189 and mm190 background sequences. 6,828 command files have been radiated to the spacecraft since launch.

Upcoming Events:
ROTOs are planned for next week in the mz183 and mz184 mini-sequences.

Wednesday 24-Jul-02

(DOY 198/19:00:00 to DOY 205/19:00:00 UTC) Launch / Days since Launch = Nov. 7, 1996 / 2086 days
Start of Mapping / Days since Start of Mapping = April 1, 1999 / 1211 days
Total Mapping Orbits = 15,094
Total Orbits = 16,777

Recent Events:
Background Sequences - The spacecraft is operating nominally in performing the beta- supplement daily recording and transmission of science data. The mm196 sequence executed successfully from 02-199 (7/18/02) through 02-201 (7/20/02). The mm197 sequence has performed well since it started on 02-202 (7/21/02). It terminates on 02-205 (7/24/02). The mm198 sequence, successfully uplinked on 02-204 (723/02), begins executing near the end of day on 02-205 (7/24/02).

Other - Seven Roll Only Targeted Observations (ROTOs) were performed successfully by the mz190 & mz191 mini-sequences. MGS has completed 282 ROTOs to date.

MGS performed another Flex Mode Test of the SA panels on DOY 199. The results are pending.

Spacecraft Health:
Spacecraft subsystems report good health and status. The science teams are investigating the cause of the PDS soft resets from last week.

Uplinks:
There have been 17 uplinks to the spacecraft during the past week, including star catalogs and ephemeris files covering 20 days, instrument command loads, the mz190 and mz191 ROTO mini-sequences, and the mm197 and mm198 background sequences. 6,898 command files have been radiated to the spacecraft since launch.

Upcoming Events:
The MGS S/C team is preparing for solar conjunction. The next 2 background sequences [mm200 (13 days, 214 -226) and mm201 (5 days, 227-231)] shall cover the solar conjunction period. Downlink during solar conjunction shall be real time, low rate data only with no playbacks. Recording will be done throughout the sequences on SSR 2A only, should a playback of recent data be necessary. The CLT will be set to 18 days. MOC will be turned off and it's heaters turned on prior to mm200. There will be no ranging during these sequences. The telecom subsystem will be configured back to normal Beta Supplement operations at the end of sequence mm201.

ROTOs are suspended until after solar conjunction.

Wednesday 31-Jul-02

(DOY 205/19:00:00 to DOY 212/19:00:00 UTC) Launch / Days since Launch = Nov. 7, 1996 / 2093 days
Start of Mapping / Days since Start of Mapping = April 1, 1999 / 1218 days
Total Mapping Orbits = 15,179
Total Orbits = 16,862

Recent Events:
Background Sequences - The spacecraft is operating nominally in performing the beta- supplement daily recording and transmission of science data. The mm198 sequence executed successfully from 02-206 (7/25/02) through 02-208 (7/27/02). The mm199 sequence has performed well since it started on 02-209 (7/28/02). It terminates on 02-213 (8/01/02). The mm200 sequence, successfully uplinked on 02-212 (7/31/02), begins executing near the end of day on 02-213 (8/01/02).

Other - Three Roll Only Targeted Observations (ROTOs) were performed successfully by the mz192 mini-sequence. MGS has completed 285 ROTOs to date.

The MOC instrument was turned off on 02-212 (7/31/02) in preparation for solar conjunction. Its replacement heaters will keep it thermally safe until it is turned back on following conjunction.

Spacecraft Health:
Spacecraft subsystems report good health and status. The signatures of the Payload Data Subsystem (PDS) software resets from several weeks ago are similar to two PDS software resets that occurred in 1999. PDS software developer's assessed the root cause for the 1999 resets to be the reprogramming of science data bandwidth. Bandwidth originally designed for the Mars Orbiter GRS and PMIRR instruments was reallocated to increase MGS MOC and TES bandwidth since GRS and PMIRR were not being flown on MGS. This reprogramming altered the PDS' finely tuned timing and marginal dynamic memory reserve such that (occasional) process memory starvation became possible. PDS' inability to complete a process due to insufficient memory is one of several triggers for a software reset.

Uplinks:
There have been 10 uplinks to the spacecraft during the past week, including instrument command loads, the mz192 ROTO mini-sequence, and the mm199 and mm200 background sequences. 6,908 command files have been radiated to the spacecraft since launch.

Upcoming Events:
MM201, the second of two solar conjunction background sequences will be uplinked on 02-214 (8/02/02). It covers the later part of solar conjuction through 02-231 (8/19/02). Downlink during solar conjunction shall be real time, low rate data only with no playbacks. Recording will be done throughout the sequences on SSR 2A only, should a playback of recent data be necessary. The Command Loss Timer (CLT) will be set to 18 days. There will be no ranging during these sequences. The telecom subsystem will be configured back to normal Beta Supplement operations at the end of sequence mm201.

Another Flex Modes Test of the solar arrays and HGA gimbal will be performed because of data losses during the playback. We are looking at opportunities during the transition between the end of solar conjunction and the start of nominal beta-supplement sequences.

ROTOs are suspended until after solar conjunction.

Wednesday 14-Aug-02

(DOY 219/19:00:00 to DOY 226/19:00:00 UTC) Launch / Days since Launch = Nov. 7, 1996 / 2107 days
Start of Mapping / Days since Start of Mapping = April 1, 1999 / 1232 days
Total Mapping Orbits = 15,351
Total Orbits = 17,034

Recent Events:
Background Sequences - The spacecraft is operating nominally in performing the beta- supplement daily recording and transmission of science data. The mm200 sequence has performed well since it started on 02-216 (8/02/02). It terminates on 02-226 (8/14/02). The mm201 sequence, successfully uplinked on 02-214 (8/02/02), begins executing near the end of day on 02-226 (8/14/02). It covers through the end of solar conjunction.

Other - The MOC has been turned off for solar conjunction so no Roll Only Targeted Observations (ROTOs) were performed. MGS has completed 285 ROTOs to date.

Spacecraft Health:
Spacecraft subsystems report good health and status.

Uplinks:
There have been no uplinks to the spacecraft during the past week due to the solar conjunction command moratorium. 6,930 command files have been radiated to the spacecraft since launch.

Upcoming Events:
The CLT will be reset to 104 hours on 02-228 (8/16/02). The mm202 background sequence will be uplinked on 02-229 (8/17/02). The mz193 Flexible Modes Test mini-sequence will be uplinked that same day. The test itself will be performed early on 02-231 (8/19/02) to characterize the dynamics of the solar arrays and HGA gimbal. The MOC will be turned on later that day, 02-231 (8/19/02). The mz203 background sequence will be uplinked on 02-232 (8/20/02).

Wednesday 21-Aug-02

(DOY 226/19:00:00 to DOY 233/19:00:00 UTC) Launch / Days since Launch = Nov. 7, 1996 / 2114 days
Start of Mapping / Days since Start of Mapping = April 1, 1999 / 1239 days
Total Mapping Orbits = 15,436
Total Orbits = 17,119

Recent Events:
Background Sequences - The spacecraft is operating nominally in performing the beta- supplement daily recording and transmission of science data. The mm201 sequence executed successfully from 02-226 (8/14/02) through 02-231 (8/19/02). The mm202 sequence has performed well since it started on 02-232 (8/20/02). It terminates on 02-234 (8/22/02). The mm203 sequence, successfully uplinked on 02-232 (8/20/02), begins executing near the end of day on 02-233 (8/21/02).

Other - The Command Loss Timer (CLT) was reset to 104 hours on 02-228 (8/16/02).

The mz193 Flexible Modes Test was performed early on 02-231 (8/19/02) to characterize the dynamics of the solar arrays and HGA gimbal. All recorded test has been played back and is in the process of being analyzed by AACS engineers.

The MOC was successfully turned-on and its replacement heater turned-off on 02-231 (8/19/02). The MOC was turned off for solar conjunction so no Roll Only Targeted Observations (ROTOs) were performed. MGS has completed 285 ROTOs to date. ROTOs resume on 02-234 (8/22/02).

Spacecraft Health:
Spacecraft subsystems report good health and status.

Uplinks:
There have been 29 uplinks to the spacecraft during the past week, including new star catalogs and ephemeris files, CLT reset commands, MOC commands, instrument command loads, the mz193 flex modes mini- sequence, the mz194 ROTO mini-sequence, and the mm202 and mm203 background sequences. 6,959 command files have been radiated to the spacecraft since launch.

Upcoming Events:
ROTOs are planned for next week in the mz194 and mz195 mini-sequences.

Wednesday 4-Sep-02

(DOY 240/19:00:00 to DOY 247/19:00:00 UTC) Launch / Days since Launch = Nov. 7, 1996 / 2128 days
Start of Mapping / Days since Start of Mapping = April 1, 1999 / 1253 days
Total Mapping Orbits = 15,607
Total Orbits = 17,290

Recent Events:
Background Sequences - The spacecraft is operating nominally in performing the beta- supplement daily recording and transmission of science data. The mm205 sequence executed successfully from 02-241 (8/29/02) through 02-243 (8/31/02). The mm206 sequence has performed well since it started on 02-244 (9/01/02). It terminates on 02-247 (9/04/02). The mm207 sequence, successfully uplinked on 02-246 (9/03/02), begins executing near the end of day on 02-247 (9/04/02).

Other - Five Roll Only Targeted Observations (ROTOs) were performed successfully by the mz196 & mz197 minisequences. MGS has completed 296 ROTOs to date.

A DSN Network Simplification Project (NSP) tracking test was performed on 02-242 (8/30/02).

A Delta Differential One-Way Range (DDOR) experiment was performed on 02-246 (9/03/01). Data from DDOR experiments are helpful in calibrating the interplanetary navigation system.

Spacecraft Health:
Spacecraft subsystems report good health and status.

Uplinks:
There have been 8 uplinks to the spacecraft during the

past week, including instrument command loads, the mz197 ROTO mini-sequence, and the mm206 and mm207 background sequences. 6,987 command files have been radiated to the spacecraft since launch.

Upcoming Events:
ROTOs are planned for next week in the mz198 and mz199 mini-sequences.

Wednesday 11-Sep-02

(DOY 247/19:00:00 to DOY 254/19:00:00 UTC) Launch / Days since Launch = Nov. 7, 1996 / 2135 days
Start of Mapping / Days since Start of Mapping = April 1, 1999 / 1260 days
Total Mapping Orbits = 15,693
Total Orbits = 17,376

Recent Events:
Background Sequences - The spacecraft is operating nominally in performing the beta- supplement daily recording and transmission of science data. The mm207 sequence executed successfully from 02-248 (9/05/02) through 02-250 (9/07/02). The mm208 sequence has performed well since it started on 02-251 (9/08/02). It terminates on 02-254 (9/11/02). The mm209 sequence, successfully uplinked on 02-253 (9/10/02), begins executing near the end of day on 02-254 (9/11/02).

Other - Three Roll Only Targeted Observations (ROTOs) were performed successfully by the mz199 minisequence. The two ROTOs contained in mz198 were not executed because mz198 could not be uplinked. Difficulties at the DSN station prevented successful uplink of the requisite payload command files. MGS has completed 299 ROTOs to date.

Flight software was successfully updated to ensure that MGS telemeters correct equator-crossing times when it is in the CSA Backup AACS mode and nadir pointed. MGS recently began managing system momentum by periodically maneuvering between the relay-16 attitude and the nadir attitude. While effectively reducing propellant consumption, this new method of managing system momentum opened a small window of vulnerability where equator-crossing times would be computed incorrectly. The flight software patch effectively eliminates this vulnerability.

Spacecraft Health:
Spacecraft subsystems report good health and status.

Uplinks:
Twenty-two of thirty command files were successfully uplinked to MGS during the past week. DSN station related timing anomalies in the uplink path resulted in MGS properly rejecting seven command files. The successful uplinks included new star catalog and ephemeris files, instrument command loads, the mi0644 equator crossing flight software patch, the mz199 ROTO mini-sequence, and the mm208 and mm209 background sequences. 7,017 command files have been radiated to the spacecraft since launch.

Upcoming Events:
ROTOs are planned for next week in the mz200 and mz201 mini-sequences. Radio Science Occultation Egress Scans will be performed on 02-273 (9/30/02) and 02-274 (10/01/02). A DDOR experiment is scheduled for 02-276 (10/03/03).

Wednesday 18-Sep-02

(DOY 254/19:00:00 to DOY 261/19:00:00 UTC) Launch / Days since Launch = Nov. 7, 1996 / 2142 days
Start of Mapping / Days since Start of Mapping = April 1, 1999 / 1267 days
Total Mapping Orbits = 15,779
Total Orbits = 17,462

Recent Events:
Background Sequences - The spacecraft is operating nominally in performing the beta- supplement daily recording and transmission of science data. The mm209 sequence executed successfully from 02-255 (9/12/02) through 02-257 (9/14/02). The mm210 sequence has performed well since it started on 02-258 (9/15/02). It terminates on 02-261 (9/18/02). The mm211 sequence, successfully uplinked on 02-260 (9/17/02), begins executing near the end of day on 02-261 (9/18/02).

Other - The three Roll Only Targeted Observations (ROTOs) reported as successful last week were in fact not successful. The images taken during those ROTOs and the first two ROTOs in mz200 were distorted because the spacecraft was slewing slowly while the images were being taken. The relative slew rate parameter had been set to orbital rate while the equator-crossing software patch was being loaded on 02-248 (9/05/02). The parameter was not reset to zero following the patch update. The problem was discovered on 02-255 (9/12/02) and corrected on 02-256 (9/13/02). The following four ROTOs in mz200 and six ROTOs in mz201 were performed successfully. MGS has successfully completed 306 ROTOs to date.

Spacecraft Health:
Spacecraft subsystems report good health and status.

Uplinks:
There have been 23 uplinks to the spacecraft during the past week, including new star catalog and ephemeris files; instrument command loads; the mi0654 slew-rate parameter update; the mz200, mz201 & mz202 ROTO mini-sequences; and the mm210 & mm211 background sequences. 7,040 command files have been radiated to the spacecraft since launch.

Upcoming Events:
ROTOs are planned for next week in the mz201, mz202, & mz203 mini- sequences. Radio Science Occultation Egress Scans will be performed on 02-273 (9/30/02) and 02-274 (10/01/02). A DDOR experiment is scheduled for 02-276 (10/03/03).

Wednesday 25-Sep-02

(DOY 261/19:00:00 to DOY 268/19:00:00 UTC) Launch / Days since Launch = Nov. 7, 1996 / 2149 days
Start of Mapping / Days since Start of Mapping = April 1, 1999 / 1274 days
Total Mapping Orbits = 15,864
Total Orbits = 17,547

Recent Events:
Background Sequences - The spacecraft is operating nominally in performing the beta- supplement daily recording and transmission of science data. The mm211 sequence executed successfully from 02-262 (9/19/02)

through 02-264 (9/21/02). The mm212 sequence has performed well since it started on 02-265 (9/22/02). It terminates on 02-268 (9/25/02). The mm213 sequence, successfully uplinked on 02-267 (9/24/02), begins executing near the end of day on 02-268 (9/25/02).

Other - Seven Roll Only Targeted Observations (ROTOs) were performed successfully by the mz201, mz202, & mz203 minisequences. MGS has completed 213 ROTOs to date.

Spacecraft Health:
Spacecraft subsystems report good health and status.

Uplinks:
There have been 16 uplinks to the spacecraft during the past week, including new star catalog and ephemeris files, instrument command loads, the mz203 & mz204 ROTO mini-sequences, and the mm212 & mm213 background sequences. 7,056 command files have been radiated to the spacecraft since launch.

Upcoming Events:
ROTOs are planned for next week in the mz204 & mz205 mini-sequences. Radio Science Occultation Egress Scans, also contained in mz205, will be performed on 02-273 (9/30/02) and 02-274 (10/01/02). The DDOR experiment scheduled for 02-276 (10/03/03) has been cancelled due to scheduling constraints.

Wednesday 2-Oct-02

(DOY 268/19:00:00 to DOY 275/19:00:00 UTC) Launch / Days since Launch = Nov. 7, 1996 / 2156 days
Start of Mapping / Days since Start of Mapping = April 1, 1999 / 1281 days
Total Mapping Orbits = 15,950
Total Orbits = 17,633

Recent Events:
Background Sequences - The spacecraft is operating nominally in performing the beta- supplement daily recording and transmission of science data. The mm213 sequence executed successfully from 02-269 (9/26/02) through 02-271 (9/28/02). The mm214 sequence has performed well since it started on 02-272 (9/29/02). It terminates on 02-275 (10/02/02). The mm215 sequence, successfully uplinked on 02-274 (10/01/02), begins executing near the end of day on 02-275 10/02/02).

Other - Nine Roll Only Targeted Observations (ROTOs) were performed successfully by the mz203, mz204, & mz205 minisequences. MGS has completed 322 ROTOs to date.

The Radio Science egress scans contained in the mz205 mini-sequence performed nominally. Radio Science Operations reported that the SNR at SPC-60 was lower than expected due to heavy rain, but all ten egress measurements were successfully acquired.

Spacecraft Health:
Spacecraft subsystems report good health and status.

Uplinks:
There have been 21 uplinks to the spacecraft during the past week, including new star catalog and ephemeris files, instrument command loads, the mz205 ROTO & Radio Science Egress Scan mini- sequence, the mz206 ROTO mini-sequence, and the mm214 & mm215 background sequences. 7,077 command files have been radiated to the spacecraft since launch.

Upcoming Events:
ROTOs are planned for next week in the mz206 & mz207 mini-sequences.

Wednesday 9-Oct-02

(DOY 275/19:00:00 to DOY 282/19:00:00 UTC) Launch / Days since Launch = Nov. 7, 1996 / 2163 days
Start of Mapping / Days since Start of Mapping = April 1, 1999 / 1288 days
Total Mapping Orbits = 16,036
Total Orbits = 17,719

Recent Events:
Background Sequences - The spacecraft is operating nominally in performing the beta- supplement daily recording and transmission of science data. The mm215 sequence executed successfully from 02-276 (10/03/02) through 02-278 (10/05/02). The mm216 sequence has performed well since it started on 02-279 (10/06/02). It terminates on 02-282 (10/09/02). The mm217 sequence, successfully uplinked on 02-281 (10/08/02), begins executing near the end of day on 02-282 10/09/02).

Other - Fifteen Roll Only Targeted Observations (ROTOs) were performed successfully by the mz206 & mz207 minisequences. MGS has completed 337 ROTOs to date.

Spacecraft Health:
Spacecraft subsystems report good health and status.

Uplinks:
There have been 13 uplinks to the spacecraft during the past week, including new star catalog and ephemeris files, instrument command loads, the mz207 & mz208 ROTO mini-sequences, and the mm216 & mm217 background sequences. 7,090 command files have been radiated to the spacecraft since launch.

Upcoming Events:
ROTOs are planned for next week in the mz208 & mz209 mini-sequences.

Wednesday 16-Oct-02

(DOY 289/19:00:00 to DOY 296/19:00:00 UTC) Launch / Days since Launch = Nov. 7, 1996 / 2177 days
Start of Mapping / Days since Start of Mapping = April 1, 1999 / 1302 days
Total Mapping Orbits = 16,206
Total Orbits = 17,889

Recent Events:
Background Sequences - The spacecraft is operating nominally in performing the beta- supplement daily recording and transmission of science data. The mm219 sequence executed successfully from 02-290 (10/17/02) through 02-292 (10/19/02). The mm220 sequence has performed well since it started on 02-293 (10/20/02). It terminates on 02-296 (10/23/02). The mm221 sequence, successfully uplinked on 02-295 (10/22/02), begins executing near the end of day on 02-296 10/23/02).

Other - Eight Roll Only Targeted Observations (ROTOs) were performed successfully by the mz210 & mz211 minisequences. MGS has completed 354 ROTOs to date.

Spacecraft Health:
Spacecraft subsystems report good health and status.

Uplinks:
There have been 17 uplinks to the spacecraft during the past week, including new star catalog and ephemeris files, instrument command loads, the mz211 & mz212 ROTO mini-sequences, and the mm220 & mm221 background sequences. 7,127 command files have been radiated to the spacecraft since launch.

Upcoming Events:
ROTOs are planned for next week in the mz211, mz212, & mz213 mini-sequences. Radio Science Occultation Egress Scans are scheduled for 02-302 (10/29/02) and 02-303 (10/30/02). The DDOR experiment slipped one day to optimize DSN coverage. It is now scheduled for 02- 305 (11/01/02).

Wednesday 30-Oct-02

(DOY 296/19:00:00 to DOY 303/19:00:00 UTC) Launch / Days since Launch = Nov. 7, 1996 / 2184 days
Start of Mapping / Days since Start of Mapping = April 1, 1999 / 1309 days
Total Mapping Orbits = 16,292
Total Orbits = 17,975

Recent Events:
Background Sequences - The spacecraft is operating nominally in performing the beta- supplement daily recording and transmission of science data. The mm221 sequence executed successfully from 02-297 (10/24/02) through 02-299 (10/26/02). The mm222 sequence has performed well since it started on 02-300 (10/27/02). It terminates on 02-303 (10/30/02). The mm223 sequence, successfully uplinked on 02-302 (10/29/02), begins executing near the end of day on 02-303 (10/30/02).

Other - Eleven Radio Science Occultation Egress Scans were performed on 02-302 (10/29/02) and 02-303 (10/30/02). Results are pending.

Twelve Roll Only Targeted Observations (ROTOs) were performed successfully by the mz211, mz212 & mz213 mini-sequences. MGS has completed 366 ROTOs to date.

Spacecraft Health:
Spacecraft subsystems report good health and performance.

Uplinks:
There have been 16 uplinks to the spacecraft during the past week, including new star catalog and ephemeris files, instrument command loads, the mz213 ROTO mini-sequence, and the mm222 & mm223 background sequences. 7,143 command files have been radiated to the spacecraft since launch.

Upcoming Events:
ROTOs are planned for next week in the mz213, mz214, & mz215 mini-sequences. A DDOR experiment will be conducted on 02- 305 (11/01/02).

Wednesday 13-Nov-02

(DOY 310/19:00:00 to DOY 317/19:00:00 UTC) Launch / Days since Launch = Nov. 7, 1996 / 2198 days
Start of Mapping / Days since Start of Mapping = April 1, 1999 / 1323 days
Total Mapping Orbits = 16,465
Total Orbits = 18,148

Recent Events:
Background Sequences - The spacecraft is operating nominally in performing the beta- supplement daily recording and transmission of science data. The mm225 sequence executed successfully from 02-311 (11/07/02) through 02-313 (11/09/02). The mm226 sequence has performed well since it started on 02-314 (11/10/02). It terminates on 02-317 (11/13/02). The mm227 sequence, successfully uplinked on 02-316 (11/12/02), begins executing near the end of day on 02-317 (11/13/02).

Twelve Roll Only Targeted Observations (ROTOs) were performed successfully by the mz216, and mz217 mini-sequences. MGS has completed 391 ROTOs to date.

Spacecraft Health:
Spacecraft subsystems report good health and performance.

Uplinks:
There have been 15 uplinks to the spacecraft during the past week, including new star catalog and ephemeris files, instrument command loads, the mz217, & mz218 ROTO mini-sequences, and the mm227 & mm225 background sequences. 7,179 command files have been radiated to the spacecraft since launch.

Other:

The MGS DSN schedule was altered this last week to accommodate Odyssey's safe mode recovery efforts. A few uplink periods were given up to Odyssey, though this did not prevent any MGS commands from being uplinked. No MGS downlink time was lost due to the schedule change.

Upcoming Events:
ROTOs are planned for next week in the mz218 & mz219 mini-sequences. A DDOR experiment will be conducted on 02-332 (11/28/02). A Nav acceptance test is scheduled for 02-333 (11/29/02). The next Radio Science Egress mini-sequence is schedule to be from 02-336 (12/02/02) to 02-337 (12/03/02).

Wednesday 27-Nov-02

(DOY 324/19:00:00 to DOY 331/19:00:00 UTC) Launch / Days since Launch = Nov. 7, 1996 / 2212 days
Start of Mapping / Days since Start of Mapping = April 1, 1999 / 1337 days
Total Mapping Orbits = 16,635
Total Orbits = 18,318

Recent Events:
Background Sequences - The spacecraft is operating nominally in performing the beta- supplement daily recording and transmission of science data. The mm229 sequence executed successfully from 02-325 (11/21/02) through 02-327 (11/23/02). The mm230 sequence has performed well since it started on 02-328 (11/24/02). It terminates on 02-331 (11/27/02). The mm231 sequence, successfully uplinked on 02-330 (11/26/02), begins executing near the end of day on 02-331 (11/27/02).

Other - Fifteen Roll Only Targeted Observations (ROTOs) were performed successfully by the mz219, mz220, and mz221 mini-sequences. MGS has completed 418 ROTOs to date.

Spacecraft Health:
Spacecraft subsystems report good health and performance.

Uplinks:
There have been 18 uplinks to the spacecraft during the past week, including new star catalog and ephemeris files, instrument command loads, the mz221, & mz222 ROTO mini- sequences, and the mm230 & mm231 background sequences. 7,214 command files have been radiated to the spacecraft since launch.

Upcoming Events:
ROTOs are planned for next week in the mz221, mz222, & mz223 mini-sequences. A DDOR experiment will be conducted on 02-332 (11/28/02). A Nav acceptance test is scheduled for 02-333 (11/29/02). The next series of Radio Science Occultation Egress measurements will occur from 02-336 (12/02/02) to 02-337 (12/03/02) as part of mz223.

Wednesday 5-Mar-03

(DOY 057/19:00:00 to DOY 064/19:00:00 UTC) Launch / Days since Launch = Nov. 7, 1996 / 2310 days
Start of Mapping / Days since Start of Mapping = April 1, 1999 / 1435 days
Total Mapping Orbits = 17,834
Total Orbits = 19,517

Recent Events:
Background Sequences - The spacecraft is operating nominally in performing the beta-supplement daily recording and transmission of science data. The mm257 sequence executed successfully from 03-058 (2/27/03) through 03-060 (3/01/03). The mm258 sequence has performed well since it started on 03-061 (3/02/03). It terminates on 03-064 (3/05/03). The mm259 sequence, successfully uplinked on 03-063 (3/04/03), begins executing near the end of day on 03-064 (3/05/03).

Other - Ten Roll Only Targeted Observations (ROTOs) were performed successfully by the mz249 and mz250 mini-sequences. MGS has completed 581 ROTOs to date.
Spacecraft Health:
Spacecraft subsystems report good health and performance. Electrical power margin is being managed by utilizing on-board solar array pointing scripts instead of auto-tracking. On 03-064 (3/05/03), the "three-part motion 25-degree offpoint"" scripts replaced the "three-part motion 0-degree offpoint"" scripts. This decreases the excess energy being produced by the solar panels thereby preserving life in the Partial Shunt Assemblies.

Uplinks:
There were 19 uplinks to the spacecraft during the past week. They included new star catalog and ephemeris files, nominal instrument command loads, three 25-degree offpoint solar array motion scripts, the mz250 & mz251 ROTO mini-sequences, and the mm258 & mm259 background sequences. 7,456 command files have been radiated to the spacecraft since launch.

Upcoming Events:
ROTOs are planned for the mz251 & mz252 mini-sequences. A Delta-DOR experiment is scheduled for 03-074 (3/15/03). Another Solar Array and HGA Flex Modes Test is scheduled for 03-076 (3/17/03).

MEDIA RELATIONS OFFICE
JET PROPULSION LABORATORY
CALIFORNIA INSTITUTE OF TECHNOLOGY
NASA PASADENA, CALIF. 91109. TELEPHONE (818) 354-5011 http://www.jpl.nasa.gov

Charli Schuler (818) 393-5467
Jet Propulsion Laboratory, Pasadena, Calif.

IMAGE ADVISORY #2003-090 June 24, 2003

NASA ORBITER EYES PHOBOS OVER MARS HORIZON

Images from the Mars Orbiter Camera aboard NASA's Mars Global Surveyor capture a faint yet distinct glimpse of the elusive Phobos, the larger and innermost of Mars' two moons. The moon, which usually rises in the west and moves rapidly across the sky to set in the east twice a day, is shown setting over Mars' afternoon horizon.

The images are available on the Internet at:

http://www.msss.com/mars_images/moc/2003/06/23/

Phobos is so close to the martian surface (less than 6,000 kilometers or 3,728 miles away), it only appears above the horizon at any instant from less than a third of the planet's surface. From the areas where it is visible, Phobos looks only half as large as Earth's full moon. Like our satellite, it always keeps the same side facing Mars. The tiny moon is also one of the darkest and mostly colorless (dark grey) objects in the solar system, so for the color image two exposures were needed to see it next to Mars. The faint orange-red hue seen in the wide-angle image is a combination of the light coming from Mars and the way the camera processes the image.

The bottom picture is a high-resolution image that shows Phobos' "trailing" hemisphere (the part facing opposite the direction of its orbit). At a range of 9,670 kilometers (6,009 miles), this image has a resolution of 35.9 meters (117.8 feet) per pixel. The image width (diagonal from lower left to upper right) is just over 24 kilometers (15 miles).

The Jet Propulsion Laboratory in Pasadena, Calif., manages Mars Global Surveyor for NASA's Office of Space Science in Washington, D.C. JPL is a division of the California Institute of Technology in Pasadena. JPL's industrial partner is Lockheed Martin Astronautics, Denver, which developed and operates the spacecraft. Malin Space Science Systems and the California Institute of Technology built the Mars Orbiter Camera, and Malin Space Science Systems operates the camera from its facilities in San Diego, Calif.

-end-

MEDIA RELATIONS OFFICE JET PROPULSION LABORATORY CALIFORNIA INSTITUTE OF TECHNOLOGY NASA PASADENA, CALIFORNIA

91109. TELEPHONE (818) 354-5011
http://www.jpl.nasa.gov

Charli Schuler (818) 393-5467 Jet Propulsion Laboratory, Pasadena, Calif.
James Hathaway (480) 965-6375 Arizona State University, Tempe, Ariz.
NEWS RELEASE: 2003-115 August 21, 2003

NEW FINDINGS COULD DASH HOPES FOR PAST OCEANS ON MARS

After a decades-long quest, scientists analyzing data from NASA's Mars Global Surveyor spacecraft have at last found critical evidence the spacecraft's infrared spectrometer instrument was built to search for: the presence of water-related carbonate minerals on the surface of Mars.

However, the discovery also potentially contradicts what scientists had hoped to prove: the past existence of large bodies of liquid water on Mars, such as oceans. How this discovery relates to the possibility of ephemeral lakes on Mars is not known at this time.

The thermal emission spectrometer on Global Surveyor found no detectable carbonate signature in surface materials at scales ranging from three to 10 kilometers (two to six miles) during its six-year Mars mapping mission. However, the sensitive instrument has detected the mineral's ubiquitous presence in martian dust in quantities between two and five percent. Planetary geologists Timothy Glotch Dr. Joshua Bandfield, and Dr. Philip Christensen of Arizona State University, Tempe, analyze the data from dust-covered areas of Mars in a report to be published Aug. 22 in the journal Science.

"We have finally found carbonate, but we've only found trace amounts in dust, not in the form of outcroppings as originally suspected. This shows that the thermal emission spectrometer can see carbonates — if they are there - and that carbonates can exist on the surface today," said Christensen, principal investigator for the instrument.

"We believe that the trace amounts that we see probably did not come from marine deposits derived from ancient martian oceans, but from the atmosphere interacting directly with dust," Christensen said. "Tiny amounts of water in Mars' atmosphere can interact with the ubiquitous dust to form the small amounts of carbonate that we see. This seems to be the result of a thin atmosphere interacting with dust, not oceans interacting with the big, thick atmosphere that many people have thought once existed there."

"What we don't see is massive regional concentrations of carbonates, like limestone," said Bandfield, who spent a year refining the techniques that allowed the group to separate carbonate's distinctive infrared signature from the spectrometer's extensive database of infrared spectra, despite the mineral's low concentrations and the masking effects of the martian atmosphere.

"We're not seeing the white cliffs of Dover or anything like that," he said. "We're not seeing high concentrations, we're just seeing ubiquitously low levels. Wherever we see the dust, we see the signature that is due to the carbonate."

Because there are known to be deposits of frozen water on Mars, the findings have important implications for Mars' past climate history.

"This really points to a cold, frozen, icy Mars that has always been that way, as opposed to a warm, humid, ocean-bearing Mars sometime in the past," said Christensen. "People have argued that early in Mars history, maybe the climate was warmer and oceans may have formed and produced extensive carbonate rock layers. If that was the case, the rocks formed in those purported oceans should be somewhere."

Although ancient carbonate rock deposits might have been buried by later layers of dust, Christensen pointed out that the global survey found no strong carbonate signatures anywhere on the planet, despite clear evidence of geological processes that have exposed ancient rocks.

Bandfield said that carbonate deposits in dust could be partially responsible for Mars' atmosphere growing even colder, to become as cold, thin and dry as it is today.

"If you store just a couple percent of carbonate in the upper crust, you can easily account for several times the Earth's atmospheric pressure," Bandfield said. "You can store a lot of carbon dioxide in a little bit of rock. If you form enough carbonates, pretty soon your atmosphere goes away. If that happens, you can no longer have liquid water on the surface because you get to the point where liquid water is not stable."

"The significance of these dramatic results may have to wait for the discoveries to be made by the Mars Exploration Rovers in 2004 and the Mars Reconnaissance Orbiter in 2006 and beyond," stated Dr. Jim Garvin, NASA's lead scientist for Mars exploration. What's important is that we have found carbon-bearing minerals at Mars, which may be linked to the history of liquid water and hence to our quest to understand whether Mars has ever been an abode for life."

The Mars Global Surveyor mission is managed for NASA's Office of Space Science, Washington, D.C. by the Jet Propulsion Laboratory, a division of the California Institute of Technology, Pasadena. Arizona State University built and operates the Thermal Emission Spectrometer on Mars Global Surveyor. Lockheed Martin Space Systems, Denver, developed and operates the spacecraft.
-end-

First MOC Public Requested Image: Caldera of Pavonis Mons

MGS MOC Release No. MOC2-481, 12 September 2003

NASA/JPL/Malin Space Science Systems

Mars Global Surveyor (MGS) first began to orbit the red planet six years ago today on 12 September 1997. More than 120,000 Mars Orbiter Camera (MOC) images have been obtained, with the high resolution camera covering about 3% of the planet. Recently, in August 2003, the MOC team began accepting public suggestions for areas on Mars to be imaged by the high resolution camera. The goal of the MOC Public Target Request effort is to cast a wide net to enhance the science return of the experiment.

On 4 September 2003, MGS MOC acquired its first images that were suggested through the public target program. Shown here are two pictures, acquired at the same time by the MOC. The first (left) is the public-requested high resolution image obtained by MOC's narrow angle camera. The second (right) is a context image taken by MOC's red wide angle camera. The white box in the context image indicates the location of the high resolution view. In both of these images, north is toward the top/upper right and sunlight illuminates them from the lower left.

The image pair shows details of the summit caldera of the martian volcano, Pavonis Mons. The caldera formed by collapse as molten rock withdrew deep within the volcano, some time in the past. The high resolution image shows that the caldera floor and walls are presently covered by a thick (perhaps a meter/yard or more) mantle of textured dust. Dark dots are boulders that are poking out from within this dust mantle in several areas on the lower caldera wall. This image partially overlaps a previous, lower-resolution view of the caldera, thus providing a close-up view at 1.5 meters (5 feet) per pixel (see E10-01691 or a smaller sub-frame in E10-01691sub for the lower-resolution image).

Pavonis Mons stands about 14 km (8.7 mi) above the martian datum (0 elevation), or roughly 6 km (3.7 mi) above its surrounding terrain. The high resolution image covers an area 1.5 km (0.9 mi) across by about 9 km (5.6 mi) long; the context frame is about 115 km (71 mi) across and down. The high resolution image is located near the equator at 0.4°N, 112.8°W.

Suggestions for MOC images of Mars are collected through the Mars Orbiter Camera Target Request Site. Each request is checked by the MOC science/operations team, and then placed in a database where the request waits until some time in the future when the spacecraft is predicted to fly over the suggested location. Because the high resolution camera field of view is so small (maximum is 3 km —1.9 mi— wide), any given request might wait for weeks, months, or even several years before it is overflown by the MGS spacecraft. Images received through this program will be placed online once per month at:
http://www.msss.com/mars_images/moc/publicresults/ — the next group of public images is anticipated to be posted in mid/late October 2003.

Beagle 2 Press Releases

Almost There

15-Dec-03 12:45 GMT

Mars Express is drawing nearer to Mars, so here we give you a brief overview of the planned activity coming up.

On December 19th, when Beagle 2 fires its spin up and eject mechanism to send it spinning on its way to Mars, we hope to bring you pictures of the tiny probe receding into the distance. Black and white and grainy maybe, but a historical moment captured in time and space.

Once separated from Mars Express, Beagle 2 becomes a spacecraft in its own right for the last 5 days of the journey. On December 20th, the mothercraft, on which Beagle 2 hitched a ride, will make a course correction to avoid a collision with Mars, thus preparing for Mars Orbit Insertion at around the same time as the lander touches down on the surface of the planet.

The spin-up and eject sequence is crucial for both the lander and orbiter. Failure to eject Beagle 2 would mean that the Mars Express would be saddled with extra mass and be unable to enter the desired science orbit, and clearly there would be no landing.

Onto Christmas Day - although the time of the landing is known, we will not be able to receive information until NASA's Odyssey orbits over the site providing a communications session. And even then, it is not certain that a signal can be received at the first attempt. Come what may there will be a press conference early on Christmas morning. The next chance to receive a message from Beagle 2 will be late on Christmas night when the giant Jodrell Bank telescope will listen out to see if Beagle 2 has switched on its transmitter.

Rest assured as soon as we know Beagle 2 is safe we will be telling everyone. So check in with early morning and late night news bulletins on Christmas day. The www.beagle2.com website will be updated as quickly as possible.

Dust storms on Mars

18-Dec-03 16:50 GMT

You may have heard there are currently dust storms on Mars and are wondering if there will be any problems for the Beagle 2 landing. We do not anticipate increased risk from dust during landing.

One storm is around the Valles Marineris area which is on the other side of the planet from Isidis Planitia (our landing site) and is therefore much less concern. A local storm is presently heading towards the Hellas Basin. But even this is several thousand kilometres from our predicted landing site.

A local dust storm can develop into a regional one or even a global one. But as global dust storms statistically happen during the summer in the southern hemisphere on Mars, and this is now coming to an end, we are hopeful that this is all a dust storm in a tea cup. We will continue to monitor events and try to work out if there are any implications for the lifetime of Beagle 2's solar arrays as the dust settles.

Separation Day Arrives for Mars Express and Beagle 2

19-Dec-03 09:30 GMT

After a joint journey of 250 million miles (400 million km), the British-built Beagle 2 spacecraft and the European Space Agency's Mars Express orbiter should now have parted and gone their separate ways.

At 8.31 GMT, software on Mars Express was scheduled to send the command for the Beagle 2 lander to separate from the orbiter. This would fire a pyrotechnic device that would slowly release a loaded spring and gently push Beagle 2 away from the mother spacecraft at around 1 ft/s (0.3 m/s). If all goes according to plan, the release mechanism will also cause Beagle 2 to rotate like a spinning top, stabilising its motion during the final stage of its flight towards Mars.

Since Beagle 2 does not have a propulsion system of its own, it must be carefully targeted at its destination. With Mars Express acting as a champion darts player aiming at a bullseye, Beagle 2 should be placed on a collision course with the planet, following a precise ballistic path that will enable it to hit a specific point at the top of the Martian atmosphere in six days' time.

Initial confirmation that the separation manoeuvre has been successful is expected at 10.40 GMT, when the European Space Operations Centre (ESOC) in Darmstadt, Germany, should receive X-band telemetry data from Mars Express. Further information from Mars Express and Beagle 2 telemetry confirming separation should be returned by 11.10 GMT. In addition, it is hoped that the orbiter's onboard Visual Monitoring Camera (VMC) will provide pictures showing the lander moving slowly away. The images are expected to be available within hours of the separation event.

However, after six months in space, during which the spacecraft were buffeted by solar storms, the manoeuvre is not without risk. Although it has been tested many times on Earth, there is always the outside possibility that something may go wrong during the all-important separation. Even if the separation is successful, Beagle 2 must rely on its own battery, which cannot last beyond 6 days, until its solar arrays are fully deployed on the surface. This means that Mars Express must release Beagle 2 at the last possible moment in order to ensure that the lander has enough power for the rest of its journey to the rust-red Martian plains.

This will be the first time that an orbiter has delivered a

lander without its own propulsion onto a planet and then attempted orbit insertion immediately afterwards. Meanwhile, Mars Express will follow Beagle 2 for a while until, three days before arrival at Mars, ground controllers will have to fire its thrusters and make it veer away to avoid crashing onto the planet.

Early on 25 December, Beagle 2 should plunge into the atmosphere before parachuting to its planned landing site, a broad basin close to the Martian equator, known as Isidis Planitia. Later that day, Mars Express should enter orbit around Mars.

Next Stop Mars!

19-Dec-03 11:28 GMT

We have separation! That was the message from the European Space Operations Centre (ESOC) in Darmstadt, Germany, to announce that the British-built Beagle 2 spacecraft is now flying independently from its Mars Express "mother ship". Initial confirmation that the separation manoeuvre has been successful came at 10.42 GMT, when Mars Express mission control at ESOC received telemetry data to indicate that electrical disconnection had taken place between Beagle 2 and the orbiter. This was followed at 11.12 GMT by confirmation that the two spacecraft had mechanically separated.

It is hoped that the orbiter's onboard Visual Monitoring Camera (VMC) has been able to capture images showing the slowly spinning Beagle 2 gradually pulling away from Mars Express. If all goes well, these images should be available early this afternoon.

"I'd like to congratulate everyone who has been a part of this project, particularly the team that built the Spin up and Eject Mechanism," said UK Science Minister, Lord Sainsbury. "This is an extraordinary example of the best of British engineering as well as the best of British science." Comparing it to a two-legged soccer or football match, both of which were being played away, Beagle 2 Lead Scientist Prof. Colin Pillinger said, "We've got a 1-0 result in the first leg, we're playing the second leg on Christmas Day."

The separation manoeuvre involved the use of a spring mechanism to give the lander a gentle push away from the orbiter. Now stabilised as it spins like a top at a rate of 14 rpm, Beagle 2 is pulling ahead of Mars Express at a rate of about 0.3 m/s (1 ft/s).

The separation marked the first key landmark at the beginning of a tense week for the Beagle 2 team. From now on, Beagle 2 will be on its own and looking after itself in terms of stability, power, thermal control and entry sequencing.

Following a carefully targeted ballistic trajectory, the 68.8 kg probe will remain switched off for most of the 5 million kilometre coast phase to Mars. Then, a few hours before entering the Martian atmosphere, an onboard timer will turn on the power and boot up Beagle's computer. Beagle 2 must rely on its own battery until its solar arrays are fully deployed on the surface.

Early on 25 December, Beagle 2 will plunge into the atmosphere at a speed of more than 20 000 km per hour (12,500 mph) before parachuting to its planned landing site, a broad basin close to the Martian equator, known as Isidis Planitia. Later that day, Mars Express should enter orbit around Mars.

The Beginning of Beagle 2's Lone Odyssey

19-Dec-03 15:03 GMT

Following the successful separation of the British-built Beagle 2 spacecraft and the European Space Agency's Mars Express orbiter earlier today, ESA has released the first images of the small lander as it begins its lone voyage to the surface of the Red Planet.

The images, taken with the Visual Monitoring Camera on board Mars Express, show Beagle 2 as a bright disk slowly drifting away from Mars Express. The spacecraft is rotating like a spinning top approximately once every 4 seconds. This motion has the effect of stabilising the spacecraft's motion during the remainder of its journey to Mars and its entry into the planet's atmosphere.

When the first picture was taken, at 8:33 GMT, Beagle 2 was about 20 metres (66 ft) away from the mother spacecraft and drifting away at a speed of about 0.3 m/s (1 ft/s).

The images show the top of the lander's cone-shaped outer shell. The heat shield that will protect the spacecraft during its headlong plunge into the planet's atmosphere on 25 December is out of sight on the far side of Beagle 2.

Farewell, Beagle 2. The bright spot on the left-hand side of the picture is the back side of Beagle 2, slowly drifting away from Mars Express. The image, taken this morning at 9:33 CET, shows the lander when it was about 20 metres away from the mother spacecraft, on its way to Mars.

Beagle 2 set for Christmas Day landing

23-Dec-03 23:31 GMT

Beagle 2 is currently travelling the last miles of its journey to the Red Planet alone and without any propulsion systems of its own. Thanks to the careful alignment of Mars Express on 19th December and accuracy of the spin-up and eject mechanism it is now believed that Beagle 2 is heading for a landing ellipse measuring 80km by 15km, much smaller than originally envisaged.

Beagle 2 was ejected successfully from Mars Express on

19th December and is presently on a collision course for Mars! It is expected to land on Christmas Day at 2.54am Greenwich Mean Time, with the aid of parachutes and airbags. The Beagle 2 team will be holding an all night vigil in London and will be eagerly awaiting a signal from the lander and news that it survived its dramatic descent.

Because Mars Express is due to be in the middle of manoeuvres to place itself in an orbit around Mars, the first attempt to contact Beagle 2 will be via NASA's Mars Odyssey orbiter (which has been circling Mars since 2001). Odyssey will pass over Beagle 2 at about 5.30am GMT, try to establish contact and hopefully beam back data to Earth at about 7am GMT.

The smaller landing ellipse than initially thought will help Odyssey position itself in the best place to pin-point Beagle 2, particularly useful as Odyssey will have about 10-20 minutes to try and pick up a signal from Beagle 2 on the surface.

More information about the first communication with Beagle 2 and answers to other frequently asked questions have been posted in the FAQ section (see the link on the right).

Beagle 2 - our thoughts are with you

25-Dec-03 02:54 GMT

At 2.47am GMT the Beagle 2 probe entered the martian atmosphere to begin the final descent to the surface, coming to rest at 2.54am.

But it will be several more hours before the waiting scientists and engineers will get the vital message confirming that all went to plan.

Beagle 2 will wait for the time that NASA's Odyssey passes overhead and send a message. This is then relayed back to Earth and the start of the message - the notes of the music composed by Blur - will tell us all is well.

Beagle starts to call home

25-Dec-03 05:50 GMT

Odyssey, the NASA orbiter, was due to pass over Beagle 2 at 5.30am GMT to try and make contact with Beagle 2.

Five minutes before Odyssey flies over the site where Beagle was scheduled to land, Beagle 2 starts to send out a message. Beagle 2 continues to call out for about 20 minutes. If we have been successful then Odyssey will be able to relay information from Beagle 2 back to Earth - but it will take some time to process any message - so we continue to wait hopefully

Beagle keeps scientists waiting

25-Dec-03 06:29 GMT

No telemetry from Beagle 2 was received during this morning's passage of NASA's Mars Odyssey over the Isidis Planitia landing site.

Professor Colin Pillinger, lead scientist for the Beagle 2 project, commented that this certainly does not mean that the probe had been damaged during its descent. There were a number of possible explanations, the most likely being that the Beagle 2 antenna was not pointing in the direction of Mars Odyssey.

The next opportunity to communicate with Beagle 2 will be late this evening (between 10pm and midnight GMT) when the Jodrell Bank Observatory will listen out for a signal from the lander.

Scientists Await First Call From Beagle

25-Dec-03 10:20 GMT

Early this morning, the Beagle 2 spacecraft landed on the surface of Mars at the end of a 250 million mile (400 million km), six-month trek to the Red Planet. Although the first attempt to use NASA's Mars Odyssey orbiter to communicate with the lander three hours later was unsuccessful, scientists and engineers are still awaiting the best Christmas present possible - the first faint signal to tell them that Beagle 2 has become only the fourth spacecraft to make a successful landing on Mars.

"This is a bit disappointing, but it's not the end of the world", said Professor Colin Pillinger, lead scientist for the Beagle 2 project. "We still have 14 contacts with Odyssey programmed into our computer and we also have the opportunity to communicate through Mars Express after 4 January."

The next window to receive confirmation that Beagle 2 has successfully landed and survived its first night on Mars will be between 10 pm and midnight (GMT) tonight, when its simple carrier signal (rather than the tune composed by Blur) may be picked up by Jodrell Bank radio observatory in Cheshire, UK. This has a much greater chance of success because the giant telescope is able to scan the entire side of the planet facing the Earth.

Another overflight by Mars Odyssey will take place around 18.15 GMT tomorrow evening, followed by daily opportunities to contact Beagle 2 via the Mars Odyssey spacecraft and the radio telescopes at Jodrell Bank and Stanford University in the United States.

There are several possible explanations for the failure of Odyssey to pick up Beagle 2's signal. Perhaps the most likely is that Beagle 2 landed off course, in an area where communication with Mars Odyssey was difficult, if not impossible. Another possibility is that the lander's antenna was not pointing in the direction of the orbiter during its brief passage over the landing site. If the onboard computer had suffered a glitch and reset Beagle 2's clock, the two spacecraft could be hailing each other at the wrong times.

The Beagle 2 lander entered the thin Martian atmosphere at 2.47 GMT today. Travelling at a speed of more than 12,500 mph (20,000 km per hour), the probe was protected from external temperatures that soared to 1,700 degrees C by a heat shield made of cork-like material.

As friction with the thin upper atmosphere slowed its descent, onboard accelerometers were used to monitor the spacecraft's progress. At an altitude of about 4.5 miles (7.1 km), Beagle's software was to order the firing of a mortar to deploy a pilot parachute, followed one minute later by deployment of the 33 ft (10 m) diameter main parachute and separation of the heat shield.

At a few hundred metres above the surface, a radar altimeter was to trigger the inflation of three gas-filled bags. Cocooned inside this protective cushion, Beagle 2 was expected to hit the rust-red surface at a speed of about 38 mph (60 km/h). As soon as the bags made contact with the surface, the main parachute was to be released so that the lander could bounce away unhindered. Like a giant beach ball, the gas bag assembly was expected to bounce along the surface for several minutes before coming to rest at 2.54am GMT.

Finally, a system of laces holding the three gas-bags onto the lander was to be cut, allowing them to roll away and drop Beagle 2 about 3 ft (1 m) onto the surface. The whole descent sequence from the top of the atmosphere to impact was to take less than seven minutes.

The "pocket watch" design of Beagle 2 ensured that it would turn upright irrespective of which way up the little lander fell. After the onboard computer sent commands to release the clamp band and open the lid, the way would be clear to deploy the four, petal-like solar panels and initiate charging of the batteries.

Confirmation of the successful landing would be provided by a musical "beeping" signal of 9 digitally encoded notes, composed by British rock group Blur. This signal should be picked up by Mars Odyssey as it passes overhead and then relayed to Earth.

The search for Beagle 2

25-Dec-03 14:55 GMT

Please find below a status report on the search for Beagle 2 and when you can expect to hear more news.

Media reports suggest that the team is conducting a continuous search for Beagle 2. This is not the case - there are only select communication windows when a search can be carried out. The next window occurs this evening when the Jodrell Bank telescope will seek out a signal from Beagle 2. The next chance to try again with Odyssey will be Boxing Day evening.

Scientists Wait For Beagle 2 To Call Home

26-Dec-03 11:15 GMT

The fate of Beagle 2 remains uncertain this morning after the giant radio telescope at Jodrell Bank in Cheshire, UK, failed in its first attempt to detect any signal from the spacecraft.

Scientists were hopeful that the 250 ft (76 m) Lovell Telescope, recently fitted with a highly sensitive receiver, would be able to pick up the outgoing call from the Mars lander between 19.00 GMT and midnight last night. An attempt to listen out for Beagle's call home by the Westerbork telescope array in the Netherlands was unfortunately interrupted by strong radio interference.

The next window of opportunity to communicate via Mars Odyssey will open at 17.53 GMT and close at 18.33 GMT this evening, when the orbiter is within range of the targeted landing site on Isidis Planitia.

Another communication session from Jodrell Bank is scheduled between 18.15 GMT and midnight tonight, when Mars will be visible to the radio telescope. It is also hoped that the Stanford University radio telescope in California will be able to listen for the carrier signal on 27 December.

The Beagle 2 team plans to continue using the Mars Odyssey spacecraft as a Beagle 2 communications relay for the next 10 days, after which the European Space Agency's Mars Express orbiter will become available.

Mars Express, which was always planned to be Beagle 2's main communication link with Earth, successfully entered orbit around the planet on 25 December and is currently being manoeuvred into its operational polar orbit.

Meanwhile, 13 more attempts to contact Mars Odyssey have been programmed into Beagle 2's computer. If there is still no contact established after that period, Beagle 2 is programmed to move into auto-transmission mode, when it will send a continuous on-off pulse signal throughout the Martian daylight hours.

The first window of opportunity to communicate with Beagle 2 took place at around 06.00 GMT yesterday, when NASA's Mars Odyssey spacecraft flew over the planned landing site. In the absence of a signal from the 33 kg lander, the mission team contacted Jodrell Bank to put their contingency plan into operation.

At present, Beagle 2 should be sending a pulsing on-off

signal once a minute (10 seconds on, 50 seconds off). Some 9 minutes later, this very slow "Morse Code" broadcast should reach Earth after a journey of some 98 million miles (157 million km).

Although the Beagle's transmitter power is only 5 watts, little more than that of a mobile phone, scientists are confident that the signal can be detected by the state-of-the-art receiver recently installed on the Lovell Telescope. However, a significant drop in signal strength would require rigorous analysis of the data before it could be unambiguously identified.

Although the ground-based radio telescopes will not be able to send any reply, the new information provided by detection of the transmission from Beagle 2 would enable the mission team to determine a provisional location for Beagle 2. This, in turn, would allow the communications antenna on Mars Odyssey to be directed more accurately towards Beagle 2 during the orbiter's subsequent overhead passes.

Still no signal from Odyssey

26-Dec-03 20:30 GMT

No signal from Beagle 2 was detected during this evening's pass by Mars Odyssey.

Jodrell Bank will be listening for a signal until midnight tonight. An update on the outcome of this session will be released as soon as any news is available.

The next opportunity to listen with Mars Odyssey will take place at about 06.15 GMT tomorrow.

Jodrell Bank doesn't find Beagle

27-Dec-03 00:25 GMT

Tonight's scan for a signal from Beagle 2 by the 250 ft (76 m) Lovell Telescope at Jodrell Bank Observatory in Cheshire, UK, was unsuccessful.

The next communication opportunity with the Mars Odyssey orbiter will take place at about 06.15 GMT this morning. The results of this session will be announced at a press briefing in the Beagle 2 Media Centre at 08.30 GMT.

During this press briefing, Professor Colin Pillinger, Beagle 2 lead scientist, will be joined by Professor David Southwood, director of science for the European Space Agency, and Professor Alan Wells from the University of Leicester Space Research Centre.

Beagle 2 Teams Continue Efforts To Communicate With The Lander

27-Dec-03 11:50 GMT

Scientists are still waiting to hear from the Beagle 2 lander on Mars. Two attempts to communicate with Beagle 2 during the last 24 hours - first with the 250 ft (76 m) Lovell Telescope at Jodrell Bank Observatory in Cheshire, UK, and then this morning with the Mars Odyssey orbiter - ended without receiving a signal. Despite this outcome, two teams at the Beagle 2 Lander Operations Control Centre in Leicester are continuing to study all possible options to establish communications with the spacecraft.

Further opportunities to scan for a signal from Beagle 2 will be undertaken over the coming days. Tonight the radio telescopes at Jodrell Bank and Stanford University in California will again listen for the carrier signal from Beagle 2, while the next Mars Odyssey pass will take place tomorrow evening at 18.57 GMT.

Meanwhile, scientists are eagerly awaiting the arrival of the European Space Agency's Mars Express spacecraft in its operational polar orbit on 4 January. Mars Express was always intended to be the prime communication relay for Beagle 2, and the lander team is hopeful that a link can be established at that time if it has not already been achieved with Mars Odyssey.

"We need to get Beagle 2 into a period when it can broadcast for a much longer period," said Professor Colin Pillinger, Beagle 2 lead scientist. "This will happen around the 4th January after the spacecraft has experienced a sufficient number of communication failures to switch to automatic transmission mode."

Both Professor Pillinger and Professor David Southwood, ESA director of science, agreed that the best chance to establish communication with Beagle 2 would now seem to be through Mars Express.

At present, Mars Express is far from the planet and preparing to fire its engines for a major trajectory change that will move it into a polar orbit around the planet. "We haven't yet played all our cards," said Professor Southwood. "With Mars Express we will be using a system that we have fully tested and understand."

"At the moment, I am frustrated rather than concerned," he added.

One possible cause of the communication failure is that the clock on Beagle 2 may have been reset as the result of a computer glitch. An attempt was made to reset the clock during this morning's Odyssey pass, the first to take place during daylight hours at the Beagle 2 landing site. The outcome of this "blind command" is still awaited.

Meanwhile, specialists at the Lander Operations Control Centre continue to investigate other potential reasons for the failure of Beagle 2 to call home, including a possible landing off course, a tilting of the spacecraft and a problem in fully opening the solar arrays which could result in a blockage of the weak signal from Beagle's antenna.

Future opportunities to communicate with Beagle 2 are listed on the Beagle 2 Web site. The results of these sessions will be announced on the Beagle 2 and PPARC Web sites as soon as they are available.

The next press briefing will be held in the Beagle 2 Media Centre at 08.30 GMT on Monday, 29 December, when Lord Sainsbury, Minister for Science and Innovation, will be a principal speaker.

Alas, no signal received by Jodrell Bank

28-Dec-03 00:20 GMT

Tonight's scan for a signal from Beagle 2 by the 250 ft (76 m) Lovell Telescope at Jodrell Bank Observatory in Cheshire, UK, was unsuccessful.

The next communication opportunity with the Mars Odyssey orbiter will take place at about 18.57 GMT this evening. The results of this session will be announced as soon as possible on the Beagle 2 and PPARC Web sites.

The next press briefing will be held in the Beagle 2 Media Centre at 08.30 GMT on Monday, 29 December, when Lord Sainsbury, Minister for Science and Innovation, will be a principal speaker.

Communication Strategy of the Beagle 2 "Think Tank"

28-Dec-03 15:45 GMT

As part of the Media update on 27th December, Professor Alan Wells (Lander Operations Control Centre, University of Leicester) outlined details of the work being undertaken by the Beagle 2 team to assess the current situation. A specialist team, titled the "Analysis and Recovery Think Tank", has been established to concentrate on understanding the reasons for Beagle 2's apparent failure to make contact with Earth, and to address the steps that may be taken to resolve these problems.

During the media briefing Professor Wells took time to explain the precise nature of the earlier attempts to make contact with the spacecraft, and detailed the remaining opportunities as follows:

A series of 15 scheduled communication sequences were programmed into Beagle 2's software prior to its separation from Mars Express on Friday 19th December. Routine communication with Mars Express before the separation confirmed that these commands had been successfully uploaded to the lander.

After touchdown the planned communication sessions should have automatically been triggered, by the onboard 'clock', to correspond with the known passing of the orbiting spacecraft Mars Odyssey (NASA), and later on in the mission, Mars Express (the ESA 'mother ship'). A number of the pre-programmed sessions were also scheduled to correspond with times when the landing site is in 'sight' of the Jodrell Bank Radio Telescope. Any signal picked up by Jodrell, or an alternative telescope, will only serve to confirm the well-being of Beagle 2 - it does not offer an opportunity for two-way communication; this can only be achieved via Mars Odyssey or Mars Express.

If the lander software is running as planned, and according to the initial timing dictated by the onboard clock, a number of the programmed communication sessions have already passed.

Sessions 1 and 3 (between 04:54 and 06:14 on Christmas Day morning and between 17:33 and 18:53 on Boxing Day evening respectively) both occurred during the martian night, and were intended to open communication with Mars Odyssey. In the planned communication sessions with an orbiting spacecraft, Beagle switches into 'listening' mode for 80 minutes. During the pass over the landing site Mars Odyssey will send out a series of 'hails' which, if picked up by Beagle, will enable the lander's receiver to lock onto the signal from the orbiter and activates the lander's transmitter and communications can proceed (any reply from Beagle would be 'headed' up with the call sign composed by Blur).

No response was received by Mars Odyssey to indicate that the 'hail' messages had reached Beagle 2 during either of the above sessions.

When a planned communication session is intended for reception by the Jodrell Bank radio telescope, Beagle 2 is configured to send a repeating unmodulated signal, transmitting for 10 seconds then remaining inactive for 50 seconds, for a duration of 80 minutes.

Sessions 2, 4 and 6 (between 22:20 and 23:40 on Christmas Day evening, between 23:00 and 24:20 on Boxing Day night and between 22:56 and 00:16 27th December respectively) coincided with opportunities to view the surface of Mars with the Jodrell Bank radio telescope. Unfortunately no signal was detected by the giant telescope.

Session 5 (between 6.17 and 7.37 on the morning of 27th December) occurred during the martian day, and Beagle 2 should have again been in its 'listening' mode at a time when Mars Odyssey was passing over the landing site. Sessions occurring in the martian daytime differ from those occurring during the night as Beagle would automatically send any stored data when sunlight is available to charge the battery. Beagle is programmed to 'concentrate' on receiving data at night time to avoid draining the battery with power intensive data transmission.

However, again the telemetry returned by Mars Odyssey

to earth did not contain any evidence of communication between the orbiter and lander.

The remaining scheduled communications sessions are as follows:

Sessions coinciding with Mars Odyssey pass:
Session 7 - 28/12/03 (18:57 - 20:17 GMT)
Session 8 - 29/12/03 (07:41 - 09:01 GMT)
Session 9 - 30/12/03 (07:24 - 08:44 GMT)
Session 10 - 30/12/03 (20:20 - 21:40 GMT)
Session 11 - 31/12/03 (09:04 - 10:24 GMT)

Please note that any signal from Beagle 2 (either directly to the Jodrell Bank radio telescope or via an orbiting spacecraft) will take at least 9 minutes to reach earth, owing to the vast distance that it must travel.

One possible explanation that has been raised for the apparent silence is the potential for incompatibility between the systems on board Beagle and those used by Mars Odyssey. The Beagle team has been in constant contact with the Jet Propulsion Lab, in Pasadena, and fellow scientists there are presently checking through data and records in order to ascertain whether there may be any problems with the transmitters and receivers aboard Odyssey. It is important to note that neither of the communication routes attempted so far has ever been tested, therefore it is possible that the best opportunity for successful communication may arise when Mars Express achieves its final orbit and can take part in the search for Beagle.

A backup has been built into the communication schedule such that if 10 scheduled sessions pass unsuccessfully then Beagle 2 will switch to an emergency mode 'search mode 1'. Planned Jodrell contact sessions are included in the count because Beagle can have no way of knowing whether these 'one-way' sessions have been successful. When the lander switches to search mode 1 it will proceed to attempt a communication with the best daytime and best nighttime orbiter pass each day - these times are calculated according to an onboard model of the orbits of both Mars Odyssey and Mars Express.

Session 10 is scheduled for the evening of 30th December, and if a regular contact has not been made by this time 'search mode 1' will be activated, increasing the opportunity for communication with a 'passing' orbiter or for detection of Beagle 2 by Radio telescope from earth.

If a further 10 communication sessions are unsuccessful, Beagle will then switch to 'search mode 2'. The second emergency mode involves the production of a signal throughout the martian day (power is still conserved during the night). With two 'search mode 1' sessions taking place each day, the adoption of search mode 2 would, in theory, begin on January 5th - soon after the date when Mars Express is first available for communication.

A further explanation for the lack of contact between Beagle and the earth is that the onboard clock may have been corrupted during the entry, descent, and landing stage of the mission. If this is indeed the case, the above scheduled communication sessions may not have corresponded accurately with the passage of Mars Odyssey over the landing site or with the viewing windows afforded to the Jodrell Bank telescope. It is possible that Beagle 2 is signalling correctly but not at a time when Mars Odyssey is passing or when Jodrell Bank can 'see' Mars. Consequently, when Beagle assumes 'search mode 1' and begins signaling more regularly it may be possible for Odyssey or Jodrell to pick up the 'additional' transmissions.

In addition, a further radio telescope at the University of Stanford, California will begin the hunt for a signal from Beagle 2 the evening of 27th December. The telescope at Stamford has a viewing window that is one hour longer than that afforded to Jodrell Bank - therefore the opportunity for a signal to be picked up, if it is being transmitted at an unexpected time, will also be increased. The Stanford telescope has previously been used to monitor faint sources of radiation in deep space. Consequently it is thought that there may be some potential for it picking up other indicators of activity on Beagle besides the expected transmitter signal. The onboard processors will produce low levels of radiation that may be weakly 'visible' to the telescope - this might be considered comparable to looking for signs of Beagle's 'heartbeat', rather than listening for its 'bark'! It was, however, noted that these systems on Beagle have been shielded to protect them from sources of external radiation hence any signal from this internal source may be extremely small. Radio telescope scientists have special ways of distinguishing small signals from other background 'noise'.

Members of the Beagle 2 team are also exploring the possibility of 'recruiting' other radio telescopes to hunt for the signal from Mars. A telescope on the other side of the Earth would allow the search to continue at different times. It is understood that staff at the Parkes telescope (Sydney, Australia) are investigating whether they have appropriate detection equipment to look for Beagle 2.

Whilst it may not be possible to establish two way communications with Beagle 2 via a radio telescope it will aid the team to pinpoint the lander's location, and provide a time reference point. If the onboard clock has been corrupted it is likely to affect the absolute timing of the communication windows rather than the relative timing.

In order to address the potential problems associated with an incorrect clock setting on Beagle 2, a 'blind command' was transmitted by Mars Odyssey during the pass over the landing site yesterday morning (27/12/03 -

session 5). The hope is that Beagle may be able to receive such signals, but is not currently able to transmit. The effect of this command would be to reset the onboard clock with the aim of resynchronising the process and prompting an opportunity for successful communication.

'Blind commands' can also be used by the 'Lander Operations Control Centre' (LOCC) to control other processes onboard Beagle 2, without the requirement for two-way communication. Commands to control the motors operating the central hinge, or the hinges controlling the solar arrays may be sent via Mars Odyssey or Mars Express to affect some minor repositioning of the lander - this approach may be used to correct any potential problems that might have occurred with the opening of the arrays, or if the lander is inappropriately positioned, for example, leaning at a restrictive angle against a rock.

Such a strategy is not without considerable risk because any command may impact other processes taking place on the lander. It is also very difficult to command a spacecraft without any information about its current status - imagine trying to drive a radio-controlled car around a track without being able to see or hear the movement, or even know what the track looks like! The LOCC team will be able to make a risk assessment by testing the implication of any proposed activity on a working replica of the Beagle lander in their control centre at the University of Leicester.

Finally Professor Wells reassured the gathered media that the team were still in good spirits, and dedicated to the challenge ahead - and promised that, 'We'll keep going until every possibility has been exhausted'.

Jodrell's last opportunity

29-Dec-03 00:20 GMT

The 250 ft (76 m) Lovell Telescope at Jodrell Bank Observatory in Cheshire, UK, was turned towards Mars between 19.30 GMT and midnight on 28th December, but no response was received from the Beagle 2 lander.

This will be Jodrell Bank's last opportunity for some time to listen for a signal from Beagle 2. The longer day on Mars means that the planet has rotated so that the Beagle 2 landing site on Isidis Planitia is no longer above the horizon at the observatory when the spacecraft should be transmitting its pulsing "Morse Code" call.

The Stanford University radio telescope in California also attempted to search for Beagle's signal on the night of 27-28th December, but no data were received.

Earlier on the evening of 28th December, no signal from Beagle 2 was received by the Mars Odyssey orbiter during its pass over the landing site. The next communication opportunity with Mars Odyssey will take place at 07.41 GMT this morning.

Other opportunities to communicate with Beagle 2, including sessions with Mars Express, are listed on the Beagle 2 Web site.

Update from the press briefing

29-Dec-03 10:00 GMT

Speakers at this morning's press briefing were Professor Colin Pillinger, Dr Mark Sims and Lord Sainsbury, Minister for Science and Innovation. A brief description of the session is given here, with more information to follow shortly.

Dr Sims spoke about the work of the LOCC recovery team, highlighting potential failure modes which could be recovered from. A run down of the forthcoming programmed communication contacts with Mars Odyssey and Mars Express was also given.

Further scenarios of what could have happened to Beagle 2 were also discussed, including the remote probability of Beagle 2 having landed in a crater, which has been recently discovered within the landing ellipse. However, the chance of Beagle 2 landing in this exact spot is thought to be very low and this is just one of the areas that is being currently investigated.

Mars Exploration and the Search for Life is a Priority Says UK Science Minister.

29-Dec-03 12:00 GMT

The latest attempts to communicate with Beagle 2 via the Lovell Telescope at Jodrell Bank and the Mars Odyssey spacecraft have been unsuccessful. However, the Beagle 2 team has not given up hope and continues to be optimistic that efforts to contact the lander will eventually be successful.

This message was also reinforced by Lord Sainsbury, UK Minister for Science and Innovation, who this morning joined members of the Beagle 2 team to answer questions about the status of the project.

"While we're disappointed that things have not gone according to plan, we are determined that the search should go on, both the search to make contact with Beagle 2 and also (the search) to answer the long term question about whether there is life on Mars," said Lord Sainsbury.

"There's clearly still a good opportunity to make contact with Beagle 2 with Mars Express when it comes into action, and that has to be the first priority at this point. I think everything is being done by the 'tiger team' in Leicester to make contact with Beagle 2 and I want to wish them every success in their efforts."

"We are looking at a number of possible failure modes

that we might do something about," said Dr. Mark Sims, Beagle 2 mission manager from the University of Leicester.

"We are working under the assumption that Beagle 2 is on the surface of Mars and for some reason cannot communicate to us. In particular, we're looking at two major issues. One is communications, and there are also related timing and software issues.

"We've got a few more Odyssey contacts, the last one being on 31 December. Then we have four contacts with Mars Express already pre-programmed into Beagle, assuming the software is running, on 6, 12, 13 and 17. The 6 and 12 are when Mars Express is manoeuvring into its final orbit, so they are not optimum for Beagle 2 communications. The 13th and 17th are very good opportunities for Mars Express."

According to Dr. Sims, one of the scenarios the team was investigating - a timer and hardware reset - now seems unlikely, and can probably be ruled out. However, other possible slips of the onboard time may have been caused by software or problems of copying data between various parts of memory. Possibly, all of the stored command times have been lost.

"None of these can yet be eliminated," he said.

After the tenth contact attempt, Beagle 2 will move into communication search mode 1 (CSM 1), taking advantage of the ability of the software on board Beagle 2 to recognise when dawn and dusk occur on Mars by measuring the current feeding from the solar arrays.

"When we get into CSM 1 mode, Beagle 2 will start putting additional contacts on its time line, independent of the clock value," said Mark Sims. "This will happen after 31 December."

The team is also looking at sending blind commands to Beagle 2. This is helped by Beagle going into CSM 1 mode.

"The team has come up with a method of fooling the receiver into accepting commands without having to talk back to the orbiter," said Dr. Sims. "We have an agreement with JPL to reconfigure Odyssey to provisionally attempt this on 31 December, the last programmed Odyssey pass."

Malin Space Science Systems has also provided the Beagle 2 team with a picture of the landing site taken by the camera on Mars Global Surveyor 20 minutes after the spacecraft's scheduled touchdown. It shows that the weather was quite good on the day Beagle landed, so it was unlikely to be a factor in the descent. The next opportunity to image the landing site with Mars Global Surveyor will not be until 5 January.

The image showing the centre of Beagle 2's landing ellipse also shows a 1 km wide crater. There is just an outside possibility that the lander could have touched down inside this crater, resulting in problems caused by steep slopes, large number of rocks or disruption to communication from the lander. This image is now available on the Beagle 2 and PPARC Web sites (see link on the right hand side).

While the Lander Operations Control Centre in Leicester continues its efforts to communicate with the Beagle 2, Lord Sainsbury took the opportunity to inform the media that the UK government is keen to continue the innovative robotic exploration effort begun with the lander.

"Long term we need to be working with ESA to ensure that in some form there is a Beagle 3 which takes forwards this technology," he said. "I very much hope that the Aurora programme, which is now being developed by ESA, will take forward this kind of robotic exploration.

"We've always recognised that Beagle 2 was a high risk project, and we must avoid the temptation in future to only do low risk projects.

"I'd like to use this opportunity to add my thanks to all those helping our efforts to make contact with Beagle 2. I think the amount of international collaboration one gets on these occasions is very, very impressive and very encouraging to the team."

"We should not ignore the importance of Mars Express, which has three British-designed instruments on board and which looks set for success," he added.

"Finally, can I use this opportunity to wish the Americans every success with its two Mars Exploration Rovers, Spirit and Opportunity."

Update on today's attempt to contact Beagle 2 via Odyssey

30-Dec-03 10:15 GMT

The sixth attempt by NASA's Mars Odyssey orbiter to communicate with Beagle 2 was made this morning, but, as on previous occasions, no data were received.

The next Mars Odyssey communication opportunity will take place at 20.20 GMT this evening. The results of this session will be announced on the Beagle 2 and PPARC websites.

Other opportunities to communicate with Beagle 2, including pre-programmed sessions with Mars Express, are listed on the Beagle 2 website.

The next Beagle 2 press briefing is scheduled to take place at the Media Centre in Camden on Sunday 4th January. Details will be confirmed on the websites at a later date.

Again, no signal via Odyssey
30-Dec-03 23:15 GMT

The seventh attempt by NASA's Mars Odyssey orbiter to communicate with Beagle 2 was made this evening, but no data were received.

The next Mars Odyssey communication opportunity will take place at 09.04 GMT tomorrow. The results of this session will be announced on the Beagle 2 and PPARC websites as soon as they are available - probably after 14.00 GMT.

Other opportunities to communicate with Beagle 2, including pre-programmed sessions with Mars Express, are listed on the Beagle 2 website.

Meanwhile, earlier today, ESA's Mars Express orbiter was successfully inserted into a polar orbit around the Red Planet. This manoeuvre means that Mars Express will be ideally placed to communicate with Beagle 2 when it passes over the landing site in Isidis Planitia in a few days' time.

The next Beagle 2 press briefing is scheduled to take place at the Media Centre in Camden on Sunday 4th January. Details will be confirmed on the websites at a later date.

Update from this morning's attempt by Odyssey to detect Beagle 2
31-Dec-03 16:05 GMT

The latest attempt to communicate with Beagle 2 by NASA's Mars Odyssey orbiter was unsuccessful.

A window of opportunity for contacting Beagle 2 was open this morning and whilst Odyssey was trying to establish contact with the lander, the Deep Space Network of telescopes back on Earth were busy involved in communication links for NASA's Mars Exploration Rover and Stardust projects, hence the delay in receiving any information back about today's attempt to contact Beagle 2.

Scientists took the opportunity this morning to upload another command to Beagle 2 to try to reset its internal clock. This time however the instructions were embedded in the "hail" command. This is designed to initiate the contact sequence with Beagle 2 and doesn't require a response from the lander to confirm the data have been received. No initial result was achieved during this pass of Odyssey but it is hoped that the command may bring success in a future communication slot.

The next opportunity for retrieval of a signal from Beagle 2 will be with Mars Express in the New Year, the date and time of which are to be confirmed.

The next Beagle 2 press briefing is scheduled to take place at the Media Centre in Camden on Sunday 4th January. Details will be confirmed on the websites at a later date.

Notice of press briefing on 4th January
02-Jan-04 18:55 GMT

The next press briefing will be held in the Beagle 2 Media Centre at 08.30 GMT on Sunday 4th January where the speakers will be Professor Colin Pillinger and Dr Mark Sims.

They will provide an update about the work of the Beagle 2 team over the last few days and speak about the opportunities that lie ahead next week with Mars Express moving into an appropriate orbit to attempt communication with Beagle 2. In addition they will provide comment on the landing of "Spirit", the first of NASA's Mars Exploration Rovers, due to land early on 4th January.

Congratulations to NASA. Beagle 2 Team Still Hopes To Repeat Mars Landing Success.
04-Jan-04 11:23 GMT

At a press briefing in London today, Professor Colin Pillinger (Open University), Beagle lead scientist, and Dr Mark Sims (University of Leicester), the mission manager, congratulated their colleagues at NASA's Jet Propulsion Laboratory on the successful landing of the Spirit rover on Mars.

"I'd like to give congratulations to NASA and the Spirit team for getting the lander down safely," said Professor Pillinger. "We wish them every luck."

Adding his congratulations, Mark Sims said, "I'd like to reiterate the international cooperation we've been getting in terms of looking for Beagle. In particular, the JPL team which has been working very strange hours supporting the Odyssey passes, Lockheed Martin, who've been running the Odyssey spacecraft, Jodrell Bank, Westerbork, the British Astronomical Association and Malin Space Science Systems. Mike Malin is looking at imaging the landing site potentially from tomorrow."

Meanwhile, the search for Beagle 2 goes on.

"We haven't in any shape or form given up on Beagle 2," said Professor Pillinger.

"We have realised that Mars Express is not in the orbit we originally expected, so our communication strategy is now different from the one that we explained at the beginning of last week."

Describing the ongoing work at the Lander Operations Control Centre, Mark Sims explained that teams from the University of Leicester, SciSys and Astrium are continuing their efforts to identify possible failure modes that can be addressed.

"We're still concentrating on both the communications and timing/software issues, and working our way through the logic and fault tree on the basis that Beagle 2 is on the surface of Mars and for some reason is failing to talk to us," said Dr. Sims.

"There are six or seven scenarios that we're still working through and we still can't eliminate any of those."

However, possible failure scenarios involving a reset of the clock hardware and a problem with a tilted antenna seem to have been ruled out. Today's successful transmission of signals from the Spirit rover via Mars Odyssey also indicates that the radio on board NASA's orbiter is working properly.

Meanwhile, an attempt to send blind commands to Beagle 2 via Mars Odyssey on 31st December also resulted in no obvious response from the lander.

There has also been no response from the Beagle 2 transceiver during 11 programmed passes. Unfortunately, the last four contact opportunities pre-programmed into Beagle 2's computer no longer coincide with Mars Express on its current orbit, so the team is now relying on the spacecraft switching to various back-up communication modes.

The mission team is now waiting for their little lander to switch to one of its backup communication modes. Beagle 2 could already be operating in 'communication search mode 1', during which it listens for 80 minutes during both the Martian day and night in an effort to establish contact with an available orbiter at Mars Odyssey overflight times.

If no link is established by this method, 'communication search mode 2' should eventually be activated. The earliest date by which this mode could become operational was 3rd January. In this mode, the receiver is on for 59 minutes out of every hour throughout the Martian day, and the spacecraft sends a carrier signal five times in each daylight hour. During the Martian night, Beagle 2's receiver will be on for one minute out of every five, but there is no carrier signal.

Although Mars Odyssey will continue to search for the lander, Mars Express will soon become the prime communication link with Beagle 2. After reaching its operational polar orbit today, ESA's orbiter should pass over the Beagle 2 landing site regularly from 7th January onwards. Various modes of communication can be attempted during passes by Mars Express, although the team anticipates starting on 7th and 8th January with the standard 'hail and command' which has been used with Mars Odyssey.

The first four passes with Mars Express (7th, 8th, 9th and 10th January) are almost directly over the landing site and only 5 to 8 minutes long, so they are not ideal for communication, whereas the opportunities on 12th and 14th January are potentially much longer.

Advance notice of Mars Express attempt to communicate with Beagle 2

05-Jan-04 17:07 GMT

The first attempt to communicate with Beagle 2 via Mars Express will be on 7th January 2004

The orbiter is scheduled to fly over the Beagle 2 site at 12.13 GMT. The result of this communication opportunity will be announced live during the media briefing by Professor Colin Pillinger and news will follow on the Beagle2.com website afterwards.

Advance notice of Mars Express attempt to communicate with Beagle 2

05-Jan-04 17:07 GMT

The first attempt to communicate with Beagle 2 via Mars Express will be on 7th January 2004.

The orbiter is scheduled to fly over the Beagle 2 site at 12.13 GMT. The result of this communication opportunity will be announced live during the media briefing by Professor Colin Pillinger and news will follow on the Beagle2.com website afterwards.

Expected time when data wil be received from Mars Express

06-Jan-04 17:42 GMT

Although the Mars Express orbiter will fly over the Beagle 2 landing site at 12.15 GMT on 7th January 2004, we have been advised by ESA that the resulting data will take some time to process and the outcome may not be known until 15.00 GMT.

In view of this, the press briefing in London and corresponding webcast will start at 14.30 (not at 11.30 as previously advertised).

Beagle 2 Fails To Call Mars Express

07-Jan-04 17:10 GMT

Today's first real opportunity for the European Space Agency's Mars Express orbiter to hear a signal from the Beagle 2 lander passed in silence.

Hopes were high that Beagle 2 would receive and respond to commands sent by Mars Express as it flew over the presumed landing site at around 12.15 GMT. Not only was Mars Express flying over Isidis Planitia at an altitude of just 220 miles (350 km), giving it an ideal listening position, but it was the first time that the primary communication link with the orbiter had been used during the Beagle 2 mission.

Speaking from the European Space Operations Centre in Darmstadt, Germany, the ESA Science Director, Professor David Southwood, said," I have, I'm afraid, to make a sad announcement, that today, when we were in conditions we thought were very good for getting direct communication between Mars Express - the 'mother ship' - and Beagle 2 - the 'baby' - we did not get any content of a signal, nor indeed a signal from the surface of Mars.

"This is not the end of the story. We have more shots to play ... but I have to say this is a setback."

"There are opportunities to contact Beagle still to come, though we've established today that it is certainly not in a particular communications mode that we had expected it to be in."

Professor Colin Pillinger, Beagle 2 lead scientist, expressed his thanks to everyone at ESOC for the efforts they had put in over the last few days.

"I think all I can say to the whole team at this stage is 'play to the final whistle'. It only takes a fraction of a second to score a goal, and that's the way we will have to look at this and not give up at this time, although it is the moment when we have to start looking at the future as well."

Efforts to contact Beagle 2 and to pin down its position on the Martian surface will continue in the weeks to come.

"We have another opportunity to look tomorrow in a more sensitive mode, the canister mode on Mars Express, which is the most sensitive mode Mars Express has for detecting an RF signal," said Dr Mark Sims, Beagle 2 mission manager.

"We have two Odyssey sessions tonight, when we will be attempting to command Beagle 2 in order to have a maximum chance of seeing data with the canister mode tomorrow. Both of those Odyssey sessions coincide with CSM I mode, both am and pm, which will be another opportunity to rule those scenarios out."

The most favourable opportunity will be on 12 January, the last Mars Express overpass that was pre-programmed into the lander before its separation from the orbiter on 19 December. However, this window will only be available if nothing has happened to reset or alter the lander's timeline.

"If we see nothing ... we're left with the scenario of Beagle 2 potentially operating but not being able to receive a signal, in which case we will have to wait till the last back-up mode in Beagle 2 becomes active, which is autotransmit," said Dr. Sims. "The latest date that will become active is 2 February."

"My personal view is that, if we have not received a signal within 5 to 10 days of that event, then we have to assume Beagle is lost."

Mars Express searching for Beagle 2

08-Jan-04 17:52 GMT

Today's pass over the Beagle 2 landing site by ESA's Mars Express orbiter occurred at about 12:50 GMT.

Mars Express was searching in 'super-sensitive' mode, which means it was gathering a large amount of data. The ground processing of this data will take several hours. The result, which is unlikely to be known until early tomorrow morning, will be posted on the Beagle 2 website.

No signal from the latest Mars Express attempt

09-Jan-04 13:34 GMT

Yesterday's pass of Mars Express over the Beagle 2 landing site failed to retrieve a signal from the lander despite Mars Express searching in a 'super-sensitive' mode.

Results have just been made available by ESA, after extensive processing of the large amounts of data received from the orbiter's canister mode search.

There are still opportunities to make contact with Beagle in the days to come. The most favourable opportunity will be on 12th January, the last Mars Express overpass that was pre-programmed into the lander before its separation from the orbiter on 19th December. However, this window will only be available if nothing has happened to reset or alter the lander's timeline.

No signal from Mars Express today

09-Jan-04 18:56 GMT

No signal from Beagle 2 was detected during today's pass by ESA's Mars Express orbiter, which flew over the landing site around 13:27 GMT.

The next opportunities for Mars Express to detect a signal from Beagle 2 are as follows: 10th January, around 14:04 GMT and 12th January, around 02:02 GMT

Again, no signal from Mars Express

10-Jan-04 19:35 GMT

No signal from Beagle 2 was detected during today's pass by ESA's Mars Express orbiter, which flew over the landing site around 14:04 GMT.

The next opportunity for Mars Express to detect a signal from Beagle 2 is: 12th January, around 02:02 GMT.

The results of future communication opportunities will be posted on the Beagle 2 and PPARC websites.

Ground control initiates radio silence to tempt Beagle 2 from its hiding place

12-Jan-04 10:33 GMT

No signal was received from Beagle 2 this morning when ESA's Mars Express orbiter passed over the landing site around 02:02 GMT.

Prof. Colin Pillinger, Beagle 2 Lead Scientist, was present at ESOC when the data came through and although the news was disappointing, Prof. Pillinger was encouraged by the continued support and determination of the team at ESA's mission control centre to continue the search.

The next phase will be to initiate a period of radio silence where no communication attempts will be made with Beagle 2 until the 22nd January. Adopting this approach will force Beagle 2 into communication search mode 2 [CSM2] where the probe will automatically transmit a signal throughout the Martian day [power is still conserved during the night].

The results from future communication attempts will be posted on the Beagle 2 and PPARC websites.

Next attempts to contact Beagle 2 planned for 24th and 25th January

20-Jan-04 17:05 GMT

The Beagle 2 team has revised its plans for trying to communicate with the lander, postponing the date for the end of radio silence by two days.

The Team has made the following announcement regarding the strategy for communication with Beagle 2 over the next 5 days:

"On 12 January we started a period when no attempts were made to contact Beagle 2. Maintaining radio silence for a period of ten days is intended to force Beagle 2 into a communication mode that should ensure that the transmitter is switched on for the majority of the daytime on Mars and thus will improve the chance of Mars Express making contact.

"During this ten-day period, Mars Express has listened for Beagle 2 but only for very short periods when Beagle 2 may not be switched on.

"The ten-day radio silence period ends on 22 January, just before a fly-over by Mars Express, but the strategy of the team is not to hail the lander immediately. Rather we are erring on the side of caution as we cannot confidently predict the precise ending of the ten-day slot. This is because the absolute accuracy of the timer on Beagle 2 could have been affected by the temperature on Mars, making the clock run slightly faster or slower than predicted. We have therefore chosen a pair of opportunities when Mars Express flies over the Beagle 2 landing site, the nights of 24 and 25 January. These two flights cover the widest possible area where Beagle 2 should be, giving us the best chance of calling the lander and getting a response from the continuous transmission. There are several other chances of just listening for Beagle 2 without calling it.

"The data will be analysed, this can take many hours, and we intend to present a complete picture of this series of attempts to contact the lander on 26 January, early afternoon. We will, at that time, outline any future communications strategy."

The results from future communication attempts will be posted on the Beagle 2 and PPARC websites.

Beagle 2 team shows support for troubled Spirit Rover

22-Jan-04 18:48 GMT

The Spirit Rover, the first of two NASA surface missions to Mars landing within a month of the Beagle 2, has stopped sending data and responding to commands

According to a spokesman from NASA's Jet Propulsion Laboratory the Mars rover, which reached the surface successfully on 3rd January, has not transmitted useful data for the last 12 hours.

As soon as they heard of the difficulty with Spirit the Beagle 2 team sent a message to Steve Squyres, Principal Investigator for the NASA project, expressing every sympathy with their predicament, and hoping they would soon be back in communication.

ESA confirmation of Martian water at pole

24-Jan-04 13:32 GMT

Beagle 2's 'mothership' Mars Express has confirmed the presence of large amounts of ice water at Mars' south pole, highlighting again the strong possibility of past life on the planet

The high resolution spectrometer on the European spacecraft, OMEGA, has sent back images showing the unmistakable signature of water at the pole. This confirmed previous data that suggested that large amounts of ice were held in the martian polar caps.

Professor Colin Pillinger, the leader of the Beagle 2 project, sent his congratulations to Jean-Pierre Bibring at the European Space Agency, whose team was responsible for the find. The ice suggests there was liquid water on Mars in the past.

Professor Pillinger noted that where there was water, there may have been life. "This discovery reemphasises the need for surface missions like Beagle 2, which can search for the signs of life directly through chemical analysis of subsurface soil and rock," he added.

Mars Express' high resolution camera has continued to return impressive data.The camera team, led by Gerhard Neukum, has been presenting some spectacularly detailed views of the martian landscape.

Disappointment in Beagle 2 search

26-Jan-04 16:09 GMT

No contact has been made with the Beagle 2 lander, despite repeated efforts over the last few days to communicate via the Mars Express and Mars Odyssey spacecraft and the Jodrell Bank radio telescope in Cheshire, UK.

At a press briefing in London this afternoon, members of the Beagle 2 team described the latest efforts to contact their missing lander.

"We haven't found Beagle 2, despite three days of intensive searching," said Professor Colin Pillinger, lead scientist for Beagle 2. "Under those circumstances, we have to begin to accept that, if Beagle 2 is on the Martian surface, it is not active.

"That isn't to say that we are going to give up on Beagle. There is one more thing that we can do - however, it is very much a last resort.We will be asking the American Odyssey spacecraft (team) tomorrow whether they will send an embedded command - a hail to Beagle with a command inside it. If it gets through, it will tell Beagle to switch off and reload the software.We are now working on the basis that there is a corrupt system and the only way we might resurrect is to send that command."

"We can also ask Mars Express to send that command. However, they cannot send it probably until the 2 or 3 February," he added.

"We'll move with the next phase in the search for Beagle 2," said Professor Pillinger. "We have discussed on our side of the house what we intend to do in the future. We are dedicated to trying to refly Beagle 2 in some shape or form, therefore we need to know how far it got because we need know which parts of this mission we don't have to study in further detail."

Detailing the efforts to contact Beagle 2 in recent days, Mark Sims, Beagle 2 Mission Manager from the University of Leicester, explained that the lander should have entered an emergency communication mode known as CSM2 no later than 22 January. In this mode, the spacecraft's receiver is switched on throughout daylight hours on Mars. The only possible explanation that no communication has been established during the last few days is that the landers battery is in a low state of charge.

Meanwhile, the academia-industry 'Tiger Team' at the National Space Centre in Leicester is beginning to concentrate on detailed analysis of the possible causes for failure of the mission and the lessons that can be learned for future missions.

The analysis of the mission now under way includes an assessment of the landing site ellipse from orbital images, reanalysis of atmospheric conditions during the entry into the Martian atmosphere on 25 December, examination of the separation from Mars Express and of the cruise phase preceding arrival at Mars.

One extremely useful piece of evidence could be provided by an image of the lander. The team is hoping that the High Resolution Stereo Camera on Mars Express or the camera on board Mars Global Surveyor may eventually be able to capture an image that reveals its location on the Martian surface.

Early Beagle 2 landing site images released

02-Feb-04 17:00 GMT

The first high resolution image of part of the Beagle 2 landing site has been released by NASA. The image covers a small strip of the ellipse that encompasses Beagle's feasible locations, but shows no obvious signs of the spacecraft.

The image was taken by the Mars Observer Camera (MOC), onboard NASA's Mars Global Surveyor orbiter, on January 5th, and is the first maximum resolution image to be obtained for the Beagle 2 landing site. Mike Malin, Principal Investigator for the MOC instrument, was able to release the image this weekend following processing by Malin Space Science Systems.

Provided are two images of the entire ellipse and surrounding area. One was obtained by the wide angle lens of the MOC instrument concurrently with the high resolution strip, and the other is a mosaic assembled from medium resolution images obtained by the camera THEMIS on NASA's other Mars orbiter, Odyssey.

This high resolution image strip acquired on January 5, 2004 at 10:01:03 UTC by the Mars Orbiter Camera (MOC) on NASA's Mars Global Surveyor spacecraft in orbit around Mars crosses the western portion of the final Beagle 2 landing ellipse. It was taken in the narrow angle mode of the MOC camera, spans a width of approximately 3 km and had a resolution of about 1.5 m per pixel. The second image shows a MOC wide angle image taken at the same time as the narrow angle view, with an overlaid outline of the landing ellipse and the narrow angle footprint. In the third image a mosaic covering all of the Beagle 2 ellipse from Mars Odyssey THEMIS observations at 15 m per pixel and 100 m per pixel, respectively, is given along with an outline of the location of the MOC narrow angle strip footprint.

What is remarkable about this MOC narrow angle image is that it is the first observation of the Beagle 2 landing area acquired after the lander's arrival at Mars on December 25, 2003. In fact, the image was requested by the Beagle 2 team to be taken in the context of a systematic search for elements of the lander such as the parachute, aeroshell and the lander itself. Dr. Mike Malin, the PI of the MOC camera, in a spirit of open cooperation quickly agreed to target this observation at the first opportunity of MGS crossing the landing ellipse in daylight, and the sequence was executed as planned and the image downlinked on January 7.

There are no obvious indications for lander elements within the high resolution MOC image. Of course, there is only a small chance that Beagle 2 actually landed in this part of the ellipse. The highly successful and astonishingly clear identification of the MER-A lander within Gusev crater as a result of another targeted observation of MOC however bodes well for eventual identification of Beagle 2 elements once more images covering the landing ellipse are acquired. This is planned to occur in a series of targeted MGS/MOC imaging activities during the course of February. As opposed to earlier imaging to identify the Viking Lander 1 and Mars Pathfinder vehicles, MER-A as well as its parachute and backshell stood out so clearly that this can only be attributed to lack of even a thin dust coating on the hardware surfaces which should just as well be the case for Beagle 2. Viking Lander-1, Mars Pathfinder and MER-A imaging was aided by prior knowledge of the 'exact' locations of these spacecraft relative to surface features as well as in the inertial frame which is not the case for Beagle 2, requiring in a worst case having to cover all of the landing ellipse with high resolution images.

Even though MOC imaging of the Beagle 2 ellipse can as yet not be performed with the image motion compensated technique (IMC) with 0.5 m along-track resolution applied for observation of the other landers - due to lack of knowledge of exact surface location and associated larger required image data volume for larger surface coverage - 'normal' MOC narrow angle imaging with 1.5 m surface resolution should suffice to show highly reflective items such as the Beagle 2 parachute or the lander itself. To gain an understanding of what may have gone wrong with Beagle 2 during its descent on Christmas Day, it will be vital to verify from MOC observations whether the parachute has deployed and thus possibly lies on the surface in a similar fashion as the MER-A chute, and whether the three airbags and the lander are visible. Depending on what we may eventually see from these observations, it may be possible to dramatically arrow the search for causes of Beagle's mishap.

Update on the hunt for Beagle 2

11-Feb-04 15:21 GMT

Whilst orbiting spacecraft continue to listen out for Beagle 2, the project has now officially moved on to assessing the possible reasons for the lack of communication.

The project team has begun an in-house investigation into all the technical aspects of Beagle 2 to establish those areas of greatest risk and what might be done to alleviate them in a future mission. It will make all the information available to the official ESA/UK government inquiry announced by Lord Sainsbury. Such an inquiry is set up as a matter of routine following loss of any spacecraft. The inquiry could be greatly aided in the search for evidence if some Beagle 2 artefact, such as the parachute, can be identified by cameras scrutinising the possible landing site.

Recognising the public interest in tracing Beagle 2, we plan to report in a few weeks time, via this web site, about the team's progress into the investigation of the fate of Beagle 2.

NATIONAL AERONAUTICS AND SPACE ADMINISTRATION

2001 Mars Odyssey Launch

Press Kit
April 2001

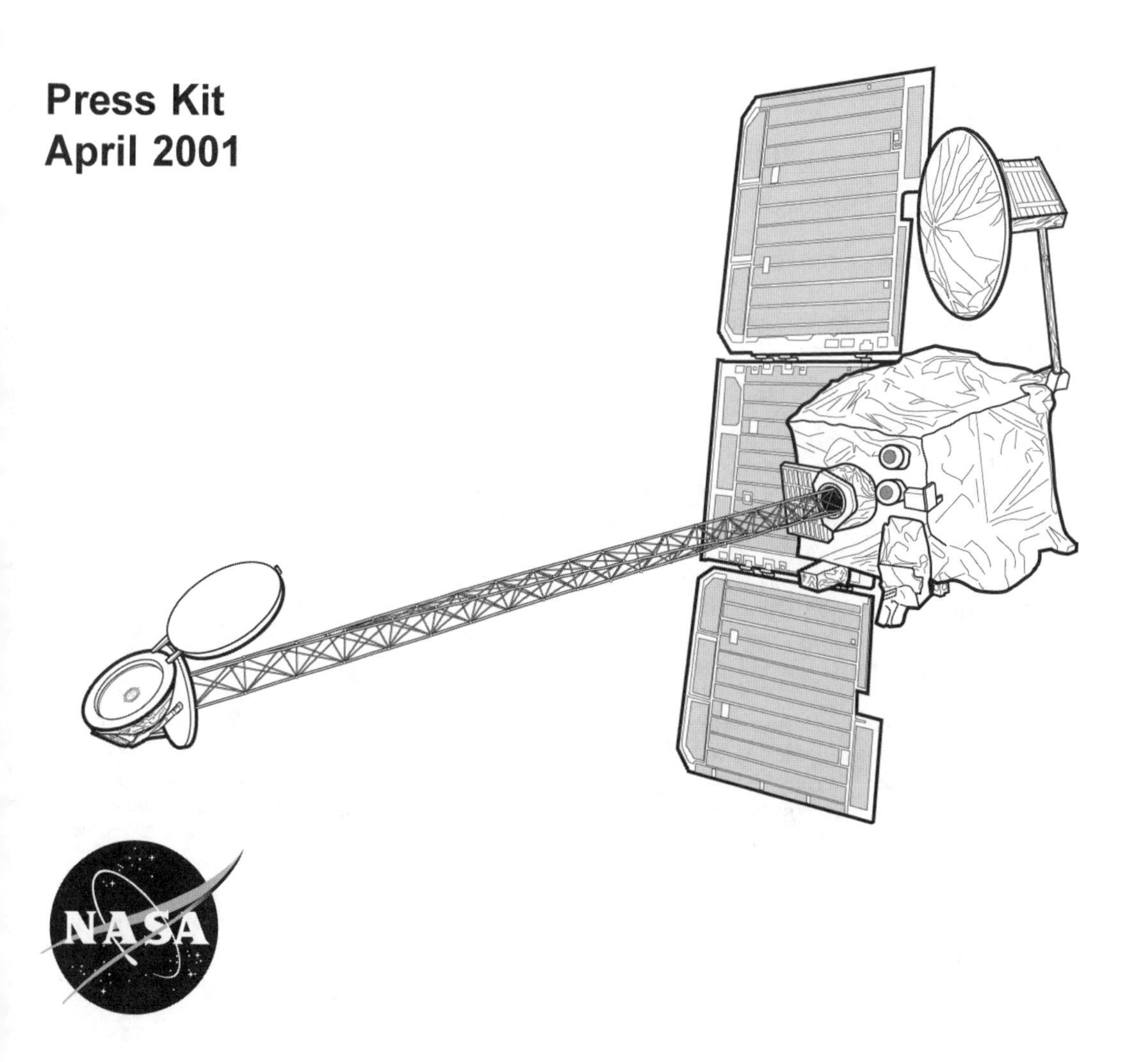

Contacts

Don Savage — Policy / Program Management 202/358-1727
Headquarters, Washington, DC

Franklin O'Donnell — 2001 Mars Odyssey Mission 818/354-5011
Jet Propulsion Laboratory, Pasadena, California

Mary Hardin — 2001 Mars Odyssey Mission 818/354-0344
Jet Propulsion Laboratory, Pasadena, California

George Diller — Launch Operations 321/867-2468
Kennedy Space Center, Florida

CONTENTS

2001 MARS ODYSSEY SET TO FIND OUT WHAT MARS IS MADE OF

When NASA's *2001 Mars Odyssey* launches in April to explore the fourth planet from the Sun, it will carry a suite of scientific instruments designed to tell us what makes up the Martian surface, and provide vital information about potential radiation hazards for future human explorers.

"The launch of *2001 Mars Odyssey* represents a milestone in our exploration of Mars – the first launch in our restructured Mars Exploration Program we announced last October," said Dr. Ed Weiler, Associate Administrator for Space Science, NASA Headquarters, Washington, D.C. "Mars continues to surprise us at every turn. We expect *Odyssey* to remove some of the uncertainties and help us plan where we must go with future missions."

Set for launch April 7 from Cape Canaveral Air Force Station, Florida, *Odyssey* is NASA's first mission to Mars since the loss of two spacecraft in 1999. Other than our Moon, Mars has attracted more spacecraft exploration attempts than any other object in the solar system, and no other planet has proved as daunting to success. Of the 30 missions sent to Mars by three countries over 40 years, fewer than one-third have been successful.

The *Odyssey* team conducted vigorous reviews and incorporated "lessons learned" in the mission plan. "The project team has looked at the people, processes, and design to understand and reduce our mission risk," said George Pace, *2001 Mars Odyssey* Project Manager at NASA's Jet Propulsion Laboratory, Pasadena, California. "We haven't been satisfied with just fixing the problems from the previous missions. We've been trying to anticipate and prevent other things that could jeopardize the success of the mission."

Odyssey is part of NASA's Mars Exploration Program, a long-term robotic exploration initiative launched in 1996 with *Mars Pathfinder* and *Mars Global Surveyor*. "The scientific trajectory of the restructured Mars Exploration Program begins a new era of reconnaissance with the *Mars Odyssey* orbiter," said Dr. Jim Garvin, Lead Scientist for NASA's Mars Exploration Program. "*Odyssey* will help identify and ultimately target those places on Mars where future rovers and landers must visit to unravel the mysteries of the red planet."

NASA's latest explorer carries three scientific instruments to map the chemical and mineralogical makeup of Mars: a thermal-emission imaging system, a gamma-ray spectrometer and a Martian radiation environment experiment. The imaging system will map the planet with high-resolution thermal images and give scientists an increased level of detail to help them understand how the mineralogy of the planet relates to the landforms. The part of *Odyssey's* imaging system that takes pictures in visible light will see objects with a clarity that fills the gaps between the *Viking* orbiter cameras of the 1970s and today's high-resolution images from *Mars Global Surveyor*.

Like a virtual shovel digging into the surface, *Odyssey*'s gamma-ray spectrometer will allow scientists to peer into the shallow subsurface of Mars, the upper few centimeters of the crust, to measure many elements, including the amount of hydrogen that exists. Since hydrogen is mostly likely present in the form of water ice, the spectrometer will be able to measure permanent ground ice and how that changes with the seasons.

"For the first time at Mars we will have a spacecraft that is equipped to find evidence for present near-surface water and to map mineral deposits from past water activity," said Dr. Steve Saunders, *2001 Mars Odyssey* Project Scientist at JPL. "Despite the wealth of information from previous missions, exactly what Mars is made of is not fully known, so this mission will give us a basic understanding about the chemistry and mineralogy of the surface."

The Martian radiation environment experiment will be the first to look at radiation levels at Mars as they relate to the potential hazards faced by future astronauts. The experiment will take data on the way to Mars and in orbit around the red planet. After completing its primary mission, the *Odyssey* orbiter will provide a communications relay for future American and international landers, including NASA's Mars Exploration Rovers, scheduled for launch in 2003.

The Jet Propulsion Laboratory, Pasadena, California, manages the 2001 Mars Odyssey mission for NASA's Office of Space Science, Washington, D.C. Principal investigators at Arizona State University, the University of Arizona and NASA's Johnson Space Center will operate the science instruments. Lockheed Martin Astronautics, Denver, Colorado, is the prime contractor for the project, and developed and built the orbiter. Mission operations will be conducted jointly from JPL, a division of the California Institute of Technology in Pasadena, and Lockheed Martin.

– End of General Release –

Media Services Information

NASA Television Transmission
NASA Television is broadcast on the satellite GE-2, transponder 9C, C band, 85 degrees west longitude, frequency 3880.0 MHz, vertical polarization, audio monaural at 6.8 MHz. The schedule for television transmissions for the 2001 Mars Odyssey launch will be available from the Jet Propulsion Laboratory, Pasadena, California; Johnson Space Center, Houston, Texas; Kennedy Space Center, Florida, and NASA Headquarters, Washington, D.C.

Status Reports
Status reports on mission activities will be issued by the Jet Propulsion Laboratory's Media Relations Office. They may be accessed on-line as noted below.

Launch Media Credentialing
Requests to cover the 2001 Mars Odyssey launch must be faxed in advance to the NASA Kennedy Space Center newsroom at 321/867-2692. Requests must be on the letterhead of the news organization and must specify the editor making the assignment to cover the launch.

Briefings
An overview of the mission will be presented in a news briefing broadcast on NASA Television originating from NASA Headquarters in Washington, D.C., at 1 p.m. EST March 19. Prelaunch briefings at Kennedy Space Center are scheduled at 1 p.m. and 2 p.m. Eastern time the day before the launch.

Internet Information
Extensive information on the 2001 Mars Odyssey mission, including an electronic copy of this press kit, press releases, fact sheets, status reports and images, is available from the Jet Propulsion Laboratory's World Wide Web home page at http://www.jpl.nasa.gov. The Mars Exploration Program maintains a home page at http://mars.jpl.nasa.gov.

Quick Facts

Spacecraft

Dimensions: Main structure 2.2 meters (7.2 feet) long, 1.7 meters (5.6 feet) tall and 2.6 meters (8.5 feet) wide; wingspan of solar array 5.7-meter (18.7-feet) tip to tip

Weight: 725 kilograms (1,598.4 pounds) total, composed of 331.8-kilogram (731.5-pound) dry spacecraft, 348.7 kilograms (768.8 pounds) of fuel and 44.5 kilograms (98.1 pounds) of science instruments
Science instruments: Thermal emission imaging system, gamma-ray spectrometer, Martian radiation environment experiment
Power: Solar array providing up to 1,500 watts just after launch, 750 watts at Mars

Launch Vehicle
Type: Delta II 7925
Weight: 230,983 kg (509,232 lbs)

Mission
Launch window: April 7 to April 27, 2001
Earth-Mars distance at launch: 125 million km (77.5 million miles)
Total distance traveled Earth to Mars: 460 million kilometers (286 million miles)
Mars arrival date: October 24, 2001
Earth-Mars distance at arrival: 150 million kilometers (93 million miles)
One-way speed of light time Mars-to-Earth at arrival: 8 minutes, 30 seconds
Primary science mapping period: January 2002 -July 2004

Program
Cost: $297 million total for 2001 Mars Odyssey
$165 million spacecraft development and science instruments
$ 53 million launch
$ 79 million mission operations and science processing

Mars at a Glance

General
One of five planets known to ancients; Mars was Roman god of war, agriculture and the state
Reddish color; at times the third brightest object in night sky after the Moon and Venus

Physical Characteristics
Average diameter 6,780 kilometers (4,217 miles); about half the size of Earth, but twice the size of Earth's Moon
Same land area as Earth
Mass 1/10th of Earth's; gravity only 38 percent as strong as Earth's
Density 3.9 times greater than water (compared to Earth's 5.5 times greater than water)
No planetwide magnetic field detected; only localized ancient remnant fields in various regions

Orbit
Fourth planet from the Sun, the next beyond Earth
About 1.5 times farther from the Sun than Earth is
Orbit elliptical; distance from Sun varies from a minimum of 206.7 million kilometers (128.4 million miles) to a maximum of 249.2 million kilometers (154.8 million miles); average distance from Sun, 227.7 million kilometers (141.5 million miles)
Revolves around Sun once every 687 Earth days
Rotation period (length of day in Earth days) 24 hours, 37 min, 23 sec (1.026 Earth days)
Poles tilted 25 degrees, creating seasons similar to Earth's

Environment
Atmosphere composed chiefly of carbon dioxide (95.3%), nitrogen (2.7%) and argon (1.6%)
Surface atmospheric pressure less than 1/100th that of Earth's average
Surface winds up to 40 meters per second (80 miles per hour)
Local, regional and global dust storms; also whirlwinds called dust devils
Surface temperature averages -53° C (-64° F); varies from -128° C (-199° F) during polar night to 27° C (80° F) at equator during midday at closest point in orbit to Sun

Features
Highest point is Olympus Mons, a huge shield volcano about 26 kilometers (16 miles) high and 600 kilometers (370 miles) across; has about the same area as Arizona
Canyon system of Valles Marineris is largest and deepest known in solar system; extends more than 4,000

kilometers (2,500 miles) and has 5 to 10 kilometers (3 to 6 miles) relief from floors to tops of surrounding plateaus

"Canals" observed by Giovanni Schiaparelli and Percival Lowell about 100 years ago were a visual illusion in which dark areas appeared connected by lines. The *Mariner 9* and *Viking* missions of the 1970s, however, established that Mars has channels possibly cut by ancient rivers

Moons

Two irregularly shaped moons, each only a few kilometers wide

Larger moon named Phobos ("fear"); smaller is Deimos ("terror"), named for attributes personified in Greek mythology as sons of the god of war

Historical Mars Missions

Mission	Country	Launch	Purpose	Results
[Unnamed]	USSR	10/10/60	Mars flyby	Did not reach Earth orbit
[Unnamed]	USSR	10/14/60	Mars flyby	Did not reach Earth orbit
[Unnamed]	USSR	10/24/62	Mars flyby	Achieved Earth orbit only
Mars 1	USSR	11/1/62	Mars flyby	Radio failed at 106 million km (65.9 million miles)
[Unnamed]	USSR	11/4/62	Mars flyby	Achieved Earth orbit only
Mariner 3	U.S.	11/5/64	Mars flyby	Shroud failed to jettison
Mariner 4	U.S.	11/28/64	Mars flyby	First successful Mars flyby 7/14/65; Returned 21 photos
Zond 2	USSR	11/30/64	Mars flyby	Passed Mars but radio failed; Returned no planetary data
Mariner 6	U.S.	2/24/69	Mars flyby	Mars flyby 7/31/69; Returned 75 photos
Mariner 7	U.S.	3/27/69	Mars flyby	Mars flyby 8/5/69; Returned 126 photos
Mariner 8	U.S.	5/8/71	Mars orbiter	Failed during launch
Kosmos 419	USSR	5/10/71	Mars lander	Achieved Earth orbit only
Mars 2	USSR	5/19/71	Mars orbiter / lander	Arrived 11/27/71; No useful data; Lander burned up due to steep entry
Mars 3	USSR	5/28/71	Mars orbiter / lander	Arrived 12/3/71; Lander operated on surface for 20 seconds before failing
Mariner 9	U.S.	5/30/71	Mars orbiter	In orbit 11/13/71 to 10/27/72; Returned 7,329 photos
Mars 4	USSR	7/21/73	Mars orbiter	Failed Mars orbiter; Flew past Mars 2/10/74
Mars 5	USSR	7/25/73	Mars orbiter	Arrived 2/12/74; Lasted a few days
Mars 6	USSR	8/5/73	Mars flyby module and lander	Arrived 3/12/74; Lander failed due to fast impact
Mars 7	USSR	8/9/73	Mars flyby module and lander	Arrived 3/9/74; Lander missed the planet
Viking 1	U.S.	8/20/75	Mars orbiter / lander	Orbit 6/19/76-1980; Lander 7/20/76-1982
Viking 2	U.S.	9/9/75	Mars orbiter / lander	Orbit 8/7/76-1987, lander 9/3/76-1980; Combined, the *Viking* orbiters and landers returned 50,000+ photos
Phobos 1	USSR	7/7/88	Mars / Phobos orbiter / lander	Lost 8/88 en route to Mars
Phobos 2	USSR	7/12/88	Mars / Phobos orbiter / lander	Lost 3/89 near Phobos
Mars Observer	U.S.	9/25/92	Mars orbiter	Lost just before Mars arrival 8/21/93
Mars Global Surveyor	U.S.	11/7/96	Mars orbiter	Arrived 9/12/97, high-detail mapping through 1/00; Now conducting second extended mission through fall 2004
Mars 96	Russia	11/16/96	orbiter and landers	launch vehicle failed
Mars Pathfinder	U.S.	12/4/96	Mars lander and rover	Landed 7/4/97; Last transmission 9/27/97
Nozomi	Japan	7/4/98	Mars orbiter	Currently in orbit around the Sun; Mars arrival delayed to 12/03 due to propulsion problem
Mars Climate Orbiter	U.S.	12/11/98	Mars orbiter	Lost upon arrival 9/23/99
Mars Polar Lander / Deep Space 2	U.S.	1/3/99	lander and soil probes	Lost on arrival 12/3/99

Why Mars?

Mars perhaps first caught public fancy in the late 1870s, when Italian astronomer Giovanni reported using a telescope to observe "canali," or channels, on Mars. A possible mistranslation of this word as "canals" may have fired the imagination of Percival Lowell, an American businessman with an interest in astronomy. Lowell founded an observatory in Arizona, where his observations of the red planet convinced him that the canals were dug by intelligent beings – a view that he energetically promoted for many years.

By the turn of the last century, popular songs envisioned sending messages between worlds by way of huge signal mirrors. On the dark side, H.G. Wells' 1898 novel *The War of the Worlds* portrayed an invasion of Earth by technologically superior Martians desperate for water. In the early 1900s, novelist Edgar Rice Burroughs, known for the *Tarzan* series, also entertained young readers with tales of adventures among the exotic inhabitants of Mars, which he called Barsoom.

Fact began to turn against such imaginings when the first robotic spacecraft were sent to Mars in the 1960s. Pictures from the first flyby and orbiter missions showed a desolate world, pocked with craters similar to those seen on Earth's Moon. The first wave of Mars exploration culminated in the *Viking* mission, which sent two orbiters and two landers to the planet in 1975. The landers included a suite of experiments that conducted chemical tests in search of life. Most scientists interpreted the results of these tests as negative, deflating hopes of identifying another world on where life might be or have been widespread.

The science community had many other reasons for being interested in Mars, apart from searching for life; the next mission on the drawing boards concentrated on a study of the planet's geology and climate. Over the next 20 years, however, new findings in laboratories on Earth came to change the way that scientists thought about life and Mars.

One was the 1996 announcement by a team from Stanford University and NASA's Johnson Space Center that a meteorite believed to have originated on Mars contained what might be the fossils of ancient bacteria. This rock and other so-called Mars meteorites discovered on several continents on Earth are believed to have been blasted away from the red planet by asteroid or comet impacts. They are thought to come from Mars because of gases trapped in the rocks that match the composition of Mars' atmosphere. Not all scientists agreed with the conclusions of the team announcing the discovery, but it reopened the issue of life on Mars.

Another development that shaped scientists' thinking was new research on how and where life thrives on Earth. The fundamental requirements for life as we know it are liquid water, organic compounds and an energy source for synthesizing complex organic molecules. Beyond these basics, we do not yet understand the environmental and chemical evolution that leads to the origin of life. But in recent years, it has become increasingly clear that life can thrive in settings much different from a tropical soup rich in organic nutrients.

In the 1980s and 1990s, biologists found that microbial life has an amazing flexibility for surviving in extreme environments – niches that by turn are extraordinarily hot, or cold, or dry, or under immense pressures – that would be completely inhospitable to humans or complex animals. Some scientists even concluded that life may have begun on Earth in heat vents far under the ocean's surface.

This in turn had its effect on how scientists thought about Mars. Life might not be so widespread that it would be found at the foot of a lander spacecraft, but it may have thrived billions of years ago in an underground thermal spring or other hospitable environment. Or it might still exist in some form in niches below the frigid, dry, windswept surface.

NASA scientists also began to rethink how to look for signs of past or current life on Mars. In this new view, the markers of life may well be so subtle that the range of test equipment required to detect it would be far too complicated to package onto a spacecraft. It made more sense to collect samples of Martian rock, soil and air to bring back to Earth, where they could be subjected to much more extensive laboratory testing.

Mars and Water

Mars today is far too cold with an atmosphere that is much too thin to support liquid water on its surface. Yet scientists studying images acquired by the *Viking* orbiters consistently uncovered landscape features that appeared to have been formed by the action of flowing water. Among those features were deep channels and curving canyons, and even landforms that resemble ancient lake shorelines. Added to this foundation is more recent evidence, especially from observations made by *Mars Global Surveyor*, that suggested widespread flowing water on the Martian surface in the planet's past. On the basis of analysis of some of the features observed by both the *Mars Pathfinder* and *Mars Global Surveyor* spacecraft, some scientists likened the action of ancient flowing water on Mars to floods with the force of thousands of Mississippi Rivers.

Continuing the saga of water in the history of Mars, in June 2000 geologists on the *Mars Global Surveyor* imaging team presented startling evidence of landscape features that dramatically resemble gullies formed by the rapid discharge of liquid water, and deposits of rocks and soils related to them. The features appear to be so young that they might be forming today. Scientists believe they are seeing evidence of a ground water supply, similar to an aquifer. Ever since the time of *Mariner 9* in the early 1970s, a large part of the focus of Mars science has been questions related to water: how much was there and where did it go (and ultimately, how much is accessible today). The spectacular images from *Mars Global Surveyor* reveal part of the answer – some of the water within the Mars "system" is stored underground, perhaps as close as hundreds of meters (or yards), and at least some of it might still be there today.

Still, there is no general agreement on what form water took on the early Mars. Two competing views are currently popular in the science community. According to one theory, Mars was once much warmer and wetter, with a thicker atmosphere; it may well have boasted lakes or oceans, rivers and rain. According to the other theory, Mars was always cold, but water trapped as underground ice was periodically released when heating caused ice to melt and gush forth onto the surface.

Even among those who subscribe to the warmer-and-wetter theory, the question of what happened to the water is still a mystery. Most scientists do not feel that the scenario responsible for Mars' climate change was necessarily a cataclysmic event such as an asteroid impact that, say, disturbed the planet's polar orientation or orbit. Many believe that the demise of flowing water on the surface could have resulted from a gradual process of climate change taking place over many millennia.

Under either the warmer-and-wetter or the always-cold scenario, Mars must have had a thicker atmosphere to support water that flowed on the surface even only occasionally. If the planet's atmosphere became thinner, then liquid water could not flow without evaporating. Mars' atmosphere today is overwhelmingly composed of carbon dioxide. Over time, carbon dioxide gas reacts with elements in rocks and becomes locked up as a kind of compound called a carbonate.

On Earth, the horizontal and vertical motions of the shifting tectonic plates that define the crust of our planet are continually plowing carbonates and other widespread minerals beneath the surface to depths at which the internal heat within Earth releases carbon dioxide, which later spews forth in volcanic eruptions. This terrestrial cycle replenishes the carbon dioxide in Earth's atmosphere. Although we are not sure Mars today harbors any active volcanoes, it clearly had abundant and widespread volcanic activity in its past. The apparent absence of a long-lasting system of jostling tectonic plates on Mars, however, suggests that a critical link in the process that leads to carbon dioxide recycling in Earth's atmosphere is missing on Mars.

These scenarios, however, are just theories. Regardless of the history and fate of the atmosphere, scientists also do not understand what happened to Mars' water. Some undoubtedly must have been lost to space. Water ice has been detected in the permanent cap at Mars' north pole. Water ice may also exist in the cap at the south pole. But much water is probably trapped under the surface – either as ice or, if near a heat source, possibly in liquid form.

NASA's next mission to the red planet, 2001 Mars Odyssey, will provide another vital piece of information to the "water puzzle" by mapping the basic chemistry and minerals that are present in the upper centimeters (or inches) of the planet's surface. *Odyssey* will be the first spacecraft to make direct observations of the element hydrogen near and within the surface of Mars, and hydrogen may provide the strongest evidence of water on or just under the Martian surface, since it is one of the key elements within the water molecule. The high-resolution imaging system on *Odyssey* will be able to identify regions such as hot springs, if any exist, which could serve as prime sites in which to refine our surface search for signs of simple biological processes.

Even if we ultimately learn that Mars never harbored life as we know it, scientific exploration of the red planet can assist in understanding life on our own home planet. Much of the evidence for the origin of life here on Earth has been obliterated by the incredible dynamics of geological processes which have operated over the past 4 billion years, such as plate tectonics and rapid weathering. Today we believe that there are vast areas of the Martian surface that date back a primordial period of planetary evolution – a time more than about 4 billion years ago that overlaps the period on Earth when pre-biotic chemical evolution first gave rise to self-replicating systems that we know of as "life."

Thus, even if life never developed on Mars – something that we cannot answer today – scientific exploration of the planet may yield absolutely critical information unobtainable by any other means about the pre-biotic chemistry that led to life on Earth. Furthermore, given the complexity we recognize in Earth's record of climate change, some scientists believe that by studying the somewhat simpler (but no less bizarre) Martian climate system, we can learn more about Earth. As such, Mars could serve as Mother Nature's great "control experiment" providing us with additional perspectives from which to understand the workings of our own home planet. The 2001 Mars Odyssey mission continues us on the path of understanding the red planet as a "system" by probing what it is made of, and where the elusive signs of surface water may have left their indelible marks.

Lessons Learned

Engineers and scientists working on the 2001 Mars Odyssey project began looking at ways to reduce risks to their mission immediately after the loss of *Mars Climate Orbiter* and *Mars Polar Lander* in 1999. In addition to the independent assessments made by the project, the team has also followed recommendations made by the NASA review boards investigating the losses and a NASA "Red Team" assigned to review the project.

Among the risk reduction actions taken are:

Identified parameters critical to mission success and did an independent verification of these parameters.
Listed both imperial and metric units on documentation for hand-off between systems and subsystems.
Added key staff at both JPL and Lockheed Martin.
Moved launch to Kennedy Space Center instead of Vandenberg Air Force Base in California to provide additional schedule margin and reduce how much the spacecraft's battery is discharged during launch.
Prepared mission fault trees and conducted mission risk reviews to formulate risk mitigation actions.
Conducted an independent verification and validation of the flight software by NASA personnel in Fairmont, West Virginia.
Conducted additional flight software tests to stress the design under off-nominal conditions.
Added check valves in the propulsion system to isolate the fuel and oxidizer until the moment of the Mars orbit insertion main engine burn.
Conducted additional pyro qualification test firings over a broader set of conditions.
Conducted additional thruster test firings to demonstrate proper operation under cold starting conditions.
Conducted life-cycling tests for assemblies in the communication system that are cycled on and off during flight.
Conducted additional measurements to assess the interference between the relay radio and the orbiter and instrument electronics.
Changed out suspect capacitors in orbiter electronics based on failures of similar capacitors on another program.
Added second- and third-shift testing to add operating time and build confidence in orbiter electronics.
Added ability to receive telemetry from spacecraft during the pressurization process prior to the Mars orbit insertion main engine burn.
Increased navigation tracking data during cruise.
Added delta differential one-way range measurements, called "delta DOR," that provide an independent measurement of the orbiter location relative to Mars.
Moved the point at which recovery from a fault would be impossible closer to Mars orbit insertion to minimize the time the system is not redundant.
Conducted additional oxidizer burn-to-depletion test to build confidence in and select parameters for the Mars orbit insertion strategy.
Raised Mars capture orbit design to a higher altitude.
Conducted additional studies to ensure that there is no fuel migration within the propulsion system that would cause excessive imbalance during the orbit insertion main engine burn.
Conducted an independent verification of Mars aerobraking by NASA Langley Research Center.
Adopted a more conservative Mars aerobraking profile to allow for dust storms and wider atmospheric variations.
Assigned clear lines of responsibility within the organization to improve communication.
Formalized operations team training.
Designated personnel to transition from development to operations.
Added a tracking station in Santiago, Chile, to fill in telemetry gaps after launch and early in cruise phase.

Where We've Been and Where We're Going

Incorporating lessons learned from past and ongoing Mars mission successes and setbacks, NASA's revamped campaign to unravel the secrets of the Red Planet moves from an era of global mapping and limited surface exploration to a much more comprehensive approach in which next-generation reconnaissance from orbit and from the surface will pave the way for multiple sample returns.

Over the next two decades, NASA's Mars Exploration Program will build upon previous scientific discoveries to establish a sustained observational presence both around and on the surface of Mars. This will be achieved from the perspective of orbital reconnaissance and telecommunication, surface-based mobile laboratories, sub-surface access and, ultimately, by means of robotic sample return missions. With international cooperation, the long-term program will maintain a science-driven, technology-enabled focus, while balancing risks against sound management principles and with attention to available resources. The strategy of the Mars Exploration Program will attempt to uncover profound new insights into Mars past environments, the history of its rocks and interior, the many roles and abundances of water and, quite possibly, evidence of past and present life.

The following are the most recently completed, ongoing and near-term future Mars missions of exploration in the

NASA program:

Mars Pathfinder (December 1996 - March 1998): The first completed mission in NASA's Discovery Program of low-cost, rapidly developed planetary missions with highly focused scientific goals, *Mars Pathfinder* far exceeded its expectations and outlived its primary design life. This lander, which released its *Sojourner* rover at the Martian surface, returned 2.3 billion bits of information, including more than 17,000 images, as well as more than 15 chemical analyses of rocks and soil and extensive data on winds and other types of weather. Investigations carried out by instruments on both the lander and the rover suggest that, in its past, Mars was warm and wet, and had liquid water on its surface and a thicker atmosphere. Engineers believed that, in October 1997, a depletion of the spacecraft's battery and a drop in the spacecraft's operating temperature were to blame for the loss of communications with *Pathfinder*. Attempts to re-establish communications with the vehicle ceased in March 1998, well beyond the mission's expected 30-day lifetime.

Mars Global Surveyor (November 1996 - January 2001 primary mapping mission): Orbiting the red planet 8,985 times so far, NASA's *Mars Global Surveyor* has collected more information than any other previous Mars mission and keeps on going into its extended mission. Sending back more than 65,000 images, 583 million topographic laser-altimeter shots and 103 million spectral measurements, *Global Surveyor*'s comprehensive observations have proven invaluable to understanding the seasonal changes on Mars. Some of the mission's most significant findings include: possible evidence for recent liquid water at the Martian surface; evidence for layering of rocks that point to widespread ponding or lakes in the planet's early history; topographic evidence for a south pole-to-north pole slope that controlled the transport of water and sediments; identification of the mineral hematite, indicating a past surface-hydrothermal environment; and extensive evidence for the role of dust in reshaping the recent Martian environment. *Global Surveyor* will continue gathering data in an extended mission approved until 2002.

Mars Exploration Rovers (2003): Identical twin rovers, able to travel almost as far in one Martian day as *Sojourner* did over its entire lifetime, will land at two separate sites and set out to determine the history of climate and water on the planet where conditions may once have been very favorable for life. By means of sophisticated sets of instruments and access tools, the twin rovers will evaluate the composition, texture and morphology of rocks and soils at a broad variety of scales, extending from those accessible to the human eye to microscopic levels. The rover science team will select targets of interest such as rocks and soils on the basis of images and infrared spectra sent back to Earth. Two different Martian landing sites will be chosen on the basis of an intensive examination of information collected by the *Mars Global Surveyor* and *Mars Odyssey* orbiters, as well as other missions.

Mars Reconnaissance Orbiter (2005): This scientific orbiter will attempt to bridge the gap between surface observations and measurements taken from orbit. It will focus on analyzing the Martian surface at new scales in an effort to follow the tantalizing hints of water from the *Mars Global Surveyor* images. For example, the *Mars Reconnaissance Orbiter* will measure thousands of Martian landscapes at 20- to 30-centimeter (8- to 12-inch) resolution, which is adequate to observe rocks the size of beach balls. In addition, maps of minerals diagnostic of the role of liquid water in their formation will be produced at unprecedented scales for thousands of potential future landing sites. Finally, a specialized, high-resolution sounding radar will probe the upper hundreds of meters (or yards) of the Martian sub-surface in search of clues of frozen pockets of water or other unique layers. Finally, the *Mars Reconnaissance Orbiter* will finish the job of characterizing the transport processes in the present-day Martian atmosphere, including the planet's annual climate cycles, using a unique infrared sounding instrument, originally carried to Mars on the ill-fated *Mars Observer*, and then again on *Mars Climate Orbiter*.

Smart Lander (2007): NASA has proposed to develop and launch a next-generation "mobile surface laboratory" with potentially long-range roving capabilities (greater than 10 kilometers (about 6 miles)) and more than a year of surface operational lifetime as a pivotal step toward a future Mars sample return mission. By providing a major leap forward in surface measurement capabilities and surface access, this mission will also demonstrate the technology needed for accurate landing and surface hazard avoidance in order to allow access to potentially compelling, but difficult to reach, landing sites. Its suite of scientific instruments could include new devices that will sample and probe the Martian subsurface in search of organic materials.

Scout Mission (2007): NASA has also proposed to create a new line of small "scout" missions that would be competitively selected from proposals submitted by the broader scientific and aerospace community. Exciting new vistas could be opened by means of this innovative approach, either through observations made from airborne vehicles, networks of small surface landers, or from highly focused orbital laboratories. NASA aims to compete these scout missions as often as possible, and potentially every four years, depending on resource availability.

Mars Sample Return (earliest launch possibility late 2011): NASA is studying additional scientific orbiters, rovers and landers, as well as approaches for returning the most promising samples of Martian materials (rocks, soils, ices and atmospheric gases / dust) back to Earth. While current schedules call for the first of several sample return missions to be launched in 2014 with a second mission in 2016, options that could move the date sooner to 2011 are presently under detailed examination. Technology development is underway for advanced capabilities, including a new generation of miniaturized surface instruments such as mass spectrometers and electron microscopes, as well as deep drilling to 20 meters (about 20 yards) or more.

Mission Overview

2001 Mars Odyssey is an orbiter carrying three packages of science experiments designed to make global observations of Mars to improve our understanding of the planet's climate and geologic history, including the search for liquid water and evidence of past life. The mission will extend across a full Martian year, or 29 Earth months.

Launch Vehicle

Odyssey will be launched on a variant of Boeing's Delta II rocket called the 7925 that includes nine strap-on solid-fuel motors. Each of the nine solid-fuel boosters is 1 meter (3.28 feet) in diameter and 13 meters (42.6 feet) long; each contains 11,765 kilograms (25,937 pounds) of a propellant called hydroxyl-terminated polybutadiene (HTPB) and provides an average thrust of 485,458 newtons (109,135 pounds) at liftoff. The casings on the solid rocket motors are made of lightweight graphite epoxy.

The main body of the first stage houses the Rocketdyne RS-27A main engine and two Rocketdyne LR101-NA-11 vernier engines. The vernier engines provide roll control during main engine burn and attitude control after main engine cutoff before the second stage separation. The RS-27A main engine burns 96,000 kilograms (211,000 pounds) of RP-1 (rocket propellant 1, a highly refined form of kerosene) as its liquid fuel and liquid oxygen as an oxidizer.

The second stage is 2.4 meters (8 feet) in diameter and 6 meters (19.7 feet) long, and is powered by an Aerojet AJ10-118K engine. The propellant is 3,929 kilograms (8,655 pounds) of a liquid fuel called Aerozine 50, a 50/50 mixture of hydrazine and unsymmetric dimethly hydrazine. The oxidizer is 2,101 kilograms (4,628 pounds) of nitrogen tetroxide. The engine is restartable and will perform two separate burns during the launch.

Orbiter Daily Launch Opportunities

The orbiter has two near-instantaneous launch opportunities each day (all times EDT)

Date	First Opportunity	Second Opportunity	Date	First Opportunity	Second Opportunity
4/7/01	11:02:22 am	11:32.22 am	4/18/01	7:58:53 am	8:37:50 am
4/8/01	10:29:00	11:29:00	4/19/01	7:38:34	8:14:26
4/9/01	9:57:36	10:57:36	4/20/01	7:29:27	8:04:03
4/10/01	9:33:01	11:01:17	4/21/01	7:23:28	7:57:21
4/11/01	9:19:24	10:27:28	4/22/01	7:15:12	7:48:06
4/12/01	9:00:53	9:57:23	4/23/01	7:07:26	7:39:28
4/13/01	8:45:05	9:35:08	4/24/01	7:00:06	7:31:25
4/14/01	8:31:00	9:16:47	4/25/01	6:53:12	7:23:49
4/15/01	8:22:12	9:05:52	4/26/01	6:46:41	7:16:43
4/16/01	8:10:05	8:51:11	4/27/01	6:40:32	7:10:01
4/17/01	7:58:53	8:37:50			

The third and final stage of the Delta II provides the final velocity required to place *Odyssey* on a trajectory to Mars. This upper stage is 1.25 meters (4.1 feet) in diameter and consists of a Star-48B solid-fuel rocket motor with 2,012 kilograms (4,431 pounds) of propellant and a system called active nutation control that provides stability after the motor ignites. A spin table attached to the top of the Delta's second stage supports, rotates and stabilizes the *Odyssey* spacecraft and Star-48B upper stage before they spin up and separate from the second stage. The *Odyssey* spacecraft is mounted to the Star-48B by a payload attachment fitting. A yo-yo despin system decreases the spin rate of the spacecraft and upper stage before they separate from each other.

During launch and ascent through Earth's atmosphere, the *Odyssey* spacecraft and Star-48B upper stage are

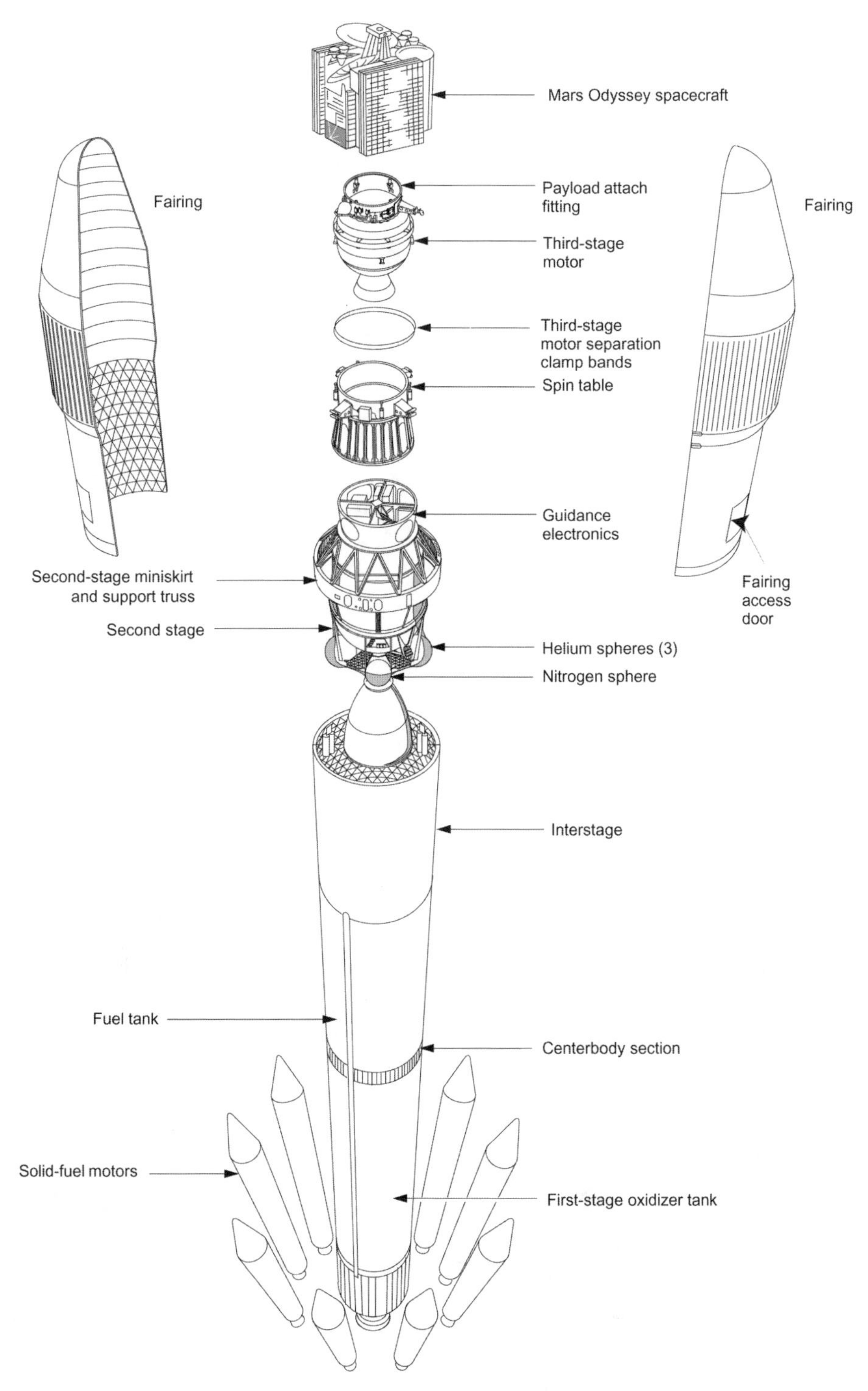

Delta II launch vehicle

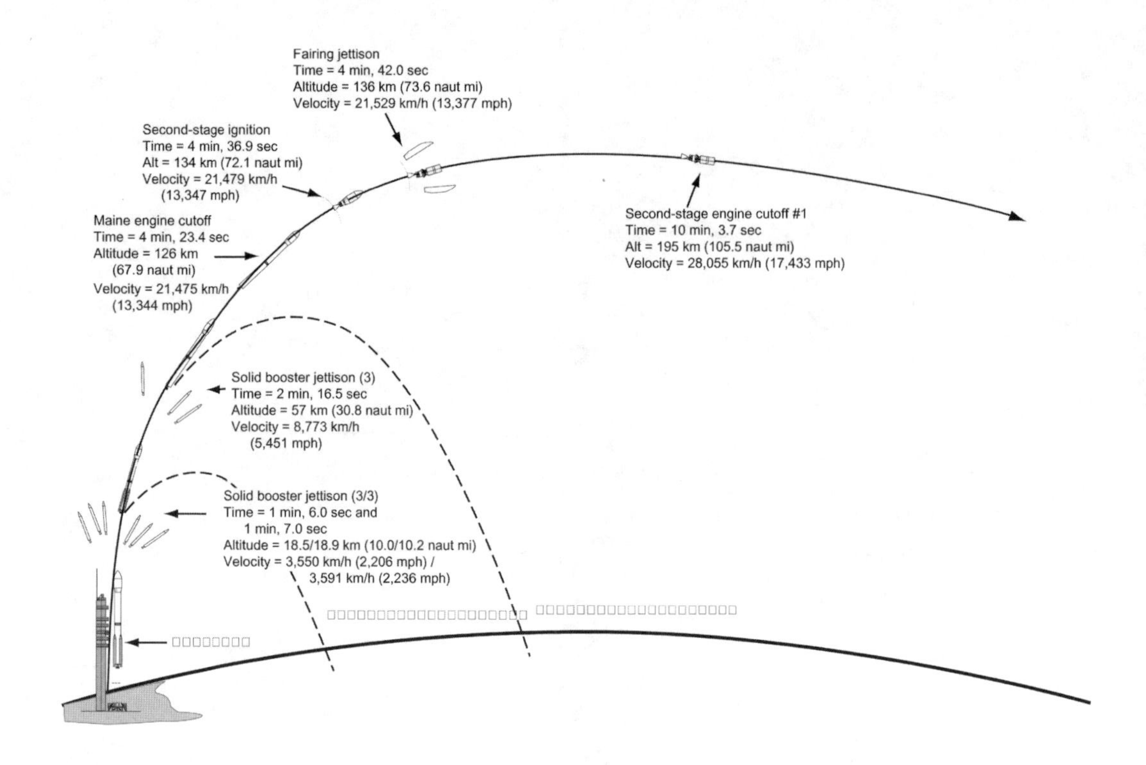

Launch boost phase

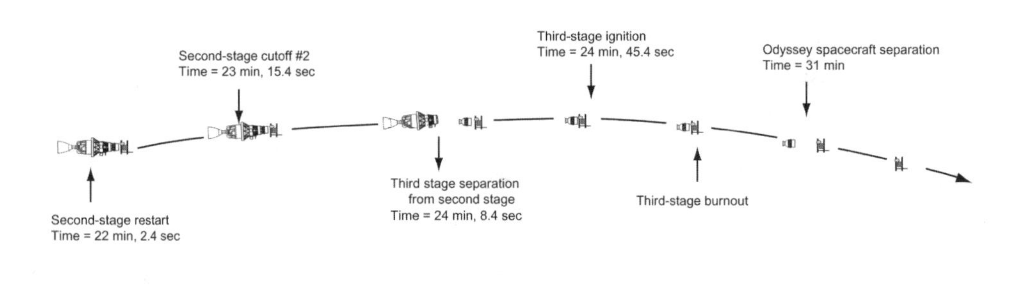

Launch injection phase

protected from aerodynamic forces by a 2.9-meter – diameter (9.5-foot) payload fairing that is jettisoned from the Delta II during second stage powered flight at an average altitude of 136 kilometers (73.6 nautical miles).

Launch Period

The orbiter launch period extends for 21 days, opening on April 7 and closing on April 27. The first 12 days of the launch period from April 7 through 18 make up what is considered the primary launch period; a secondary launch period runs from April 19 through 27. If *Odyssey* is launched during the secondary period, science data return at Mars may need to be reduced slightly because of higher arrival speeds and a longer aerobraking periods. Arrival dates at Mars vary with launch dates, and range from October 17 to 28, 2001.

Daily Windows

Two nearly instantaneous launch opportunities occur each day during the launch period each is separated by 30 to 90 minutes depending on the day. On April 7 the first is at 11:02 a.m. EDT and the second is at 11:32 a.m. EDT. The opportunities become earlier each day through the launch period.

Liftoff

Odyssey will lift off from Space Launch Complex 17 at Cape Canaveral Air Station, Florida. Sixty-six seconds after launch, the first three solid rocket boosters will be discarded followed by the next three boosters one second later. The final three boosters are jettisoned two minutes, 11 seconds after launch. About four minutes, 24 seconds after liftoff, the first stage will stop firing and be discarded eight seconds later. About five seconds later, the second stage engine ignites. The fairing or nose cone will be discarded four minutes, 41 seconds after launch. The first burn of the second stage engine occurs at 10 minutes, three seconds after launch.

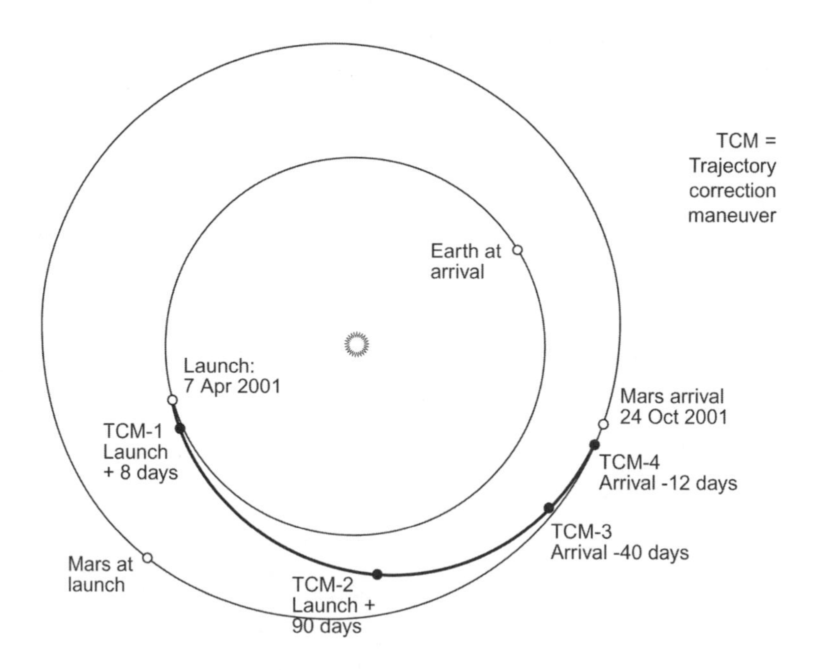

Interplanetary trajectory

At this point the vehicle is in low Earth orbit at an altitude of 189 kilometers (117 miles). Depending on the actual launch day and time the vehicle will then coast for several minutes, once it is in the correct point in its orbit, the second stage will be restarted at 24 minutes, 32 seconds after launch.

Small rockets will then be fired to spin up the third stage on a turntable attached to the second stage. The third stage will separate and ignite its motor, sending the spacecraft out of Earth orbit. A nutation control system (a thruster on an arm mounted on the side of the third stage) will be used to maintain stability during this the third stage burn. After that, the spinning upper stage and the attached *2001 Mars Odyssey* spacecraft must be despun so that the spacecraft can be separated and acquire its proper cruise orientation. This is accomplished by a set of weights that are reeled out from the side of the spinning vehicle on flexible lines, much as spinning ice skaters slow themselves by extending their arms. Odyssey will separate from the Delta third stage about 33 minutes after launch.

Any remaining spin will be removed using the orbiter's onboard thrusters.

About 36 minutes after launch the solar array is unfolded and about eight minutes later it is locked in place. Then the spacecraft turns to its initial communication attitude and the transmitter is turned on. About one hour after launch the 34-meter-diameter (112 foot) antenna at the Deep Space Network complex near Canberra, Australia will acquire Odyssey's signal.

Interplanetary Cruise

The interplanetary cruise phase is the period of travel from the Earth to Mars and lasts about 200 days. It begins with the first contact by the DSN after launch and extends until seven days before Mars arrival. Primary activities during the cruise include check out of the spacecraft in its cruise configuration, checkout and monitoring of the spacecraft and the science instruments and navigation activities necessary to determine and correct *Odyssey*'s flight path to Mars.

There are science activities planned for the cruise phase including payload health and status checks, instrument calibrations, as well as data taking by some of the science instruments as spacecraft limitations allow.

Odyssey's flight path to Mars is called a Type I trajectory, that takes it less than 180 degrees around the Sun. During the first two months of cruise, only the Deep Space Network station in Canberra will be capable of viewing the spacecraft. Late in May California's Goldstone station will come into view, and by early June the Madrid station will also be able to track the spacecraft. The project has also added the use of a tracking station in Santiago, Chile, to fill in tracking coverage during the first seven days following launch.

The orbiter will transmit to Earth using its medium-gain antenna and it will receive commands on its low-gain antenna during the early portion on its flight. At some point during the first 30 days after launch, the orbiter will be commanded to receive and transmit through its high-gain antenna. Cruise command sequences are generated and uplinked approximately once every four weeks during one of the regularly scheduled Deep Space Network passes.

The spacecraft will determine its orientation in space chiefly via a star camera and a device called an inertial measurement unit. The spacecraft will fly with its medium or high gain antenna pointed toward the Earth at all times while keeping the solar panels pointed toward the Sun. The spacecraft is stabilized in three axes and will not spin to maintain its orientation, or "attitude."

The spacecraft's orientation will be controlled by reaction wheels, devices with spinning wheels similar to gyroscopes. These devices will be occasionally "desaturated," meaning that their momentum will be unloaded by firing the spacecraft's thrusters.

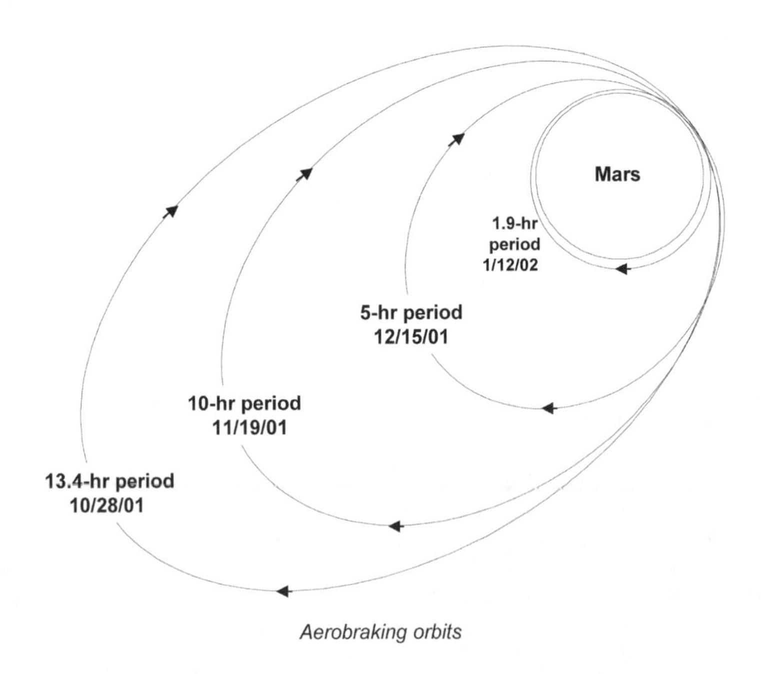

Aerobraking orbits

During interplanetary cruise, *Odyssey* is scheduled to fire its thrusters a total of five times to adjust its flight path. The first of these trajectory correction maneuvers is scheduled for eight days after launch, and will correct launch injection errors and adjust the Mars arrival aim point. It will be followed by a second maneuver 90 days after launch.

The remaining three trajectory correction maneuvers will be used to direct the spacecraft to the proper aim point at Mars. These maneuvers are scheduled at 90 days after launch, 12 days before arrival and seven hours (October 24) before arrival. The spacecraft will communicate with Deep Space Network antennas continuously for 24 hours around all of the trajectory correction maneuvers. Maneuvers will be conducted in what engineers are calling a "constrained turn-and-burn" mode in which the spacecraft will turn to the desired burn attitude and fire the thrusters, while remaining in contact with Earth.

Navigation tracking during cruise involves the collection of two-way doppler and ranging data. In order to provide additional information for navigation, the project has added a program of delta differential one-way range measurements, called delta DOR, that will be taken periodically during cruise and Mars approach. Delta DOR measurements are interferometric measurements between two radio sources. In this case, one of the radio sources is the DOR tones or telemetry signal coming from *Odyssey*. The second source will be either a known, stable natural radio source like a quasar or the telemetry signal from the *Mars Global Surveyor* spacecraft. Each source is recorded simultaneously at two radio antennas. The triangulation achieved through this method provides navigators with much more refined knowledge of the spacecraft's position. With this information, spacecraft operators can more precisely adjust *Odyssey*'s flight path. Delta DOR measurements will be collected and processed for system testing during early and mid-cruise and weekly during the Mars approach phase to provide additional data to the navigation team. For the first 14 days after launch, the Deep Space Network will continuously track the spacecraft. During the quiet phase of cruise when spacecraft activity is at a minimum, only three 8-hour passes per day are scheduled. Continuous tracking will resume for the final 50 days before Mars arrival.

Science instruments will be powered on, tested and calibrated during cruise. The thermal emission imaging system will take a picture of the Earth-Moon system about 12 days after launch if the spacecraft is operating normally. Star calibration imaging is also planned 45 days after launch, while a Mars approach image is planned about 12 days before arrival if the Earth-Moon calibration image is not taken.

Two calibration periods are planned for the gamma-ray spectrometer during cruise. Each of the spectrometer's three sensors may be operated during the calibration periods, depending upon spacecraft power capabilities. The Mars radiation environment experiment is designed to collect radiation data constantly during cruise to help determine what the radiation environment is like on the way to Mars.

A test of the orbiter's UHF radio system is planned between 60 and 80 days after launch. The 45-meter (150-foot) antenna at California's Stanford University will be used to test the UHF system ability to receive and transmit. The UHF system will be used during *Odyssey*'s relay phase to support future landers, it is not used as part of the orbiter's science mission.

Mars Orbit Insertion

Odyssey will arrive at Mars on October 24, 2001. As it nears its closest point to the planet over the northern hemisphere, the spacecraft will fire its 640-Newton main engine for approximately 22 minutes to allow itself to be captured into an elliptical, or egg-shaped, orbit. If the launch occurs early in the period, *Odyssey* will loop around the planet every 17 hours. About three orbits after insertion, the spacecraft will fire its thrusters in what is called a period reduction maneuver so that it orbits the planet approximately once every 11 hours.

Aerobraking

Aerobraking is the transition from the initial elliptical orbit to the science orbit where *Odyssey* will circle Mars at a uniform altitude. It is a technique that slows the spacecraft down by using frictional drag as it flies through the upper part of the planet's atmosphere.

During each of its long, elliptical loops around Mars, the orbiter will pass through the upper layers of the atmosphere each time it makes its closest approach to the planet. Friction from the atmosphere on the spacecraft and its wing-like solar array will cause the spacecraft to lose some of its momentum during each close approach, known as an "a drag pass." As the spacecraft slows during each close approach, the orbit will gradually lower and circularize.

Aerobraking will occur in three primary phases that engineers call walk-in, the main phase and walk-out. The walk-in phase occurs during the first four to eight orbits following Mars arrival. The main aerobraking phase begins once the point of the spacecraft's closest approach to the planet, know as the orbit's "periapsis," has been lowered to within about 100 kilometers (60 miles) above the Martian surface. As the spacecraft's orbit is reduced and circularized during approximately 273 drag passes in 76 days, the periapsis will moved northward, almost directly over Mars' north pole. Small thruster firings when the spacecraft is at its most distant point from the planet will keep the drag pass altitude at the desired level to limit heating and dynamic pressure on the orbiter. The walk-out phase occurs during the last few days of aerobraking when the period of the spacecraft's orbit is the shortest.

The aerobraking drag pass events will be executed by stored onboard command sequences. The drag pass sequence begins with the heaters for the thrusters being warmed up for about 20 minutes. The transmitter is turned off to conserve power during the drag pass. The spacecraft then turns to the aerobraking attitude under reaction wheel control.

Following aerobraking walk-out, the orbiter will be in an elliptical orbit with a periapsis near an altitude of 120 kilometers (75 miles) and an "apoapsis" – the farthest point from Mars – near a desired 400-kilometer (249-mile) altitude. Periapsis will be near the equator. A maneuver to raise the periapsis will be performed to achieve the final 400-kilometer (249-mile) circular science orbit.

The transition from aerobraking to the beginning of the science orbit will take about one week. The high-gain antenna will be deployed during this time and the spacecraft and science instruments will be checked out.

NASA's Langley Research Center in Hampton, Virginia, will provide aerobraking support to JPL's navigation team during mission operations. Langley's role includes performing independent verification and validation, developing simulation tools and assisting the navigation team with trade studies and performance analysis.

Mapping Orbit

The science mission begins about 45 days after the spacecraft is captured into orbit about Mars. The primary science phase will last for 917 Earth days. The science orbit inclination is 93.1 degrees, which results in a nearly Sun-synchronous orbit. The orbit period will be just under two hours. Successive ground tracks are separated in longitude by approximately 29.5 degrees and the entire ground track nearly repeats every two sols, or Martian days.

During the science phase, the thermal emission imaging system will take multispectral thermal-infrared images to make a global map of the minerals on the Martian surface, and will also acquire visible images with a resolution of about 18 meters (59 feet). The gamma-ray spectrometer will take global measurements during all Martian seasons. The Martian radiation environment experiment will be operated throughout the science phase to collect data on the planet's radiation environment. Opportunities for science collection will be assigned on a time-phased basis depending on when conditions are most favorable for specific instruments.

Relay Phase

The relay phase begins at the end of the first Martian year in orbit (about two Earth years). During this phase the orbiter will provide communication support for U.S. and international landers and rovers.

Spacecraft

The shape of *2001 Mars Odyssey* is anything but uniform, but its size can most easily be visualized by mentally placing the spacecraft inside of a box. Pictured this way, the box would measure 2.2 meters (7.2 feet) long, 1.7 meters (5.6 feet) tall and 2.6 meters (8.5 feet) wide. At launch *Odyssey* weighs 725.0 kilograms (1598.4 pounds), including the 331.8-kilogram (731.5-pound) dry spacecraft with all of its subsystems, 348.7 kilograms (768.8 pounds) of fuel and 44.5 kilograms (98.1 pounds) of instruments.

The framework of the spacecraft is composed mostly of aluminum and some titanium. The use of titanium, a lighter and more expensive metal, is an efficient way of conserving mass while retaining strength. *Odyssey*'s metal structure is similar to that used in the construction of high-performance and fighter aircraft.

Most systems on the spacecraft are fully redundant. This means that, in the event of a device failure, there is a backup system to compensate. The main exception is a memory card that collects imaging data from the thermal emission imaging system.

Command and Data Handling

All of *Odyssey*'s computing functions are performed by the command and data handling subsystem. The heart of this subsystem is a RAD6000 computer, a radiation-hardened version of the PowerPC chip used on most models of Macintosh computers. With 128 megabytes of random access memory (RAM) and three megabytes of non-volatile memory, which allows the system to maintain data even without power, the subsystem runs *Odyssey*'s flight software and controls the spacecraft through interface electronics.

Interface electronics make use of computer cards to communicate with external peripherals. These cards slip into slots in the computer's main board, giving the system specific functions it would not have otherwise. For redundancy purposes, there are two identical strings of these computer and interface electronics, so that if one fails the spacecraft can switch to the other.

Communication with *Odyssey*'s sensors that measure the spacecraft's orientation in space, or "attitude," and its science instruments is done via another interface card. A master input / output card collects signals from around the spacecraft and also sends commands to the electrical power subsystem. The interface to *Odyssey*'s telecommunications subsystems exists through another card called the uplink / downlink card.

There are two other boards in the command and data handling subsystem, both internally redundant. The module interface card controls when the spacecraft switches to backup hardware and serves as the spacecraft's time clock. A converter card takes electricity produced by the power subsystem and converts it into the proper voltages for the rest of the command and data handling subsystem components.

The last interface card is a single, non-redundant, one-gigabyte mass memory card that is used to store imaging data. The entire command and data handling subsystem weighs 11.1 kilograms (24.5 pounds).

Telecommunications

Odyssey's telecommunications subsystem is composed of both a radio system operating in the X-band microwave frequency range and a system that operates in the ultra high frequency (UHF) range. It provides communication capability throughout all phases of the mission. The X-band system is used for communications between Earth and the orbiter, while the UHF system will be used for communications between *Odyssey* and future Mars landers. The telecommunication subsystem weighs 23.9 kilograms (52.7 pounds).

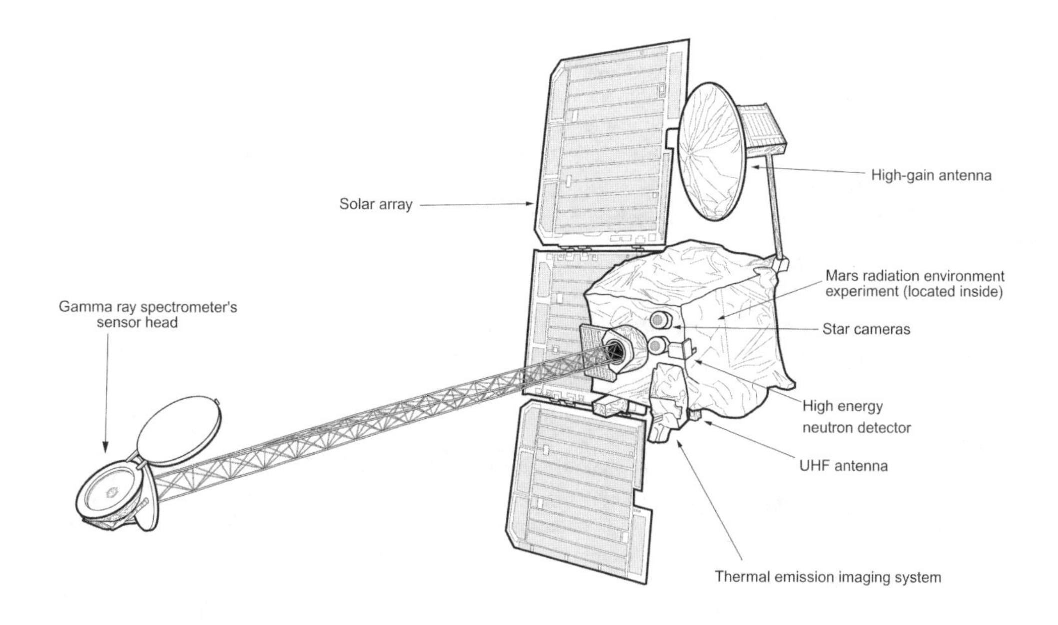

2001 Mars Odyssey spacecraft

Electrical Power

All of the spacecraft's power is generated, stored and distributed by the electrical power subsystem. The system obtains its power from an array of gallium arsenide solar cells on a panel measuring seven square meters (75 square feet). A power distribution and drive unit contains switches that send power to various electrical loads around the spacecraft. Power is also stored in a 16-amp-hour nickel-hydrogen battery.

The electrical power subsystem operates the gimbal drives on the high-gain antenna and the solar array. It contains also a pyro initiator unit, which fires pyrotechnically actuated valves, activates burn wires, and opens and closes thruster valves. The electrical power subsystem weighs 86.0 kilograms (189.6 pounds).

Guidance, Navigation and Control

Using three redundant pairs of sensors, the guidance, navigation and control subsystem determines the spacecraft's orientation, or "attitude." A Sun sensor is used to detect the position of the Sun as a backup to the star camera. A star camera is used to look at star fields. Between star camera updates, a device called the inertial measurement unit collects information on spacecraft orientation.

This system also includes the reaction wheels, gyro-like devices used along with thrusters to control the spacecraft's orientation. Like most spacecraft, *Odyssey*'s orientation is held fixed in relation to space ("three-axis stabilized") as opposed to being stabilized via spinning. There are a total of four reaction wheels, with three used for primary control and one as a backup. The guidance, navigation and control subsystem weighs 23.4 kilograms (51.6 pounds).

Propulsion

The propulsion subsystem features sets of small thrusters and a main engine. The thrusters are used to perform *Odyssey*'s attitude control and trajectory correction maneuvers, while the main engine is used to place the spacecraft in orbit around Mars.

The main engine, which uses hydrazine propellant with nitrogen tetroxide as an oxidizer, produces a minimum thrust of 65.3 kilograms of force (144 pounds of force). Each of the four thrusters used for attitude control produce a thrust of 0.1 kilogram of force (0.2 pound of force). Four 2.3-kilogram-force (5.0-pound-force) thrusters are used for turning the spacecraft.

In addition to miscellaneous tubing, pyro valves and filters, the propulsion subsystem also includes a single gaseous helium tank used to pressurize the fuel and oxidizer tanks. The propulsion subsystem weighs 49.7 kilograms (109.6 pounds).

Structures

The spacecraft's structure is divided into two modules. The first is a propulsion module, containing tanks, thrusters and associated plumbing. The other, the equipment module, is composed of an equipment deck, which supports engineering components and the radiation experiment, and a science deck connected by struts. The top side of the science deck supports the thermal emission imaging system, gamma-ray spectrometer, the high-energy neutron detector, the neutron spectrometer and the star cameras, while the underside supports engineering components and the gamma-ray spectrometer's central electronics box. The structures subsystem weighs 81.7 kilograms (180.1 pounds).

Thermal control

The thermal control subsystem is responsible for maintaining the temperatures of each component on the spacecraft to within their allowable limits. It does this using a combination of heaters, radiators, louvers, blankets and thermal paint. The thermal control subsystem weighs 20.3 kilograms (44.8 pounds).

Mechanisms

There are a number of mechanisms used on *Odyssey*, several of which are associated with its high-gain antenna. Three retention and release devices are used to lock the antenna down during launch, cruise and aerobraking. Once the science orbit is attained at Mars, the antenna is released and deployed with a motor-driven hinge. The antenna's position is controlled with a two-axis gimbal assembly.

There are also four retention and release devices used for the solar array. The three panels of the array are folded together and locked down for launch. After deployment, the solar array is also controlled using a two-axis gimbal assembly.

The last mechanism is a retention and release device for the deployable 6-meter (19.7-feet) boom for the gamma-ray spectrometer. All of the mechanisms combined weigh 24.2 kilograms (53.4 pounds).

Flight Software

Odyssey receives its commands via radio from Earth and translates them into spacecraft actions. The flight software is capable of running multiple concurrent sequences, as well as executing immediate commands as they are received.

The software responsible for the data collection is extremely flexible. It collects data from the science and engineering devices and puts them in a variety of holding bins. The choice of which channel is routed to which holding bin, and how often it is sampled, is easily modified via ground commands.

The flight software is also responsible for a number of autonomous functions, such as attitude control and fault protection, which involves frequent internal checks to determine if a problem has occurred. If the software senses a problem, it will automatically perform a number of preset actions to resolve the problem and put the spacecraft in a safe standby awaiting further direction from ground controllers.

Science Objectives

One of the chief scientific goals that *2001 Mars Odyssey* will focus on is mapping the chemicals and minerals that make up the Martian surface. As on Earth, the geology and elements that form the Martian planet chronicle its history. And while neither elements, the building blocks of minerals, nor minerals, the building blocks of rocks, can convey the entire story of a planet's evolution, both contribute significant pieces to the puzzle. These factors have profound implications for understanding the evolution of Mars' climate and the role of water on the planet, the potential origin and evidence of life, and the possibilities that may exist for future human exploration.

Other major goals of the Odyssey mission are to:

Determine the abundance of hydrogen, most likely in the form of water ice, in the shallow subsurface
Globally map the elements that make up the surface
Acquire high-resolution thermal infrared images of surface minerals
Provide information about the structure of the Martian surface
Record the radiation environment in low Mars orbit as it relates to radiation-related risk to human exploration

During the 917-day science mission, *Odyssey* will also serve as a communication relay for U.S. or international scientific orbiters and landers in 2003 and 2004. After this period, the orbiter will be available as a communication relay for an additional 457 days, making for a total mission duration of 1,374 days, or two Martian years. Science operations may still continue during the communication relay-only phase depending on remaining orbiter resources.

The orbiter carries three science instruments: a thermal infrared imaging system, a gamma-ray spectrometer and a radiation environment experiment. These are all calibrated during the spacecraft's cruise phrase on its way to Mars. Opportunities for data collection are assigned on a time-phased basis depending on when conditions are most favorable for specific instruments.

Thermal Emission Imaging System
This instrument is responsible for determining Mars' surface mineralogy. Unlike our eyes, which can only detect visible light waves, a small portion of the electromagnetic spectrum, the instrument can see in both visible and infrared, thus collecting imaging data that has been previously invisible the scientists.

In the infrared spectrum, the instrument uses 10 spectral bands to help detect minerals within the Martian terrain. These spectral bands, similar to ranges of colors, serve as signatures, or spectral fingerprints, of particular types of geological materials.

Minerals, such as carbonates, silicates, hydroxides, sulfates, hydrothermal silica, oxides and phosphates, all show up as different colors in the infrared spectrum. This multispectral method allows researchers to detect in particular the presence of minerals that form in water and understand those minerals in their proper geological context.

Remote-sensing studies of natural surfaces, together with laboratory measurements, have demonstrated that 10 spectral bands are sufficient to detect minerals at abundances of five to 10 percent. In addition, the use of 10 infrared spectral bands can determine the absolute mineral abundance in a specific location within 15 percent.

The instrument's multispectral approach will also provide data on localized deposits associated with hydrothermal and subsurface water and enable 100-meter (328-feet) resolution mapping of the entire planet. In essence, this allows a broad geological survey of the planet for the purpose of identifying minerals, with 100 meters (328 feet) of Martian terrain captured in each pixel, or single point, of every image. It will also allow the instrument to search for thermal spots during the night that could result in discovering hot springs on Mars.

Using visible imaging in five spectral bands, the experiment will also take 18-meter (59-feet) resolution mineralogical and structural measurements specifically to determine the geological record of past liquid environments. More than

15,000 images each 20 by 20 kilometers (12 by 12 miles) will be acquired for Martian surface studies. These more detailed data will be used in conjunction with mineral maps to identify potential future Martian landing sites. These images will provide an important bridge between the data acquired by the *Viking* missions and the high-resolution images captured by *Mars Global Surveyor*.

The instrument weighs 11.2 kilograms (24.7 pounds); is 54.5 centimeters (21.5 inches) long, 34.9 centimeters (13.7 inches) tall and 28.6 centimeters (11.3 inches) wide; and runs on 17 watts of electrical power.

The principal investigator for the instrument is Dr. Philip Christensen of Arizona State University in Tempe.

Gamma-Ray Spectrometer

This instrument plays a lead role in determining the elemental makeup of the Martian surface. Using a gamma-ray spectrometer and two neutron detectors, the experiment detects and studies gamma-rays and neutrons emitted from the planet's surface.

When exposed to cosmic rays, all chemical elements emit gamma-rays with distinct signatures. This spectrometer looks at these signatures, or energies, coming from the elements present in the Martian soil. By measuring gamma-rays coming from the Martian surface, it is possible to calculate how abundant various elements are and how they are distributed around the planet's surface.

By measuring neutrons, it is possible to calculate Mars' hydrogen abundance, thus inferring the presence of water. The neutron detectors are sensitive to concentrations of hydrogen in the upper meter of the surface.

Gamma-rays, emitted from the nuclei of atoms, show up as sharp emission lines on the instrument's spectrum. While the energy represented in these emissions determines which elements are present, the intensity of the spectrum reveals the elements' concentrations. The spectrometer will send a reading to Earth every 20 seconds. This data will be collected over time and used to build up a full-planet map of elemental abundances and their distributions.

The spectrometer's data, collected at 300-kilometer (186-mile) resolution, will enable researchers to address many questions and problems regarding Martian geoscience and life science, including crust and mantle composition, weathering processes and volcanism. The spectrometer is expected to add significantly to the growing understanding of the origin and evolution of Mars and of the processes shaping it today and in the past.

The gamma-ray spectrometer consists of two main components: the sensor head and the central electronics assembly. The sensor head is separated from the rest of the *Odyssey* spacecraft by a 6-meter (20-feet) boom, which will be extended after *Odyssey* has entered the mapping orbit at Mars. This is done to minimize interference from any gamma-rays coming from the spacecraft itself. The initial spectrometer activity, lasting between 15 and 40 days, will perform an instrument calibration before the boom is deployed. After 100 days in orbit, the boom will deploy and remain in this position for the duration of the mission. The two neutron detectors – the neutron spectrometer and the high-energy neutron detector – are mounted on the main spacecraft structure and will operate continuously throughout the mission.

The instrument weighs 30.2 kilograms (66.6 pounds) and uses 32 watts of power. Along with its cooler, the gamma-ray spectrometer measures 46.8 centimeters (18.4 inches) long, 53.4 centimeters (21.0 inches) tall and 60.4 centimeters (23.8 inches) wide. The neutron spectrometer is 17.3 centimeters (6.8 inches) long, 14.4 centimeters (5.7 inches) tall and 31.4 centimeters (12.4 inches) wide. The high-energy neutron detector measures 30.3 centimeters (11.9 inches) long, 24.8 centimeters (9.8 inches) tall and 24.2 centimeters (9.5 inches) wide. The instrument's central electronics box is 28.1 centimeters (11.1 inches) long, 24.3 centimeters (9.6 inches) tall and 23.4 centimeters (9.2 inches) wide.

The principal investigator for the gamma-ray spectrometer is Dr. William Boynton of the University of Arizona.

Martian Radiation Environment Experiment

This instrument characterizes aspects of the radiation environment both on the way to Mars and in the Martian orbit. Since space radiation presents an extreme hazard to crews of interplanetary missions, the experiment will attempt to predict anticipated radiation doses that would be experienced by future astronauts and help determine possible effects of Martian radiation on human beings.

Space radiation comes from two sources – energetic particles from the Sun and galactic cosmic rays from beyond our solar system. Both kinds of radiation can trigger cancer and cause damage to the central nervous system. A spectrometer inside the instrument will measure the energy from these radiation sources. As the spacecraft orbits the red planet, the spectrometer sweeps through the sky and measures the radiation field.

The instrument, with a 68-degree field of view, is designed to continuously collect data during *Odyssey*'s cruise from Earth to Mars. It can stores large amounts of data for downlink whenever possible, and will operate throughout the entire science mission.

The instrument weighs 3.3 kilograms (7.3 pounds) and uses 7 watts of power. It measures 29.4 centimeters (11.6 inches) long, 23.2 centimeters (9.1 inches) tall and 10.8 centimeters (4.3 inches) wide.

The principal investigator for the radiation environment experiment is Dr. Gautum Badhwar of NASA's Johnson Space Center.

Program / Project Management

The *2001 Mars Odyssey* mission is managed by the Jet Propulsion Laboratory, Pasadena, California, for NASA's Office of Space Science, Washington, D.C. At NASA Headquarters, Dr. Edward Weiler is the Associate Administrator for Space Science, G. Scott Hubbard is the Mars Program Director, Dr. Jim Garvin is the Lead Scientist for the Mars Exploration Program, Mark Dahl is the 2001 Mars Odyssey Program Executive, and Dr. Michael Meyer is the 2001 Mars Odyssey Program Scientist.

At the Jet Propulsion Laboratory, Dr. Firouz Naderi is the Mars Program Manager, Dr. Dan McCleese is Mars Program Scientist, George Pace is the 2001 Mars Odyssey Project Manager and Dr. R. Stephen Saunders is the 2001 Mars Odyssey Project Scientist.

At Lockheed Martin Astronautics, Denver, Colorado, Robert L. Berry is the company's Mars Program Director.

NASA's *Odyssey* to Mars

2001 Mars Odyssey is part of NASA's Mars Exploration Program, a long-term effort of robotic exploration of the red planet. The opportunity to go to Mars comes around every 26 months, when the alignment of Earth and Mars in their orbits around the sun allows spacecraft to travel between the two planets with the least amount of energy. *2001 Mars Odyssey* launched on April 7, 2001, and arrived at Mars on October 24, 2001, 0230 Universal Time (October 23, 7:30 pm PDT/ 10:30 EDT).

Odyssey's primary science mission will take place February 2002 through August 2004. For the first time, the mission will map the amount and distribution of chemical elements and minerals that make up the Martian surface. The spacecraft will especially look for hydrogen, most likely in the form of water ice, in the shallow subsurface of Mars. It will also record the radiation environment in low-Mars orbit to determine the radiation related risk to any future human explorers who may one day go to Mars. All of these objectives support the four science goals of the Mars Exploration Program.

The three primary instruments carried by *2001 Mars Odyssey* are:

THEMIS (Thermal Emission Imaging System), for determining the distribution of minerals, particularly those that can only form in the presence of water;

GRS (Gamma Ray Spectrometer), for determining the presence of 20 chemical elements on the surface of Mars, including hydrogen in the shallow subsurface (which acts as a proxy for determining the amount and distribution of possible water ice on the planet); and,

MARIE (Mars Radiation Environment Experiment), for studying the radiation environment.

During and after its science mission, the *Odyssey* orbiter will also support other missions in the Mars Exploration program. It will provide the communications relay for U.S. and international landers, including the next mission in NASA's Mars Program, the Mars Exploration Rovers to be launched in 2003. Scientists and engineers will also use *Odyssey* data to identify potential landing sites for future Mars missions.

The name *2001 Mars Odyssey* was selected as a tribute to the vision and spirit of space exploration as embodied in the works of renowned science fiction author Arthur C. Clarke. Evocative of one of his most celebrated works, the name speaks to our hopes for the future and of the fundamental human desire to explore the unknown despite great dangers, the risk of failure and the daunting, enormous depths of space.

The 2001 Mars Odyssey mission makes use of many innovative technologies, but the most important among them are the three instrument packages that will carry out science investigations once the spacecraft arrives at Mars. All three involve the use of spectrometers.

2001 Mars Odyssey makes use of several kinds of spectrometers:

THEMIS: The Thermal Emission Imaging System is a camera that images Mars in the visible and infrared parts of the spectrum in order to determine the distribution of minerals on the surface of Mars.

GRS: The Gamma Ray Spectrometer uses the gamma-ray part of the spectrum to look for the presence of 20 elements from the periodic table (e.g., carbon, silicon, iron, magnesium, etc.). Its neutron detectors look for water and ice in the soil by measuring neutrons.

MARIE: The Martian Radiation Experiment is designed to measure the radiation environment of Mars using an energetic particle spectrometer.

Spectrometers

Spectrometers are instruments that allow scientists to collect data that would otherwise be invisible to us. Our eyes are sophisticated detectors that can reveal much of the world around us, but they are only sensitive to a very small part of the electromagnetic spectrum that characterizes light.

We call the part of the electromagnetic spectrum that we can see "visible" or "optical" light. To fully appreciate the complexity of the world around us, however, we need to rely on human-made devices to provide views of the "invisible" world – that is, the parts of the electromagnetic spectrum we cannot see without the aid of technology: gamma-rays, X-rays, ultraviolet waves, infrared waves, microwaves, and radio waves.

All of these different types of energy are "light," even if our human eyes can only "see" part of it. The only difference between them is wavelength (the distance between the peaks of each). Wavelengths get larger as we move across the electromagnetic spectrum from gamma-rays to radio waves. The wavelength of visible light is about $^{1}/_{10}{}^{th}$ of a micrometer, but the full electromagnetic spectrum includes both shorter and longer wavelengths.

Familiar ways of studying "invisible light" (with wavelengths greater or less than what our eye can see) include x-rays for medical diagnosis and radar for guiding airplanes. By using spectrometers at Mars, scientists can learn a great deal about the planet's composition (what it is made of) and its radiation environment. None of this knowledge would be possible if we only relied on our eyes.

How Spectrometers Work

Spectrometers can essentially spread light out into its wavelengths to create spectra, which look something like rainbow-colored bars. Within these spectra, scientists can study the emission and absorption lines that provide "fingerprints" of any atoms and molecules that may be present. Each atom has a unique fingerprint because they each can only emit or absorb certain energies or wavelengths. That is why the location and spacing of spectral lines – the fingerprint – is unique for each atom. Spectrometers are the instruments that engineers build to detect these kinds of fingerprints.

Odyssey Science Objectives

Mars Odyssey's job is to map chemical elements and minerals on the surface of Mars, look for water in the shallow subsurface, and analyze the radiation environment to determine its potential effects on human health.

Odyssey's science investigations directly support the Mars Exploration Program's overall science strategy of " Following the Water". The four science goals that support this strategy for discovery are:

Goal 1: Determine whether Life ever arose on Mars

While Odyssey does not carry instruments for detecting life on Mars, data from the mission will help us understand whether the environment of Mars was – or is – conducive to life. One of the fundamental requirements for life as we know it is the presence of liquid water. For the first time at Mars, we'll have a spacecraft that is equipped to find evidence of present near-surface water and to map mineral deposits from past water activity. It will also be able to identify regions with hot springs, if any exist, which would be prime areas in which to search for signs of simple life in future missions.

Goal 2: Characterize the Climate of Mars

Mars today is far too cold with an atmosphere that is far too thin to support liquid water on the surface. However, much of the water on Mars is probably trapped under the surface, either as ice or possibly in liquid form if any exists near a heat source on the planet. Odyssey will allow scientists to measure the amount of permanent ground ice and how it changes with the seasons. In addition, Odyssey's studies of the geologic landforms and minerals – especially those that formed in the presence of water – will help us understand the role of water in the evolution of the Martian climate since the planet first formed some 4.5 billion years ago.

Goal 3: Characterize the Geology of Mars

For the first time, 2001 Mars Odyssey will determine the chemical elements (e.g., carbon, silicon, iron, etc.) and minerals that make up the planet Mars, as well as help explain how the planet's landforms developed over time. The chemical elements are the building blocks of minerals, minerals are the building blocks of rocks, and all of these relate to the structure and landforms of the Martian surface. This understanding in turn provides clues to the geological and climatic history of Mars and the potential for finding past or present life.

Goal 4: Prepare for Human Exploration

The Mars Radiation Environment Experiment will give us a first look at the radiation levels at Mars as they relate to the potential hazards faced by possible future astronaut crews. The experiment will take data on the way to Mars and in orbit, so that future mission designers will know better how to outfit human explorers for their journey to the red planet.

Innovative applications of technology in the 2001 Mars Odyssey mission enable us to meet these science goals.

The 2001 Mars Odyssey Mission Summary

2001 Mars Odyssey is an orbiter carrying science experiments designed to make global observations of Mars to improve our understanding of the planet's climate and geologic history, including the search for water and evidence of life-sustaining environments. The mission will extend for more than a full Martian year (2½ Earth years).

Mars Odyssey was launched April 7, 2001 on a Delta II rocket from Cape Canaveral, Florida, and reached Mars on October 24, 2001, 0230 Universal Time (October 23, 7:30 pm PDT/ 10:30 EDT). The spacecraft's main engine fired to brake the spacecraft's speed and allowed it to be captured into orbit around Mars. Odyssey used a technique called "aerobraking" that gradually brought the spacecraft closer to Mars with each orbit. By using the atmosphere of Mars to slow down the spacecraft in its orbit rather than firing its engine or thrusters, Odyssey was able to save more than 200 kilograms (440 pounds) of propellant.

Aerobraking ended in January, and began its science mapping mission in February. The primary science mission will continue through August 2004. The spacecraft will also serve as a communications relay for U.S. and international spacecraft scheduled to arrive at Mars in 2003 and 2004.

Mars Odyssey's THEMIS Begins Posting Daily Images

News Release
James Hathaway, (480) 965-6375
Hathaway@asu.edu
March 27, 2002

Need to get away to someplace exotic? Mars is now open for daily sightseeing.

Beginning March 27, 2002, recent images of Mars taken by the Thermal Emission Imaging System on NASA's Mars Odyssey spacecraft will be available to the public on the Internet. A new, "uncalibrated" image taken by the visible light camera will be posted at 10 A.M. EST daily, Monday through Friday. The pictures can be viewed and downloaded at http://themis.asu.edu/latest.html .

The images will show 22 kilometer-wide strips of the martian surface at a resolution of 18 meters. Though the images will not yet be fully calibrated for scientific use, they give the public an unprecedented opportunity to get a close look at many of Mars' unusual geological features. The visible light camera's resolution is about eight to 16 times better than most of the images taken by NASA's Viking missions, which completed the first global map of the martian surface.

"We want to generate a steady flow of images so we can share some of the excitement of what we're seeing with the public," said Greg Mehall, THEMIS mission manager at Arizona State University. "We're seeing a lot of very interesting things, since much of Mars has never been viewed so closely before."

Though the posted images have undergone only minimal image processing, the team wanted to share them with the public as soon as possible. "They're still pretty spectacular to look at," Mehall said. "And we want people to feel they are getting a first look at the images with us."

THEMIS began mapping Mars from an orbit of 420 kilometers in mid-February, taking images in both infrared and visible light The instrument is expected to take as many as 15,000 visible light images through the course of the mission.

The Jet Propulsion Laboratory, a division of the California Institute of Technology in Pasadena, manages the 2001 Mars Odyssey mission for NASA's Office of Space Science in Washington. Investigators at Arizona State University in Tempe, the University of Arizona in Tucson and NASA's Johnson Space Center, Houston, operate the science instruments. Additional science partners are located at the Russian Aviation and Space Agency and at Los Alamos National Laboratories, New Mexico.

Lockheed Martin Astronautics, Denver, is the prime contractor for the project, and developed and built the orbiter. Mission operations are conducted jointly from Lockheed Martin and from JPL. Additional information about the 2001 Mars Odyssey is available on the Internet at: http://mars.jpl.nasa.gov/odyssey/
ASU

Mars Rover Landing Sites Chosen

April 11 2003

NASA has selected the two landing sites for the Mars Exploration Rovers, and the winners are Gusev crater and Meridiani Planum. THEMIS data was extensively used in the landing site selection process. Onboard each rover will be one of our instruments (MiniTES) from Arizona State University. The twin rovers are due to touchdown on Mars next January 4 and January 25, 2004.

Gusev Crater

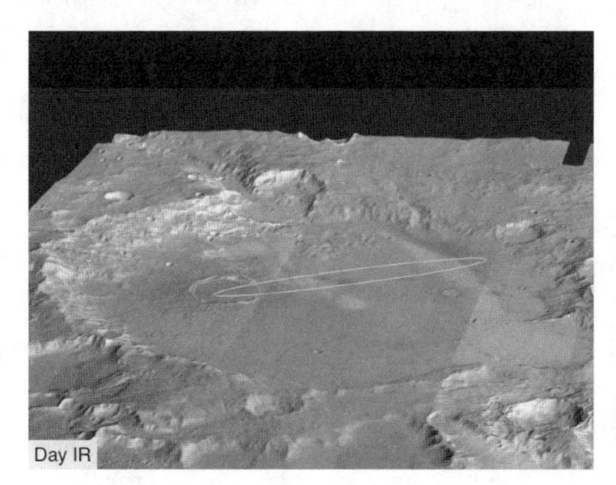

The temperature of Gusev Crater, Mars. A mosaic of THEMIS daytime infrared images of Gusev Crater has been draped over the martian topography determined from the Mars Global Surveyor MOLA instrument. The daytime temperatures range from approximately -45 ° C (black) to -5 ° C (white). The temperature differences in these daytime images are due primarily to lighting effects, where sunlit slopes are warm (bright) and shadowed slopes are cool (dark). Gusev crater is one of the selected landing sites for the Mars Exploration Rovers, to be launched to Mars this summer. The large ancient river channel of Ma'Adim that once flowed into Gusev can be seen at the top of the mosaic. The resolution of the THEMIS data is 100 m per pixel. This THEMIS mosaic covers an area approximately 180 km on each side centered near 14° S, 175° E, looking toward the south in this simulated view. These images were acquired using the THEMIS infrared Band 9 centered at 12.6 μm Image Credit: NASA/JPL/ASU

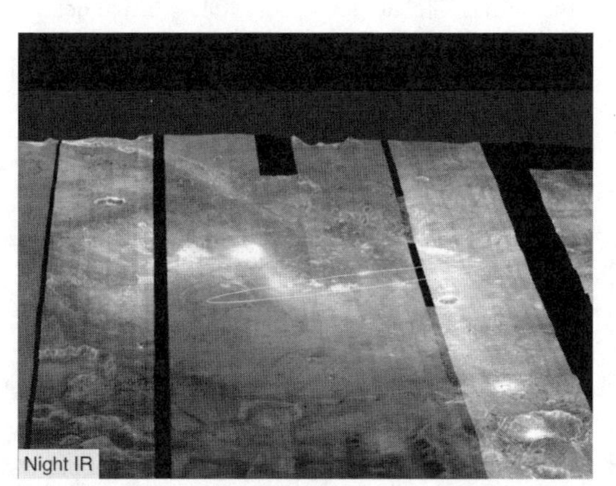

Gusev Crater at night. A mosaic of THEMIS nighttime infrared images of Gusev Crater has been draped over the martian topography determined from the Mars Global Surveyor MOLA instrument. The nighttime temperature differences are due to differences in the abundance of rocky materials that retain their heat at night and stay relatively warm (bright). Fine grained dust and sand (dark) cools off more rapidly at night. This THEMIS mosaic covers an area approximately 180 km on each side centered near 14° S, 175° E, looking toward the south in this simulated view. These images were acquired using the THEMIS infrared Band 9 centered at 12.6 μm.
Image Credit: NASA/JPL/ASU

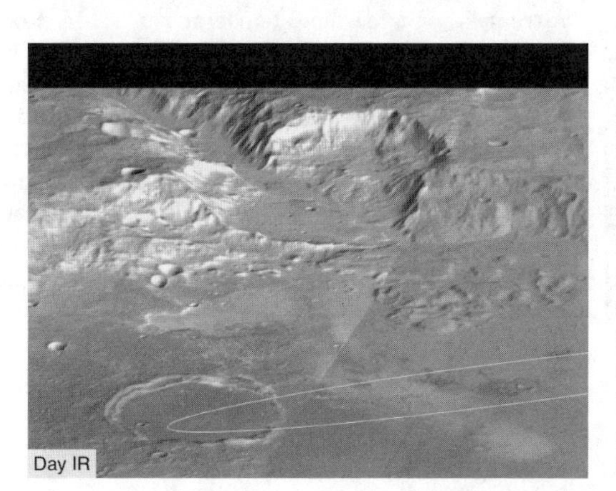

Close up view of Gusev crater landing ellipse. A mosaic of THEMIS daytime infrared images of Gusev crater has been draped over the martian topography determined from the Mars Global Surveyor MOLA instrument. The daytime temperatures range from approximately -45 ° C (black) to -5 ° C (white). The temperature differences in these daytime images are due primarily to lighting effects, where sunlit slopes are warm (bright) and shadowed slopes are cool (dark). Gusev crater is one of the selected landing sites for the Mars Exploration Rovers, to be launched to Mars this summer. The resolution of the THEMIS data is 100 m per pixel. The length of the landing ellipse is approximately 100 km and the width is approximately 20 km.
Image Credit: NASA/JPL/ASU

Meridiani Planum

The temperature of Meridiani Planum, Mars. A mosaic of THEMIS daytime infrared images of Meridiani Planum has been draped over the martian topography determined from the Mars Global Surveyor MOLA instrument. The daytime temperatures range from approximately -45 ° C (black) to -5 ° C (white). The temperature differences in these daytime images are due primarily to lighting effects, where sunlit slopes are warm (bright) and shadowed slopes are cool (dark). Meridiani Planum is one of the selected landing sites for the Mars Exploration Rovers, to be launched to Mars this summer. The resolution of the THEMIS data is 100 m per pixel. This THEMIS mosaic covers an area approximately 250 km on each side centered near 0°, 355° E, looking toward the north in this simulated view. These images were acquired using the THEMIS infrared Band 9 centered at 12.6 μm. Image Credit: NASA/JPL/ASU

Meridiani Planum at night. A mosaic of THEMIS nighttime infrared images of Meridiani Planum has been draped over the martian topography determined from the Mars Global Surveyor MOLA instrument. The nighttime temperature differences are due to differences in the abundance of rocky materials that retain their heat at night and stay relatively warm (bright). Fine grained dust and sand (dark) cools off more rapidly at night. This THEMIS mosaic covers an area approximately 250 km on each side centered near 0°, 355° E, looking toward the north in this simulated view. These images were acquired using the THEMIS infrared Band 9 centered at 12.6 μm. Image Credit: NASA/JPL/ASU Image Credit: NASA/JPL/ASU

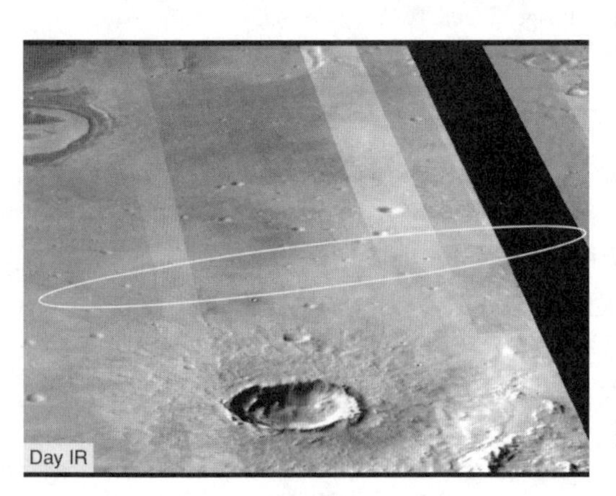

Close up view of Meridiani Planum landing ellipse. A mosaic of THEMIS daytime infrared images of Meridiani Planum has been draped over the martian topography determined from the Mars Global Surveyor MOLA instrument. The daytime temperatures range from approximately -45 ° C (black) to -5 ° C (white). The temperature differences in these daytime images are due primarily to lighting effects, where sunlit slopes are warm (bright) and shadowed slopes are cool (dark). Meridiani Planum is one of the selected landing sites for the Mars Exploration Rovers, to be launched to Mars this summer. The resolution of the THEMIS data is 100 m per pixel. The length of the landing ellipse is approximately 120 km and the width is approximately 20 km.

Isidis Planitia

The temperature of Isidis Planitia, Mars. A mosaic of THEMIS daytime infrared images of Isidis Planitia has been draped over the martian topography determined from the Mars Global Surveyor MOLA instrument. The daytime temperatures range from approximately -45 ° C (black) to -5 ° C (white). The temperature differences in these daytime images are due primarily to lighting effects, where sunlit slopes are warm (bright) and shadowed slopes are cool (dark). Isidis Planitia was a potential landing site for the Mars Exploration Rovers, to be launched to Mars this summer. The resolution of the THEMIS data is 100 m per pixel. This THEMIS mosaic covers an area approximately 250 km on each side centered near 5° N, 88° E, looking toward the south in this simulated view. These images were acquired using the THEMIS infrared Band 9 centered at 12.6 μm.
Image Credit: NASA/JPL/ASU

Isidis Planitia at night. A mosaic of THEMIS nighttime infrared images of Isidis Planitia has been draped over the martian topography determined from the Mars Global Surveyor MOLA instrument. The nighttime temperature differences are due to differences in the abundance of rocky materials that retain their heat at night and stay relatively warm (bright). Fine grained dust and sand (dark) cools off more rapidly at night. This THEMIS mosaic covers an area approximately 250 km on each side centered near 14° S, 175° E, looking toward the south in this simulated view. These images were acquired using the THEMIS infrared Band 9 centered at 12.6 μm. Image Credit: NASA/JPL/ASU

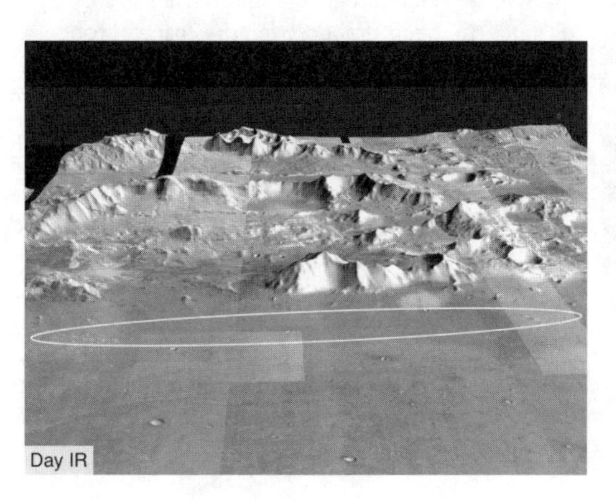

Close up view of Isidis Planitia landing ellipse. A mosaic of THEMIS daytime infrared images of Isidis Planitia has been draped over the martian topography determined from the Mars Global Surveyor MOLA instrument. The daytime temperatures range from approximately -45 ° C (black) to -5 ° C (white). The temperature differences in these daytime images are due primarily to lighting effects, where sunlit slopes are warm (bright) and shadowed slopes are cool (dark). Isidis Planitia was one of the candidate landing sites for the Mars Exploration Rovers, to be launched to Mars this summer. The resolution of the THEMIS data is 100 m per pixel. The length of the landing ellipse is approximately 135 km and the width is approximately 20 km.
Image Credit: NASA/JPL/ASU

Elysium Planitia

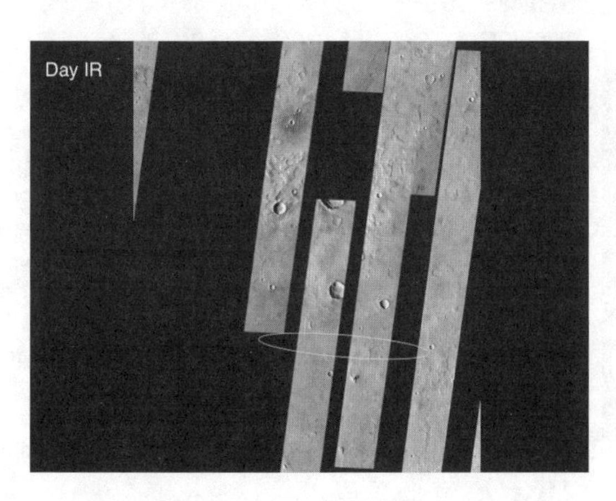

The temperature of Elysium Planitia, Mars. A mosaic of THEMIS daytime infrared images of Elysium Planitia has been draped over the martian topography determined from the Mars Global Surveyor MOLA instrument. The daytime temperatures range from approximately -45 ° C (black) to -5 ° C (white). The temperature differences in these daytime images are due primarily to lighting effects, where sunlit slopes are warm (bright) and shadowed slopes are cool (dark). Elysium Planitia was a potential landing site for the Mars Exploration Rovers, to be launched to Mars this summer. The resolution of the THEMIS data is 100 m per pixel. This THEMIS mosaic covers an area approximately 500 km on each side centered near 13° N, 123° E, looking toward the south in this simulated view. These images were acquired using the THEMIS infrared Band 9 centered at 12.6 μm. Image Credit: NASA/JPL/ASU

Elysium Planitia at night. A mosaic of THEMIS nighttime infrared images of Elysium Planitia has been draped over the martian topography determined from the Mars Global Surveyor MOLA instrument. The nighttime temperature differences are due to differences in the abundance of rocky materials that retain their heat at night and stay relatively warm (bright). Fine grained dust and sand (dark) cools off more rapidly at night. This THEMIS mosaic covers an area approximately 500 km on each side centered near 13° N, 123° E, looking toward the south in this simulated view. These images were acquired using the THEMIS infrared Band 9 centered at 12.6 μm.
Image Credit: NASA/JPL/ASU

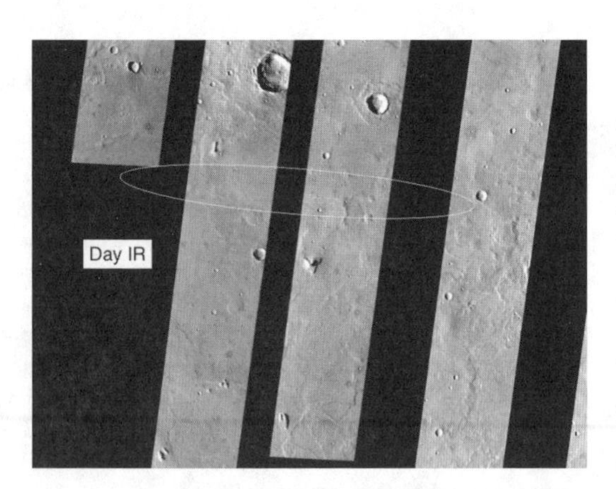

Close up view of Elysium Planitia landing ellipse. A mosaic of THEMIS daytime infrared images of Elysuim Planitia has been draped over the martian topography determined from the Mars Global Surveyor MOLA instrument. The daytime temperatures range from approximately -45 ° C (black) to -5 ° C (white). The temperature differences in these daytime images are due primarily to lighting effects, where sunlit slopes are warm (bright) and shadowed slopes are cool (dark). Elysium Planitia was one of the candidate landing sites for the Mars Exploration Rovers, to be launched to Mars this summer. The resolution of the THEMIS data is 100 m per pixel. The length of the landing ellipse is approximately 165 km and the width is approximately 15 km.

NASA's Mars Odyssey Points to Melting Snow as Cause of Gullies

February 19, 2003

MEDIA RELATIONS OFFICE JET PROPULSION LABORATORY CALIFORNIA INSTITUTE OF TECHNOLOGY
NATIONAL AERONAUTICS AND SPACE ADMINISTRATION PASADENA, CALIFORNIA 91109. TELEPHONE (818) 354-5011
http://www.jpl.nasa.gov

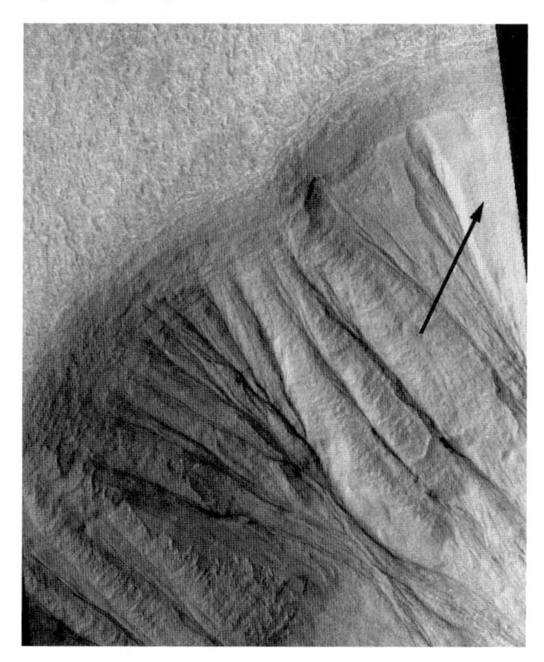

Gullies on martian crater, seen by Mars Global Surveyor

Images from the visible light camera on NASA's Mars Odyssey spacecraft, combined with images from NASA's Mars Global Surveyor, suggest melting snow is the likely cause of the numerous eroded gullies first documented on Mars in 2000 by Global Surveyor.

The now-famous martian gullies were created by trickling water from melting snow packs, not underground springs or pressurized flows, as had been previously suggested, argues Dr. Philip Christensen, the principal investigator for Odyssey's camera system and a professor from Arizona State University in Tempe. He proposes gullies are carved by water melting and flowing beneath snow packs, where it is sheltered from rapid evaporation in the planet's thin atmosphere. His paper is in the electronic February 19 issue of Nature.

Looking at an image of an impact crater in the southern mid-latitudes of Mars, Christensen noted eroded gullies on the crater's cold, pole-facing northern wall and immediately next to them a section of what he calls "pasted-on terrain." Such unique terrain represents a smooth deposit of material that Mars researchers have concluded is "volatile" (composed of materials that evaporate in the thin Mars atmosphere), because it characteristically occurs only in the coldest, most sheltered areas. The most likely composition of this slowly evaporating material is snow. Christensen suspected a special relationship between the gullies and the snow.

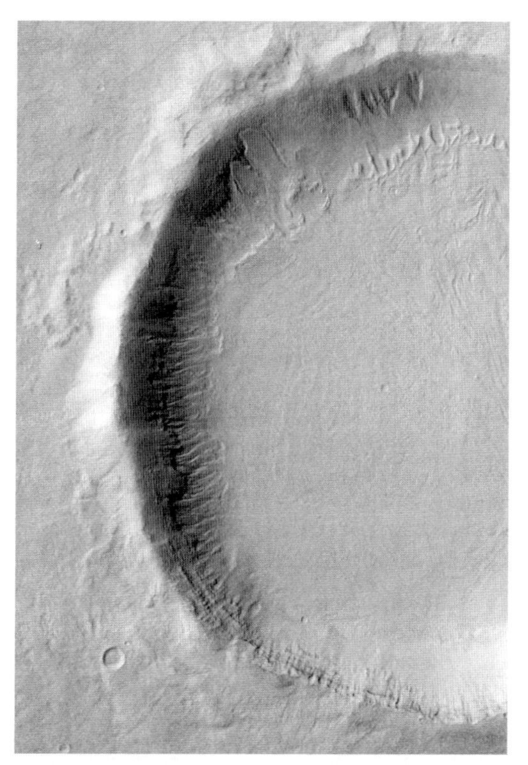

Gullies on martian crater, seen by Odyssey's Themis instrument

"The Odyssey image shows a crater on the pole-facing side has this 'pasted-on' terrain, and as you come around to the west there are all these gullies," said Christensen. "I saw it and said 'Ah-ha!' It looks for all the world like these gullies are being exposed as this terrain is being removed through melting and evaporation."

Eroded gullies on martian crater walls and cliff sides were first observed in images taken by Mars Global Surveyor in 2000. There have been other scientific theories offered to explain gully formation on Mars, including seeps of ground water, pressurized flows of ground water (or carbon dioxide), and mudflows caused by collapsing permafrost deposits, but no explanation to date has been universally accepted. The scientific community has remained puzzled, yet has been eagerly pursuing various possibilities.

"The gullies are very young," Christensen said. "That's always bothered me, because how is it that Mars has groundwater close enough to the surface to form these gullies, and yet the water has stuck around for billions of years? Second, you have craters with rims that are raised, and the gullies go almost to the crest of the rim. If it's a leaking subsurface aquifer, there's not much subsurface up there. And, finally, why do they occur preferentially on the cold face of the slope at mid-latitudes? If it's melting groundwater causing the flow, that's the coldest place, and the least likely place for that to happen."

Christensen points out that finding water erosion under melting snow deposits answers many of these problems, "Snow on Mars is most likely to accumulate on the pole-facing slopes, the coldest areas. It accumulates and drapes the landscape in these areas during one climate period, and then it melts during a warmer one. Melting begins first in the most exposed area right at the crest of the ridge. This explains why gullies start so high up." Once he started to think about snow, Christensen began finding a large number of other images showing a similar relationship between "pasted on" snow deposits and gullies in the high resolution images taken by the camera on Global Surveyor. Yet it was the unique mid-range resolution of the visible light camera in Mars Odyssey's thermal emission imaging system that was critical for the insight, because of its wide field of view.

"It was almost like finding a Rosetta Stone. The basic idea comes out of having a regional view, which Odyssey's camera system gives. It's a kind of you-can't-see-the forest-for-the-trees problem. An Odyssey image made it all suddenly click, because the resolution was high enough to identify these features and yet low enough to show their relationship to each other in the landscape," he said.

"Christensen's new hypothesis was made possible by NASA's tandem of science orbiters currently laying the groundwork for locating the most interesting areas for future surface exploration by roving laboratories, such as the Mars Exploration Rovers, scheduled for launch in May and June of this year," said Dr. Jim Garvin, NASA's lead scientist for Mars Exploration in Washington, D.C.

The Jet Propulsion Laboratory manages the Mars Exploration Program for NASA's Office of Space Science in Washington, D.C.

The new images are available online at http://photojournal.jpl.nasa.gov/catalog/PIA04408 and http://photojournal.jpl.nasa.gov/catalog/PIA04409 .

More information about the 2001 Mars Odyssey mission is available on the Internet at http://mars.jpl.nasa.gov/odyssey/ .

NASA's Mars Odyssey Changes Views about Red Planet

March 13, 2003

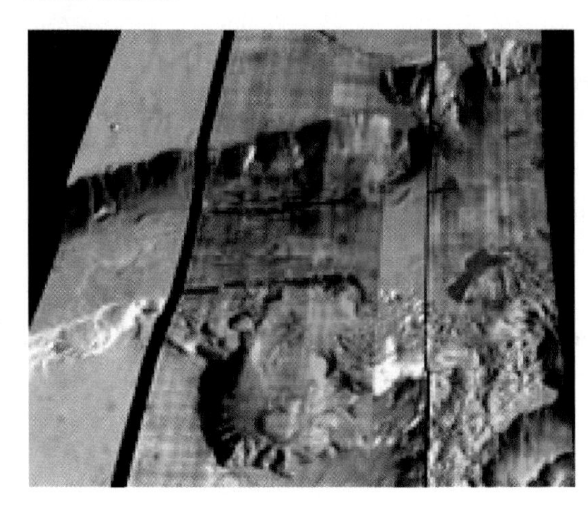

False-color infrared image of Ganges Chasma

NASA's Mars Odyssey spacecraft has transformed the way scientists are looking at the red planet.

"In just one year, Mars Odyssey has fundamentally changed our understanding of the nature of the materials on and below the surface of Mars," said Dr. Jeffrey Plaut, Odyssey's project scientist at NASA's Jet Propulsion Laboratory, Pasadena, Calif.

During its first year of surveying the martian surface, Odyssey's camera system provided detailed maps of minerals in rocks and soils. "A wonderful surprise has been the discovery of a layer of olivine-rich rock exposed in the walls of Ganges Chasm. Olivine is easily destroyed by liquid water, so its presence in these ancient rocks suggests that this region of Mars has been very dry for a very long time," said Dr. Philip Christensen, principal investigator for Odyssey's thermal emission imaging system at Arizona State University, Tempe.

"Infrared images have provided a remarkable new tool for mapping the martian surface. The temperature differences we see in the day and night images have revealed complex patterns of rocks and soils that show the effects of lava flows, impact craters, wind and possibly water throughout the history of Mars," Christensen said.

Odyssey has measured radiation levels at Mars that are substantially higher than in low-Earth orbit. "The martian radiation environment experiment has confirmed expectations that future human explorers of Mars will face significant long-term health risks from space radiation," said Dr. Cary Zeitlin, principal investigator for the martian radiation environment experiment, National Space Biomedical Research Institute, Houston. "We've also observed solar particle events not seen by near-Earth radiation detectors."

The gamma ray spectrometer suite, which early in the mission discovered vast amounts of hydrogen in the form of water ice trapped beneath the martian surface, has also begun to map the elemental composition of the surface.

"We are just now getting our first look at global elemental composition maps, and we are seeing Mars in a whole new light, gamma ray 'light,' and that's showing us aspects of the surface composition never seen before," said Dr. William Boynton, team leader for the gamma ray spectrometer suite at the University of Arizona, Tucson.

JPL, a division of the California Institute of Technology in Pasadena, Calif., manages the 2001 Mars Odyssey mission for NASA's Office of Space Science in Washington, D.C. Investigators at Arizona State University, the University of Arizona, and NASA's Johnson Space Center, Houston, built and operate the science instruments.

Additional science partners are located at the Russian Aviation and Space Agency and at Los Alamos National Laboratories, New Mexico. Lockheed Martin Astronautics, Denver, the prime contractor for the project, developed and built the orbiter. Mission operations are conducted jointly from Lockheed Martin and from JPL.

Odyssey Thermal Data Reveals a Changing Mars

June 05, 2003

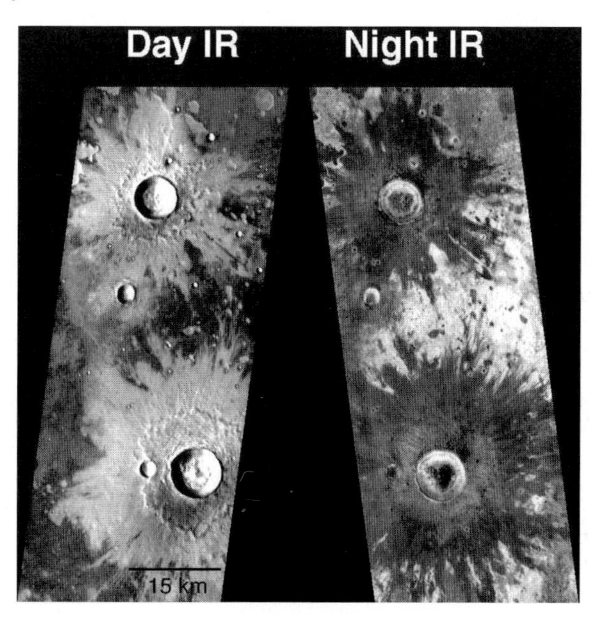

The first overview analysis of a year's worth of high-resolution infrared data gathered by the Thermal Emission Imaging System (THEMIS) on NASA's Mars Odyssey spacecraft is opening Mars to a new kind of detailed geological analysis and revealing a dynamic planet that has experienced dramatic environmental change.

The report by THEMIS's science team will appear in an upcoming issue of Science and will be released on June 5 in the magazine's online preview, Science Express.

"THEMIS is creating a set of data that is going to revolutionize our mapping of the planet and our idea of the planet's geology," said lead author and THEMIS Principal Investigator Philip Christensen, Korrick Professor of Geological Sciences at Arizona State University. "It will keep Mars scientists busy for the next 20 years trying to understand the processes that have produced this landscape."

THEMIS is providing planetary geologists with detailed temperature and infrared radiation images of the martian surface. The images reveal geological details that were impossible to detect even with the high-resolution Mars Orbital Camera on NASA's Mars Global Surveyor and that have 300 times higher resolution than MGS's Thermal Emission Spectrometer. Among the significant findings noted in the report is the detection of layers in the martian surface that indicate major changes in past environmental conditions.

"With a visible light camera, I can take a picture of a lava flow, but even with the highest resolution cameras that we have today the smallest thing we can see is the size of a bus and in order to do geology I need to have more detail," said Christensen.

"The camera on Mars Global Surveyor takes exquisite images that show layers, but it doesn't tell me anything about composition – is it a layer of boulders with a layer of sand on top? I have no way of knowing. With the THEMIS temperature data, I can actually get an idea because the layers vary – and each layer has remarkably different physical properties."

Daytime and nighttime temperature data can allow scientists to distinguish between solid rock and a variety of loose materials, from boulders to sand and dust. As any beach-goer knows, fine-grained sand heats up more rapidly at the surface than solid stone (which transmits more heat inward) but it also cools off more rapidly at night, when solid materials retain heat.

"We have seen layers, each with dramatically different physical properties, in places like Terra Meridiani," Christensen said. "Why do the physical properties in the different layers change? They change because the environment in which those rocks were deposited changed.

"It's very difficult to say exactly what happened in any particular place, but what we've found is that in many places on Mars it hasn't just been the same old thing happening for year after year for billions of years. These data have been so remarkable and so different from all of our previous experience that it has taken time to sift through the images and figure out what we're seeing."

Among the details that have stood out so far are kilometer-wide stretches of bare bedrock that Christensen notes were unexpected, given the Mars' known dustiness. Large areas of exposed rock indicate that strong environmental forces are currently at work, "scouring" from the surface any past sediment as well as any new material that might be falling from the atmosphere.

Also unexpected is the finding that accumulations of loose rock are common on martian hillsides, indicating recent processes of weathering continuing to affect the planet. " If those rocks had been made a billion years ago, they'd be covered with dust," Christensen pointed out. "This shows a dynamic Mars – it's an active place."

However, despite Odyssey's past findings of significant martian ice deposits, there are also indications that, in many places on the planet, water may not be one of the active causes behind the observed geological features.

Analyzing the spectra from the ten different bands of infrared light the instrument can detect, the THEMIS team has begun to identify specific mineral deposits, including a significant layer of the mineral olivine near the bottom of a four-and-a-half kilometer deep canyon known as Ganges Chasma. Olivine, Christensen notes, is significant because it decomposes rapidly in the presence of water.

"This gives us an interesting perspective of water on Mars," he said. "There can't have been much water – ever – in this place. If there was groundwater present when it was deep within the surface, the olivine would have disappeared. And since the canyon has opened up, if there had ever been water at the surface it would be gone too. This is a very dry place, because it's been exposed for hundreds of millions of years. We know that some places on Mars have water, but here we see that some really don't."

Overall, Christensen notes that the emerging diversity and complexity of the planet point to the likelihood of future surprises and keep enlarging the possibilities for discovery on Mars.

"With Odyssey, we are looking at Mars in its entirety, in context. It's remarkable how much this has already changed our view of the complexity and richness of the planet. We discovered that it has a really dynamic geologic history. It has far more ice and water than we thought – we're seeing snow and gullies, layers – and there are also processes involving volcanoes, impact craters and wind. It's a fascinating place."

In addition to Christensen, the authors on the paper include Joshua L. Bandfield, James F. Bell, Noel Gorelick, Victoria E. Hamilton, Anton Ivanov, Bruce M. Jakosky, Hugh H. Kieffer, Melissa D. Lane, Michael C. Malin, Timothy McConnochie, Alfred S. McEwen, Harry Y. McSween, Greg L. Mehall, Jeffery E. Moersch, Kenneth H Nealson, James W. Rice, Mark I. Richardson, Steven W. Ruff, Michael D. Smith, Timothy N. Titus, and Michael B Wyatt.

Link to Science Magazine online site
http://www.sciencemag.org/sciencexpress/recent.shtml

The Jet Propulsion Laboratory, a division of the California Institute of Technology in Pasadena, manages the 2001 Mars Odyssey mission for NASA's Office of Space Science in Washington. Investigators at Arizona State University in Tempe, the University of Arizona in Tucson and NASA's Johnson Space Center, Houston, operate the science instruments. Additional science partners are located at the Russian Aviation and Space Agency and at Los Alamos National Laboratories, New Mexico. Lockheed Martin Astronautics, Denver, is the prime contractor for the project, and developed and built the orbiter. Mission operations are conducted jointly from Lockheed Martin and from JPL. Additional information about the 2001 Mars Odyssey is available on the Internet at: http://mars.jpl.nasa.gov/odyssey/

Mars Odyssey Orbiter Watches a Frosty Mars

June 26, 2003

NASA's Mars Odyssey detected water ice in the northern hemisphere. During the winter months, the icy soil is covered by a thick layer of carbon dioxide ("dry ice") frost obscuring the water ice signature. Detailed caption link

NASA's Mars Odyssey spacecraft is revealing new details about the intriguing and dynamic character of the frozen layers now known to dominate the high northern latitudes of Mars. The implications have a bearing on science strategies for future missions in the search of habitats.

Odyssey's neutron and gamma-ray sensors have tracked seasonal changes as layers of "dry ice" (carbon-dioxide frost or snow) accumulate during northern Mars' winter and then dissipate in the spring, exposing a soil layer rich in water ice – the Martian counterpart to permafrost.

Researchers used measurements of martian neutrons combined with height measurements from the laser altimeter on another NASA spacecraft, Mars Global Surveyor, to monitor the amount of dry ice during the northern winter and spring seasons.

"Once the carbon-dioxide layer disappears, we see even more water ice in northern latitudes than Odyssey found last year in southern latitudes," said Odyssey's Dr. Igor Mitrofanov of the Russian Space Research Institute (IKI), Moscow, lead author of a paper in the June 27 issue of the journal Science. "In some places, the water ice content is more than 90 percent by volume," he said. Mitrofanov and co-authors used the changing nature of the relief of these regions, measured more than 2 years ago by the Global Surveyor's laser altimeter science team, to explore the implications of the changes.

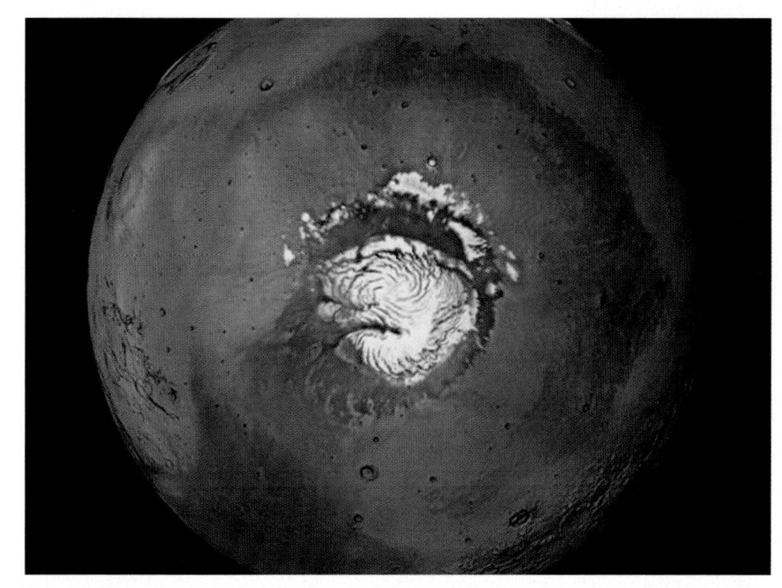

MARS' NORTHERN HEMISPHERE

This is a mosaic of the northern hemisphere of Mars as seen Viking.

Credit: NASA/JPL/GSFC

Mars Odyssey's trio of instruments, called the gamma-ray spectrometer suite, can identify elements in the top meter (3 feet) or so of Mars' surface. Mars Global Surveyor's laser altimeter is precise enough to monitor meter-scale changes in the thickness of the seasonal frost, which can accumulate to depths greater than a meter. The new findings show a correlation in the springtime between Odyssey's detection of dissipating carbon dioxide in latitudes poleward of 65 degrees north and Global Surveyor's measurement of the thinning of the frost layer in prior years.

"Odyssey's high-energy neutron detector allows us to measure the thickness of carbon dioxide at lower latitudes, where Global Surveyor's altimeter does not have enough sensitivity," Mitrofanov said. "On the other hand, the neutron detector loses sensitivity to measure carbon-dioxide thickness greater than one meter (3 feet), where the altimeter obtained reliable data. Working together, we can examine the whole range of dry ice snow accumulations."

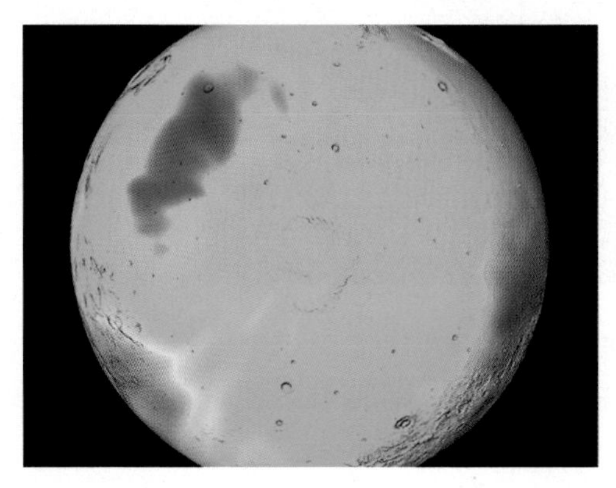

WATER ICE/WINTER OBSERVATIONS

Blue colors indicate water ice observed by Odyssey during the northern winter months, when carbon dioxide covers the surface.

Credit: NASA/JPL/GSFC/IKI

"The synergy between the measurements from our two 'eyes in the skies of Mars' has enabled these new findings about the nature of near-surface frozen materials, and suggests compelling places to visit in future missions in order to understand habitats on Mars," said Dr. Jim Garvin, NASA's Lead Scientist for Mars Exploration.

Another report, to be published in the Journal of Geophysical Research – Planets, combines measurements from Odyssey and Global Surveyor to provide indications of how densely the winter layer of carbon dioxide frost or snow is packed at northern latitudes greater than 85 degrees. The Odyssey data are used to estimate the mass of the deposit, which can then be compared with the thickness to obtain a density. The dry ice layer appears to have a fluffy texture, like freshly fallen snow, according to the report by Dr. William Feldman of Los Alamos National Laboratory, N.M., and 11 co-authors. The study also found that once the dry ice disappears, the remaining surface near the pole is composed almost entirely of water ice.

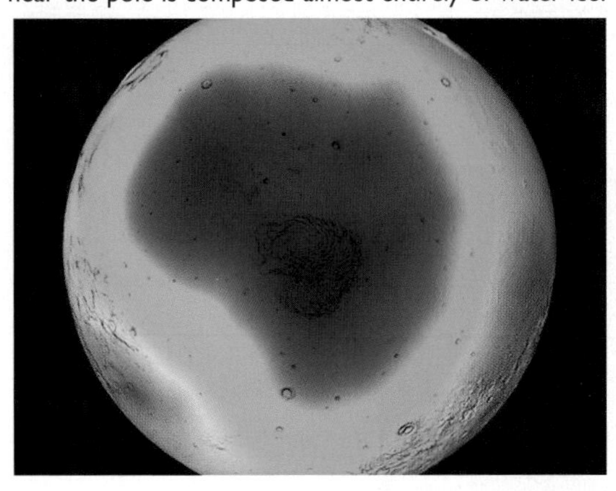

WATER ICE/SUMMER OBSERVATIONS

Blue colors show the water ice that is revealed during the martian summer.

Credit: NASA/JPL/GSFC/IKI

"Mars is constantly changing," said Dr. Jeffrey Plaut, Mars Odyssey project scientist at NASA's Jet Propulsion Laboratory, Pasadena, Calif. With Mars Odyssey, we plan to examine these dynamics through additional seasons, to watch how the winter accumulations of carbon dioxide on each pole interact with the atmosphere in the current climate regime."

Mitrofanov's co-authors include researchers at the Institute for Space Research of the Russian Academy of Science, Moscow; MIT, Cambridge, MA; NASA's Goddard Space Flight Center, Greenbelt, Md.; TechSource, Santa Fe, N.M.; and NASA Headquarters, Washington. Feldman's co-authors include researchers at New Mexico State University, Las Cruces; Cornell University, Ithaca, N.Y.; and Observatoire Midi-Pyrenees, Toulouse, France.

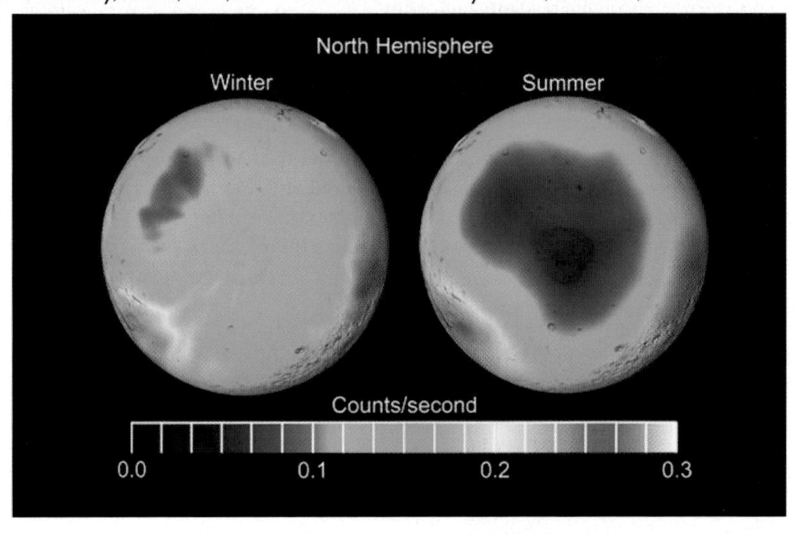

WATER ICE REVEALED AS DRY ICE DISSIPATES

A comparison of the water ice detected by Odyssey in northern summer and winter seasons. In some places the water ice content is more than 90 percent by volume.

Credit: NASA/JPL/GSFC/IKI

JPL manages the Mars Odyssey and Mars Global Surveyor missions for NASA's Office of Space Science in Washington. Investigators at Arizona State University, the University of Arizona, and NASA's Johnson Space Center, Houston, built and operate Odyssey science instruments. The Russian Aviation and Space Agency supplied the high-energy neutron detector and Los Alamos National Laboratory supplied the neutron spectrometer. NASA's Goddard Space Flight Center supplied Global Surveyor's laser altimeter. Information about NASA's Mars exploration program is available online at: http://mars.jpl.nasa.gov

About the images: The images and animations were rendered by the Scientific Visualization Studio at the Goddard Space Flight Center by data provided by the Mars Odyssey and MGS Mars Orbital Laser Altimeter (MOLA) Science teams.

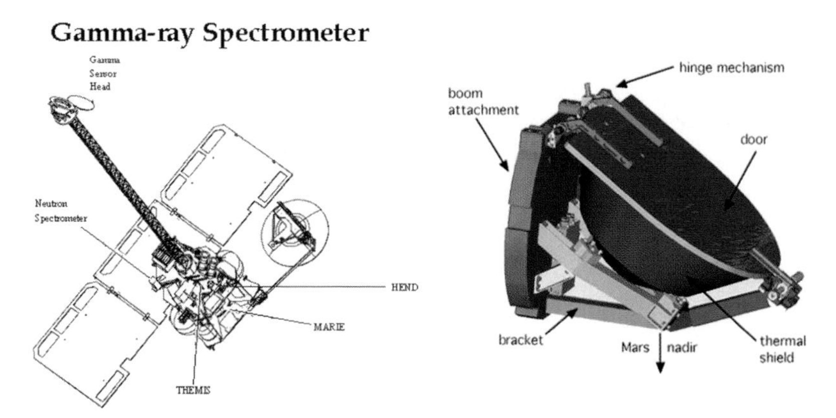

GRS: The Gamma Ray Spectrometer

The gamma ray spectrometer will be able to measure the abundance and distribution of about 20 primary elements of the periodic table, including silicon, oxygen, iron, magnesium, potassium, aluminum, calcium, sulfur, and carbon. Knowing what elements are at or near the surface will give detailed information about how Mars has changed over time. To determine the elemental makeup of the Martian surface, the experiment uses gamma ray spectrometer and two neutron detectors.

How GRS Works

When exposed to cosmic rays (charged particles in space that come from the stars, including our sun), chemical elements in soils and rocks emit uniquely identifiable signatures of energy in the form of gamma rays. The gamma ray spectrometer looks at these signatures, or energies, coming from the elements present in the Martian soil.

By measuring gamma rays coming from the Martian surface, it is possible to calculate how abundant various elements are and how they are distributed around the planet's surface. Gamma rays, emitted from the nuclei of atoms, show up as sharp emission lines on the instrument's spectrum. While the energy represented in these emissions determines which elements are present, the intensity of the spectrum reveals the elements concentrations. The spectrometer is expected to add significantly to the growing understanding of the origin and evolution of Mars and the processes shaping it today and in the past.

How are gamma rays and neutrons produced by cosmic rays? Incoming cosmic rays – some of the highest-energy particles – collide with atoms in the soil. When atoms are hit with such energy, neutrons are released, which scatter and collide with other atoms. The atoms get " excited" in the process, and emit gamma rays to release the extra energy so they can return to their normal rest state. Some elements like potassium, uranium, and thorium are naturally radioactive and give off gamma rays as they decay, but all elements can be excited by collisions with cosmic rays to produce gamma rays. The HEND and Neutron Spectrometers on GRS directly detect scattered neutrons, and the Gamma Sensor detects the gamma rays.

How GRS Will Help Detect Water

By measuring neutrons, it is possible to calculate the abundance of hydrogen on Mars, thus inferring the presence of water. The neutron detectors are sensitive to concentrations of hydrogen in the upper meter of the surface. Like a virtual shovel "digging into" the surface, the spectrometer will allow scientists to peer into this shallow subsurface of Mars and measure the amount of hydrogen that exists there. Since hydrogen is most likely present in the form of water ice, the spectrometer will be able to measure directly the amount of permanent ground ice and how it changes with the seasons.

GRS will supply data similar to that of the successful Lunar Prospector mission, which told us how much hydrogen, and thus water, is likely on the moon.

The gamma ray spectrometer consists of four main components: the gamma sensor head, the neutron spectrometer, the high energy neutron detector, and the central electronics assembly. The sensor head is separated from the rest of the Odyssey spacecraft by a 6.2-meter (20-ft) boom, which will be extended after Odyssey has entered the mapping orbit at Mars. This maneuver is done to minimize interference from any gamma rays coming from the spacecraft itself. The initial spectrometer activity, lasting between 15 and 40 days, will perform an instrument calibration before the boom is deployed. After about 100 days of the mapping mission, the boom will deploy and remain in this position for the duration of the mission. The two neutron detectors – the neutron spectrometer and the high-energy neutron detector – are mounted on the main spacecraft structure and will operate continuously throughout the mapping mission.

GRS

The Gamma Ray Spectrometer weighs 30.5 kilograms (67.2 pounds) and uses 32 watts of power. Along with its cooler, the Gamma Ray Spectrometer measures 46.8 centimeters (18.4 inches) by 53.4 centimeters (21.0 inches) by 60.4 centimeters (23.8 inches).

The neutron spectrometer is 17.3 centimeters (6.8 inches) by 14.4 centimeters (5.7 inches) by 31.4 centimeters (12.4 inches).

The high-energy neutron detector measures 30.3 centimeters (11.9 inches) by 24.8 centimeters (9.8 inches) by 24.2 centimeters (9.5 inches). The instrument's central electronics box is 28.1 inches (11.1 inches) by 24.3 centimeters (9.6 inches) by 23.4 centimeters (9.2 inches).

THEMIS

The Thermal Emission Imaging System weighs 11.2 kilograms (24.7 pounds). It is 54.5 centimeters (21.5 inches) by 37 centimeters (14.6 inches) by 28.6 centimeters (11.3 inches). THEMIS runs on 14 watts of electrical power.

THEMIS: The Thermal Emission Imaging System

By looking at the visible and infrared parts of the spectrum, THEMIS will determine the distribution of minerals on the surface of Mars and help understand how the mineralogy of the planet relates to the landforms.

How THEMIS Works in the Infrared

During the Martian day, the sun heats the surface. Surface minerals radiate this heat back to space in characteristic ways that can be identified and mapped by the instrument. At night, since it maps heat, the imager will search for active thermal spots and may discover "hot springs" on Mars.

In the infrared spectrum, the instrument uses 9 spectral bands to help detect minerals within the Martian terrain. These spectral bands, similar to ranges of colors, can obtain the signatures (spectral "fingerprints") of particular types of geological materials. Minerals, such as carbonates, silicates, hydroxides, sulfates, hydrothermal silica, oxides and phosphates, all show up as different colors in the infrared spectrum. This multi-spectral method allows researchers to detect in particular the presence of minerals that form in water and to understand those minerals in their proper geological context. THEMIS' infrared capabilities will significantly improve the data from TES, a similar instrument on Mars Global Surveyor.

The instrument's multi-spectral approach will also provide data on localized deposits associated with hydrothermal and subsurface water and enable 100-meter (328-feet) images of Martian terrain to be captured in each pixel, or single point, of every image.

ASTER, an Earth orbiting instrument on the Terra spacecraft, has used a similar approach to map the distribution of minerals here on Earth.

Variations in the thermal infrared "color" in the right-hand image are due to differences in the kinds of minerals that make up rocks and soil. In the visible part of the spectrum that our eyes can see (left-hand image), it would not be apparent what minerals are present.

How THEMIS Works in the Visible

Using visible imaging in five spectral bands, the experiment will also take 18-meter-resolution (59-foot) images to determine the geological record of past liquid environments on Mars. More than 15,000 images – each 20X20 kilometers (12X12 miles) – will be acquired for Martian surface studies. These more detailed data will be used in conjunction with mineral maps to identify potential landing sites for future Mars missions.

The part of the imaging system that takes pictures in visible light will be able to show objects about as big as a semi-truck. This resolution will help fill in the gap between large-scale geological images from the Viking orbiters in the 1970s and the very high-resolution images from the currently orbiting Mars Global Surveyor.

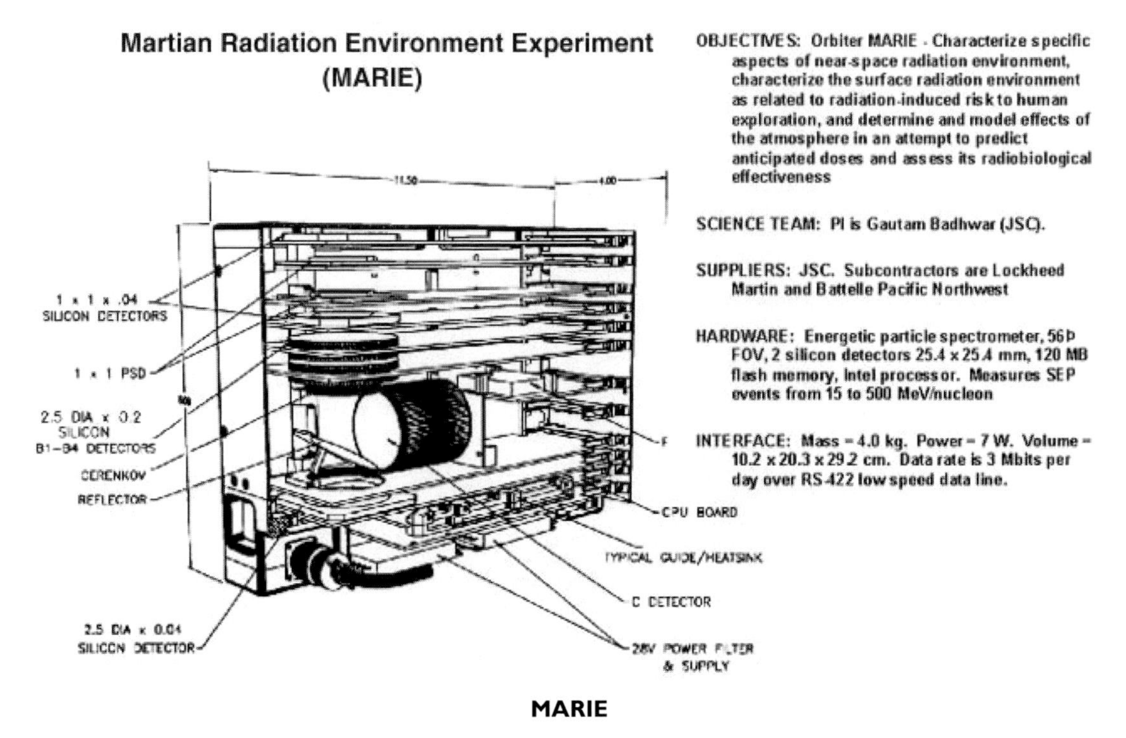

MARIE

The Mars Radiation Environment Experiment weighs 3.3 kilograms (7.3 pounds) and uses 7 watts of power. It measures 29.4 centimeters (11.6 inches) by 23.2 centimeters (9.1 inches) by 10.8 centimeters (4.3 inches).

Led by NASA's Johnson Space Center, this science investigation is designed to characterize aspects of the radiation environment both on the way to Mars and in the Martian orbit.

Since space radiation presents an extreme hazard to crews of interplanetary missions, the experiment will attempt to predict anticipated radiation doses that would be experienced by future astronauts and help determine possible effects of Martian radiation on human beings.

Space radiation comes from cosmic rays emitted by our local star, the sun, and from stars beyond our solar system as well. Space radiation can trigger cancer and cause damage to the central nervous system. Similar instruments are flown on the Space Shuttles and on the International Space Station (ISS), but none have ever flown outside of Earth's protective magnetosphere, which blocks much of this radiation from reaching the surface of our planet.

How the Instrument Works

A spectrometer inside the instrument will measure the energy from these radiation sources. As the spacecraft orbits the red planet, the spectrometer sweeps through the sky and measures the radiation field.

The instrument, with a 68-degree field of view, is designed to collect data continuously during Odyssey's cruise from Earth to Mars. It can store large amounts of data for downlink whenever possible, and will operate throughout the entire science mission.

Los Alamos Releases New Maps of Mars Water

July 24, 2003

LOS ALAMOS NATIONAL LABORATORY A DEPARTMENT OF ENERGY/UNIVERSITY OF CALIFORNIA LABORATORY Communications and External Relations Division Public Affairs Office CONTACTS: Nancy Ambrosiano, 505-667-0471, nwa@lanl.gov 03-101
Jim Danneskiold, 505-667-1640, slinger@lanl.gov

LOS ALAMOS, N.M., July 24, 2003 – "Breathtaking" new maps of likely sites of water on Mars showcase their association with geologic features such as Valles Marineris, the largest canyon in the solar system.

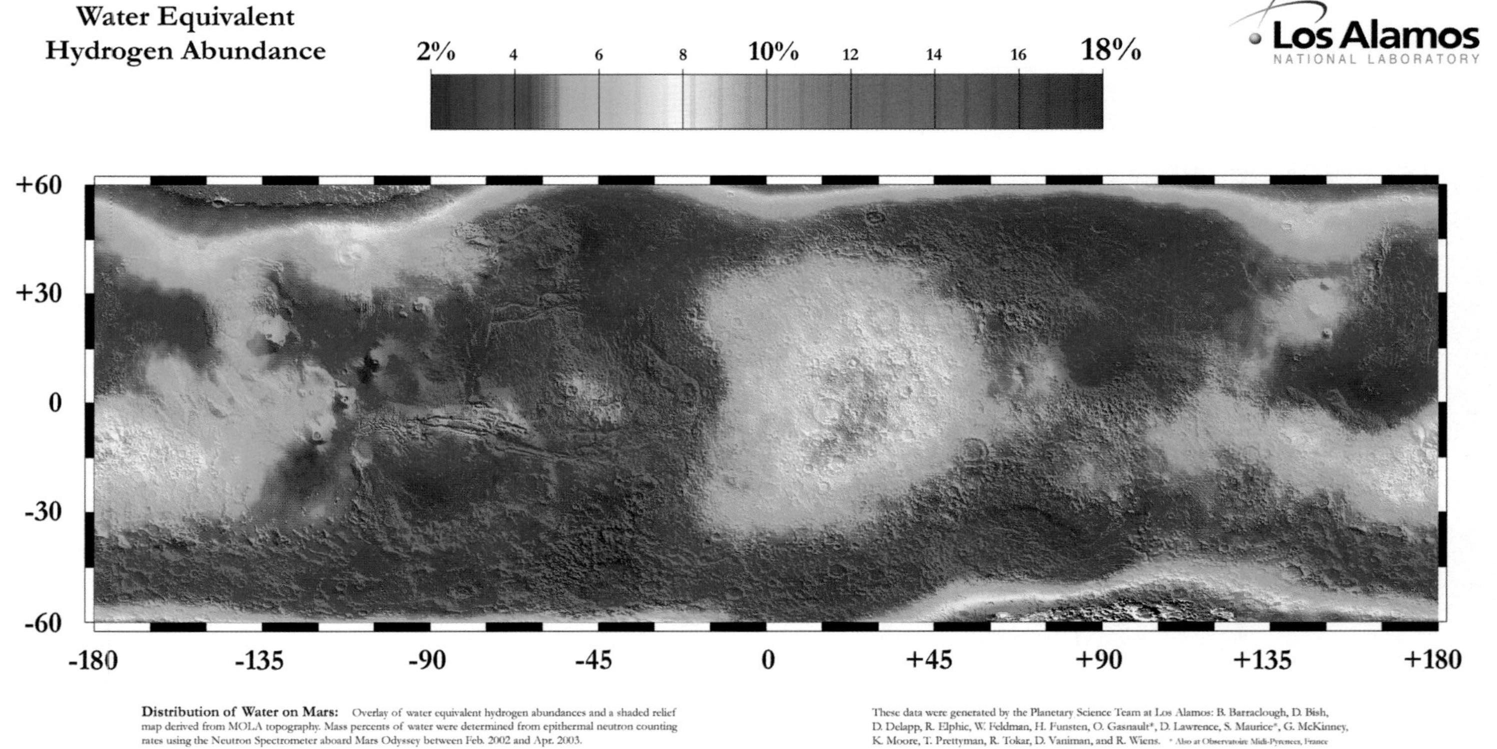

Distribution of Water on Mars: Overlay of water equivalent hydrogen abundances and a shaded relief map derived from MOLA topography. Mass percents of water were determined from epithermal neutron counting rates using the Neutron Spectrometer aboard Mars Odyssey between Feb. 2002 and Apr. 2003.

Reference: Feldman W. C., T. H. Prettyman, S. Maurice, J. J. Plaut, D. L. Bish, D. T. Vaniman, M. T. Mellon, A. E. Metzger, S. W. Squyres, S. Karunatillake, W. V. Boynton, R. C. Elphic, H. O. Funsten, D. J. Lawrence, and R. L. Tokar, The global distribution of near-surface hydrogen on Mars, *JGR-planets*, submitted July 2003.

These data were generated by the Planetary Science Team at Los Alamos: B. Barraclough, D. Bish, D. Delapp, R. Elphic, W. Feldman, H. Funsten, O. Gasnault*, D. Lawrence, S. Maurice*, G. McKinney, K. Moore, T. Prettyman, R. Tokar, D. Vaniman, and R. Wiens. * Also at Observatoire Midi-Pyrenees, France

The neutron spectrometer aboard Mars Odyssey, a component of the Gamma-ray Spectrometer suite of instruments, was designed and built by the Los Alamos National Laboratory and is operated by the University of Arizona in Tucson. The Mars Odyssey mission is managed by the Jet Propulsion Laboratory.

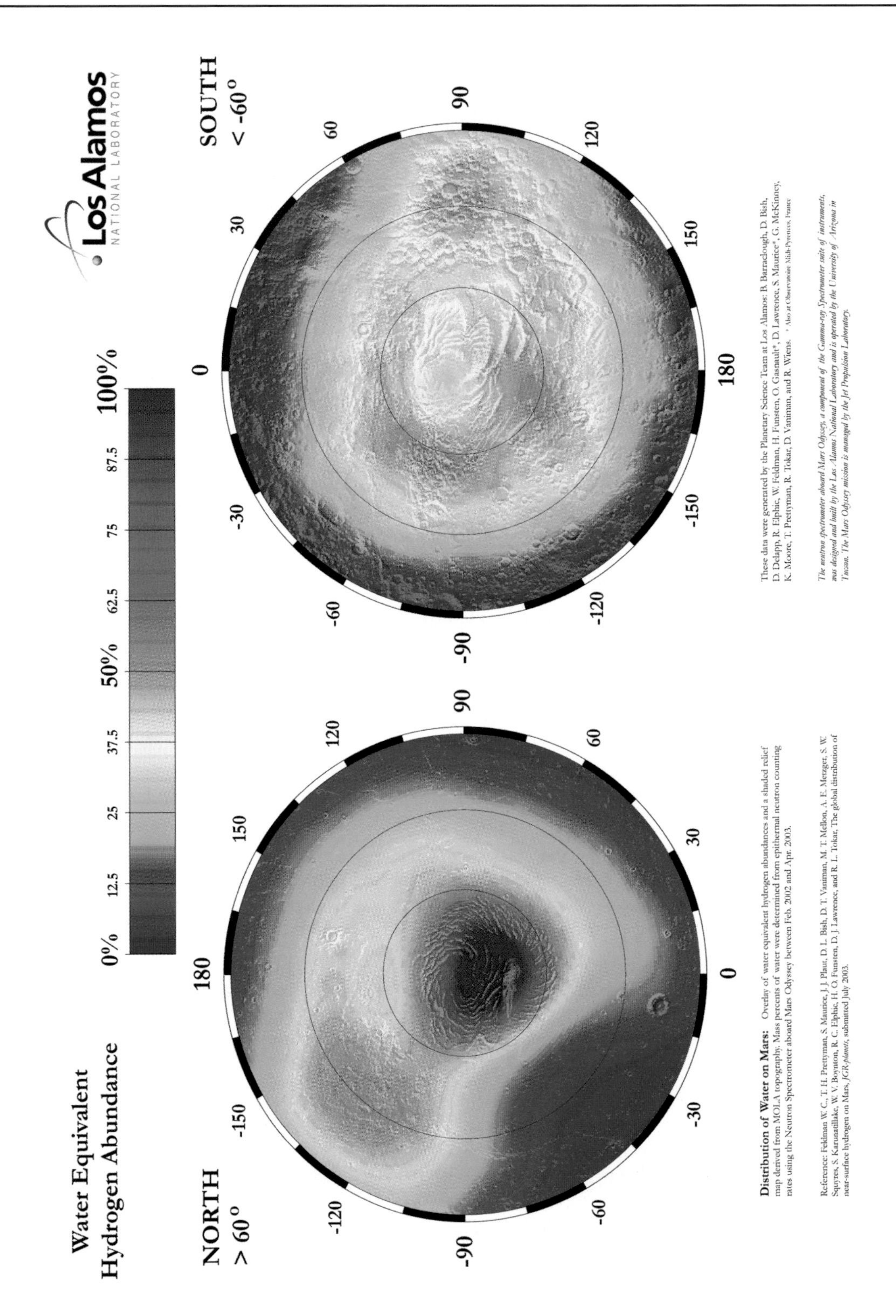

Distribution of Water on Mars

Overlay of water equivalent hydrogen abundances and a shade relief map derived from MOLA topography. Mass percents of water were determined from epithermal neutron counting rates using the Neutron Spectrometer aboard Mars Odyssey between February 2002 and April 2003.

The maps detail the distribution of water-equivalent hydrogen as revealed by Los Alamos National Laboratory-developed instruments aboard NASA's Mars Odyssey spacecraft. In an upcoming talk at the Sixth International Conference on Mars at the California Institute of Technology, in Pasadena, Los Alamos space scientist Bill Feldman and coworkers will offer current estimates of the total amount of water stored near the Martian surface. His presentation will be at 1:20 p.m., Friday, July 25.

For more than a year, Los Alamos' neutron spectrometer has been carefully mapping the hydrogen content of the planet's surface by measuring changes in neutrons given off by soil, an indicator of hydrogen likely in the form of water ice. The new color maps are available at http://www.lanl.gov/worldview/news/photos/mars.shtml online.

"The new pictures are just breathtaking, the water-equivalent hydrogen follows the geographic features beautifully," said Feldman. "There's a lane of hydrogen-rich material following the western slopes of the biggest volcanoes in the solar system, a maximum reading sits right on Elysium mons, and another maximum is in the deepest canyon in the solar system."

The new maps combine images from the Mars Orbiter Laser Altimeter (MOLA) on the Mars Global Surveyor with Mars Odyssey spectrometer data through more than half a Martian year of 687 Earth days. From about 55 degrees latitude to the poles, Mars boasts extensive deposits of soils that are rich in water ice, bearing an average of 50 percent water by mass. In other words, Feldman said, a typical pound of soil scooped up in those polar regions would yield an average of half a pound of water if it were heated in an oven.

The tell-tale traces of hydrogen, and therefore the presence of hydrated minerals, also are found in lower concentrations closer to Mars' equator, ranging from two to 10 percent water by mass. Surprisingly, two large areas, one within Arabia Terra, the 1,900-mile-wide Martian desert, and another on the opposite side of the planet, show indications of relatively large concentrations of sub-surface hydrogen.

Scientists are attracted to two possible theories of how all that water got into the Martian soils and rocks.

The vast water icecaps at the poles may be the source. The thickness of the icecaps themselves may be enough to bottle up geothermal heat from below, increasing the temperature at the bottom and melting the bottom layer of the icecaps, which then could feed a global water table.

On the other hand, there is evidence that about a million years or so ago, Mars' axis was tilted about 35 degrees, which might have caused the polar icecaps to evaporate and briefly create enough water in the atmosphere to make ice stable planetwide. The resultant thick layer of frost may then have combined chemically with hydrogen-hungry soils and rocks.

"We're not ready yet to precisely describe the abundance and stratigraphy of these deposits, but the neutron spectrometer shows water ice close to the surface in many locations, and buried elsewhere beneath several inches of dry soils," Feldman said. "Some theories predict these deposits may extend a half mile or more beneath the surface; if so, their total water content may be sufficient to account for the missing water budget of Mars."

In fact, a team of Los Alamos scientists has begun a research project to interpret the Mars Odyssey data and their ramifications for the history of Mars' climate. The project is funded through the Laboratory Directed Research and Development program – which funds innovative science with a portion of the Laboratory's operating budget – and seeks to develop a global Martian hydrology model, using vast amounts of remote sensing data, topography maps and experimental results on water loading of minerals.

Members of the Planetary Science team at Los Alamos working with Feldman on the Odyssey project include Bruce Barraclough, David Bish, Dorothea Delapp, Richard Elphic, Herbert Funsten, Olivier Gasnault, David Lawrence, G. McKinney, Kurt Moore, Robert Tokar, Thomas Prettyman, David Vaniman and Roger Wiens as well as Sylvestre Maurice of the Observatoire Midi-Pyrénées (France), S.W. Squyres of Cornell University, and Jeff Plaut of the Jet Propulsion Laboratory.

Los Alamos' neutron spectrometer, a more sensitive version of the instrument that found water ice on the moon five years ago, is one component of the gamma-ray spectrometer suite of instruments aboard Odyssey. W.T. Boynton of the University of Arizona leads the gamma-ray spectrometer team.

The neutron spectrometer looks for neutrons generated when cosmic rays slam into the nuclei of atoms on the planet's surface, ejecting neutrons skyward with enough energy to reach the Odyssey spacecraft 250 miles above the surface.

Elements create their own unique distribution of neutron energy – fast, thermal or epithermal – and these neutron flux signatures are shaped by the elements that make up the soil and how they are distributed. Thermal neutrons are low-energy neutrons in thermal contact with the soil; epithermal neutrons are intermediate, scattering down in energy after bouncing off soil material; and fast neutrons are the highest-energy neutrons produced in the interaction between high-energy galactic cosmic rays and the soil.

By looking for a decrease in epithermal neutron flux, researchers can locate hydrogen. Hydrogen in the soil efficiently absorbs the energy from neutrons, reducing their flux in the surface and also the flux that escapes the surface to space where it is detected by the spectrometer. Since hydrogen is likely in the form of water ice at high latitudes, the spectrometer can measure directly, a yard or so deep into the Martian surface, the amount of ice and how it changes with the seasons.

The Los Alamos expertise in neutron spectroscopy stems from longtime nuclear nonproliferation work at the Laboratory, funded by the U.S. Department of Energy's National Nuclear Security Administration. The ability to measure and detect signatures of nuclear materials is a vital component of the Laboratory's mission to reduce the threats from weapons of mass destruction.

Mars Odyssey was launched from Cape Canaveral Air Force Station in April 2001 and arrived in Martian orbit in late October 2001. During the rest of the spacecraft's 917-day science mission, Los Alamos' neutron spectrometer will continue to improve the hydrogen map and solve more Martian moisture mysteries.

Jet Propulsion Laboratory, a division of the California Institute of Technology in Pasadena, manages the Mars Odyssey mission for NASA's Office of Space Science in Washington, D.C. Investigators at Arizona State University in Tempe, the University of Arizona in Tucson and NASA's Johnson Space Center, Houston, operate the science instruments. Additional science partners are located at the Russian Aviation and Space Agency and at Los Alamos National Laboratories, New Mexico. Lockheed Martin Astronautics, Denver, the prime contractor for the project, developed and built the orbiter. Mission operations are conducted jointly from Lockheed Martin and from JPL.

Los Alamos National Laboratory is operated by the University of California for the National Nuclear Security Administration (NNSA) of the U.S. Department of Energy and works in partnership with NNSA's Sandia and Lawrence Livermore national laboratories to support NNSA in its mission.

Los Alamos develops and applies science and technology to ensure the safety and reliability of the U.S. nuclear deterrent; reduce the threat of weapons of mass destruction, proliferation and terrorism; and solve national problems in defense, energy, environment and infrastructure.

Mars Odyssey Mission Status

November 26, 2003

The martian radiation environment experiment on NASA's 2001 Mars Odyssey orbiter has collected data continuously from the start of the Odyssey mapping mission in March 2002 until late last month. The instrument has successfully monitored space radiation to evaluate the risks to future Mars-bound astronauts. Its measurements are the first of their kind to be obtained during an interplanetary cruise and in orbit around another planet.

On Oct. 28, 2003, during a period of intense solar activity, the instrument stopped working properly. Controllers' efforts to restore the instrument to normal operations have not been successful. These efforts will continue for the next several weeks or months.

The martian radiation environment experiment detects energetic charged particles, including galactic cosmic rays and particles emitted by the Sun in coronal mass ejections. The dose equivalent from galactic cosmic rays as measured by the instrument agrees well with predictions based on modeling. Validation of radiation models is a crucial step in predicting radiation-related health risks for crews of future missions.

"Even if the instrument provides no additional data in the future, it has been a great success at characterizing the radiation environment that a crewed mission to Mars would need to anticipate," said Dr. Jeffrey Plaut, project scientist for Mars Odyssey at NASA's Jet Propulsion Laboratory, Pasadena, Calif.

JPL manages the Mars Odyssey and Global Surveyor missions for NASA's Office of Space Science, Washington, D.C. Investigators at Arizona State University, Tempe; University of Arizona, Tucson; NASA's Johnson Space Center,

Houston; the Russian Aviation and Space Agency, Moscow; and Los Alamos National Laboratory, Los Alamos, N.M., built and operate Odyssey science instruments. Information about NASA's Mars exploration program is available on the Internet at: http://mars.jpl.nasa.gov .

Contact: Guy Webster (818) 354-6278 JPL
2003-156

Odyssey Studies Changing Weather And Climate On Mars

December 08, 2003

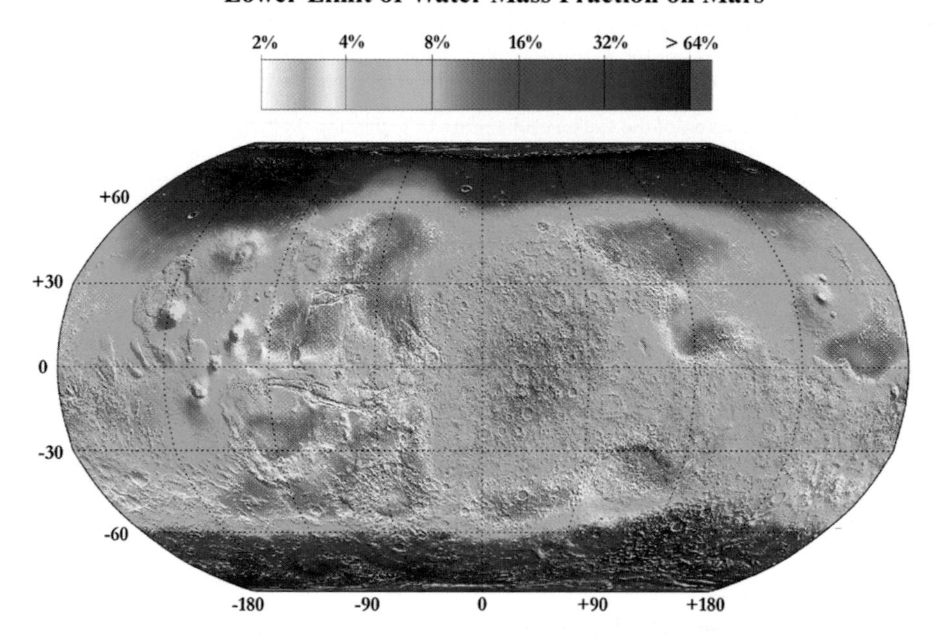

Water mass map from neutron spectrometer 08 Dec 2003

This map shows the estimated lower limit of the water content of the upper meter of Martian soil. The estimates are derived from the hydrogen abundance measured by the neutron spectrometer component of the gamma ray spectrometer suite on NASA's Mars Odyssey spacecraft.

The highest water mass fractions, exceeding 30 percent to well over 60 percent, are in the polar regions, beyond about 60 degrees latitude north or south. Farther from the poles, significant concentrations are in the area bound in longitude by minus 10 degrees to 50 degrees and in latitude by 30 degrees south to 40 degrees north, and in an area to the south and west of Olympus Mons (30 degrees to 0 degrees south latitude and minus 135 degrees to 110 degrees longitude).

NASA's Jet Propulsion Laboratory, a division of the California Institute of Technology in Pasadena, manages the 2001 Mars Odyssey mission for the NASA Office of Space Science in Washington. Investigators at Arizona State University in Tempe, the University of Arizona in Tucson and NASA's Johnson Space Center, Houston, operate the science instruments. The gamma-ray spectrometer was provided by the University of Arizona in collaboration with the Russian Aviation and Space Agency, which provided the high-energy neutron detector, and the Los Alamos National Laboratories, New Mexico, which provided the neutron spectrometer. Lockheed Martin Space Systems, Denver, is the prime contractor for the project, and developed and built the orbiter. Mission operations are conducted jointly from Lockheed Martin and from JPL.

Credit: NASA/JPL/Los Alamos National Laboratory

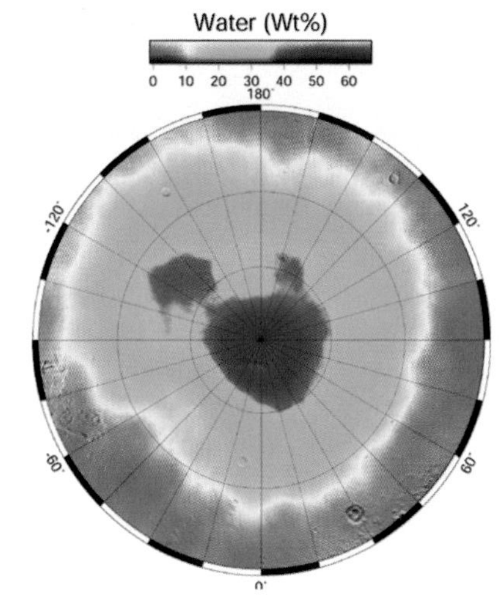

North polar water ice by weight 08 Dec 2003

This map shows the percent of water by weight in near-surface materials of Mars' north polar region. It is derived from the gamma ray spectrometer component of the gamma ray spectrometer suite of instruments on NASA's Mars Odyssey spacecraft.

Significant concentrations of water (greater than 20 percent) are poleward of 55 degrees north latitude. The highest concentration, greater than 50 percent, is between 75 degrees north and the pole. Another area with a high concentration of water by weight is in the north polar plains between longitudes minus 105 degrees and minus 140 degrees, and between latitudes 60 degrees and 75 degrees.

NASA's Jet Propulsion Laboratory, a division of the California Institute of Technology in Pasadena, manages the 2001 Mars Odyssey mission for the NASA Office of Space Science in Washington. Investigators at Arizona State University in Tempe, the University of Arizona in Tucson and NASA's Johnson Space Center, Houston, operate the science instruments. The gamma-ray spectrometer was provided by the University of Arizona in collaboration with the Russian Aviation and Space Agency, which provided the high-energy neutron detector, and the Los Alamos National Laboratories, New Mexico, which provided the neutron spectrometer. Lockheed Martin Space Systems, Denver, is the prime contractor for the project, and developed and built the orbiter. Mission operations are conducted jointly from Lockheed Martin and from JPL.

Credit: NASA/JPL/University of Arizona

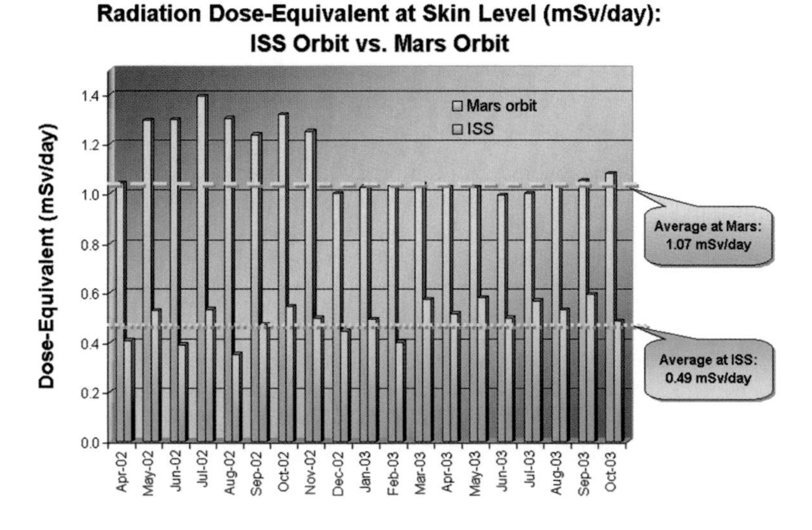

Radiation environment at Mars and Earth 08 Dec 2003

This graphic shows the radiation dose equivalent as measured by Odyssey's martian radiation environment experiment at Mars and by instruments aboard the Earth-orbiting International Space Station (ISS), for the 18-month period from April 2002 through October 2003. The accumulated total in Mars orbit is just over two times larger than that aboard the Space Station. The bars where the Mars instrument's measurements are well above the average (as shown by the orange line) are months when there was significant solar activity, which increases the dose equivalent. Dose equivalent is expressed in units of milliSieverts per day.

NASA's Jet Propulsion Laboratory manages the 2001 Mars Odyssey mission for NASA's Office of Space Science, Washington. The radiation experiment was provided by the Johnson Space Center, Houston, Texas. Lockheed Martin Space Systems, Denver, Colo., is the prime contractor for the project, and developed and built the orbiter. Mission operations are conducted jointly from Lockheed Martin and from JPL, a division of the California Institute of Technology in Pasadena.

Credit: NASA/JPL/JSC

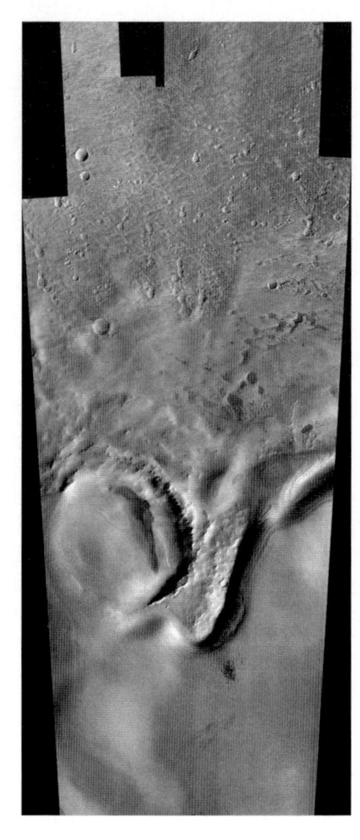

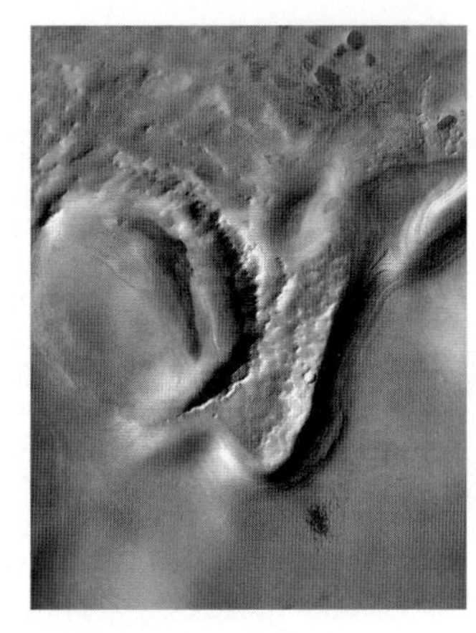

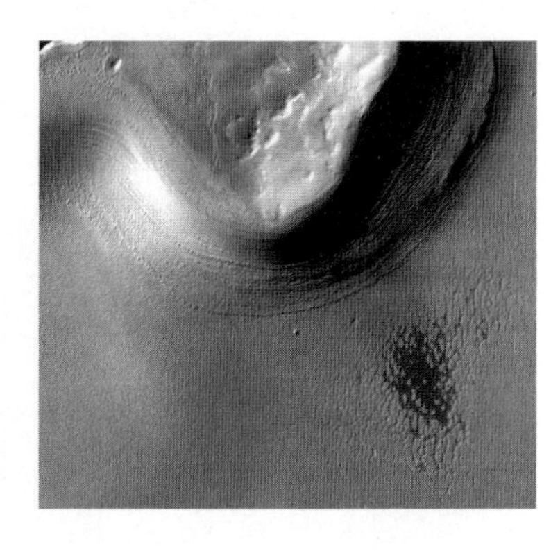

Mars south polar layered deposits 08 Dec 2003

Remarkable layered deposits covering older, cratered surfaces near Mars' south pole dominate this mosaic of images taken by the camera on NASA's Mars Odyssey spacecraft between Nov. 8 and Nov. 26, 2003. The margin of these layered deposits appears to be eroding poleward, exposing a series of layers in the retreating cliff.

The mosaic, stitched from eight visible wavelength images from Odyssey's thermal emission imaging system, covers an area more than 325 kilometers (200 miles) long and 100 kilometers (62 miles) wide. The pictured area lies between 78 degrees and 82 degrees south latitude and between 90 degrees and 104 degrees east longitude.

A zoom shows details in an area about 75 kilometers (47 miles) by 100 kilometers (62 miles), centered at about 80 degrees south latitude and 99 degrees east longitude. An older impact crater in the left part of the scene is filled with younger deposits from the layered terrain.

A further zoom emphasizes a small, fresh crater about 350 meters (1,150 feet) in diameter near the center of the scene. The adjacent cliff contains numerous individual layers. An unusual set of small mesas, seen in the lower right part of the image, is being eroded from the polar layered material. The images making up this mosaic have a spatial resolution of 36 meters (118 feet) per pixel, allowing detection of features as small as 75 to 100 meters (246 to 328 feet) across.

NASA's Jet Propulsion Laboratory manages the 2001 Mars Odyssey mission for the NASA Office of Space Science, Washington. The thermal emission imaging system on Odyssey was developed by Arizona State University, Tempe, in collaboration with Raytheon Santa Barbara Remote Sensing. Lockheed Martin Space Systems, Denver, is the prime contractor for the Odyssey project, and developed and built the orbiter. Mission operations are conducted jointly from Lockheed Martin and from JPL, a division of the California Institute of Technology in Pasadena.

Credit: NASA/JPL/Arizona State University

Mars may be going through a period of climate change, new findings from NASA's Mars Odyssey orbiter suggest.

Odyssey has been mapping the distribution of materials on and near Mars' surface since early 2002, nearly a full annual cycle on Mars. Besides tracking seasonal changes, such as the advance and retreat of polar dry ice, the orbiter is returning evidence useful for learning about longer-term dynamics.

The amount of frozen water near the surface in some relatively warm low-latitude regions on both sides of Mars' equator appears too great to be in equilibrium with the atmosphere under current climatic conditions, said Dr. William Feldman of Los Alamos National Laboratory, N.M. He is the lead scientist for an Odyssey instrument that assesses water content indirectly through measurements of neutron emissions.

"One explanation could be that Mars is just coming out of an ice age," Feldman said. "In some low-latitude areas, the ice has already dissipated. In others, that process is slower and hasn't reached an equilibrium yet. Those areas are like the patches of snow you sometimes see persisting in protected spots long after the last snowfall of the winter."

Frozen water makes up as much as 10 percent of the top meter (three feet) of surface material in some regions close to the equator. Dust deposits may be covering and insulating the lingering ice, Feldman said. He and other

Odyssey scientists described their recent findings today at the fall meeting of the American Geophysical Union in San Francisco.

"Odyssey is giving us indications of recent global climate change in Mars," said Dr. Jeffrey Plaut, project scientist for the mission at NASA's Jet Propulsion Laboratory, Pasadena, Calif.

High latitude regions of Mars have layers with differing ice content within the top half meter (20 inches) or so of the surface, researchers conclude from mapping of hydrogen abundance based on gamma-ray emissions.

"A model that fits the data has three layers near the surface," said Dr. William Boynton of the University of Arizona, Tucson, team leader for the gamma-ray spectrometer instrument on Odyssey. "The very top layer would be dry, with no ice. The next layer would contain ice in the pore spaces between grains of soil. Beneath that would be a very ice rich layer, 60 to nearly 100 percent water ice."

Boynton interprets the iciest layer as a deposit of snow or frost, mixed with a little windblown dust, from a cold climate era. The middle layer could be the result of changes brought in a warmer era: The ice down to a certain depth dissipates into the atmosphere. The dust left behind collapses into a soil layer with limited pore space for returning ice.

Information from the gamma-ray spectrometer alone is not enough to tell how recently the climate changed from colder to warmer, but an estimated range might come from collaborations with climate modelers, Boynton said.

Other Odyssey instruments are providing other pieces of the puzzle. Images from the orbiter's camera system have been combined into the highest resolution complete map ever made of Mars' south polar region. "We can now accurately count craters in the layered materials of the polar regions to get an idea how old they are," said Dr. Phil Christensen of Arizona State University, Tempe, principal investigator for the camera system.

Temperature information from the camera system's infrared imaging has produced a surprise about dark patches that dot bright expanses of seasonal carbon dioxide ice. "Those dark features look like places where the ice has gone away, but thermal infrared maps show that even the dark areas have temperatures so low they must be carbon dioxide ice." Christensen said. "One possibility is that the ice is clear in these areas and we're seeing down through the ice to features underneath."

Odyssey's high-energy neutron detector continues to monitor seasonal changes in the amount of carbon dioxide ice deposited in polar regions, allowing tests of atmosphere circulation models, said Dr. Igor Mitrofanov of the Institute for Space Research, Moscow, Russia.

Measurements by an instrument for monitoring the radiation environment at Mars show the level of radiation hazard that Mars-bound astronauts might face, including levels during a period of unusually intense solar activity, said Dr. Cary Zeitlin of the National Space Biomedical Research Institute, Houston.

JPL manages Mars Odyssey for NASA's Office of Space Science, Washington. Investigators at Arizona State University, Tempe; University of Arizona, Tucson; NASA's Johnson Space Center, Houston; the Russian Aviation and Space Agency, Moscow; and Los Alamos National Laboratory, Los Alamos, N.M., built and operate Odyssey science instruments. Information about the mission is available on the Internet at: http://mars.jpl.nasa.gov/odyssey.

Contact: Guy Webster (818) 354-6278 Jet Propulsion Laboratory, Pasadena, Calif.
Donald Savage (202) 358-1727 NASA Headquarters, Washington
RELEASE: 2003-164

Mars May Be Emerging From an Ice Age

December 17, 2003

NASA's Mars Global Surveyor and Mars Odyssey missions have provided evidence of a relatively recent ice age on Mars. In contrast to Earth's ice ages, a Martian ice age waxes when the poles warm, and water vapor is transported toward lower latitudes. Martian ice ages wane when the poles cool and lock water into polar icecaps.

The "pacemakers" of ice ages on Mars appear to be much more extreme than the comparable drivers of climate change on Earth. Variations in the planet's orbit and tilt produce remarkable changes in the distribution of water ice from Polar Regions down to latitudes equivalent to Houston or Egypt. Researchers, using NASA spacecraft data and analogies to Earth's Antarctic Dry Valleys, report their findings in Thursday's edition of the journal Nature.

"Of all the solar system planets, Mars has the climate most like that of Earth. Both are sensitive to small changes in orbital parameters," said planetary scientist Dr. James Head of Brown University, Providence, R.I., lead author of the study. "Now we're seeing that Mars, like Earth, is in a period between ice ages," he said.

Discoveries on Mars, since 1999, of relatively recent water carved gullies, glacier-like flows, regional buried ice and

possible snow packs created excitement among scientists who study Earth and other planets. Information from the Mars Global Surveyor and Odyssey missions provided more evidence of an icy recent past.

Head and co-authors from Brown (Drs. John Mustard and Ralph Milliken), Boston University (Dr. David Marchant) and Kharkov National University, Ukraine (Dr. Mikhail Kreslavsky) examined global patterns of landscape shapes and near-surface water ice the orbiters mapped. They concluded a covering of water ice mixed with dust mantled the surface of Mars to latitudes as low as 30 degrees, and is degrading and retreating. By observing the small number of impact craters in those features and by backtracking the known patterns of changes in Mars' orbit and tilt, they estimated the most recent ice age occurred just 400 thousand to 2.1 million years ago, very recent in geological terms. "These results show Mars is not a dead planet, but it undergoes climate changes that are even more pronounced than on Earth," Head said.

Marchant, a glacial geologist, who spent 17 field seasons in the Mars-like Antarctic Dry Valleys, said, "These extreme changes on Mars provide perspective for interpreting what we see on Earth. Landforms on Mars that appear to be related to climate changes help us calibrate and understand similar landforms on Earth. Furthermore, the range of microenvironments in the Antarctic Dry Valleys helps us read the Mars record."

Mustard said, "The extreme climate changes on Mars are providing us with predictions we can test with upcoming Mars missions, such as Europe's Mars Express and NASA's Mars Exploration Rovers. Among the climate changes that occurred during these extremes is warming of the poles and partial melting of water at high altitudes. This clearly broadens the environments in which life might occur on Mars."

According to the researchers, during a Martian ice age, polar warming drives water vapor from polar ice into the atmosphere. The water comes back to ground at lower latitudes as deposits of frost or snow mixed generously with dust. This ice rich mantle, a few meters thick, smoothes the contours of the land. It locally develops a bumpy texture at human scales, resembling the surface of a basketball, and also seen in some Antarctic icy terrains. When ice at the top of the mantling layer sublimes back into the atmosphere, it leaves behind dust, which forms an insulating layer over remaining ice. On Earth, by contrast, ice ages are periods of polar cooling. The buildup of ice sheets draws water from liquid water oceans, which Mars lacks.

"This exciting new research really shows the mettle of NASA's 'follow-the-water' strategy for studying Mars," said Dr. Jim Garvin, NASA's lead scientist for Mars exploration. "We hope to continue pursuing this strategy in January, if the Mars Exploration Rovers land successfully. Later, the 2005 Mars Reconnaissance Orbiter and 2007 Phoenix near-polar lander will be able to directly follow up on these astounding findings by Professor Head and his team."

Global Surveyor has been orbiting Mars since 1997, Odyssey since 2001. NASA's Jet Propulsion Laboratory, a division of the California Institute of Technology, Pasadena, manages both missions for the NASA Office of Space Science, Washington. Information about NASA's Mars missions is available on the Internet at: http://mars.jpl.nasa.gov

Contact: Mark Nickel (401) 863-2476 Brown University, Providence, R.I.
Guy Webster (818) 354-6278 Jet Propulsion Laboratory, Pasadena, Calif.
Donald Savage (202) 358-1727 NASA Headquarters, Washington
RELEASE: 2003-415

Anniversary Party for Odyssey at Mars

January 11, 2004

As we celebrate Spirit's success, another of our robotic friends is celebrating an anniversary of sorts. Last week, NASA's Mars Odyssey orbiter reached an important milestone: a full Mars year (687 Earth days) of science mapping. During this martian year, it has:

* shown us where water ice lies buried beneath the surface;

* analyzed "what Mars is made of" by identifying minerals and chemical elements; and,

* studied the martian radiation environment to help us understand potential health effects on future human explorers.

"Before you send any landers to Mars, you want to look at the planet as a whole. We call that 'global reconnaissance,'" said Bob Mase, Odyssey Mission Manager.

To help rovers like Spirit and Opportunity be successful, orbiters must do the crucial work of mapping the planet, identifying scientifically interesting locations and classifying potential hazards at the landing sites. Odyssey, like its predecessor, Mars Global Surveyor, is a valuable asset in the convoy of martian spacecraft NASA continues to send to the red planet.

"We are both limited in what we can do. Orbiters can't scrape rocks and look at them microscopically, and rovers cannot traverse and image the entire planet. So, the two types of missions really complement one another," Mase said, from the desk where he monitors the spacecraft.

Beyond science studies of their own, orbiters have an important communications role to play. Not since Viking has NASA employed both orbiters and landed vehicles together. Today, the Odyssey and Mars Global Surveyor orbiters are helping the Spirit rover "talk" to ground controllers at JPL.
"It's difficult to communicate from the surface of Mars directly to Earth," Mase said. "You'd need a big antenna and a lot of power. It turns out that the rovers can more efficiently send the information up to the orbiters, which are better equipped to relay the data back to Earth."

With this additional role, Odyssey team members may not have had much time to party on Odyssey's one-martian-year anniversary. They are receiving a pretty good gift though. During Spirit's time on Mars, 75% of the rover's pictures and data have come to Earth through Odyssey.

Mars Odyssey

NATIONAL AERONAUTICS AND SPACE ADMINISTRATION

Mars Exploration Rover Launches

Press Kit
June 2003

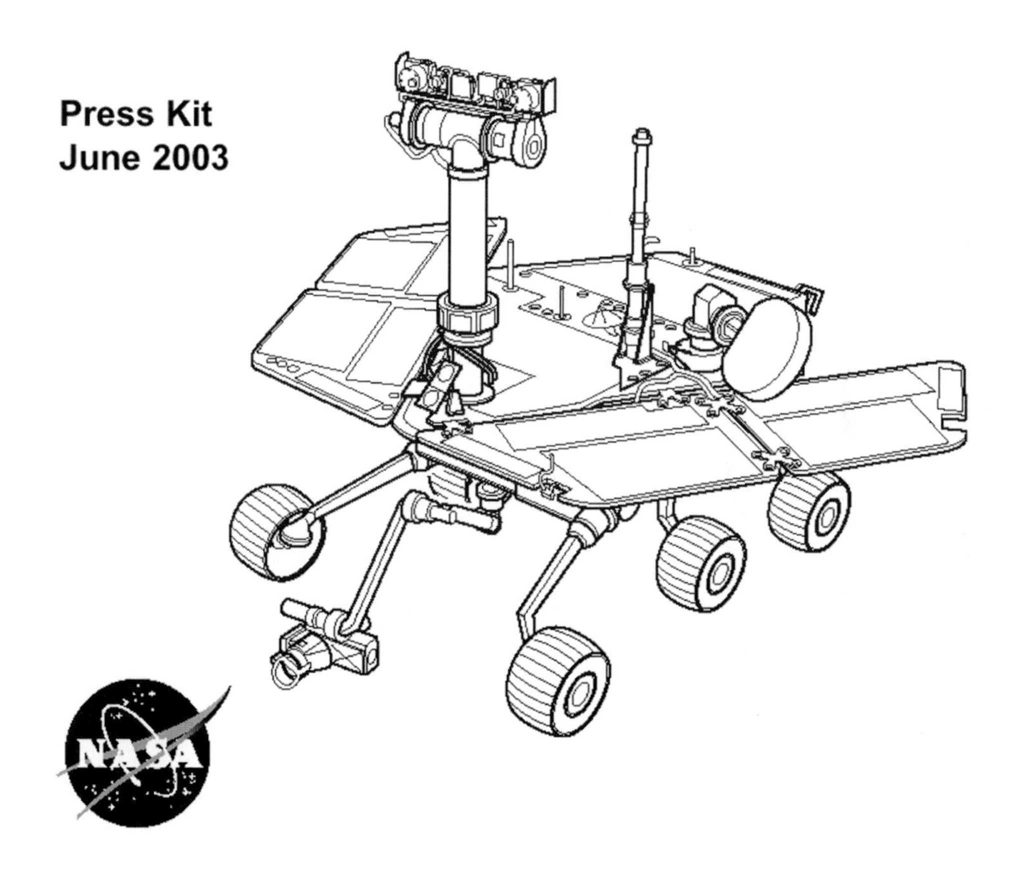

Media Contacts

Donald Savage HeadquartersWashington, D.C.	Policy / Program Management	202/358-1547 donald.savage@hq.nasa.gov
Guy Webster Jet Propulsion Laboratory, Pasadena, California	Mars Exploration Rover Mission	818/354-5011 guy.webster@jpl.nasa.gov
David Brand Cornell University, Ithaca, New York	Science Payload	607/255-3651 deb27@cornell.edu
George Diller Kennedy Space Center, Florida	Launch	321/867-2468 george.diller-1@ksc.nasa.gov

GENERAL RELEASE: NASA PREPARES TWO ROBOT ROVERS FOR MARS EXPLORATION

NASA's Mars Exploration Rover project kicks off by launching the first of two unique robotic geologists, as early as June 8. The identical rolling rovers see sharper images, can explore farther and examine rocks better than anything that's ever landed on Mars. The second rover mission, bound for a different site on Mars, will launch as soon as June 25.

"The instrumentation onboard these rovers, combined with their great mobility, will offer a totally new view of Mars, including a microscopic view inside rocks for the first time," said Dr. Ed Weiler, associate administrator for space science, NASA Headquarters, Washington.

"However, missions to Mars have proven to be far more hazardous than missions to other planets. Historically, two out of three missions, from all countries who have tried to land on Mars, ended in failure. We have done everything we can to ensure our rovers have the best chance of success, and today I gave the order to proceed to launch," Weiler said.

The first rover will arrive at Mars on January 4, 2004, the second on January 25. Plans call for each to operate for at least three months. These missions continue NASA's quest to understand the role of water on Mars. "We will be using the rovers to find rocks and soils that could hold clues about wet environments of Mars' past," said Dr. Cathy Weitz, Mars Exploration Rover program scientist at NASA Headquarters. "We'll analyze the clues to assess whether those environments may have been conducive to life."

First, the rovers have to safely reach Mars. "The rovers will use innovations to aid in safe landings, but risks remain," said Peter Theisinger, Mars Exploration rover project manager at NASA's Jet Propulsion Laboratory, Pasadena, California.

The rovers will bounce to airbag-cushioned landings at sites offering a balance of favorable conditions for safe landings and interesting science. The designated site for the first mission is Gusev Crater. The second rover will go to a site called Meridiani Planum. "Gusev and Meridiani give us two different types of evidence about liquid water in Mars' history," said Dr. Joy Crisp, Mars Exploration Rover project scientist at JPL. "Gusev appears to have been a crater lake. The channel of an ancient riverbed indicates water flowed right into it. Meridiani has a large deposit of gray hematite, a mineral that usually forms in a wet environment," Crisp said.

The rovers, working as robotic field geologists, will examine the sites for clues about what happened there. "The clues are in the rocks, but you can't go to every rock, so you split the job into two pieces," said Dr. Steve Squyres of Cornell University, Ithaca, N.Y., principal investigator for the package of science instruments on the rovers.

First, a panoramic camera at human-eye height, and a miniature thermal emission spectrometer, with infrared vision, help scientists identify the most interesting rocks. The rovers can watch for hazards in their way and maneuver around them. Each six-wheeled robot has a deck of solar panels, about the size of a kitchen table, for power. The rover drives to the selected rock and extends an arm with tools on the end. Then, a microscopic imager, like a geologist's hand lens, gives a close-up view of the rock's texture. Two spectrometers identify the composition of the rock. The fourth tool substitutes for a geologist's hammer. It exposes the fresh interior of a rock by scraping away the weathered surface layer.

Both rover missions will lift off from Cape Canaveral Air Force Station, Florida, on Delta II launch vehicles. Launch opportunities begin for the first mission at 2:06 p.m. EDT June 8 and for the second mission at 12:38 a.m. EDT June 25, and repeat twice daily for up to 21 days for each mission.

"We see the twin rovers as stepping stones for the rest of the decade and to a future decade of Mars exploration that will ultimately provide the knowledge necessary for human exploration," said Orlando Figueroa, director of the Mars Exploration Program at NASA Headquarters.

JPL, a division of the California Institute of Technology in Pasadena, manages the Mars Exploration Rover project for NASA's Office of Space Science, Washington.

For information about the Mars Exploration Rover project on the Internet, visit:
http://mars.jpl.nasa.gov/mer

NASA will feature live webcasts of the launches on the Internet at:
http://www.jpl.nasa.gov/webcast/mer

Cornell University's web site on the science payload is at:
http://athena.cornell.edu

Media Services Information

NASA Television Transmission
NASA Television is broadcast on the satellite AMC-2, transponder 9C, C band, 85 degrees west longitude, frequency 3880.0 MHz, vertical polarization, audio monaural at 6.8 MHz. The schedule for Mars arrival television transmissions will be available from the Jet Propulsion Laboratory, Pasadena, California; and NASA Headquarters, Washington.

Launch Media Credentialing
News media representatives who would like to cover the launch in person must be accredited through the NASA Kennedy Space Center newsroom. Journalists may contact the newsroom at 321/867-2468 for more information.

Briefings
An extensive schedule of news and background briefings will beheld at JPL during the landing period, with later briefings originating jointly from JPL and NASA Headquarters. A schedule of briefings is available on the Internet at JPL's Mars News site (below).

Internet Information
Extensive information on the Mars Exploration Rover project including an electronic copy of this press kit, press releases, fact sheets, status reports, briefing schedule and images, is available from the Jet Propulsion Laboratory's Mars Exploration Rover newsroom website: http://www.jpl.nasa.gov/mer. The Mars Exploration Rover project also maintains a web site at: http://mars.jpl.nasa.gov/mer. Cornell University's web site on the science payload is at: http://athena.cornell.edu.

Quick Facts

Spacecraft
Cruise vehicle dimensions: 2.65 meters (8.7 feet) diameter, 1.6 meters (5.2 feet) tall
Rover dimensions: 1.5 meter (4.9 feet) high by 2.3 meters (7.5 feet) wide by 1.6 meter (5.2 feet) long
Weight: 1,062 kilograms (2,341 pounds) total at launch, consisting of 174-kilogram (384-pound) rover, 365-kilogram (805-pound) lander, 198-kilogram (436-pound) backshell and parachute, 90-kilogram (198-pound) heat shield and 183-kilogram (403-pound) cruise stage, plus 52 kilograms (115 pounds) of propellant
Power: Solar panel and lithium-ion battery system providing 140 watts on Mars surface
Science instruments: Panoramic cameras, miniature thermal emission spectrometer, Mössbauer spectrometer, alpha particle X-ray spectrometer, microscopic imager, rock abrasion tool, magnet arrays

Rover A Mission
Launch vehicle: Delta II 7925
Launch period: June 8-24, 2003
Earth-Mars distance at launch: 105 million kilometers (65 million miles)
Mars landing: January 4, 2004, at about 2 p.m. local Mars time (8:11 p.m. January 3 PST)
Landing site: Gusev Crater, possible former lake in giant impact crater
Earth-Mars distance on landing day: 170.2 million kilometers (105.7 million miles)
One-way speed-of-light time Mars-to-Earth on landing day: 9.46 minutes
Total distance traveled Earth to Mars (approximate): 500 million kilometers (311 million miles)
Near-surface atmospheric temperature at landing site: -100 C (-148 F) to 0 C (32 F)
Primary mission: 90 Mars days, or "sols" (equivalent to 92 Earth days)

Rover B Mission
Launch vehicle: Delta II 7925H (larger solid-fuel boosters than 7925)
Launch period: June 25-July 15, 2003
Earth-Mars distance at launch: 89 million kilometers (55 million miles)
Mars landing: January 25, 2004, at about 1:15 p.m. local Mars time (8:56 p.m. January 24 PST)
Landing site: Meridiani Planum, where mineral deposits suggest wet past
Landing time: Approximately 1:15 p.m. local Mars time (8:56 p.m. PST)
Earth-Mars distance on landing day: 198.7 million kilometers (123.5 million miles)
One-way speed-of-light time Mars-to-Earth on landing day: 11 minutes
Total distance traveled Earth to Mars (approximate): 491 million kilometers (305 million miles)
Near-surface atmospheric temperature at landing site: -100 C (-148 F) to 0 C (32 F)
Primary mission: 90 Mars days, or "sols" (equivalent to 92 Earth days)

Program
Cost: Approximately $800 million total, consisting approximately of $625 million spacecraft development and science instruments; $100 million launch; $75 million mission operations and science processing

Mars at a Glance

General
One of five planets known to the ancients; Mars was the Roman god of war, agriculture and the state
Yellowish brown to reddish color; occasionally the third brightest object in the night sky after the Moon and Venus

Physical Characteristics
Average diameter 6,780 kilometers (4,212 miles); about half the size of Earth, but twice the size of Earth's Moon
Same land area as Earth, reminiscent of a rocky desert
Mass $^1/_{10}$th of Earth's; gravity only 38 percent as strong as Earth's
Density 3.9 times greater than water (compared to Earth's 5.5 times greater than water)
No planet-wide magnetic field detected; only localized ancient remnant fields in various regions

Orbit
Fourth planet from the Sun, the next beyond Earth
About 1.5 times farther from the Sun than Earth is
Orbit elliptical; distance from Sun varies from a minimum of 206.7 million kilometers (128.4 millions miles) to a maximum of 249.2 million kilometers (154.8 million miles); average distance from the Sun 227.7 million kilometers (141.5 million miles)
Revolves around Sun once every 687 Earth days
Rotation period (length of day) 24 hours, 39 min, 35 sec (1.027 Earth days)
Poles tilted 25 degrees, creating seasons similar to Earth's

Environment
Atmosphere composed chiefly of carbon dioxide (95.3%), nitrogen (2.7%) and argon (1.6%)
Surface atmospheric pressure less than $^1/_{100}$th that of Earth's average
Surface winds up to 80 miles per hour (40 meters per second)
Local, regional and global dust storms; also whirlwinds called dust devils
Surface temperature averages -53° C (-64° F); varies from -128° C (-199° F) during polar night to 27° C (80° F) at equator during midday at closest point in orbit to Sun

Features
Highest point is Olympus Mons, a huge shield volcano about 26 kilometers (16 miles) high and 600 kilometers (370 miles) across; has about the same area as Arizona
Canyon system of Valles Marineris is largest and deepest known in solar system; extends more than 4,000 kilometers (2,500 miles) and has 5 to 10 kilometers (3 to 6 miles) relief from floors to tops of surrounding plateaus
"Canals" observed by Giovanni Schiaparelli and Percival Lowell about 100 years ago were a visual illusion in which dark areas appeared connected by lines. The *Mariner 9* and *Viking* missions of the 1970s, however, established that Mars has channels possibly cut by ancient rivers

Moons
Two irregularly shaped moons, each only a few kilometers wide
Larger moon named Phobos ("fear"); smaller is Deimos ("terror"), named for attributes personified in Greek mythology as sons of the god of war

Historical Mars Missions

Mission	Country	Launch Date	Purpose	Results
[Unnamed]	USSR	10/10/60	Mars flyby	Did not reach Earth orbit
[Unnamed]	USSR	10/14/60	Mars flyby	Did not reach Earth orbit
[Unnamed]	USSR	10/24/62	Mars flyby	Achieved Earth orbit only
Mars 1	USSR	11/1/62	Mars flyby	Radio failed at 106 million km (65.9 million miles)
[Unnamed]	USSR	11/4/62	Mars flyby	Achieved Earth orbit only
Mariner 3	U.S.	11/5/64	Mars flyby	Shroud failed to jettison
Mariner 4	U.S.	11/28/64	Mars flyby	First successful Mars flyby 7/14/65; Returned 21 photos
Zond 2	USSR	11/30/64	Mars flyby	Passed Mars but radio failed; Returned no planetary data
Mariner 6	U.S.	2/24/69	Mars flyby	Mars flyby 7/31/69; Returned 75 photos
Mariner 7	U.S.	3/27/69	Mars flyby	Mars flyby 8/5/69; Returned 126 photos
Mariner 8	U.S.	5/8/71	Mars orbiter	Failed during launch
Kosmos 419	USSR	5/10/71	Mars lander	Achieved Earth orbit only
Mars 2	USSR	5/19/71	Mars orbiter / lander	Arrived 11/27/71; No useful data; Lander burned up due to steep entry
Mars 3	USSR	5/28/71	Mars orbiter / lander	Arrived 12/3/71; Lander operated on surface for 20 seconds before failing
Mariner 9	U.S.	5/30/71	Mars orbiter	In orbit 11/13/71 to 10/27/72; Returned 7,329 photos
Mars 4	USSR	7/21/73	Mars orbiter	Failed Mars orbiter; Flew past Mars 2/10/74
Mars 5	USSR	7/25/73	Mars orbiter	Arrived 2/12/74; Lasted a few days
Mars 6	USSR	8/5/73	Mars flyby module and lander	Arrived 3/12/74; Lander failed due to fast impact
Mars 7	USSR	8/9/73	Mars flyby module and lander	Arrived 3/9/74; Lander missed the planet
Viking 1	U.S.	8/20/75	Mars orbiter / lander	Orbit 6/19/76-1980; Lander 7/20/76-1982
Viking 2	U.S.	9/9/75	Mars orbiter / lander	Orbit 8/7/76-1987, lander 9/3/76-1980; Combined, the *Viking* orbiters and landers returned 50,000+ photos
Phobos 1	USSR	7/7/88	Mars / Phobos orbiter / lander	Lost 8/88 en route to Mars
Phobos 2	USSR	7/12/88	Mars / Phobos orbiter / lander	Lost 3/89 near Phobos
Mars Observer	U.S.	9/25/92	Mars orbiter	Lost just before Mars arrival 8/21/93
Mars Global Surveyor	U.S.	11/7/96	Mars orbiter	Arrived 9/12/97, high-detail mapping through 1/00; Now conducting second extended mission through fall 2004
Mars 96	Russia	11/16/96	orbiter and landers	launch vehicle failed
Mars Pathfinder	U.S.	12/4/96	Mars lander and rover	Landed 7/4/97; Last transmission 9/27/97
Nozomi	Japan	7/4/98	Mars orbiter	Currently in orbit around the Sun; Mars arrival delayed to 12/03 due to propulsion problem
Mars Climate Orbiter	U.S.	12/11/98	Mars orbiter	Lost upon arrival 9/23/99
Mars Polar Lander / Deep Space 2	U.S.	1/3/99	lander and soil probes	Lost on arrival 12/3/99
Mars Odyssey	U.S.	3/7/01	Mars orbiter	Arrived 10/24/01; currently conducting prime mission studying global composition, ground ice, thermal imaging
Mars Express / Beagle 2	European	6/2/03	Mars orbiter / lander	Due to enter orbit 12/03; Landing 12/25/03

Mars: The Water Trail

Thirty-eight years ago, on the eve of the first spacecraft flyby of Mars, everything we knew about the Red Planet was based on what sparse details could be gleaned by peering at it from telescopes on Earth. Since the early 1900s, popular culture had been enlivened by the notion of a habitable neighboring world criss-crossed by canals and, possibly, inhabited by advanced life forms that might have built them – whether friendly or not. Astronomers were highly skeptical about the canals, which looked more dubious the closer they looked. About the only hard information they had on Mars was that they could see it had seasons with ice caps that waxed and waned, along with seasonally changing surface markings. By breaking down the light from Mars into colors, they learned that its atmosphere was thin and dominated by an unbreathable gas known as carbon dioxide.

The past four decades have completely revolutionized that view. First, hopes of a lush, Earth-like world were deflated when *Mariner 4*'s flyby in 1965 revealed large impact craters, not unlike those on Earth's barren, lifeless Moon. Those holding out for Martians were further discouraged when NASA's two Viking landers were sent to the surface in 1976 equipped with a suite of chemistry experiments that turned up no conclusive sign of biological activity. Mars, as we came to know it was cold, nearly airless and bombarded by hostile radiation from both the Sun and from deep space.

But along the way since then, new possibilities of a more hospitable Martian past have emerged. Mars is a much more complex body than Earth's Moon. Scientists scrutinizing pictures from the Viking orbiters have detected potential signs of an ancient coastline that may have marked the edges of a long-lost sea. Today's Mars Global Surveyor and Mars Odyssey orbiters have revealed many features that strongly appear to have been shaped by running water that has since disappeared, perhaps buried as layers of ice just under the planet's surface.

Although it appears unlikely that complex organisms similar to Earth's could have existed in any recent time on Mars' comparatively hostile surface, scientists are intrigued by the possibility that life in some form, perhaps very simple microbes, may have gained a foothold in ancient times, when Mars may have been warmer and wetter. It is not unthinkable that life in some form could persist today in underground springs warmed by heat vents around smoldering volcanoes, or even beneath the thick ice caps. To investigate those possibilities, scientists must start by learning more about the history of water on Mars – how much there was and when, in what form it existed, and how long it lasted.

One of the most promising ways to answer those questions is to look at the diverse clues that water has left on Mars. Besides the water-carved land forms visible for decades from orbiting spacecraft, many details of the story of water on the Red Planet are locked up in the rocks littered across its surface. Rocks are made up of building blocks known as minerals, each of which tells the story of how it came to be a part of a any given rock. Some types of minerals, for example, are known to form on Earth only submerged under water, while others are profoundly altered when hot water runs through them, leaving behind residues. Up until now, it has been very difficult to get to know the minerals in Martian rocks because we have not had the tools to unravel their mineralogies. By understanding Mars' rocks in a more complete manner, scientists can gain a better view into the history of liquid water on the planet. Like their predecessor mission, *Mars Pathfinder*, the Mars Exploration Rovers will pursue this goal by placing robotic geologists on the planet's surface – ideally suited to "reading the rocks" to understand the still mysterious history of water, and even of life-friendly ancient environments.

Myths and Reality

Mars caught public fancy in the late 1870s, when Italian astronomer Giovanni Schiaparelli reported using a telescope to observe "canali," or channels, on Mars. A possible mistranslation of this word as "canals" may have fired the imagination of Percival Lowell, an American businessman with an interest in astronomy. Lowell founded an observatory in Arizona, where his observations of the Red Planet convinced him that the canals were dug by intelligent beings – a view that he energetically promoted for many years.

By the turn of the last century, popular songs envisioned sending messages between worlds by way of huge signal mirrors. On the dark side, H.G. Wells' 1898 novel *The War of the Worlds* portrayed an invasion of Earth by technologically superior Martians desperate for water. In the early 1900s, novelist Edgar Rice Burroughs, known for the *Tarzan* series, also entertained young readers with tales of adventures among the exotic inhabitants of Mars, which he called Barsoom.

Fact began to turn against such imaginings when the first robotic spacecraft were sent to Mars in the 1960s. Pictures from the 1965 flyby of *Mariner 4* and the 1969 flybys of *Mariner 6* and *Mariner 7* showed a desolate world, pocked with impact craters similar to those seen on Earth's Moon. *Mariner 9* arrived in 1971 to orbit Mars for the first time, but showed up just as an enormous dust storm was engulfing the entire planet. When the storm died down, *Mariner 9* revealed a world that, while partly crater-pocked like Earth's Moon, was much more geologically complex, complete with gigantic canyons, volcanoes, dune fields and polar ice caps. This first wave of Mars exploration culminated in the *Viking* mission, which sent two orbiters and two landers to the planet in 1975. The landers included a suite of experiments that conducted chemical tests in direct search of life. Most scientists interpreted the results of these tests as negative, deflating hopes of identifying another world on where life might be or have been wide-spread. However, *Viking* left a huge legacy of information about Mars that fed a hungry science community for two decades.

The science community had many other reasons for being interested in Mars, apart from the direct search for life; the next mission on the drawing boards concentrated on a study of the planet's geology and climate using advanced orbital reconnaissance. Over the next 20 years, however, new findings in laboratories on Earth came to change the way that scientists thought about life and Mars.

One was the 1996 announcement by a team from Stanford University and NASA's Johnson Space Center that a meteorite believed to have originated on Mars contained what might be the fossils of ancient bacteria. This rock and other likely Mars meteorites discovered on several continents on Earth are believed to have been blasted off the Red Planet by asteroid or comet impacts. They are presently believed to have come from Mars because of gases trapped in them that unmistakably match the composition of Mars' atmosphere as measured by the *Viking* landers. Many scientists questioned the conclusions of the team announcing the discovery of possible life in one Martian meteorite, but if nothing else the mere presence of organic compounds in the meteorites increases the odds of life forming at an earlier time on a far wetter Mars.

Another development that shaped scientists' thinking was spectacular new findings on how and where life thrives on Earth. The fundamental requirements for life as we know it today are liquid water, organic compounds and an energy source for synthesizing complex organic molecules. Beyond these basics, we do not yet understand the environmental and chemical evolution that leads to the origin of terrestrial life. But in recent years, it has become increasingly clear that life can thrive in settings much different – and more harsh – from a tropical soup rich in organic nutrients.

In the 1980s and 1990s, biologists found that microbial life has an amazing flexibility for surviving in extreme environments – niches that by turn are extraordinarily hot, or cold, or dry, or under immense pressures – that would be completely inhospitable to humans or complex animals. Some scientists even concluded that life may have begun on Earth in heat vents far under the ocean's surface.

This in turn had its effect on how scientists thought about Mars. Martian life might not be so widespread that it would be readily found at the foot of a lander spacecraft, but it may have thrived billions of years ago in an underground thermal spring or other hospitable environment. Or it might still exist in some form in niches below the currently frigid, dry, windswept surface, perhaps entombed in ice or in liquid water aquifers.

After years of studying pictures from the *Viking* orbiters, scientists gradually came to conclude that many features they saw suggested that Mars may have been warm and wet in an earlier era. And two currently operating orbiters – *Mars Global Surveyor* and *Mars Odyssey* – are giving scientists yet new insights into the planet. *Global Surveyor*'s camera detected possible evidence for recent liquid water in a large number of settings, while *Odyssey*'s camera system has found large amounts of ice mixed in with Mars surface materials at high latitudes, as well as potential evidence of ancient snow packs.

The Three Ages of Mars

Based on what they have learned from spacecraft missions, scientists view Mars as the "in-between" planet of the inner solar system. Small rocky planets such as Mercury and Earth's Moon apparently did not have enough internal heat to power volcanoes or to drive the motion of tectonic plates, so their crusts grew cold and static relatively soon after they formed, when the solar system condensed into planets about 4.6 billion years ago. Devoid of atmospheres, they are riddled with craters that are relics of impacts during a period of bombardment when the inner planets were sweeping up remnants of small rocky bodies that failed to "make it as planets" in the solar system's early times.

Earth and Venus, by contrast, are larger planets with substantial internal heat sources and significant atmospheres. Earth's surface is continually reshaped by tectonic plates sliding under and against each other and materials spouting forth from active volcanoes where plates are ripped apart. Both Earth and Venus have been paved over so recently that both lack any discernible record of cratering from the era of bombardment in the early solar system.

Mars appears to stand between those sets of worlds, on the basis of current yet evolving knowledge. Like Earth and Venus, it possesses a myriad of volcanoes, although they probably did not remain active as long as counterparts on Earth and Venus. On Earth, a single "hot spot" or plume might form a chain of middling-sized islands such as the Hawaiian Islands as a tectonic plate slowly slides over it. On Mars there are apparently no such tectonic plates, at least as far as we know today, so when volcanoes formed in place they had the time to become much more enormous than the rapidly moving volcanoes on Earth. Overall, Mars appears to be neither as dead as Mercury and our Moon, nor as active as Earth and Venus. As one scientist quips, "Mars is a warm corpse if not a fire-breathing dragon." Thanks to the ongoing observations by the *Global Surveyor* and *Odyssey* orbiters, however, this view of Mars is still evolving.

Mars almost resembles two different worlds that have been glued together. From latitudes around the equator to the south are ancient highlands pockmarked with craters from the solar system's early era, yet riddled with channels that attest to the flow of water. The northern third of the planet, however, overall is sunken and much smoother at kilometer (mile) scales. There is as yet no general agreement on how the northern plains got to be that way. At one end of the spectrum is the theory that it is the floor of an ancient sea; at the other, the notion that it is merely the end product of innumerable lava flows. New theories are emerging thanks to the discoveries of *Mars Odyssey*, and some scientists believe a giant ice sheet may be buried under much of the relatively smooth northern plains. Many scientists suspect that some unusual internal process not yet fully understood may have caused the northern plains to sink to relatively low elevations in relation to the southern uplands.

Scientists today view Mars as having had three broad ages, each named for a geographic area that typifies it:

The Noachian Era is the name given to the time spanning perhaps the first billion years of Mars' existence after the planet was formed 4.6 billion years ago. In this era, scientists suspect that Mars was quite active, with periods of warm and wet environment, erupting volcanoes and some degree of tectonic activity. The planet may have had a thicker atmosphere to support running water, and it may have rained and snowed.

In the Hesperian Era, which lasted for about the next 500 million to 1.5 billion years, geologic activity was slowing down and near-surface water perhaps was freezing to form surface and buried ice masses. Plunging temperatures probably caused water pooled underground to erupt when heated by impacts in catastrophic floods that surged across vast stretches of the surface – floods so powerful that they unleashed the force of thousands of Mississippi Rivers. Eventually, water became locked up as permafrost or subsurface ice, or was partially lost into outer space.

The Amazonian Era is the current age that began around 2 billion to 3 billion years ago. The planet is now a dry, desiccating environment with only a modest atmosphere in relation to Earth. In fact, the atmosphere is so thin that water can exist only as a solid or a gas, not as a liquid.

Apart from that broad outline, there is lively debate and disagreement on the details of Mars' history. How wet was the planet, and how long ago? What eventually happened to all of the water? That is all a story that is still being written.

In addition to studying the planet from above with orbiting spacecraft, NASA's Mars Exploration Program is putting robotic geologists on the surface in the form of instrumented rovers. Both of the landing sites selected for the Mars Exploration Rovers show evidence of water activity in their past. The rovers will look at rocks to understand the types of minerals that they are made of, and hence the environments in which they formed. This, in turn, will offer clues about the environment in which the rocks formed. Some types of rocks, for example, might be of types that form in running water, whereas others might be typical of the sediments that form on the beds of lakes.

Even if we ultimately learn that Mars never harbored life as we know it here on Earth, scientific exploration of the Red Planet can assist in understanding the history and evolution of life on our own home world. Much, if not all, of the evidence for the origin of life here on Earth has been obliterated by the incredible pace of weathering and global tectonics that have operated over billions of years. Mars, by comparison, is a composite world with some regions that may have histories similar to Earth's crust, while others serve as a frozen gallery of the solar system's early days.

Thus, even if life never developed on Mars – something that we cannot answer today – scientific exploration of the planet may yield critical information unobtainable by any other means about the prebiotic chemistry that led to life on Earth. Mars as a fossil graveyard of the chemical conditions that fostered life on Earth is an intriguing possibility.

Where We've Been and Where We're Going

Building on scientific discoveries and lessons learned from past and ongoing missions, NASA's Mars Exploration Program will establish a sustained observational presence both around and on the surface of Mars in coming years. This will include orbiters that view the planet from above and act as telecommunications relays; surface-based mobile laboratories; robots that probe below the planet's surface; and, ultimately, missions that return soil and rock samples to Earth. With international cooperation, the long-term program will be guided by compelling questions that scientists are interested in answering about Mars, developing technologies to make missions possible within available resources. The program's strategy is to seek to uncover profound new insights into Mars' past environments, the history of its rocks and interior, the many roles and abundances of water and, quite possibly, evidence of past and present life.

The following are the most recently completed, ongoing and near-term future Mars missions of exploration in the NASA program:

Mars Pathfinder (December 1996 - March 1998): The first completed mission in NASA's Discovery Program of low-cost, rapidly developed planetary missions with highly focused scientific goals, *Mars Pathfinder* far exceeded its expectations and outlived its primary design life. This lander, which released its *Sojourner* rover at the Martian surface, returned 2.3 billion bits of information, including more than 17,000 images and more than 15 chemical analyses of rocks and soil and extensive data on winds and other types of weather. Investigations carried out by instruments on both the lander and the rover suggest that, in its past, Mars was warm and wet, with liquid water on its surface and a thicker atmosphere. The lander and rover functioned far beyond their planned lifetimes (30 days for the lander and 7 days for the rover), but eventually, after about three months on the Martian surface, depletion of the lander's battery and a drop in the lander's operating temperature are thought to have ended the mission.

Mars Global Surveyor (November 1996 - present): During its primary mapping mission from March 1999 through January 2001, NASA's *Mars Global Surveyor* collected more information than any other previous Mars mission. Today the orbiter continues to gather data in a second extended mission. As of May 1, 2003, it has completed more than 20,000 orbits of Mars and returned more than 137,000 images, 671 million laser-altimeter shots and 151 million spectrometer measurements. Some of the mission's most significant findings include: evidence of possibly recent liquid water at the Martian surface; evidence for layering of rocks that points to widespread ponds or lakes in the planet's early history; topographic evidence that most of the southern hemisphere is higher in elevation than most of the northern hemisphere, so that any downhill flow of water and sediments would have tended to be northward; identification of gray hematite, a mineral suggesting a wet environment when it was formed; and extensive evidence for the role of dust in reshaping the recent Martian environment. Global Surveyor provided valuable details for evaluating the risks and attractions of potential landing sites for the Mars Exploration Rover missions, and it will serve as a communications relay for the rovers as they descend to land on Mars and afterwards.

Mars Climate Orbiter and *Mars Polar Lander* (1998-99): These spacecraft were both lost upon Mars arrival.

Mars Odyssey (April 2001 - present): This orbiter's prime mapping mission began in March 2002. Its suite of gamma-ray spectrometer instruments has provided strong evidence for large quantities of frozen water mixed into the top layer of soil in the 20 percent of the planet near its north and south poles. By one estimate – likely an underestimate – the amount of water ice near the surface, if melted, would be enough water to fill Lake Michigan twice. *Odyssey*'s infrared camera system has also provided detailed maps of minerals in rocks and soils. A layer of olivine-rich rock in one canyon near Mars' equator suggests that site has been dry for a long time, since olivine is easily weathered by liquid water. Nighttime infrared imaging by *Odyssey*'s camera system provides information about how quickly or slowly surface features cool off after sunset, which gives an indication of where the surface is rocky and where it is dusty. *Odyssey*'s observations have helped evaluate potential landing sites for the Mars Exploration Rovers. When the rovers reach Mars, radio relay via *Odyssey* will be one way they will return data to Earth.

Mars Reconnaissance Orbiter (2005): This mission is being developed to provide detailed information about thousands of sites on Mars, connecting the big-picture perspective of an orbiter with a level of local detail that has previously come only from landing a spacecraft on the surface. The spacecraft's telescopic camera will reveal Martian landscapes in resolution fine enough to show rocks the size of a desk. Maps of surface minerals will be produced in unprecedented detail for thousands of potential future landing sites. Scientists will search in particular for types of minerals that form in wet environments. A radar instrument on the orbiter will probe hundreds of meters (or yards) below Mars' surface for layers of frozen or melted water, and other types of geologic layers. Another instrument will document atmospheric processes changing with Mars' seasons, and study how water vapor enters, moves within and leaves the atmosphere.

Mars Scouts (2007 and later): *Mars Scouts* are competitively proposed missions intended to supplement and complement, at relatively low cost, the core missions of NASA's Mars Exploration Program. From 25 original proposals, NASA selected four candidate *Scout* missions in late 2002 for further study. One will be chosen in August 2003 as the first *Mars Scout*, for launch in 2007. The four finalists include an orbiter, a lander, an airplane, and a quick dip into Mars' atmosphere to fetch dust and gas samples back to Earth. Mars Volcanic Emission and Life Scout consists of an orbiter for exploring Mars' atmosphere for emissions that could be related to active volcanism or microbial activity. *Phoenix* is a surface laboratory that proposes to land in Mars' northern plains to investigate water ice, organic molecules and climate. The Aerial Regional-scale Environmental Study proposes to fly a rocket-propelled aircraft through Mars' atmosphere to measure water vapor and other gases near the surface for improved understanding of the chemical evolution of the planet and potential biological activity. The Sample Collection for Investigation of Mars would swoop close enough to the Martian surface to grab a sampling of atmospheric dust and gas and return them back to Earth. A second round of Scout solicitation in the future will select a handful of additional *Mars Scout* missions, one of which would fly in 2011.

Mars Science Laboratory (2009): NASA proposes to develop and launch a roving science laboratory that would operate on Mars for more than a year and travel for at least several kilometers or miles. The mission would mark major advances in measurement capabilities and surface access. The rover will examine the potential of the Red Planet as a habitat for extant or extinct life. It would also demonstrate technologies for accurate landing and surface-hazard avoidance that will be necessary for sending future missions to sites that are scientifically compelling but difficult to reach. This mission is designed to make the transition from a program in which we "follow the water" to one in which we "follow the clues to search for the missing carbon" – and hence to perform the first indirect life detection in a generation on the Martian surface.

The Next Decade of Mars Exploration: For the second decade of this century, NASA proposes additional reconnaissance orbiters, rovers and landers, and the first mission to return samples of Martian rock and soil to Earth. The flexible program includes many options. Scientists and mission planners foresee technology development for advanced capabilities, such as Mars ascent vehicle, automatic rendezvous in Mars orbit and planetary protection.

Science Investigations

The Mars Exploration Rover mission seeks to determine the history of climate and water at sites on Mars where conditions may once have been favorable to life. Each rover is equipped with a suite of science instruments that will be used to read the geologic record at each site, to investigate what role water played there, and to determine how suitable the conditions would have been for life.

Science Objectives
Based on priorities of the overall Mars Exploration Program, the following science objectives were developed for the 2003 rovers:

Search for and characterize a diversity of rocks and soils that hold clues to past water activity (water-bearing minerals and minerals deposited by precipitation, evaporation, sedimentary cementation, or hydrothermal activity).

Investigate landing sites, selected on the basis of orbital remote sensing, that have a high probability of containing physical and/or chemical evidence of the action of liquid water.

Determine the spatial distribution and composition of minerals, rocks and soils surrounding the landing sites.

Determine the nature of local surface geologic processes from surface morphology and chemistry.

Calibrate and validate orbital remote sensing data and assess the amount and scale of heterogeneity at each landing site.

For iron-containing minerals, identify and quantify relative amounts of specific mineral types that contain water or hydroxyls, or are indicators of formation by an aqueous process, such as iron-bearing carbonates.

Characterize the mineral assemblages and textures of different types of rocks and soils and put them in geologic context.

Extract clues from the geologic investigation, related to the environmental conditions when liquid water was present and assess whether those environments were conducive for life.

Science Instruments

The package of science instruments on the rovers is collectively known as the Athena science payload. Led by Dr. Steven Squyres, professor of astronomy at Cornell University, Ithaca, New York, the Athena package was originally proposed to fly under different Mars lander and rover mission concepts before being finalized as the science payload for the Mars Exploration Rovers.

The package consists of two instruments designed to survey the landing site, as well as three other instruments on an arm designed for close-up study of rocks. Also on the arm is a tool that can scrape away the outer layers of rocks. Those instruments are supplemented by magnets and calibration targets that will enable other studies.

The two instruments that will survey the general site are:

Panoramic Camera will view the surface using two high-resolution color stereo cameras to complement the rover's navigation cameras. Delivering panoramas of the Martian surface with unprecedented detail, the instrument's narrow-angle optics provide angular resolution more than three times higher than that of the *Mars Pathfinder* cameras. The camera's images will help scientists decide what rocks and soils to analyze in detail, and will provide information on surface features, the distribution and shape of nearby rocks, and the presence of features carved by ancient waterways.

The Mini-Thermal Emission Spectrometer is an instrument that sees infrared radiation emitted by objects. It will determine from afar the mineral composition of Martian surface features and allow scientists to select specific rocks and soils to investigate in detail. Observing in the infrared allows scientists to see through dust that coats many rocks, allowing the instrument to recognize carbonates, silicates, organic molecules and minerals formed in water. Infrared data will also help scientists assess the capacity of rocks and soils to hold heat over the wide temperature range of a Martian day. Besides studying rocks, the instrument will be pointed upward to make the first-ever high-resolution temperature profiles through the Martian atmosphere's boundary layer. The data from the instrument will be complement that obtained by the thermal emission spectrometer on the *Mars Global Surveyor* orbiter.

The instruments on the rover arm are:

The Microscopic Imager is a combination of a microscope and a camera. It will produce extreme close-up views (at a scale of hundreds of microns) of rocks and soils examined by other instruments on the rover arm, providing context for the interpretation of data about minerals and elements. The imager will help characterize sedimentary rocks that formed in water, and thus will help scientists understand past watery environments on Mars. This instrument will also yield information on the small-scale features of rocks formed by volcanic and impact activity as well as tiny veins of minerals like the carbonates that may contain microfossils in the famous Mars meteorite, ALH84001. The shape and size of particles in the Martian soil can also be determined by the instrument, which provides valuable clues about how the soil formed.

Because many of the most important minerals on Mars contain iron, the Mössbauer Spectrometer is designed to determine with high accuracy the composition and abundance of iron-bearing minerals that are difficult to detect by other means. Identification of iron-bearing minerals will yield information about early Martian environmental conditions. The spectrometer is also capable of examining the magnetic properties of surface materials and identifying minerals formed in hot, watery environments that could preserve fossil evidence of Martian life. The instrument uses two pieces of radioactive cobalt-57, each about the size of a pencil eraser, as radiation sources. The instrument is provided by Germany.

The Alpha Particle X-ray Spectrometer will accurately determine the elements that make up rocks and soils. This information will be used to complement and constrain the analysis of minerals provided by the other science instruments. Through the use of alpha particles and X-rays, the instrument will determine a sample's abundances of all major rock-forming elements except hydrogen. Analyzing the elemental make-up of Martian surface materials will provide scientists with information about crustal formation, weathering processes and water activity on Mars. The instrument uses small amounts of curium-244 for generating radiation. It is provided by Germany.

The arm-mounted instruments will be aided by a Rock Abrasion Tool that will act as the rover's equivalent of a geologist's rock hammer. Positioned against a rock by the rover's instrument arm, the tool uses a grinding wheel to remove dust and weathered rock, exposing fresh rock underneath. The tool will expose an area 4.5 centimeters (2 inches) in diameter, and grind down to a depth of as much as 5 millimeters (0.2 inch).

In addition, the rovers are equipped with the following that work in conjunction with science instruments:

Each rover has three sets of Magnet Arrays that will collect airborne dust for analysis by the science instruments. Mars is a dusty place, and some of that dust is highly magnetic. Magnetic minerals carried in dust grains may be freeze-dried remnants of the planet's watery past. A periodic examination of these particles and their patterns of accumulation on magnets of varying strength can reveal clues about their mineralogy and the planet's geologic history. One set of magnets will be carried by the rock abrasion tool. As it grinds into Martian rocks, scientists will have the opportunity to study the properties of dust from these outer rock surfaces. A second set of two magnets is mounted on the front of the rover for the purpose of gathering airborne dust. These magnets will be reachable for analysis by the Mössbauer and alpha particle X-ray spectrometers. A third magnet is mounted on the top of the rover deck in view of the panoramic camera. This magnet is strong enough to deflect the paths of wind-carried, magnetic dust. The magnet arrays are provided by Denmark.

Calibration Targets are reference points that will help scientists fine-tune observations not only from imagers but also other science instruments. The Mössbauer spectrometer, for example, uses as a calibration target a thin slab of rock rich in magnetite. The alpha particle X-ray spectrometer uses a calibration target on the interior surfaces of doors designed to protect its sensor head from Martian dust. The miniature thermal emission spectrometer has both an internal target located in the mast assembly as well as an external target on the rover's deck.

The panoramic camera's calibration target is, by far, the most unique the rover carries. It is in the shape of a Sundial and is mounted on the rover deck. The camera will take pictures of the sundial many times during the mission so that scientists can make adjustments to the images they receive from Mars. They will use the colored blocks in the corners of the sundial to calibrate the color in images of the Martian landscape. Pictures of the shadows that are cast by the sundial's center post will allow scientists to properly adjust the brightness of each camera image. Children provided artwork for the sides of the base of the sundial.

Landing Sites

Selection of the landing sites for the two Mars Exploration Rovers involved over two years of intensive study by more than 100 scientists and engineers. Their job was to find sites that offered both excellent chances for a safe landing and outstanding science after the landings are achieved.

To qualify for consideration, candidate sites had to be near Mars' equator, low enough in elevation (so the spacecraft would pass through enough atmosphere to slow them), not too rugged, not too rocky and not too dusty. In all, 155 potential sites met the initial safety constraints. The two that made the final cut satisfied all of the safety criteria; they also show powerful evidence of past liquid water, but in two very different ways:

Gusev Crater, named after the 19th century Russian astronomer Matvei Gusev, is an impact crater about 150 kilometers (95 miles) in diameter and about 15 degrees south of Mars' equator. It lies near the transition between the planet's ancient highlands to the south and smoother plains to the north.

What makes Gusev an attractive landing site is a 900-kilometer-long (550-mile) meandering valley that enters the crater from the southeast. Called Ma'adim Vallis (from the Hebrew name for Mars), this valley is believed to have been eroded long ago by flowing water. The water likely cut through the crater's rim and filled much of the crater, creating a large lake not unlike current crater lakes here on Earth such as Lake Bosumtwi in Ghana). The lake is gone now, but the floor of Gusev Crater may contain water-laid sediments that still preserve a record of what conditions were like in the lake when the sediments were deposited.

Are lake sediments still preserved at Gusev Crater, or have they been buried by younger geologic materials? If sediments can be found, what do they reveal about the conditions that existed in the lake? Did the lake create an environment that would have been suitable for life? Are there other clues at Gusev that can reveal more about whether Mars had a warmer, wetter past?

Meridiani Planum is near the Martian equator, halfway around the planet from Gusev. The region of the planet in which it lies has been known as Meridiani since the earliest days of telescopic study of Mars, because it lies near the planet's meridian, or line of zero longitude. "Planum" means plains, and the name fits: Meridiani Planum is one of the smoothest, flattest places on Mars.

The scientific appeal of Meridiani comes not from its smooth landscape, but from its strange mineral composition. Looking down from orbit, the thermal emission spectrometer instrument on the *Mars Global Surveyor* spacecraft has shown that Meridiani Planum is rich in an iron oxide mineral called gray hematite. Gray hematite is found on Earth, where it usually – though not always – forms in association with liquid water.

Did the formation of the telltale mineral hematite at Meridiani involve liquid water? If it did, what was the process? Was the water in a lake? Was it percolating through rocks, perhaps at high temperatures? Was it present only as a trace on the surfaces of rocks? If water was present, were the conditions at Meridiani favorable for life? Or did the hematite form by some other process that didn't involve water at all? And what other clues does Meridani Planum hold regarding past conditions on Mars?

Rover A – the first to launch and land – will go to Gusev, while Rover B will go to Meridiani.

Mission Overview

NASA's Mars Exploration Rover Project will deliver two mobile laboratories to the surface of Mars for robotic geological fieldwork, including the examination of rocks and soils that may reveal a history of past water activity.

Sequences of launch, cruise and arrival operations will dispatch each rover to a different area of the planet three weeks apart to explore those areas for about three months each.

The two identical rovers can recognize and maneuver around small obstacles on their way to target rocks selected by scientists from images sent by the rovers. They will conduct unprecedented studies of Mars geology, such as the first microscopic observations of rock samples. They will provide "ground truth" characterization of the landing vicinities that will help to calibrate observations from instruments that view the planet from above on Mars orbiters.

NASA selected the sites to be explored, Gusev Crater and Meridiani Planum, from 155 potential locations as the two offering the best combination of safe landing potential and scientific appeal in assessing whether liquid water on Mars has ever made environments conducive to life.

While the rovers and the instruments they carry are the centerpieces of the project, each rover mission also depends on the performance of other components: the launch vehicle; a cruise stage; a system for entering Mars' atmosphere, descending through it and landing; a versatile system for deep-space communications; Earth facilities for data processing; and an international team of engineers, scientists and others.

Launch Vehicle
The two rover spacecraft will be lofted on three-stage Delta II rockets from Florida's Cape Canaveral Air Station. Rover A will launch on a version of the Delta II known as model 7925, a vehicle with a history of more than 40 successful launches, including those of the *Mars Global Surveyor*, *Mars Pathfinder* and *Mars Odyssey* missions. Rover B will use a newer, slightly more powerful version called model 7925H; the H identifies the vehicle as a heavy lifter.

Both of the Deltas feature a liquid-fueled first stage with nine strap-on solid-fuel boosters; a second-stage liquid-fueled engine; and a third stage solid-fuel rocket. The difference between the two versions is in the size of the strap-on boosters. With their payloads on top, each launch vehicle stands 39.6 meters (130 feet) tall.

The first stage of the Delta II uses a Rocketdyne RS-27A main engine. The engine provides nearly 890,000 newtons (200,000 pounds) of thrust by reacting RP-1 fuel (thermally stable kerosene) with liquid oxygen. The nine boosters for the first rover mission are 1,016 millimeters (40 inches) in diameter and fueled with enough hydroxyl-terminated polybutadiene solid propellant to provide about 446,000 newtons (100,000 pounds) of thrust apiece. The nine for the second mission are each 1,168 millimeters (46 inches) in diameter, with about 25 percent more thrust.

The Delta's second stage is powered by a restartable Aerojet AJ10-118K engine, which produces about 44,000 newtons (9,900 pounds) of thrust. The engine uses a fuel called Aerozine 50, which is a mixture of hydrazine and dimethyl hydrazine, reacted with nitrogen tetroxide as an oxidizer.

A Star-48B solid-fuel rocket made by Thiokol powers the third stage. It adds a final kick of about 66,000 newtons (14,850 pounds), using a propellant made primarily of ammonium perchlorate and aluminum.

Launch Timing

Rover A will be launched between June 8 and June 24, 2003, followed by Rover B between June 25 and July 15, 2003. To allow changeover of ground equipment at the launch pads, the two missions must be launched at least 10 days apart, so if Rover A launches at the end of its launch period Rover B's launch will be slipped accordingly. Rover A will lift off from Cape Canaveral's Space Launch Complex 17A, while Rover B will use the station's Space Launch Complex 17B.

Each mission has a total of two nearly instantaneous launch opportunities each day. On the first day of Rover A's launch period, June 8, the first opportunity is at 2:05:55 p.m. Eastern Daylight Time. On the first day of Rover B's launch period, June 25, the first opportunity is at 12:38:16 a.m. EDT. Opportunities for both missions occur a few minutes earlier each day as the launch period progresses.

Launch Sequences

When each Delta II launches, its first-stage engine and six of its nine strap-on boosters ignite at the moment of lift-off. The remaining three boosters will ignite following burnout of the first six. The boosters' spent casings will be jettisoned in sets of three between 1 and 3 minutes after lift-off.

About 4 minutes and 23 seconds into the flight, the main engine will cut off. Within the following 20 seconds, the first stage will separate from the second, the second stage will ignite, and the nose cone, or "fairing," will fall away. At about 10 minutes after lift-off for Rover A and 9 minutes after lift-off for Rover B, the second-stage engine will temporarily stop firing.

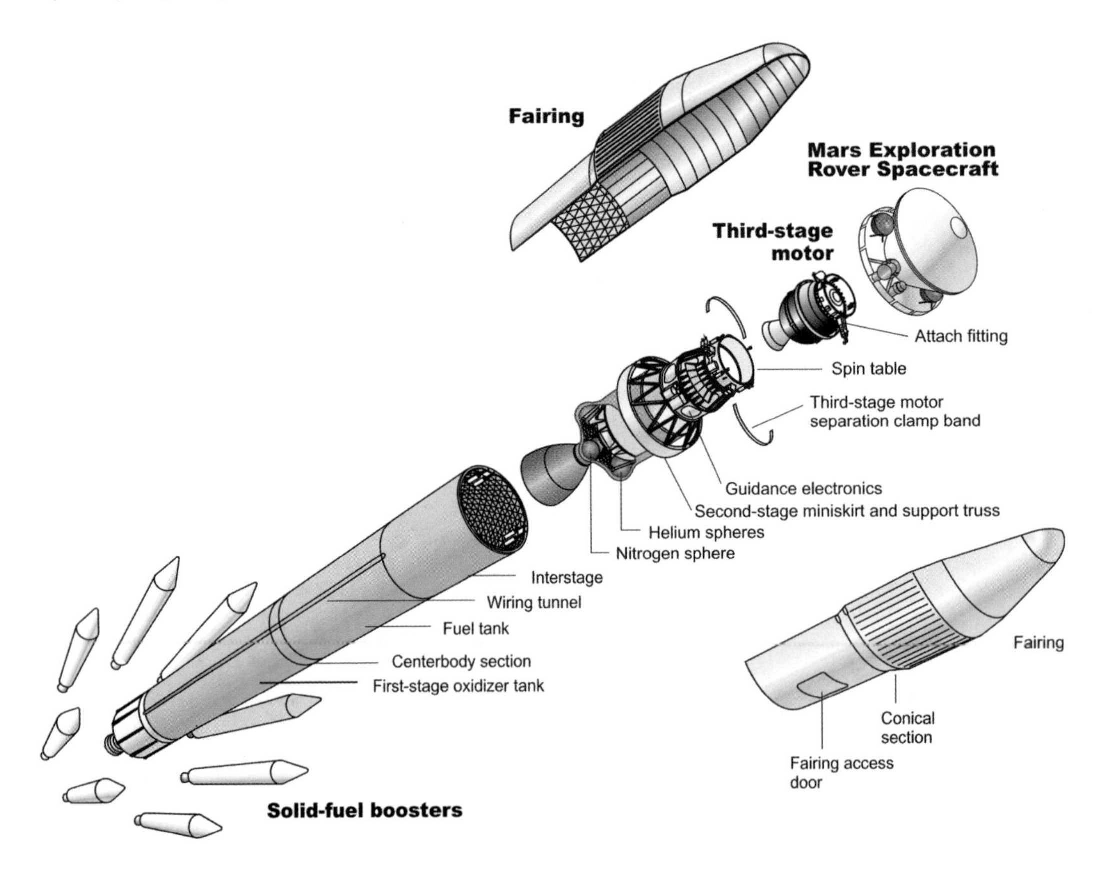

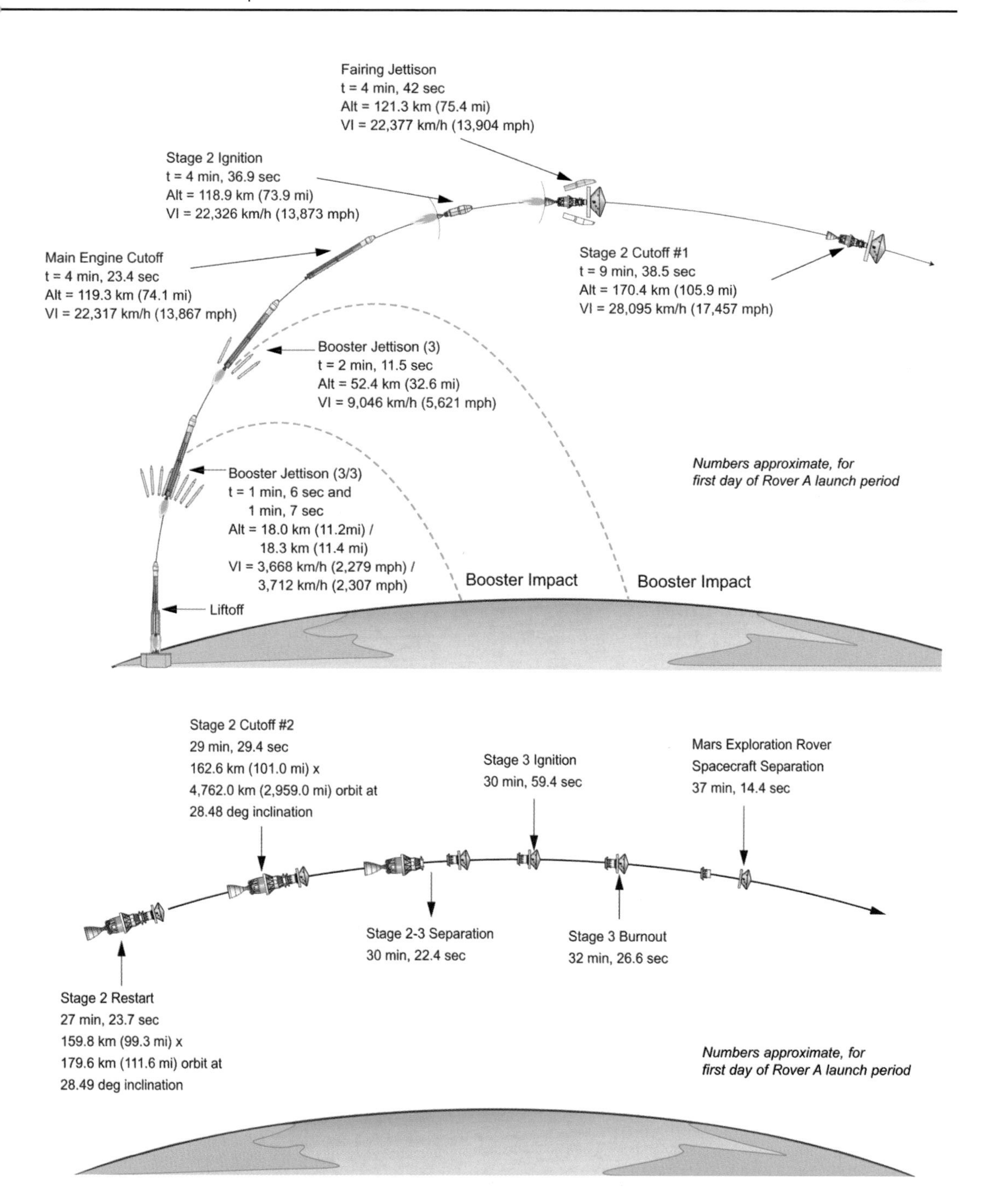

At this point, the spacecraft with the second and third stages of the Delta still attached will be in a circular parking orbit 167 kilometers (104 miles) above Earth. Before completion of even one orbit, however, the Delta's second stage will reignite to begin pushing the spacecraft out onto its interplanetary trajectory toward Mars. This begins about 14 to 19 minutes after lift-off for Rover A, depending on the date and time of launch, and about 59 to 67 minutes after lift-off for Rover B. The second burn of the Delta's second stage will last 2 to 3 minutes.

Small rockets will be fired to spin the Delta's third stage to about 63 rotations per minute on a turntable attached to the second stage. The third stage will then separate from the second, firing its engine for about 87 seconds to finish putting the spacecraft on course for Mars. To reduce the spacecraft's spin rate after the third-stage engine finishes firing, a set of yo-yo-like weights will reel out on flexible lines. These work like a twirling ice-skater's arms, slowing the spin as they are extended.

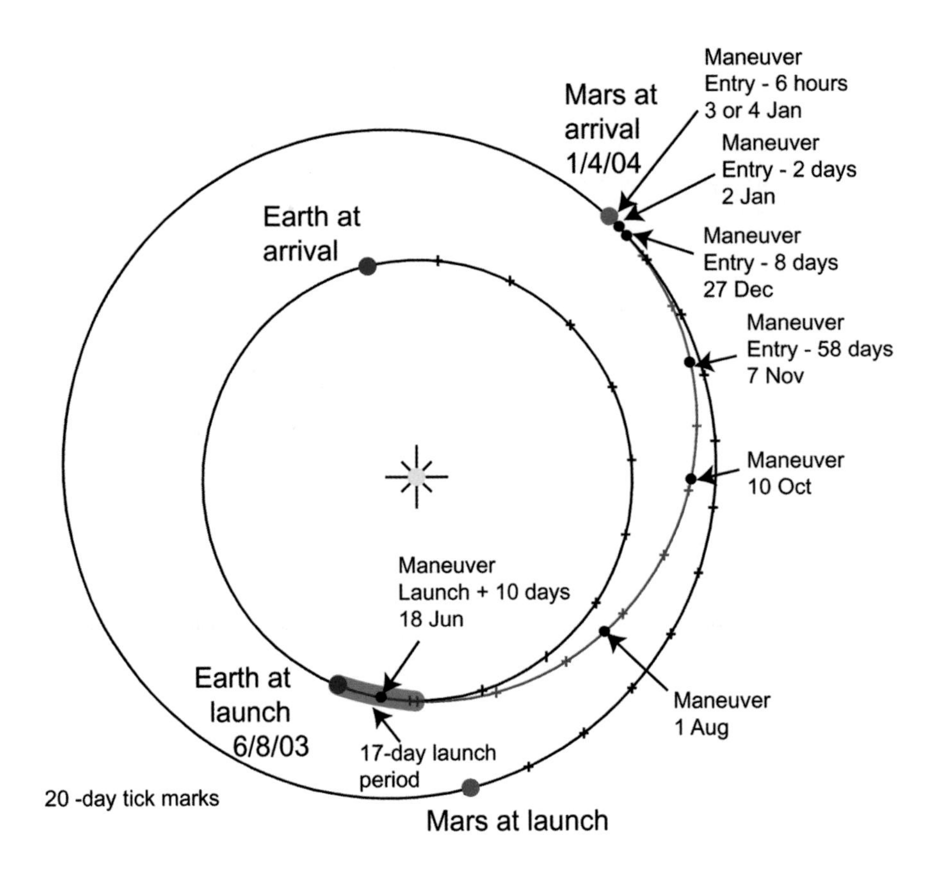

Mars at arrival 1/4/04
Maneuver Entry - 6 hours 3 or 4 Jan
Maneuver Entry - 2 days 2 Jan
Earth at arrival
Maneuver Entry - 8 days 27 Dec
Maneuver Entry - 58 days 7 Nov
Maneuver 10 Oct
Maneuver Launch + 10 days 18 Jun
Earth at launch 6/8/03
Maneuver 1 Aug
17-day launch period
20 -day tick marks
Mars at launch

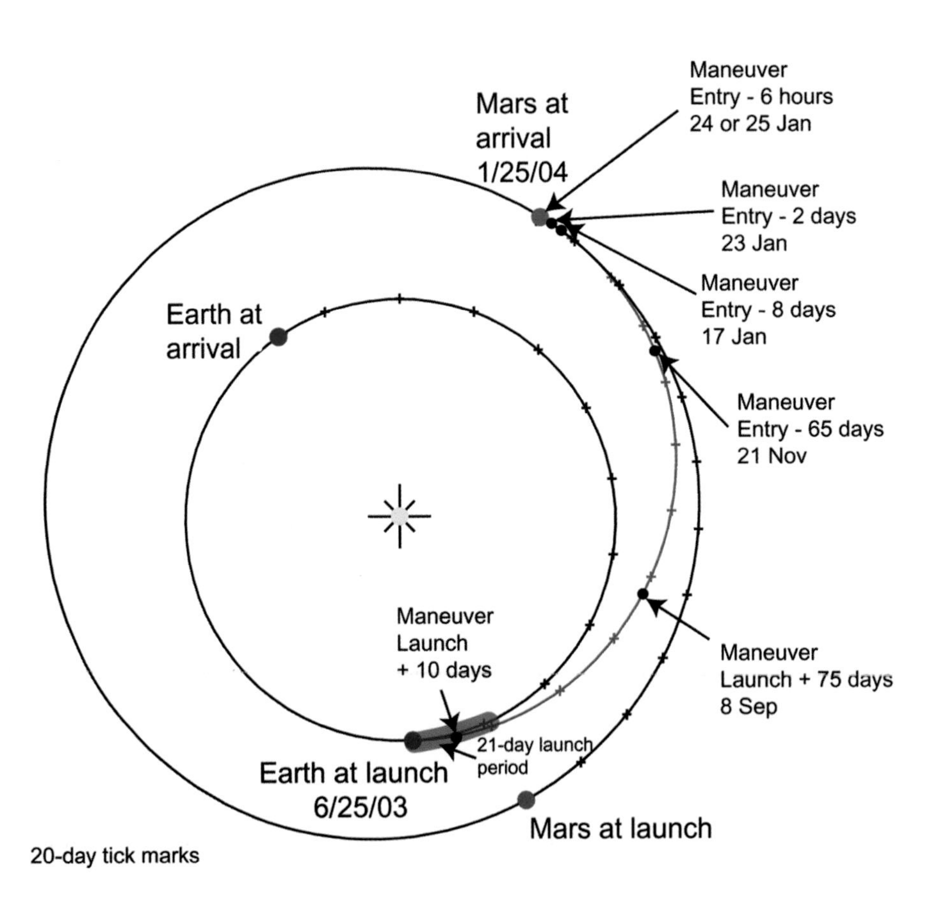

Maneuver Entry - 6 hours 24 or 25 Jan
Mars at arrival 1/25/04
Maneuver Entry - 2 days 23 Jan
Maneuver Entry - 8 days 17 Jan
Earth at arrival
Maneuver Entry - 65 days 21 Nov
Maneuver Launch + 10 days
Maneuver Launch + 75 days 8 Sep
21-day launch period
Earth at launch 6/25/03
Mars at launch
20-day tick marks

The spacecraft will shed the burned-out Delta third stage about 34 to 39 minutes after lift-off for Rover A and 78 to 87 minutes after lift-off for Rover B. Springs will push the third stage away, exposing an antenna on the rover spacecraft's cruise stage. Radio transmissions from the Delta II will enable controllers on the ground to monitor critical events throughout the launch sequence. However, communications with each Mars Exploration Rover spacecraft cannot begin until after it separates from the Delta's third stage. NASA's Deep Space Network will begin receiving radio signals from Rover A about 50 minutes after launch, using the network's Canberra, Australia, antenna complex. For Rover B, initial radio contact with the Deep Space Network is expected at the Goldstone, California, antenna complex one minute or more after third-stage separation.

Interplanetary Cruise and Approach to Mars

No matter which day in its launch period Rover A leaves Earth, it will reach Mars on January 4, 2004, so the journey may last anywhere from 194 to 210 days. Similarly, Rover B has a fixed appointment for arriving on January 25, 2004, so the duration of its journey to Mars will be from 194 days to 214 days, depending on its launch date.

Engineers refer to the first few months of each trip as the cruise phase, while the final 45 days before arrival are known as the approach phase. During both phases, each spacecraft is connected to a cruise stage that will be jettisoned in the final minutes of the flight. Solar panels on the cruise stage will provide electricity for the spacecraft in flight.

Thrusters on the cruise stage will be fired to adjust the spacecraft's flight path three times during the cruise phase and up to three more times during the final eight days of the approach phase. The first one or two of these maneuvers will commit the spacecraft to a specific target area on Mars. An additional thruster firing may be added during Rover A's cruise stage to allow ground controllers to retarget the landing site later if this is deemed necessary. Later maneuvers for both missions will refine the targeting based on calculations using frequently updated determinations of the spacecraft's position and course. The final trajectory correction maneuver, which is optional, is scheduled just six hours before landing.

Like NASA's *Mars Odyssey* orbital mission, the Mars Exploration Rover project will combine two traditional tracking schemes with a relatively new triangulation method to improve navigational precision. One of the traditional methods is ranging, which measures the distance to the spacecraft by timing precisely how long it takes for a radio signal to travel to the spacecraft and back. The other is doppler, which measures the spacecraft's speed relative to Earth by the amount of shift in the pitch of a radio signal from the craft.

The newer method, called delta differential one-way range measurement, adds information about the location of the spacecraft in directions perpendicular to the line of sight. Pairs of antennas at Deep Space Network sites on two different continents simultaneously receive signals from the spacecraft, then use the same antennas to observe natural radio waves from a known celestial reference point, such as a quasar. Successful use of this triangulation method is expected to shave several kilometers or miles off the amount of uncertainty in delivering the rovers to their targeted landing sites.

The months in which the rovers travel from Earth to Mars will also provide time for testing critical procedures, equipment and software in preparation for arrival.

Entry, Descent and Landing

The Mars Exploration Rovers will use the same airbag-cushioned landing scheme that successfully delivered *Mars Pathfinder* to the Red Planet in 1997.

About 70 minutes before entering Mars' atmosphere, each rover spacecraft will turn to orient its heat shield forward. From that point until the rover deploys its own solar panels after landing, five batteries mounted on the lander will power the spacecraft.

The planned sequence of events for entering the atmosphere, descending and landing is essentially the same for each of the two rover missions. Fifteen minutes before atmospheric entry, the protective aeroshell encasing the lander and rover will separate from the cruise stage, whose role will at that point be finished. Each cruise stage will ultimately impact Mars.

Each spacecraft will hit the top of the atmosphere, about 128 kilometers (80 miles) above Mars' surface, at a flight path angle of about 11.5 degrees and a velocity of about 5.4 kilometers per second (12,000 miles per hour). Although Mars has a much thinner atmosphere than Earth does, the friction of traveling through it will heat and slow the spacecraft dramatically. The surface of the heat shield is expected to reach a temperature of 1,447° C (2,637° F). By 4 minutes after atmospheric entry, speed will have decreased to about 430 meters per second (960

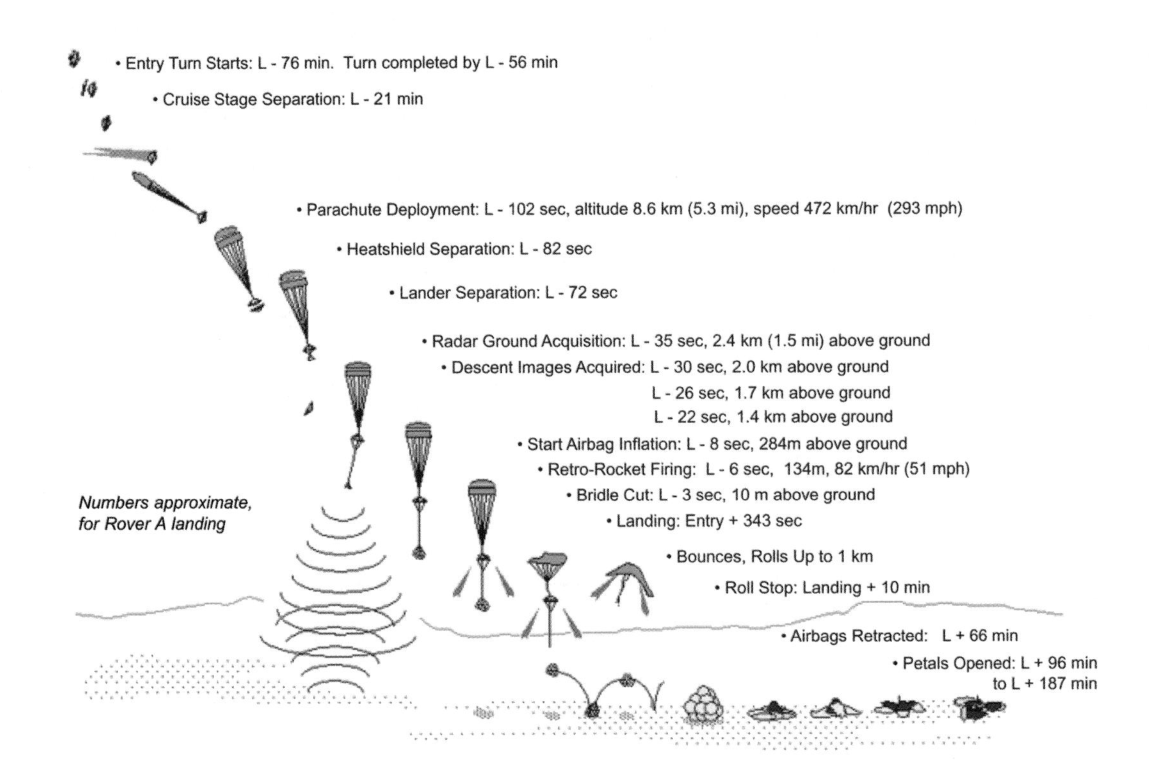

miles per hour). At that point, about 8.5 kilometers (5.3 miles) above the ground, the spacecraft will deploy its parachute.

Within 2 minutes, the spacecraft will be bouncing on the surface, but those minutes will be packed with challenging events crucial to the mission's success.

Twenty seconds after parachute deployment, the spacecraft will jettison the bottom half of its protective shell, the heat shield, exposing the lander inside. Ten seconds later, the backshell, still attached to the parachute, will begin lowering the lander on a tether-like bridle about 20 meters (66 feet) long. Spooling out the bridle to full length will take 10 seconds. Almost immediately, a radar system on the lander will begin sending pulses toward the ground to measure its altitude. Radar will detect the ground when the craft is about 2.4 kilometers (1.5 miles) above the surface, approximately 35 seconds before landing.

The Mars Exploration Rover design has two new tools, absent on *Mars Pathfinder*, to avoid excessive horizontal speed during ground impact in case of strong winds near the surface. One is a downward-looking camera mounted on the lander. Once the radar has sensed the surface, this camera will take three pictures of the ground about 4 seconds apart and automatically analyze them to estimate the spacecraft's horizontal velocity. The other innovation is a set of three small transverse rockets mounted on the backshell that can be fired in any combination to reduce horizontal velocity or counteract effects of side-to-side swinging under the parachute and bridle.

Eight seconds before touchdown, gas generators will inflate the lander's airbags. Two seconds later, the three main deceleration rockets on the backshell – and, if needed, one or two of the transverse rockets – will ignite. After 3 more seconds, when the lander should be about 15 meters (50 feet) above ground and have zero vertical velocity, its bridle will be cut, releasing it from the backshell and parachute. The airbag-protected lander will then be in free fall for a few seconds as it drops toward the ground.

The first bounce may take the airbag-protected lander back up to 15 meters (49 feet) or more above the ground. Bouncing and rolling could last several minutes. By comparison, the airbag-cushioned *Mars Pathfinder* bounced about 15 times, as high as 15 meters (49 feet), before coming to a rest 2½ minutes later about a kilometer (0.6 miles) from its point of initial impact.

Twelve minutes after landing, motors will begin retracting the airbags, a process likely to take about an hour. Then the lander petals will open. No matter which of the four petals is on the bottom when the folded-up lander stops rolling, the petal-opening action will set all four face up, with the rover's base petal in the center.

Mars Surface Operations

Opening of the four-sided lander will uncover the rover tucked snugly inside. Each rover's first action will be to unfold its solar-array panels. Then, still in a crouch, it will take images of the immediate surroundings with four hazard-identification cameras mounted below the plane of the solar panels.

Since the rovers rely on sunlight to generate electrical power, their operations on the surface will run on a schedule timed to the length of the Martian day. A Martian day, or "sol," lasts 24 hours, 39 minutes and 35 seconds.

Each rover will need to spend several sols completing housekeeping tasks before moving off its lander. Before the first Martian night, each rover may deploy its main antenna and the mast on which its panoramic camera and navigation camera are mounted. The navigation camera will take the first panorama of the landing site. Once transmitted to Earth during the following sol, the panorama and initial imaging by the rover's hazard-identification cameras will help mission engineers identify the safest route for the rover's later departure from the lander.

The rover will rise up from its crouching position and stand up at its full height while still on the lander base petal. From this height, it will take a 360-degree high-resolution, stereo, color panorama with its panoramic camera and a matching 360-degree panorama with its miniature thermal infrared spectrometer. Scientists will rely heavily on those images to decide which rocks the rover should go examine.

Unlike *Mars Pathfinder*, when each Mars Exploration Rover rolls off its lander, the lander's role in the mission will have ended. A new chapter in Mars exploration will begin.

In the next few sols after roll-off, the rover will finish checking and calibrating its science instruments and move to whichever nearby rock or patch of soil the science team has selected as the first target by analyzing the panoramic and infrared images taken earlier. The rover will examine each target up close, then begin moving on the following sol toward its next target. It may travel as much as about 40 meters (44 yards) in a sol, but is likely to cover less than that on most travel days as it maneuvers itself to avoid hazards on the way.

To coordinate their work with the rovers, flight team engineers and scientists operating the rovers from NASA's Jet Propulsion Laboratory in Pasadena, California, will be living on a Martian schedule, too. The 40-minute difference from Earth's day length means that, by about two weeks after the rovers land on Mars, team members' wake-up times and meal times will have shifted by about 9 hours. After the second rover reaches Mars, its team will be working on a different Martian schedule than the first rover's team because the two chosen landing sites are about halfway around Mars from each other. When it's noon at Meridiani, it's midnight at Gusev. Each rover will typically transmit each sol's accumulation of data in the Martian afternoon. The flight team will analyze that data, refine plans for the next sol's rover activity, and send updated commands to the rover the next Martian morning.

Each rover has a prime-mission goal of operating for at least 90 Martian sols (92 Earth days) after landing, though environmental conditions such as dust storms could cut the mission shorter.

Mars' distance from the Sun varies much more than Earth's does, and Mars will have passed the closest point to the Sun in its 23-month elliptical orbit about 5 months before the rovers arrive. The distance between Mars and the Sun will therefore increase by about 7 percent between mid-January and mid-April 2004, resulting in two principal consequences for how long the rovers can keep working. The rovers land at the end of summer in Mars' southern hemisphere, and with the onset of autumn the decreasing intensity of solar radiation reaching their solar panels will lessen the amount of electrical power produced. Also, colder nights will increase the need for electrically powered heating to keep the batteries warm enough to work. On top of those factors, a less predictable but possibly most important element limiting the rovers' lifetime will be the accumulation of dust on their solar panels.

Communications

Like all of NASA's interplanetary missions, the Mars Exploration Rover project will rely on the agency's Deep Space Network to track and communicate with both spacecraft. During the critical minutes of arrival at Mars, the two rovers will communicate essential spacecraft-status information throughout their atmospheric entry, descent and landing. On the surface of Mars, the rovers will be capable of communicating either directly with Earth or through Mars orbiters acting as relays. The distance between Earth and Mars will increase by about 65 percent between mid-January and mid-April 2004, reducing the rate at which data can be transmitted across space.

The Deep Space Network, which will be 40 years old on December 24, 2003, transmits and receives radio signals through large dish antennas at three sites spaced approximately one-third of the way around the world from each other. This configuration ensures that spacecraft remain in view of one antenna complex or another as Earth rotates. The antenna complexes are at Goldstone in California's Mojave Desert; near Madrid, Spain; and near Canberra, Australia. Each complex is equipped with one antenna 70 meters (230 feet) in diameter, at least two antennas 34 meters (112 feet) in diameter, and smaller antennas. All three complexes communicate directly with the control hub at NASA's Jet Propulsion Laboratory, Pasadena, California. The network served more than 25 spacecraft in 2002.

The network has been preparing to deal with an extraordinary level of demand for interplanetary communications in late 2003 and early 2004. Several missions besides the Mars Exploration Rovers will be conducting critical events. Among others, the European Space Agency's *Mars Express* will enter Mars orbit after dropping the *Beagle 2* lander to the surface; Japan's *Nozomi* orbiter will be arriving at Mars; NASA's *Stardust* spacecraft will fly by a comet; and NASA's *Cassini* spacecraft will be nearing its mid-2004 arrival at Saturn. The Deep Space Network is upgrading antenna capabilities at all three complexes and is completing the construction of a new 34-meter antenna at the Madrid complex. That new antenna alone will add about 70 hours of spacecraft-tracking time per week during the periods when Mars is in view of Madrid.

During each Mars Exploration Rover mission's early cruise phase, a low-gain antenna mounted on the cruise stage will provide the communications link with Earth. A low-gain antenna does not need to be pointed as precisely as a higher-gain antenna. During early cruise it would be difficult to keep an antenna pointed at Earth and the solar panels oriented toward the Sun, due to the Sun-Earth angle at that stage of the mission. Later in the cruise toward Mars, the angle between the Sun and Earth will shrink, making it possible for the spacecraft to switch to a more directional medium-gain antenna, also mounted on the cruise stage.

Data transmission is most difficult during the critical sequence of atmospheric entry, descent and landing activities, but communication from the spacecraft is required during this period in order to diagnose any potential problems that may occur.

Minutes before the spacecraft turns to point its heat shield forward in preparation for entering Mars' atmosphere, the cruise stage's low-gain antenna will take over again, which will reduce the data transmission rate to 10 bits per second, less than 2 percent of the mid-gain antenna's rate. Through this antenna, and later through other low-gain antennas on the backshell, lander and rover, transmissions during the next hour or more will consist of simple signal tones coded to indicate the accomplishment of critical activities. For example, a change in tone will tell controllers when the spacecraft has successfully jettisoned its cruise stage about 15 minutes before hitting the atmosphere. During the descent through the atmosphere, about 36 ten-second signal tones will be transmitted.

Before its first night on the surface of Mars, each rover may deploy its high-gain antenna for use the following morning. The rovers will be able to communicate directly with Earth at transmission rates greater than 11,000 bits per second using this antenna.

About a minute before each lander drops to the Martian surface, another important communication method – relay through Mars orbiter spacecraft – will begin to be used. An antenna mounted on each lander will transmit status information to the orbiting *Mars Global Surveyor* from the time the descending lander emerges from the backshell until ground impact. If that antenna survives the first bounce, it will continue to relay information for a few minutes as the lander bounces and rolls to a stop. The orbit of *Mars Global Surveyor* will be adjusted in preceding weeks to place it over the landing vicinity during those crucial minutes to receive the transmissions. The orbiter will later transmit the data to Earth.

Throughout each rover's surface mission, a rover-mounted antenna will be able to communicate with *Mars Global Surveyor* and *Mars Odyssey* for several minutes once or twice per sol while each of the two orbiters pass overhead via a UHF link at 128,000 bits per second. Plans call for using direct-to-Earth communications for transmissions critical to mission success, but about half the total data returned from the rovers could be relayed via the orbiters. One engineering goal for the project is to demonstrate relay capability at least once with the European Space Agency's *Mars Express* orbiter, which is due to begin circling Mars in December 2003.

Planetary Protection Requirements

In the study of whether Mars has had environments conducive to life, precautions are taken against introducing microbes from Earth. The United States is a signatory to an international treaty that stipulates that exploration must be conducted in a manner that avoids harmful contamination of celestial bodies.

The primary strategy for preventing contamination of Mars with Earth organisms is to be sure that the hardware intended to reach the planet is clean. Each Mars Exploration Rover spacecraft must comply with requirements to carry a total of no more than 300,000 bacterial spores on any surface from which the spores could get into the Martian environment. Technicians assembling the spacecraft and preparing them for launch frequently cleaned surfaces by wiping them with an alcohol solution. The planetary protection team carefully sampled the surfaces and performed microbiology tests to demonstrate that each spacecraft meets requirements for biological cleanliness. Components tolerant of high temperature, such as the parachute and thermal blanketing, were heated to 110° C (230° F) or hotter to kill microbes. The core box of each rover, containing the main computer and other key electronics, is sealed and vented through high-efficiency filters that keep any microbes inside. Some smaller electronics compartments are also isolated in this manner.

Another type of precaution is to be sure that other hardware doesn't go to Mars accidentally. When the Delta's third stage separates from the spacecraft, the two objects are traveling on nearly identical trajectories. To prevent the possibility of the Delta's third stage hitting Mars, that shared course is deliberately set so that the spacecraft would miss Mars if not for its first trajectory correction maneuver, about 10 days later. The NASA planetary protection officer is responsible for the establishment and enforcement of the agency's planetary protection regulations.

Launch Safety

The rovers use small amounts of radioactive materials in two science instruments and to prevent electronics from getting too cold during Martian nights. NASA has safely used radioactive materials for four decades in a variety of scientific instruments and for spacecraft heating or electrical power when necessary.

There is little radiological danger to the public from a Mars Exploration Rover 2003 launch accident. Analysis performed for the mission's environmental impact statement indicates that the chance of an accident occurring during launch is about 1 in 30. Most accidents would not present a threat to the radioisotope heater units onboard the spacecraft because of the rugged design of the units. There are also small-quantity radioactive sources on board the spacecraft (curium-244 and cobalt-57) that are used for instrument calibration or science experiments. Since these small sources of curium-244 and cobalt-57 have relatively low melting temperatures compared to the plutonium dioxide in the radioisotope heater units, these radioactive materials would likely be released in an early launch accident (i.e., the first 23 seconds of launch). The chance of an early launch accident that releases any radioactive material is about 1 in 1,030.

If a launch-area accident resulting in the release of radioactive sources were to occur, spectators and people off-site in the downwind direction could be exposed to small quantities of radionuclides. The person with the highest exposure would typically receive less than a few tens of millirem. (The average annual dose from naturally occurring sources of radiation in the United States is about 300 millirem per year.) No health consequences would be expected with this level of radiation exposure.

Precautionary measures include deployment of radiological monitoring teams and remote air monitoring stations at strategic locations at the launch site. A radiological control center at Kennedy Space Center would coordinate any local emergency actions required during the pre-launch or early launch phases of the mission in the event of a launch mishap.

In the event of a radiological release, federal, state and county agencies would determine an appropriate course of action for any areas outside the Cape Canaveral Air Station-Kennedy Space Center site based on actual monitoring information.

Spacecraft

Each of the two Mars Exploration Rover spacecraft resembles a nested set of Russian dolls. The rover will travel to Mars tucked inside a folded-up lander wrapped in airbags. The lander in turn will be encased in a protective aeroshell. Finally, a disc-shaped cruise stage is attached to the aeroshell on one side and to the Delta II launch vehicle on the other.

Cruise stage

The cruise stage provides capabilities needed during the seven-month passage to Mars but not later in the mission, such as a propulsion system for trajectory correction maneuvers. Approximately 2.6 meters (8.5 feet) in diameter and 1.6 meters (5.2 feet) tall, the disc-shaped cruise stage is outfitted with solar panels and antennas on one face, and with fuel tanks and the aeroshell on the other. Around the rim sit thrusters, a star scanner and a Sun sensor.

The propulsion system uses hydrazine propellant stored in two titanium tanks. Since the entire spacecraft spins at about 2 rotations per minute, fuel in the tanks is pushed outward toward outlets and through fuel lines to two clusters of thrusters. Each cluster has four thrusters pointing in different directions.

The star scanner and Sun sensor help the spacecraft determine its orientation. Since the rover's solar arrays are tucked away inside the aeroshell for the trip, the cruise stage needs its own for electrical energy. The arrays can generate more than 600 watts when the spacecraft is about as far from the Sun as Earth is, and about half that much when it nears Mars.

The cruise stage also carries a system for carrying excess heat away from the rover's computer with a pumped freon loop and rim-mounted radiators.

Entry, Descent and Landing System
The system for getting each rover safely through Mars' atmosphere and onto the surface relies on an aeroshell, a parachute and airbags. The aeroshell has two parts: a heat shield that faces forward and a backshell. Both are based on designs used successfully by NASA's *Viking* Mars landers in 1976 and *Mars Pathfinder* in 1997.

The parachute is attached to the backshell and opens to about 15 meters (49 feet) in diameter. The parachute design was tested under simulated Martian conditions in a large wind tunnel at NASA's Ames Research Center near Sunnyvale, California.

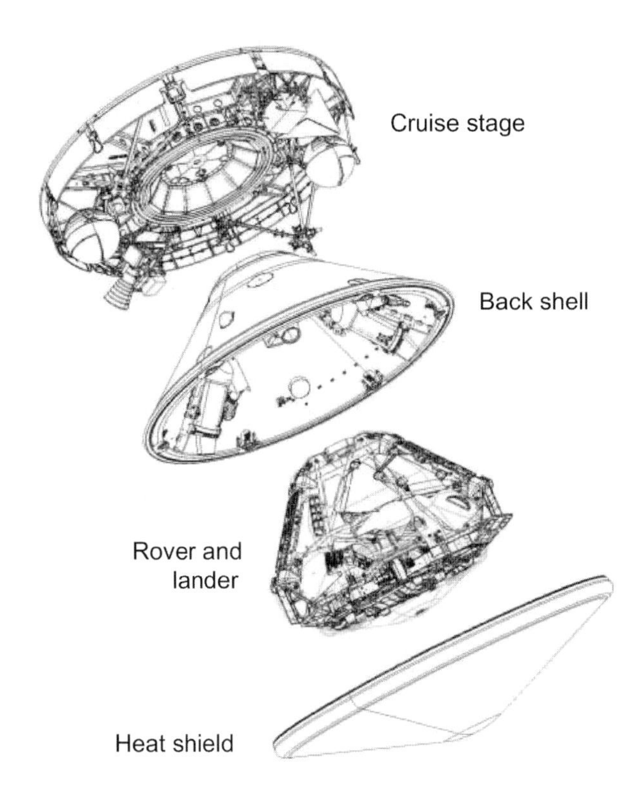

The backshell carries a deceleration meter used to determine the right moment for deploying the parachute. Solid-fuel rockets mounted on the underside of the shell reduce vertical velocity and any excessive horizontal velocity just before landing.

The airbags, based on *Pathfinder*'s design, cushion the impact of the lander on the surface. Each of the four faces of the folded-up lander is equipped with an envelope of six airbags stitched together. Explosive gas generators rapidly inflate the airbags to a pressure of about 6900 Pascal (one pound per square inch). Each airbag has double bladders to support impact pressure and, to protect the bladders from sharp rocks, six layers of a special cloth woven from polymer fiber that is five times stronger than steel. The fiber material, Vectran, is used in the strings of archery bows and tennis racquets.

Lander
The lander, besides deploying the airbags, can set the rover right-side-up, if necessary, and provides an adjustable platform from which the rover can roll onto Mars' surface. It also carries a radar altimeter used for timing some descent events, as well as two antennas.

The lander's basic structure is four triangular petals made of graphite-epoxy composite material. Three petals are each attached with a hinge to an edge of the central base petal. The rover stays fastened to the base petal during the flight and landing. When folded up, the lander's petals form a tetrahedral box around the stowed rover. Any of the petals could end up on the bottom when the airbag-cushioned bundle rolls to a stop after landing. Electric motors at the hinges have enough torque to push the lander open, righting the rover, if it lands on one of the side petals.

Other motors retract the deflated airbags. An apron made out of the same type of tough fabric as the airbags stretches over ribs and cables connected to the petals, providing a surface that the rover can drive over to get off the lander. The side petals can also be adjusted up or down from the plane of the base petal to accommodate uneven terrain and improve the rover's path for driving off of the lander.

Nearly 4 million people have a special connection to the Mars Exploration Rover project by having their names recorded on each mission's lander. Each of the two landers carries a digital versatile disc, or DVD, containing millions of names of people around the world collected during a "Send Your Name to Mars" campaign, which ended in November 2002.

Rover

At the heart of each Mars Exploration Rover spacecraft is its rover. This is the mobile geological laboratory that will study the landing site and travel to examine selected rocks up close.

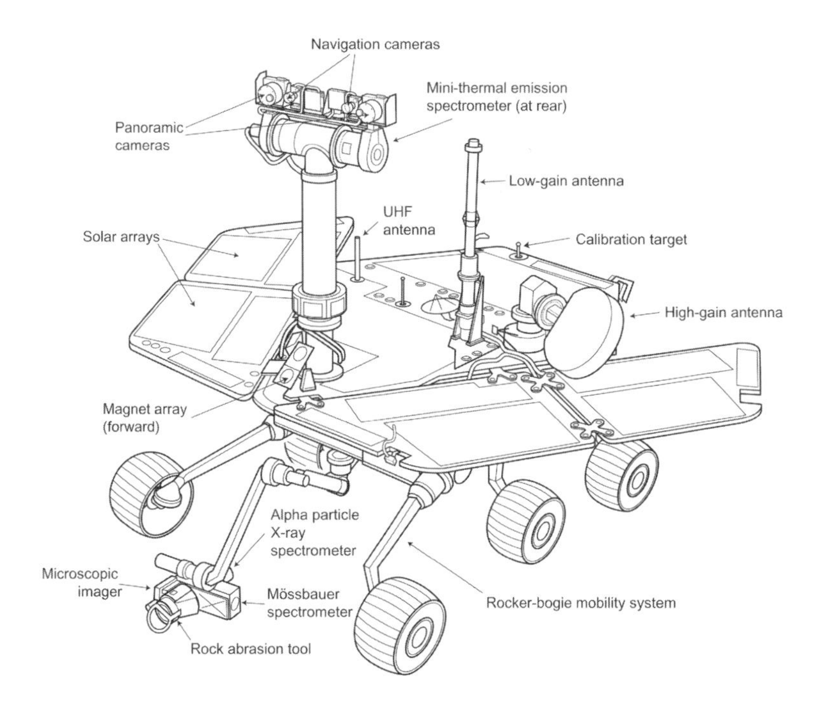

The Mars Exploration Rovers differ in many ways from their only predecessor, *Mars Pathfinder's Sojourner* rover. *Sojourner* was about 65 centimeters (2 feet) long and weighed 10 kilograms (22 pounds). Each Mars Exploration Rover is 1.6 meter (5.2 feet) long and weighs 174 kilograms (384 pounds). *Sojourner* traveled a total distance equal to the length of about one football field during its 12 weeks of activity on Mars. Each Mars Exploration Rover is expected to travel six to 10 times that distance during its three-month prime mission. *Pathfinder*'s lander, not *Sojourner*, housed that mission's main telecommunications, camera and computer functions. The Mars Exploration Rovers carry equipment for those functions onboard and do not interact with their landers any further once they roll off.

On each Mars Exploration Rover, the core structure is made of composite honeycomb material insulated with a high-tech material called aerogel. This core body, called the warm electronics box, is topped with a triangular surface called the rover equipment deck. The deck is populated with three antennas, a camera mast and a panel of solar cells. Additional solar panels are connected by hinges to the edges of the triangle. The solar panels fold up to fit inside the lander for the trip to Mars, and deploy to form a total area of 1.3 square meters (14 square feet) of three-layer photovoltaic cells. Each layer is of different materials: gallium indium phosphorus, gallium arsenide and germanium. The array can produce nearly 900 watt-hours of energy per Martian day, or sol. However, by the end of the 90-sol mission, the energy generating capability is reduced to about 600 watt-hours per sol because of accumulating dust and the change in season. The solar array repeatedly recharges two lithium-ion batteries inside the warm electronics box.

Doing sport utility vehicles one better, each rover is equipped with six-wheel drive. A rocker-bogie suspension system, which bends at its joints rather than using any springs, allows rolling over rocks bigger than the wheel diameter of 26 centimeters (10 inches). The distribution of mass on the vehicle is arranged so that the center of mass is near the pivot point of the rocker-bogie system. That enables the rover to tolerate a tilt of up to 45 degrees in any direction without overturning, although onboard computers are programmed to prevent tilts of more than 30 degrees. Independent steering of the front and rear wheels allows the rover to turn in place or drive in gradual arcs.

The rover has navigation software and hazard-avoiding capabilities it can use to make its own way toward a destination identified to it in a daily set of commands. It can move at up to 5 centimeters (2 inches) per second on flat hard ground, but under automated control with hazard avoidance, it travels at an average speed about one-fifth of that.

Two stereo pairs of hazard-identification cameras are mounted below the deck, one pair at the front of the rover and the other at the rear. Besides supporting automated navigation, the one on the front also provides imaging of what the rover's arm is doing. Two other stereo camera pairs sit high on a mast rising from the deck: the panoramic camera included as one of the science instruments, and a wider-angle, lower-resolution navigation camera pair. The mast also doubles as a periscope for another one of the science instruments, the miniature thermal emission spectrometer.

The rest of the science instruments are at the end of an arm, called the "instrument deployment device," which tucks under the front of the rover while the vehicle is traveling. The arm extends forward when the rover is in position to examine a particular rock or patch of soil.

Batteries and other components that are not designed to survive cold Martian nights reside in the warm electronics box. Nighttime temperatures may fall as low as minus 105° C (minus 157° F). The batteries need to be kept above minus 20° C (minus 4° F) for when they are supplying power, and above 0° C (32° F) when being recharged. Heat inside the warm electronics box comes from a combination of electrical heaters, eight radioisotope heater units and heat given off by electronics components.

Each radioisotope heater unit produces about one watt of heat and contains about 2.7 grams (0.1 ounce) of plutonium dioxide as a pellet about the size and shape of the eraser on the end of a standard pencil. Each pellet is encapsulated in a metal cladding of platinum-rhodium alloy and surrounded by multiple layers of carbon-graphite composite material, making the complete unit about the size and shape of a C-cell battery. This design of multiple protective layers has been tested extensively, and the heater units are expected to contain their plutonium dioxide under a wide range of launch and orbital-reentry accident conditions. Other spacecraft, including *Mars Pathfinder's Sojourner* rover, have used radioisotope heater units to keep electronic systems warm and working.

The computer in each Mars Exploration Rover runs with a 32-bit Rad 6000 microprocessor, a radiation-hardened version of the PowerPC chip used in some models of Macintosh computers, operating at a speed of 20 million instructions per second. Onboard memory includes 128 megabytes of random access memory, augmented by 256 megabytes of flash memory and smaller amounts of other non-volatile memory, which allows the system to retain data even without power.

Program / Project Management

The Mars Exploration Rover Project is managed by the Jet Propulsion Laboratory, Pasadena, California, for NASA's Office of Space Science, Washington, D.C. At NASA Headquarters, Dr. Edward Weiler is associate administrator for space science, Orlando Figueroa is Mars program director, Dr. Jim Garvin is the lead scientist for the Mars Exploration Program, David Lavery is Mars Exploration Rover program executive and Dr. Catherine Weitz is Mars Exploration Rover program scientist.

At the Jet Propulsion Laboratory, Dr. Firouz Naderi is the Mars program manager, Dr. Dan McCleese is Mars chief scientist, Peter Theisinger is Mars Exploration Rover project manager and Dr. Joy Crisp is Mars Exploration Rover project scientist.

At Cornell University, Ithaca, New York, Dr. Steve Squyres is principal investigator for Mars Exploration Rover's Athena suite of science instruments.

MEDIA RELATIONS OFFICE
JET PROPULSION LABORATORY
CALIFORNIA INSTITUTE OF TECHNOLOGY
NATIONAL AERONAUTICS AND SPACE ADMINISTRATION
PASADENA, CALIF. 91109 TELEPHONE (818) 354-5011

http://www.jpl.nasa.gov
Contact: Mary Hardin (818) 354-0344
FOR IMMEDIATE RELEASE

NASA'S MARS ROVER TEST DRIVE RACKS UP MILES AND SMILES

April 29, 1999

It is the ultimate test drive for the newest otherworldly vehicle. A few practice spins around an ancient lake bed in the Mojave desert this week with the next-generation Mars rover are helping NASA scientists and engineers learn more about driving the real thing on Mars.

"It's pretty exciting out here. We want to rack up a lot of miles and see how far this rover can go," said Dr. Raymond Arvidson, a geologist from Washington University in St. Louis, MO, and mission director for the field tests. "We are doing an 'end-to-end' test, using systems similar to what we will use on Mars. These test drives will help ensure that we will have a successful Mars rover mission."

Future robotic rovers on Mars will need to find the best rocks to bring back to Earth, samples that are likely to contain the evidence scientists need to prove that life once existed on the red planet. The rovers are being built and tested by NASA's Jet Propulsion Laboratory, Pasadena, CA.

To find the best sample, scientists need a good retriever. This week they're testing the work horse, er dog, named FIDO — Field Integrated Design and Operations — that is helping them figure out how to use the kinds of instruments the next Mars rovers will need to fetch the most scientifically interesting rocks. FIDO is designed to test the advanced technology of the Athena flight rover and science payload that will be launched as part of NASA's Mars Sample Return missions in 2003 and 2005.

"No place on Earth is like Mars, but our field site on an ancient lake bed in the Mojave Desert comes close. So far we've been able to use the rover's mini-corer to drill a rock sample and we've used the microscopic camera to look inside the hole," Arvidson said. "We're practicing looking for rocks that contain carbonate minerals. If we find those kinds of rocks on Mars it may tell us if the early planet had a carbon dioxide atmosphere."

"We've had a fantastic week. In just a few days, we've shown that we can find good rocks, drill samples out of them, and take the samples back to a lander. That's a huge step forward in preparing to bring the first samples back from Mars," said Dr. Steven Squyres, principal investigator for the Athena rover payload from Cornell University, Ithaca, NY.

"FIDO's advanced technology includes the ability to navigate over distances on its own and avoid natural obstacles without receiving directions from a controller," said Dr. Eric Baumgartner, a robotics engineer at JPL and mission engineer for the desert field tests. "The rover also uses a robot arm to manipulate science instruments and it has a new mini-corer or drill to extract and cache rock samples. There are also several camera systems onboard that allow the rover to collect science and navigation images by remote-control."

FIDO is about the size of a coffee table and weighs as much as a St. Bernard, about 70 kilograms (150 pounds). It is approximately 85 centimeters (about 33 inches) wide, 105 centimeters (41 inches) long, and 55 centimeters (22 inches) high. The rover moves up to 300 meters an hour (less than a mile per hour) over smooth terrain, using its onboard stereo vision systems to detect and avoid obstacles as it travels "on-the-fly." During these tests, FIDO is powered by both solar panels that cover the top of the rover and by replaceable, rechargeable batteries.

"FIDO is about six times the size of Mars Pathfinder's Sojourner and is far more capable of performing its job without frequent human help," Dr. Paul S. Schenker, who directs FIDO rover development at JPL as part of the NASA Exploration Technology Program. "FIDO navigates continuously using on-board computer vision and autonomous controls, and has similar capabilities for eye-to-hand coordination of its robotic science arm and mast. The rover has six wheels that are all independently steered and can drive forward or backward allowing FIDO to turn or back up with the use of its rear-mounted cameras."

In addition to testing FIDO, the scientists and engineers are supporting students from four schools around the country in designing and carrying out their own mission with the rover. This is the first time students have been able to remotely operate a NASA/JPL rover. The students, from Los Angeles, Phoenix, Ithaca, NY, and St, Louis, (LAPIS), form an integrated mission team and are responsible for planning, conducting and archiving a two-day mission using FIDO.

"It is important to excite young people about space exploration and discovery and these tests provide an excellent educational opportunity," Arvidson said. "We're including high school students in the FIDO tests as a pilot experiment in which the students gain a sense of participation in the field trials by planning their own mission segments and working with us to implement the rover's assignments."

The FIDO rover development and the Mars Sample Return 2003/2005 missions are managed by NASA's Jet Propulsion Laboratory for NASA's Office of Space Science Washington, DC. JPL is a division of the California Institute of Technology, Pasadena, CA.

More information about FIDO is available at: http://wundow.wustl.edu/rover

April 1999 Mojave Field Test

FIDO uses a robot arm to manipulate science instruments and it has a new mini-corer or drill to extract and cache rock samples. Several camera systems onboard allow the rover to collect science and navigation images by remote-control. The rover is about the size of a coffee table and weighs as much as a St. Bernard, about 70 kilograms (150 pounds). It is approximately 85 centimeters (about 33 inches) wide, 105 centimeters (41 inches) long, and 55 centimeters (22 inches) high. The rover moves up to 300 meters an hour (less than a mile per hour) over smooth terrain, using its onboard stereo vision systems to detect and avoid obstacles as it travels "on-the-fly." During these tests, FIDO is powered by both solar panels that cover the top of the rover and by replaceable, rechargeable batteries.

NASA Identifies Two Options for 2003 Mars Missions Decision in July

May 12, 2000

Donald Savage Headquarters, Washington, DC (Phone: 202/358-1547) RELEASE: 00-81

In 2003, NASA may launch either a Mars scientific orbiter mission or a large scientific rover which will land using an airbag cocoon like that on the successful 1997 Mars Pathfinder mission. The two concepts were selected from dozens of options that had been under study. NASA will make a decision on the options, including whether or not to proceed to launch, in early July.

Two teams, one centered at NASA's Jet Propulsion Laboratory (JPL), Pasadena, CA, and the other at Lockheed Martin Astronautics, Denver, CO, will conduct separate, intensive two- month studies to further define the concepts. In the studies the teams also will evaluate risk, cost, and readiness for flight, allowing 36 months of development leading to a May 2003 launch date.

The reports will be submitted for review to Mars Program Director Scott Hubbard at NASA Headquarters, Washington, DC. Dr. Ed Weiler, Associate Administrator for Space Science at NASA Headquarters, will make the final decision of which mission - if any - to launch in the 2003 opportunity. If selected, the cost of the 2003 mission will be about the same as the successful 1997 Mars Pathfinder mission (adjusted for inflation).

"Our budget will support only one of these two outstanding missions for the 2003 launch opportunity, and it will be a very tough decision to make," said Dr. Weiler. "Following this decision, later in the year we will have a more complete overall Mars exploration program to present to the American public which will represent the most exciting, most scientifically rich program of exploration we have ever undertaken of the planet Mars."

"These two mission concepts embody the requirements we have learned through the hard lessons of two recent Mars mission failures, and either one will extend the tremendous scientific successes we have had with the Mars Global Surveyor and Mars Pathfinder," said Hubbard.

The Mars Surveyor Orbiter is a multi-instrument spacecraft similar in size to the currently operating Mars Global Surveyor. It is designed to recapture all the lost science capability of the Mars Climate Orbiter mission as well as to seek new evidence of water-related materials. The orbiter's mission will be to study the martian atmosphere and trace the signs of ancient and modern water. Its instruments potentially will include a very high- resolution imaging system, a moderate-to-wide-angle multicolor camera, an atmospheric infrared sounder, a visible-to-near- infrared imaging spectrometer, an ultraviolet spectrometer, and possibly a magnetometer and laser altimeter. Telecommunications relay equipment that could be used to support Mars missions for 10 years also would be included.

The rover is a based on the Athena rover design, which already has been operated in field tests and previously was considered for the cancelled 2001 lander mission. The concept being proposed for the 2003 mission involves packaging the 286- pound (130-kilogram) rover in a system similar to the 1997 Mars Pathfinder structure, which would be cushioned on landing by airbags. Unlike the 1997 mission, however, the four-petal, self- righting enclosure would serve only as a means to deliver the rover to the surface and not function as a science or support station.

After landing, the Mars Mobile Lander would serve as a self- contained mission, communicating directly with Earth or with an orbiting spacecraft band as the rover traverses the martian terrain. The rover would be capable of travelling up to 100 yards (100 meters) a day, providing unprecedented measurements of the mineralogy and geochemistry of the martian surface, particularly of rocks, using a newly developed suite of instruments optimized to search for clues about ancient water on Mars. The mobile surface-laboratory will be able to gain access to a broad diversity of rocks and fine-scale materials for the first time on the surface of Mars, in its search for evidence of water-related materials. The rover's mission would last for at least 30 days on the surface.

"We are opening up a new frontier on the Red Planet, and we can't afford to overlook anything," Weiler added. "We have to make sure we plan it well, provide our people with the tools they need, and do whatever it takes to ensure the best possible chances for success."

Back to the Future on Mars

Above: This artist's rendering shows a view of NASA's Mars 2003 Rover as it sets off roam the surface of the red planet. The rover is scheduled for launch in June 2003 and will arrive in January 2004, shielded in its landing by an airbag shell. The airbag/lander structure, which has no scientific instruments of its own, is shown to the right in this image, behind the rover. The rover will carry five scientific instruments and rock abrading device. The Panoramic Camera and the Miniature Thermal Emission Spectrometer are located on the large mast shown on the front of the rover. The camera will be supplied by NASA's Jet Propulsion Laboratory, Pasadena, Calif.; and the spectrometer will be supplied by Arizona State University in Tempe. The payload also includes magnetic targets, provided by the Niels Bohr Institute in Copenhagen, Denmark, that will collect magnetic dust for further study by the science instruments.

July 28, 2000
NASA announces plans for a Mars rover in 2003 with a second rover under consideration.

In 2003, NASA plans to launch a relative of the now-famous 1997 Mars Pathfinder rover. Using drop, bounce, and roll technology, this larger cousin is expected to reach the surface of the Red Planet in January, 2004 and begin the longest journey of scientific exploration ever undertaken across the surface of that alien world.

Dr. Edward Weiler, Associate Administrator, Office of Space Science, NASA Headquarters, Washington, DC., announced today that the Mars Rover was his choice from two mission options which had been under study since March.

The Rock Abrasion Tool is located on a robotic arm that can be deployed to study rocks and soil.(In this view, the robotic arm is tucked under the front of the rover.) The tool, provided by Honeybee Robotics Ltd., New York, N.Y., will grind away the outer surfaces of rocks, which may be dusty and weathered, allowing the science instruments to determine the nature of rock interiors. The three instruments that will study the abraded rocks are a Mossbauer Spectrometer, provided by the Johannes Gutenberg-University Mainz, Germany; an Alpha-Proton X-ray Spectrometer provided by Max Planck Institute for Chemistry, also in Mainz, Germany; and a Microscopic Imager, supplied by JPL. The payload also includes magnetic targets, provided by the Niels Bohr Institute in Copenhagen, Denmark, that will collect magnetic dust for further study by the science instruments.

In a landing similar to that of the 1997 Mars Pathfinder spacecraft, a parachute will deploy to slow the spacecraft down and airbags will inflate to cushion the landing. Petals of the landing structure will unfold to release the rover, which will drive off to begin its exploration. JPL manages the Mars 2003 Rover for NASA's Office of Space Science, Washington, D.C. JPL is a division of the California Institute of Technology in Pasadena. Cornell University, Ithaca, NY is the lead institution for the science payload.

"Today I am announcing that we have selected the Mars Exploration Program Rover rather than the orbiter option, which was an extremely difficult decision to make," said Weiler. "At the same time, we want to look into what could be an amazing opportunity, as well as a challenge, by sending two such rovers to two very different locations on Mars in 2003 rather than just one."

"We are evaluating the implications of a two-rover option, Weiler added. "I intend to make a decision in the next few weeks so that, if the decision is to proceed with two rovers, we can meet the development schedule for a 2003 launch."

With far greater mobility and scientific capability than the 1997 Mars Pathfinder Sojourner rover, this new robotic explorer will be able to trek up to110 yards (100 meters) across the surface each Martian day, which is 24 hrs. 37 min. The Mars rover will carry a sophisticated set of instruments that will allow it to search for evidence of liquid water that may have been present in the planet's past, as well as study the geologic building blocks on the surface.

"This mission will give us the first ever robot field geologist on Mars. It not only has the potential for breakthrough scientific discoveries, but also gives us necessary experience in full-scale surface science operations which will benefit all future missions," said Scott Hubbard, Mars Program Director at NASA Headquarters. "A landed mission in 2003 also allows us to take advantage of a very favorable alignment between Earth and Mars."

After launch atop a Delta II rocket, and a cruise of seven and a half months, the spacecraft should enter the Martian atmosphere January 20, 2004. In a landing similar to that of the Pathfinder spacecraft, a parachute will deploy to slow the spacecraft down, and airbags will inflate to cushion the landing. Upon reaching the surface the spacecraft will bounce about a dozen times and could roll as far as a half-mile (about one kilometer). When it comes to a stop, the airbags will deflate and retract, and the petals will open, bringing the lander to an upright position and revealing the rover.

Parents and Educators: Please visit Thursday's Classroom for lesson plans and activities related to this story.

Where the Pathfinder mission consisted of a lander, with science instruments and camera, as well as the small Sojourner rover, the Mars 2003 mission features a design that is dramatically different. This new spacecraft will consist entirely of the large, long-range rover, which comes to the surface inside a Pathfinder landing system, making it essentially a mobile scientific lander.

Immediately after touchdown, the rover is expected to provide a virtual tour of the landing site by sending back a high resolution 360-degree, panoramic, color and infrared image. It will then leave the petal structure behind, driving off as scientists command the vehicle to go to rock and soil targets of interest.

Above: This 360 degree image shows in colorful detail the surroundings of the Sagan Memorial Station at the Mars Pathfinder landing site. Like Pathfinder, the Mars 2003 lander will send back a panoramic color image soon after it reaches Mars.

This rover will be able to travel almost as far in one Martian day as the Sojourner rover did over its entire lifetime. Rocks and soils will be analyzed with a set of five instruments. A special tool called the "RAT," or Rock Abrasion Tool, will also be used to expose fresh rock surfaces for study.

The rover will weigh about 300 pounds (nearly 150 kilograms) and has a range of up to about 110 yards (100 meters) per sol, or Martian day. Surface operations will last for at least 90 sols, extending to late April 2004, but could continue longer, depending on the health of the rover.

"By studying a diverse array of martian materials, including the interiors of rocks, the instruments aboard the Rover will reveal the secrets of past martian environments, possibly providing new perspectives on where to focus the quest for signs of past life," said Dr. Jim Garvin, NASA Mars Program Scientist at NASA Headquarters. "Furthermore, the Rover offers never-before-possible opportunities for discoveries about the martian surface at scales ranging from microscopic to that of gigantic boulders. This is a key stepping stone to the future of our Mars exploration program."

One aspect of the Mars Rover's mission is to determine history of climate and water at a site or sites on Mars where conditions may once have been warmer and wetter and thus potentially favorable to life as we know it here on Earth.

The exact landing site has not yet been chosen, but is likely to be a location such as a former lakebed or channel deposit - a place where scientists believe there was once water. A site will be selected on the basis of intensive study of orbital data collected by the Mars Global Surveyor spacecraft, as well as the Mars 2001 orbiter, and other missions.

Left :This is a close-up view of the arm on NASA's Mars 2003 Rover that contains several of the scientific instruments. The Microscopic Imager is being extended toward the rock, the Alpha-Proton X-ray Spectrometer (APXS) is pointing back toward the rover body, the Mossbauer spectrometer is pointing away from the viewer (i.e., toward the rover's left front wheel), and the Rock Abrasion Tool is pointing toward the viewer. The rover is set for launch in June 2003 and will arrive at Mars in January 2004. JPL will manage the Mars 2003 Rover for NASA's Office of Space Science, Washington, D.C. JPL is a division of the California Institute of Technology, Pasadena, Calif. Cornell University, Ithaca, NY is the lead institution for the science payload.

The alternative mission, which had been under consideration for the 2003 opportunity, was a Mars scientific orbiter, which featured a camera capable of imaging objects as small as about two feet (60 cm) across, an imaging spectrometer designed to search for mineralogical evidence of the role of ancient water in martian history, and other science objectives.

Teams at NASA's Jet Propulsion Laboratory (JPL), Pasadena, CA, and Lockheed Martin Astronautics, Denver, CO, conducted separate, intensive, two-month studies of the missions.

"Both teams did an absolutely superb job in preparing these proposals in a very compressed time frame," said Dr. Weiler. "They both deserve a lot of credit for what they were able to achieve."

"This project can be accommodated within the President's budget request for NASA and we will spend the next few weeks refining our budget estimates and other requirements, plus the impacts and the consequences of sending two rovers to Mars instead of one," said Hubbard. "When we have fully addressed all of the issues, which may take several weeks, we will announce our final plans."

July 27, 2000 Mars 2003 Rover

This artist's rendering shows a side view of NASA's Mars 2003 Rover as it sets off on its exploration of the red planet. The rover is scheduled for launch in June 2003 and will arrive at Mars in January 2004 with an airbag-shielded landing shell.

The Mars 2003 Rover will carry five scientific instruments and a rock abrading tool. The instruments include a Panoramic Camera and a Miniature Thermal Emission Spectrometer, both on the large mast shown on the front of the rover. A Mossbauer Spectrometer, an Alpha-Proton X-ray Spectrometer, and a Microscopic Imager are located on a robotic arm that is tucked under the front of the rover, as is a Rock Abrasion Tool that will grind away the outer surfaces of rocks to determine the nature of rock interiors.

NASA's Jet Propulsion Laboratory, Pasadena, Calif., manages the Mars 2003 Rover for NASA's Office of Space Science, Washington, D.C. JPL is a division of the California Institute of Technology in Pasadena. Cornell University, Ithaca, NY is the lead institution for the science payload.

NASA releases video made by Cornell undergraduate Dan Maas to dramatize plans for two-rover space mission in 2003

FOR RELEASE: Aug. 10, 2000
Contact: David Brand Office: (607) 255-3651 E-Mail: deb27@cornell.edu

An image of the Rover from the Dan Maas video, with the collapsed lander in the background NASA/JPLAdditional pictures and high-resolution versions are available from the JPL picture archive.

ITHACA, N.Y. — When NASA today announced its intention to send two rover exploration vehicles to Mars on its previously announced 2003 space shot, it introduced the ambitious venture with a two-minute, computer-generated video that dramatizes the mission with startling clarity and accuracy.

The video is the work of Dan Maas, a 19-year-old undergraduate at Cornell University enrolled in the university's College Scholar program for independent, interdisciplinary study.

Maas has been perfecting his artistic and technical ability to depict the drama of Mars landing missions for the past two years working with Steven Squyres, Cornell professor of astronomy. Squyres, who will be the principal investigator on the Athena science cargoes to be carried by the new, long-range rovers, calls Dan's work "sensational."

When, earlier this summer, space agency officials saw a previous Mars-landing video made by Maas, they requested that he make a video to herald the 2003 mission. Squyres found Maas working at a summer job at the University of Southern California Institute for Creative Technologies, a new graphics research lab in Los Angeles, and asked him to return to Ithaca.

Back home, the student produced the NASA video in just three days. "Thankfully I was able to re-use certain elements, such as the Martian landscape, from an earlier video. That saved a great deal of time and effort," says Maas. "It was quite a stretch, though, to accomplish all of that rendering before the NASA deadline — I had all of our home computers and two laptops churning out the frames around the clock."

The movie opens in true Hollywood style, with the rover's antenna slowly appearing over a Martian ridge. The vehicle then descends a slope and after maneuvering its way around boulders approaches the edge of a crater, where its microscopic imager takes a fine-scale picture of the soil. Then it heads away into the distance as Maas's "camera" swings up toward the sky — and a second space capsule. The capsule makes a fiery descent, then a parachute is deployed and airbags inflate to cushion the landing, which is made "bouncing ball"- style, first used in the successful 1997 Pathfinder mission. Then the second rover emerges and begins exploring an ancient lakebed.

Maas, who entered Cornell's College of Arts and Sciences at the age of 16, has been producing digital animations since he was 10, although his interest in film goes even farther back. His father, James, the noted Cornell professor of psychology, recalls giving him a home-built toy film-editing machine for his third birthday. At the age of 16, Dan Maas started his own company, Digital Cinema, to provide animations for television commercials, and at the age of 17, he went to Los Angeles to intern at one of the leading digital animation studios. In the meantime, he has been studying theater arts at Cornell under David Feldshuh, professor of theatre, film and dance, and taking courses in math and physics.

Somehow Maas also found time to work and study in Cornell's astronomy department, which is where he met Squyres, who as a seasoned mission scientist had not been impressed by previous videos that had attempted to depict space missions. "They were dry as dust," he says. But when Squyres discovered Maas's abilities to create such computer-generated scenes as a helicopter being struck by a missile or a prison guard tower blowing up, he signed on the then-freshman student. "The job interview consisted of two words: 'You're hired,' " he recalls.

Maas begins each video by hand sketching a storyboard, with each panel depicting a specific scene from the Mars mission, which he transfers to the computer with a wash of color. Then, using a program called Lightwave, he begins creating the images in three-dimensional detail. Later, using another program, Digital Fusion, he creates special effects, such as graininess to simulate the look of film, and lens flare — the bright flash caused by the sun.

Almost none of Maas's scenes contain actual photographic images. Instead he uses a wealth of material — conversations with Squyres and engineers, blueprints, images from NASA web sites — to create his computer-generated space flight. He has even visited the Jet Propulsion Lab (JPL) in Pasadena to talk to the engineers managing the rover mission.

Take, for example, his depiction of the 2003 rover vehicle. First, Maas created digital geometric models of the chassis, wheels and sensors using blueprints provided by JPL as a guide. Maas then worked through each shot in the sequence, selecting camera angles and blocking out the rover's movements. Next, Squyres and Maas together refined the shots to combine scientific accuracy with cinematic excitement. The final step was rendering — or automatically computing — the nearly 3,000 individual frames of animation that make up the finished video, a process that took several hundred hours of non-stop calculation on seven computers.

Of course, there's much more to it than that. There is, for example, the considerable artistry involved in turning a simple shape into a convincing image of a rover wheel. "Basically, I think my job title is digital artist," says Maas. "It takes both technical skill and a strong grasp of traditional filmmaking techniques to make the most of my tools."

For Squyres, what impresses most is the Cornell student's "general cinematographer's sense" and "his fanatical attention to detail." The end product, he says, is a piece of work you look at and you think is real.

The Mars 2003 rover project will be managed at JPL for NASA's Office of Space Science.

NASA Outlines Mars Exploration Program for Next Two Decades

October 26, 2000

Donald Savage Headquarters, Washington, DC
(Phone: 202/358-1547)
RELEASE: 00-171

By means of orbiters, landers, rovers and sample return missions, NASA's revamped campaign to explore Mars, announced today, is poised to unravel the secrets of the Red Planet 's past environments, the history of its rocks, the many roles of water and, possibly, evidence of past or present life.

Six major missions are planned in this decade as part of a scientific tapestry that will weave a tale of new understanding of Earth's sometimes enigmatic and surprising neighbor.

The missions are part of a long-term Mars exploration program which has been developed over the past six months. The new program incorporates the lessons learned from previous mission successes and failures, and builds on scientific discoveries from past missions. The NASA-led effort to define the program well into the next decade focused on the science goals, management strategies, technology development and resource availability in an effort to design and implement missions which would be successful and provide a balanced program of discoveries. International participation, especially from Italy and France, will add significantly to the plan. The next step will be an 18-month programmatic systems engineering study to refine the costs and technology needs.

In addition to the previously announced 2001 Mars Odyssey orbiter mission and the twin Mars Exploration Rovers in 2003, NASA plans to launch a powerful scientific orbiter in 2005. This mission, the Mars Reconnaissance Orbiter, will focus on analyzing the surface at new scales in an effort to follow the tantalizing hints of water from the Mars Global Surveyor images and to bridge the gap between surface observations and measurements from orbit. For example, the Reconnaissance Orbiter will measure thousands of Martian landscapes at 8-to-12-inch (20-to-30-cm) resolution, good enough to observe rocks the size of beach balls.

NASA proposes to develop and to launch a long-range, long-duration mobile science laboratory that will be a major leap in surface measurements and pave the way for a future sample return mission. NASA is studying options to launch this mobile science labroatory mission as early as 2007. This capability will also demonstrate the technology for accurate landing and hazard avoidance in order to reach what may be very promising but difficult-to-reach scientific sites.

NASA also proposes to create a new line of small "Scout" missions which would be selected from proposals from the science community, and might involve airborne vehicles or small landers, as an investigation platform. Exciting new vistas could be opened up by this approach either through the airborne scale of observation or by increasing the number of sites visited. The first Scout mission launch is planned for 2007.

In the second decade, NASA plans additional science orbiters, rovers and landers, and the first mission to return the most promising Martian samples to Earth. Current plans call for the first sample return mission to be launched in 2014 and a second in 2016. Options which would significantly increase the rate of mission launch and/or accelerate the schedule of exploration are under study, including launching the first sample return mission as early as 2011. Technology development for advanced capabilities such as miniaturized surface science instruments and deep drilling to several hundred feet will also be carried out in this period.

Mars missions can be launched every 26 months during advantageous alignments - called launch opportunities - of the Earth and Mars, which facilitate the minimum amount of fuel needed to make the long trip.

The agency's Mars Exploration Program envisions significant international participation, particularly by France and Italy. In cooperation with NASA, the French and Italian Space Agencies plan to conduct collaborative scientific orbital and surface investigations and to make other major contributions to sample collection/return systems, telecommunications assets and launch services. Other nations also have expressed interest in participating in the program.

"We have developed a campaign to explore Mars unparalleled in the history of space exploration. It will change and adapt over time in response to what we find with each mission. It's meant to be a robust, flexible, long-term program that will give us the highest chances for success," said Scott Hubbard, Mars Program Director at NASA Headquarters, Washington, DC. "We're moving from the early era of global mapping and limited surface exploration to a much more intensive approach. We will establish a sustained presence in orbit around Mars and on the surface with long-duration exploration of some of the most scientifically promising and intriguing places on the planet."

"The scientific strategy developed for the new program is that of first seeking the most compelling places from above, before moving to the surface to investigate Mars," said Dr. Jim Garvin, NASA Mars Exploration Program Scientist at Headquarters. "The new program offers opportunities for competitively selected instruments and investigations at every step, and endeavors to keep the public informed on each mission via higher bandwidth telecommunication on the web."

"NASA's new Mars Exploration Program may well prove to be a watershed in the history of Mars exploration," said Dr. Ed Weiler, NASA's Associate Administrator for Space Science. "With this new strategy, we're going to dig deep into the details of Mars' mineralogy, geology and climate history in a way we've never been able to do before. We also plan to 'follow the water' so that in the not-to-distant future we may finally know the answers to the most far-reaching questions about the Red Planet we humans have asked over the generations: Did life ever arise there, and does life exist there now?"

NASA Exercises Delta II Contract Option for Mars 2003 Rover

December 21, 2000
George H. Diller Kennedy Space Center 321/867-2468
KSC Release No. 111-00

NASA today announced that it is exercising a contract option with the Boeing Company for a Delta II vehicle to launch the Mars Exploration Rover 2 (MER-B). The spacecraft is scheduled for launch at the beginning of a 21-day planetary window that opens on June 27, 2003. This firm-fixed price option is covered under the NASA Launch Services contract (NAS10-00-001) officially awarded by NASA's Kennedy Space Center on June 16,2000 to:

Delta Launch Services, Inc., 5301 Bolsa Avenue, Huntington Beach, CA 92647-2099

NASA's total launch services budget for the MER-B campaign is approximately $68 million dollars.

The goals of the Mars Exploration Rover 2003 mission are to land a roving vehicle on Mars for science observations that will help determine the water, climatic and geological history of a site on Mars where conditions may have been favorable to the preservation of evidence of life or associated pre-biotic processes.

NASA's Mars Program is managed by the NASA Headquarters Office of Space Science, Washington, D.C. The MER-B spacecraft project is managed for NASA by the Jet Propulsion Laboratory, Pasadena, CA.

Mars Explorer Rover 2003 Landing Sites

February 8, 2001

Interested in how the landing sites for Mars missions get picked? The first Mars 2003 Landing Site Workshop was held January 24th and 25th, to evaluate landing sites best suited to safely achieving the mission's science objectives. The mission will attempt to determine the history of water and climate in locations where conditions may have been favorable for life. More info about the Workshop and potential landing sites at:

http://marsoweb.nas.nasa.gov/landingsites/mer2003/

Can Liquid Water Exist on Present-Day Mars?

March 27, 2001

NASA Astrobiology Institute By staff writer, Science Communications

In 1998, NASA's Associate Administrator Wesley Huntress, Jr., stated, "Wherever liquid water and chemical energy are found, there is life. There is no exception."

Could there, then, be life on Mars? In the mid-1970s, the Viking Lander mission's Gas Exchange Experiment detected strong chemical activity in the martian soil. Liquid water seems to be the one element needed for the equation of life on Mars. The presence of water there, however, is still hotly contested.

Many scientists believe that liquid water does not and cannot exist on the surface of Mars today. Although surface water may have been plentiful in Mars' past, they say, the current conditions of freezing temperatures and a thin atmosphere mean that any water on Mars would have to be deep underground. Moreover, if any water ice existing on Mars were somehow warmed, it still wouldn't melt into water. The thin martian atmosphere instead would cause the ice to sublime directly into water vapor.

But Dr. Gilbert Levin of Spherix, Inc., and his son, Dr. Ron Levin of MIT's Lincoln Laboratory, believe differently. They say that liquid water-in limited amounts and for limited times-can exist on the surface of present-day Mars. They have based their theory on data collected from the Viking landers and on the 1997 Mars Pathfinder mission.

This father-son team has suggested a diurnal water cycle on Mars: water vapor in the air freezes out by night, then during the day the ice melts. As the day progresses, the heat of the Sun causes this liquid water to evaporate back into the air.

It has already been established from Viking photographs that a thin frost does form overnight on certain areas of the martian surface. Unlike many scientists, the Levins believe that this frosty layer does not instantly revert back into water vapor when the Sun rises. They suggest that, in the early hours of the martian morning, the atmosphere more than one meter above the martian surface remains too cold to hold water vapor. So the moisture stays on the ground.

Data from the Mars Pathfinder support this theory, as the Pathfinder temperature readings noted that temperatures one meter above the surface were often dozens of degrees colder than the temperatures closer to the ground.

This layer of cold air, say the Levins, provides a form of insulation, trapping the water moisture below. Since the atmosphere is too cold to hold the water as vapor and the ground is warm enough to melt the ice, the water melts into a liquid. This liquid water, the Levins believe, remains on the surface until the temperature of the atmosphere rises enough to allow the water to evaporate. In this way, they argue, the martian soil becomes briefly saturated with liquid water every day.

"The meteorological data fully confirm the presence of liquid water in the topsoil each morning," says Gilbert Levin. "The black-and-white as well as the color images show slick areas that may well be moist patches."

Such a scenario is certainly possible, admits Christopher McKay. McKay is a planetary scientist at NASA Ames Research Center in Mountain View, CA, and a member of the NASA Astrobiology Institute.

"At the surface the frost may melt to form a very short-lived layer of liquid," says McKay. "The experiments show that this is the case." But, he cautions, "how long it persists is not yet accurately determined.

"There have been several attempts to look at the problem of frost evaporation and melting on Mars theoretically," says McKay. But Levin's analysis, he says, is "badly flawed. The way to address this question," he says, "is with experiment."

The Levins look to tests conducted in Death Valley, CA, for support of their theory. Soil samples taken from the top one to two millimeters of the Californian sand dunes and analyzed by soil scientists from NASA's Jet Propulsion Laboratory were reported to contain 0.9% moisture, comparable to the moisture levels found in the martian soil by the Viking mission.

These desert samples from California also contained aerobic microorganisms. No clear evidence has yet been found, however, that there is life in the topmost layer of the martian soil.

Mars may, indeed, contain such forms of microorganic life. The Levins point to a study published in the Federation of European Microbiological Societies Reviews in 1997 by Elena Vorobyova, et al., entitled "The Deep Cold Biosphere: Facts and Hypothesis." This study reported that permafrost conditions provide a constant and stable environment to permit microbial communities to survive for millions of years.

The Levins cite this research as direct evidence for adaptive physiological and biochemical processes in microorganisms during long exposure to cold. While these findings refer to terrestrial microorganisms, the Levins believe they might also apply to Mars.

McKay does not believe these analogies to terrestrial environments prove anything about Mars, however. "Mars is still much drier and much colder than even the Atacama Desert in Chile or the dry valleys of Antarctica," argues McKay. "And Death Valley is not that dry. It rains there 25 millimeters a year."

Gilbert Levin is a long-time proponent of life on Mars. He worked on the Viking missions in the mid-1970s and steadfastly believes that the Viking Lander's Labeled Release (LR) experiment proved that primitive life does exist on present-day Mars.

The LR experiment dropped liquid nutrient into a sample of martian soil, then measured the gases that were released by the mixture. If martian bacteria had consumed the nutrients and had begun to multiply, certain gases would have been released. When the LR experiment was conducted on both Viking Landers, some of the gases emitted seemed to suggest that microbes were ingesting the released nutrients. But, overall, the results were ambiguous.

Many in the scientific community believe that the LR results can be explained non-biologically. One such explanation is that the LR experiment showed the surface of Mars to contain oxides. When the nutrients mixed with the oxides, a chemical reaction-not a biological one-occurred. Moreover, these oxides would actually prevent life from forming on the martian surface.

Gilbert Levin isn't swayed by this reasoning. After examining all the non-biological possibilities and looking at the new findings about life in extreme environments on Earth, Levin now firmly believes that the LR experiment did find microbial life on Mars.

His new model for the formation of liquid water, he argues, "removes the final constraint preventing acceptance of the biological interpretation of the Viking LR Mars data as having detected living microorganisms in the soil of Mars. It comes at a time when a growing body of evidence from the Earth and space are supporting the presence of life not only on Mars, but on many celestial bodies."

For McKay, the Viking experiments do not prove-or even suggest-that life could exist on the surface of Mars. "I support a chemical explanation for the Labled Release experiment and the other Viking instruments, such as the Gas Chromatograph/Mass Spectrometer and the Gas Exchange experiment," he says.

The Gas Chromatograph/Mass Spectrometer (GCMS) was designed to measure organic compounds in the martian soil. Organic compounds are present in space (for example, in meteorites), but the GCMS found no trace of them on the surface of Mars. Gilbert Levin believes, however, that the GCMS instrument sent to Mars could easily have missed biologically significant amounts of organic matter in the soil, as it had in a number of tests on Earth.

The Gas Exchange (GEX) experiment submerged a sample of martian soil in a nutrient mixture, and incubated the soil for 12 days in a simulated martian atmosphere. Gases emitted by organisms consuming the nutrients would have been detected by the gas chromatograph.

While the GEX experiment did detect some gases, it also got results with the control sample-soil that had been heated to sterilize it of any possible life. In other words, non-biological processes may have been at work. Subsequent laboratory experiments on Earth demonstrated that similar results were obtained when water was added to highly-reactive oxidizing compounds, such as the oxides or superoxides now believed to be present in martian soil.

"A biology explanation [for the Viking test results] is inconsistent, ecologically, with what we know about Mars' surface environment," says McKay.

What Next?

In 2003, NASA will send two rovers to Mars to hunt for signs of water in the rocks and surface soil. In the same year, the European Space Agency will launch Mars Express, which will include a lander. The Lander, dubbed Beagle 2, will contain a scientific payload dedicated to detecting signs of biogenic activity on Mars-the first such payload to be sent to Mars since Viking.

Rehearsal Readies Scientists for NASA's Next Mars Landing

August 19, 2002

With less than a year to go before the launch of NASA's Mars Exploration Rover mission, scientists have spent the last few weeks at a high-tech summer camp, rehearsing their roles for when the spacecraft take center stage.

"The purpose of this test is really to teach the science team how to remotely conduct field geology using a rover, rather than to test the rover hardware," said Dr. John Callas, science manager for the Mars Exploration Rover mission at NASA's Jet Propulsion Laboratory, Pasadena, Calif. "We sent one of our engineering development rovers out to a distant, undisclosed desert location, with the science team back at JPL planning the operations and sending commands, just as they'll do when the actual rovers are on Mars."

Fido rover in desert in the southwest

The 10-day blind test, which ran from Aug. 10 to 19, used the Field Integrated Design Operations testbed, called Fido, which is similar in size and capability to the Mars Exploration Rovers. Although important differences exist, the similarities are great enough that the same types of challenges exist in commanding these rovers in complex realistic terrain as are expected for the rovers on Mars.

"The scientific instruments on this test rover are similar to the Athena science payload that will be carried by the Mars Exploration Rovers," said Dr. Steve Squyres, principal investigator for the Mars Exploration Rover mission at Cornell University, Ithaca, N.Y. "We're using the test rover now to learn how to do good field geology with a robot. When we get to real Mars rover operations in 2004, we'll be able to use everything we're learning now to maximize our science return."

"The test rover has received and executed daily commands via satellite communications between JPL and the remote desert field site. Each day, they have sent images and science data to JPL that reveal properties of the desert geology," said Dr. Eddie Tunstel, the rover's lead engineer at JPL.

Scientists gather in JPL's practice mission control room during Fido tests

The Mars Exploration Rovers will be launched in May and June 2003. Upon their arrival at Mars in January 2004, they will spend at least three months conducting surface operations, exploring Mars for evidence of past water interaction with the surface and looking for other clues to the planet's past.

The science team of more than 60 scientists from around the world will tell the rovers what to do and where to go from the mission control room at JPL. This month's test is one of several training operations that are planned before landing.

The rovers are currently being built at JPL and will be shipped to the Kennedy Space Center in Florida early next year to begin preparations for launch. Shortly before the launch, NASA will select the landing sites.

More information about the rover mission is available at
http://www.jpl.nasa.gov/news/fact_sheets/mars03rovers.pdf or http://mars.jpl.nasa.gov/mer.
A description of the Fido rover is available at http://mars.jpl.nasa.gov/mer/fido or http://fido.jpl.nasa.gov.
More information about the Mars Exploration Program is available at http://mars.jpl.nasa.gov.

The Mars Exploration Rover mission is managed by JPL for NASA's Office of Space Science, Washington, D.C. JPL is a division of the California Institute of Technology in Pasadena.

NASA Selects LEGO Company to Run Mars Rover Naming Contest

November 04, 2002

NASA announced a contest, which will give American school kids a chance to make history, by naming two rovers being launched to explore Mars.

The robotic explorers are part of NASA's upcoming Mars Exploration Rover (MER) mission. The twin rovers will land at two different locations on the mysterious red planet to explore the surface in search of answers about the history of water on Mars.

The NASA "Name the Rovers" contest is a collaborative effort between NASA and the LEGO Company. The LEGO Company will manage the contest in conjunction with The Planetary Society. The contest provides students with the unique opportunity to suggest a name for each of the two Mars-bound rovers, temporarily known as MER-A and MER-B. The rovers are scheduled to launch in May and June 2003 respectively. The rovers are scheduled to land on Mars in January 2004.

"We are very excited about providing students with an opportunity to actively participate in the next mission to Mars," said Dave Lavery, Program Executive for Solar System Exploration at NASA Headquarters. "We are eagerly looking for some really creative and innovative ideas from the students as they compete to name the next Mars rovers and become part of history."

"The LEGO Company is dedicated to furthering hands-on discovery, playful learning, and the boundless frontiers of imagination. Space exploration also embraces these qualities, which is why the LEGO Company is so pleased to partner with NASA," said Brad Justus, LEGO Senior Vice President. "By involving children actively in the Mars mission, through the 'Name the Rovers' contest and with other related activities, we hope to help excite and inspire the next generation of space explorers," he said.

The NASA "Name the Rovers" contest is open to students 5 to 18 years of age who attend a U.S. school and are enrolled in the Fall 2002 school season.

Submissions must include suggested names for both rovers and a 50-500 word essay justifying why the students believe the names should be chosen. The contest has many educational benefits and encourages students to do research for their essays and to learn more about Mars and space exploration.

The contest is open for submissions through January 31, 2003. NASA will announce the contest winners prior to launching the rovers in the spring of 2003.

Information about the contest is available at: http://www.nametherovers.org
Information about the Mars Exploration Rovers is at: http://mars.jpl.nasa.gov/mer

NASA Twins Plan Martian Ramble

December 09, 2002

With just over a year to go before NASA's twin Mars Exploration Rovers land on the red planet, members of the science team are previewing the mission's goals and candidate landing sites at a special session of the American Geophysical Union meeting in San Francisco.

"The twin rovers will be able to travel the distance of several football fields during their missions. They will carry sophisticated instruments that effectively make them robotic geologists, acting as the eyes and hands of the science team on Earth," said Dr. Mark Adler, mission manager at NASA's Jet Propulsion Laboratory, Pasadena, Calif. "We are very busy at JPL building and testing the two rovers and the spacecraft that will land them safely on Mars."

Remote sensing instruments will be mounted on a rover mast, including high-resolution color stereo panoramic cameras and an infrared spectrometer for determining the mineralogy of rocks and soils. When interesting scientific targets are identified, the rovers will drive over to them and perform detailed investigations with instruments mounted on a robotic arm.

Rover instruments include a microscopic imager, to see micron-size particles and textures; an alpha-particle/x-ray spectrometer, for measuring elemental composition; and a Moessbauer spectrometer for determining the mineralogy of iron-bearing rocks. Each rover will carry a rock abrasion tool, the equivalent of a geologist's rock hammer, to remove the weathered surfaces from rocks and analyze their interior.

"All the instruments on the payload are undergoing intensive calibration and test activities in preparation for flight," said Dr. Steve Squyres, principal investigator for the science payload at Cornell University, Ithaca, N.Y.

"Once at Mars, the instruments will be used, together with the rover's ability to traverse long distances, to study the geologic history of the two landing sites," Squyres explained. The scientific focus of the mission is to investigate what role water played there, and to determine how suitable the conditions would have been for life.

NASA scientists are in the process of picking the landing site for each rover. Four sites look the most promising. "Three of the sites, Terra Meridiani, known as the Hematite site, Gusev, and Isidis show evidence for surface processes involving water. These sites appear capable of addressing the science objectives of the rover missions: to determine if water was present on Mars and whether there are conditions favorable to the preservation of evidence for ancient life," said Dr. Matt Golombek, landing site scientist at JPL. The fourth site, Elysium, appears to contain ancient terrain, which may hold clues to Mars' early climate when conditions may have been wetter.

The launch period for the first rover opens May 30, 2003, and the second rover's launch period opens June 25, 2003. The first rover will reach Mars January 4, 2004, and the second arrives January 25, 2004. Each rover will have a primary mission lasting at least three months on the martian surface.

The Jet Propulsion Laboratory manages the Mars Exploration Rover mission for NASA's Office of Space Science, Washington, D.C. JPL is a division of the California Institute of Technology, Pasadena.

Homestretch for NASA & LEGO "Name the Rovers Contest"

January 14, 2003

NASA is reminding America's school kids that time is running out on a chance to make history by naming two rovers being launched to explore Mars. The NASA "Name the Mars Rovers" contest closes January 31, 2003, so there is still time to submit the winning entries.

The robotic explorers are part of NASA's upcoming Mars Exploration Rover (MER) mission. The twin rovers will land at two different locations on the mysterious red planet to explore the surface in search of answers about the history of water on Mars.

The NASA "Name the Rovers" contest is a collaborative effort between NASA and the LEGO Company. The LEGO Company is managing the contest in conjunction with The Planetary Society. The contest provides students with the unique opportunity to suggest a name for each of the two Mars-bound rovers, temporarily known as MER-A and MER-B. The rovers are scheduled to launch in May and June 2003 respectively. The rovers are scheduled to land on Mars in January 2004.

"We are eagerly looking forward to some really creative and innovative ideas from students, as they compete to name the next Mars rovers and become part of history," said Dave Lavery, Program Executive for Solar System Exploration at NASA Headquarters.

"The LEGO Company is dedicated to furthering hands-on discovery, playful learning, and the boundless frontiers of imagination. Space exploration also embraces these qualities, which is why the LEGO Company is so pleased to partner with NASA," said Brad Justus, LEGO Senior Vice President. "By involving children actively in the Mars mission, through the 'Name the Rovers' contest and with other related activities, we hope to help excite and inspire the next generation of space explorers," he said.

The NASA "Name the Rovers" contest is open to students 5 to 18 years of age who attend a U.S. school and are enrolled in the current school season.

Submissions must include suggested names for both rovers and a 50-500 word essay justifying why the students believe the names should be chosen. The contest has many educational benefits and encourages students to do research for their essays, learn more about Mars and space exploration.

The contest is open for submissions through January 31, 2003. NASA will announce the contest winners prior to launching the rovers in the spring of 2003.

Information about the contest is available at: http://www.nametherovers.org
Information about the Mars Exploration Rovers is at: http://mars.jpl.nasa.gov/mer

MEDIA RELATIONS OFFICE
JET PROPULSION LABORATORY CALIFORNIA INSTITUTE OF TECHNOLOGY
NATIONAL AERONAUTICS AND SPACE ADMINISTRATION PASADENA, CALIF. 91109. TELEPHONE (818) 354-5011 http://www.jpl.nasa.gov

Guy Webster (818) 354-6278Jet Propulsion Laboratory, Pasadena, Calif.
Donald Savage (202) 358-1547 NASA Headquarters, Washington, D.C.
RELEASE: 2003-051

NASA ROVERS SLATED TO EXAMINE TWO INTRIGUING SITES ON MARS

April 11, 2003

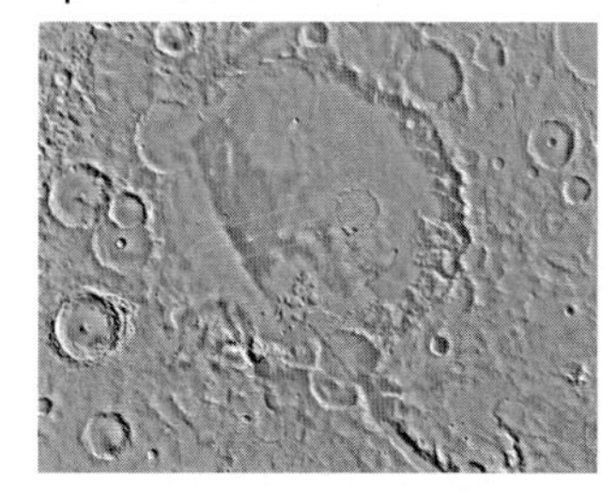

The designated landing site for the first Mars Exploration Rover mission is Gusev Crater, seen here in its geological context from NASA Viking images.

NASA has chosen two scientifically compelling landing sites for twin robotic rovers to explore on the surface of Mars early next year. The two sites are a giant crater that appears to have once held a lake, and a broad outcropping of a mineral that usually forms in the presence of liquid water.

Each Mars Exploration Rover will examine its landing site for geological evidence of past liquid water activity and past environmental conditions hospitable to life.

"Landing on Mars is very difficult, and it's harder on some parts of the planet than others," said Dr. Ed Weiler, NASA associate administrator for space science in Washington, D.C. "In choosing where to go, we need to balance science value with engineering safety considerations at the landing sites. The sites we have chosen provide such balance."

The first rover, scheduled for launch May 30, will be targeted to land at Gusev Crater, 15 degrees south of Mars' equator. The second, scheduled to launch June 25, will be targeted to land at Meridiani Planum, an area with deposits of an iron oxide mineral (gray hematite) about two degrees south of the equator and halfway around the planet from Gusev.

Which rover is targeted to a specific site is still considered tentative, while further analyses and simulations are conducted. NASA can change the order as late as approximately one month after the launch of the first rover. The first mission will parachute to an airbag-cushioned landing on Jan. 4, 2004, and the second on Jan. 25, 2004.

"A tremendous amount of effort has gone into evaluating possible landing sites in the past two years, to maximize the probability of mission success," said Peter Theisinger, Mars Exploration Rover project manager at NASA's Jet Propulsion Laboratory, Pasadena, Calif.

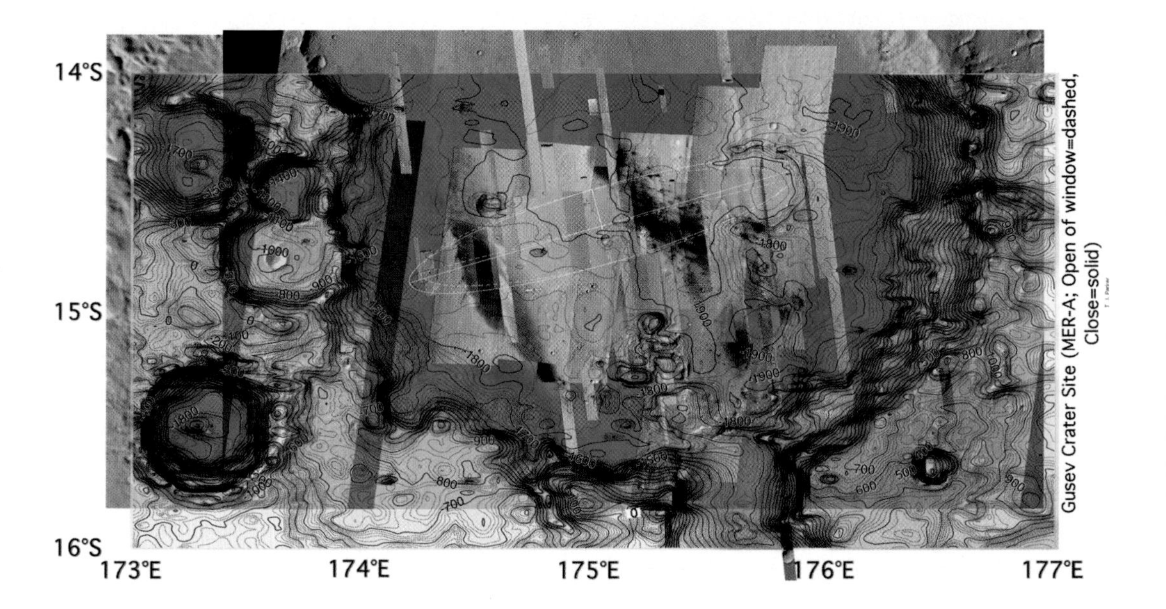

Details of the Gusev Crater designated landing site are added with topographic information and higher-resolution imaging from instruments on the Mars Global Surveyor and Mars Odyssey orbiters.

Images and measurements from two NASA spacecraft orbiting Mars provided scientists and engineers evaluating potential landing sites with details of candidate site topography, composition, rockiness and geological context.

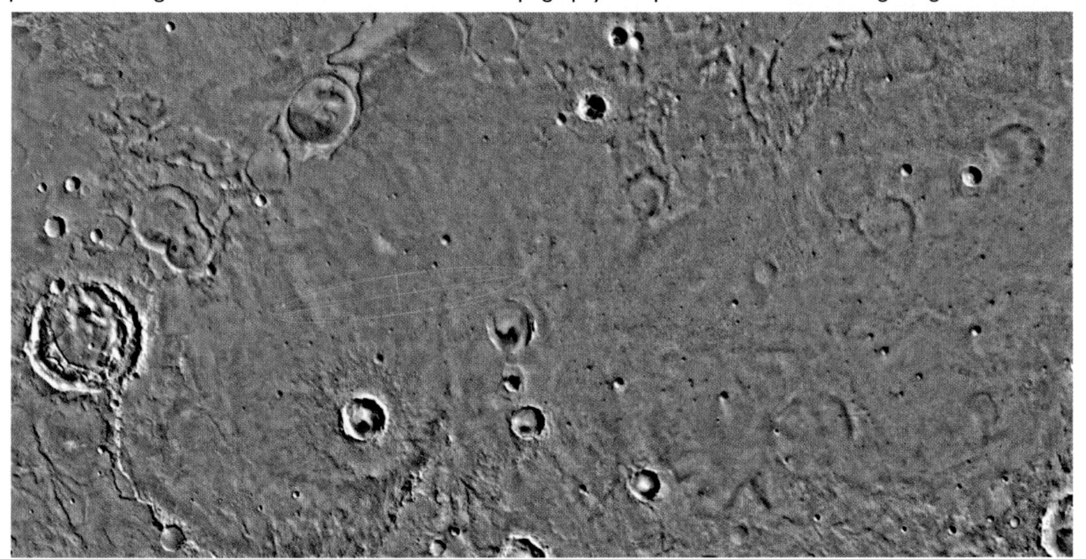

The designated landing site for the second Mars Exploration Rover mission is Meridiani Planum, seen here in its geological context from NASA Viking images.

"Meridiani and Gusev both show powerful evidence of past liquid water, but in very different ways," said Dr. Steve Squyres, principal investigator for the rovers' science toolkit and a geologist at Cornell University, Ithaca, N.Y. "Meridiani has a chemical signature of past water. Gray hematite is usually, but not always, produced in an environment where there is liquid water. At Gusev, you've got a big hole in the ground with a dry riverbed going right into it. There had to have been a lake in Gusev Crater at some point. They are fabulous sites, and they complement each other because they're so different."

Mars Exploration Rover site selection began with identifying all areas on Mars that fit a set of engineering-driven requirements, said JPL's Dr. Matt Golombek, co-chair of a landing-site steering committee. To qualify, candidate sites had to be near the equator, low in elevation, not too steep, not too rocky and not too dusty, among other criteria; 155 potential sites were studied. A series of public meetings evaluated the merits of potential landing sites. More than 100 Mars scientists participated in the meetings.

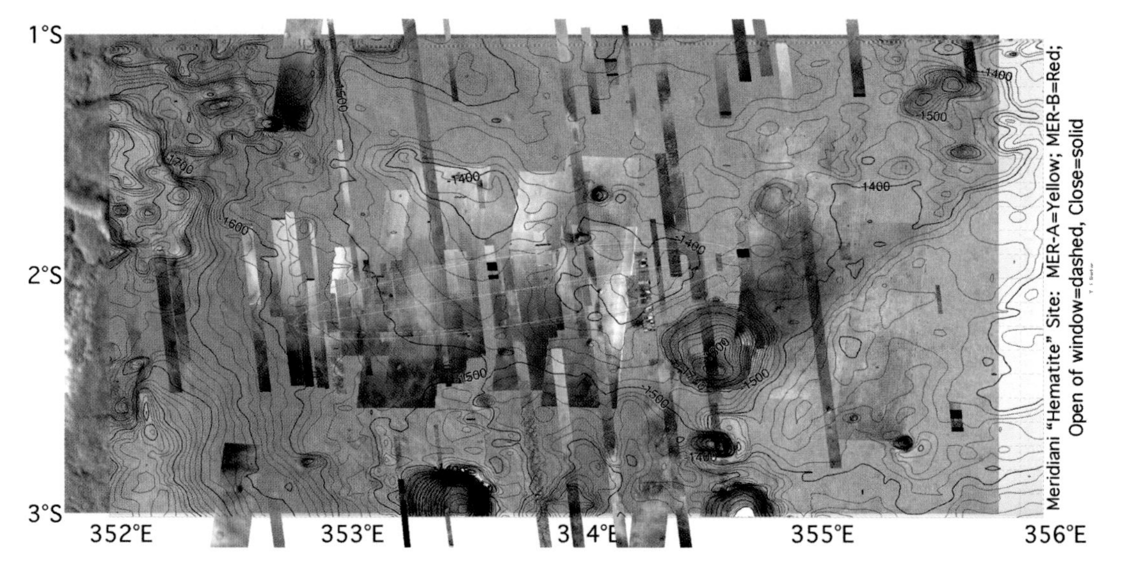

Details of the Meridiani Planum designated landing site are added with topographic information and higher-resolution imaging from instruments on the Mars Global Surveyor and Mars Odyssey orbiters.

"These two landing sites have been studied more than anywhere else on Mars. Both sites have specific scientific hypotheses that can be tested using the instruments on board each rover. It should be a very busy and exciting time after landing for the scientists analyzing the wealth of new data from the ground," said Dr. Cathy Weitz, Mars Exploration Rover program scientist at NASA Headquarters.

"Clearly there is tremendous interest in the science community in what these missions can accomplish and eagerness to help see that the rovers go to the best possible sites," said the National Air and Space Museum's Dr. John Grant, the steering committee's other co-chair.

Once they reach their landing sites, each rover's prime mission will last at least 90 martian days (92 Earth days). The rovers are solar-powered, and in approximately 90 days, dust accumulating on the solar arrays likely will be diminishing the power supply.

The twin Mars Exploration Rover spacecraft are at NASA's Kennedy Space Center, Fla., in preparation for launch. JPL built the rovers and manages the project for NASA's Office of Space Science, Washington D.C. JPL is a division of the California Institute of Technology in Pasadena.

Information about the contest is available at: http://mars.jpl.nasa.gov/mer/
For more information about NASA on the Internet, visit http://www.nasa.gov

NASA Will Send Two Robotic Geologists to Roam on Mars

June 04, 2003

NASA's Mars Exploration Rover project kicks off by launching the first of two unique robotic geologists on June 8. The identical rolling rovers can see sharper images, explore farther and examine rocks better than anything that's ever landed on Mars. The second rover mission, bound for a different site on Mars, will launch as soon as June 25.

"The instrumentation onboard these rovers, combined with their great mobility, will offer a totally new view of Mars, including a microscopic view inside rocks for the first time," said Dr. Ed Weiler, associate administrator for space science at NASA Headquarters, Washington, D.C. However, missions to Mars have proven to be far more hazardous than missions to other planets. Historically, two out of three missions, from all countries that have tried to land on Mars, ended in failure. We have done everything we can to ensure our rovers have the best chance of success."

The first Mars Exploration Rover will arrive at Mars on Jan. 4, 2004; the second on Jan. 25. Plans call for each to operate for at least three months.

These missions continue NASA's quest to understand the role of water on Mars. "We will be using the rovers to find rocks and soils that could hold clues about wet environments of Mars' past," said Dr. Cathy Weitz, Mars Exploration Rover program scientist at NASA Headquarters. "We'll analyze the clues to assess whether those environments may have been conducive to life."

First, the rovers have to safely reach Mars. "The rovers will use innovations to aid in a safe landing, but risks remain," said Peter Theisinger, Mars Exploration Rover project manager at NASA's Jet Propulsion Laboratory, Pasadena, Calif.

The rovers will bounce to airbag-cushioned landings at sites offering a balance of favorable conditions for safe landings and interesting science. The designated site for the first mission is Gusev Crater. The second rover will go to a site called Meridiani Planum. "Gusev and Meridiani give us two different types of evidence about liquid water in Mars' history," said Dr. Joy Crisp, Mars Exploration Rover project scientist at JPL. "Gusev appears to have been a crater lake. The channel of an ancient riverbed indicates water flowed right into it. Meridiani has a large deposit of gray hematite, a mineral that usually forms in a wet environment."

The rovers, working as robotic field geologists, will examine the sites for clues about what happened there. "The clues are in the rocks, but you can't go to every rock, so you split the job into two pieces," said Dr. Steve Squyres of Cornell University, Ithaca, N.Y., principal investigator for the package of science instruments on the rovers.

First, a panoramic camera at human-eye height, and a miniature thermal emission spectrometer with infrared vision help scientists identify the most interesting rocks. The rovers can watch for hazards and maneuver around them. Each six-wheeled robot has a deck of solar panels, about the size of a kitchen table, for power. The rover drives to the selected rock and extends an arm with tools on the end. Then, a microscopic imager, like a geologist's hand lens, gives a close-up view of the rock's texture. Two spectrometers identify the composition of the rock. The fourth tool substitutes for a geologist's hammer. It exposes the fresh interior of a rock by scraping away the weathered surface layer.

Both rover missions will lift off from Cape Canaveral Air Force Station, Fla., on Delta II launch vehicles. Launch opportunities begin for the first mission at 2:06 p.m. (Eastern Daylight Time) June 8 and for the second mission at 12:38 a.m. June 25, and repeat twice daily for up to 21 days for each mission.

"We see the twin rovers as stepping stones for the rest of the decade and to a future decade of Mars exploration that will ultimately provide the knowledge necessary for human exploration," said Orlando Figueroa, director of the Mars Exploration Program at NASA Headquarters.

JPL, a division of the California Institute of Technology in Pasadena, manages the Mars Exploration Rover Project for NASA's Office of Space Science, Washington, D.C.

Additional information about the project is online at http://mars.jpl.nasa.gov/mer/ .

A press kit for the mission is available at http://www.jpl.nasa.gov/news/press_kits/merlaunch.pdf

NASA Television will broadcast both launches live. NASA Television is offered by some cable providers and is available via the AMC-2 satellite, transponder 9C, located at 85 degrees west longitude, vertical polarization, frequency 3880.0 megahertz. JPL will carry live webcasts of the launches at http://www.jpl.nasa.gov/webcast/mer/ .
June 08, 2003

Girl With Dreams Names Mars Rovers 'Spirit' and 'Opportunity'

New names for Mars Exploration Rovers: Spirit and Opportunity

Twin robotic geologists NASA is sending to Mars will embody in their newly chosen names — Spirit and Opportunity — two cherished attributes that guide humans to explore.

NASA Administrator Sean O'Keefe and 9-year-old Sofi Collis, who wrote the winning essay in a naming contest, unveiled the names this morning at NASA's Kennedy Space Center. "Now, thanks to Sofi Collis, our third grade explorer-to-be from Scottsdale, Ariz., we have names for the rovers that are extremely worthy of the bold mission they are about to undertake," O'Keefe said.

Sofi read her essay: "I used to live in an orphanage. It was dark and cold and lonely. At night, I looked up at the sparkly sky and felt better. I dreamed I could fly there. In America, I can make all my dreams come true. Thank you for the 'Spirit' and the 'Opportunity.'"

Hers was selected from nearly 10,000 entries in the contest sponsored by NASA and the Lego Co., a Denmark-based toymaker, with collaboration from the Planetary Society, Pasadena, Calif..

Sofi Collis with a model of Mars Exploration Rover

Collis was born in Siberia. At age two, she was adopted by Laurie Collis and brought to the United States. "She has in her heritage and upbringing the soul of two great spacefaring countries," O'Keefe said. "One of NASA's goals is to inspire the next generation of explorers. Sofi is a wonderful example of how that next generation also inspires us."

Collis' dream of flying now takes the form of wanting to become an astronaut. Meanwhile, she enjoys playing with her older sister, swimming, reading Harry Potter stories, and her family's three dogs and one cat.

Lego President Kjeld Kirk Kristiansen, commenting on the naming contest, said, "The early days of space exploration stimulated the creativity of an entire generation, expanded our imagination and encouraged us to push our limits, making us better and braver human beings. With this project, the Lego Co. wants to bring part of that magic back. Everything we do is aimed at giving children that same power to create, and by involving children in the Name the Rovers Contest and other related playful learning activities, we hope to motivate and inspire the next generation of explorers."

Eleven miles from today's naming ceremony, Spirit, formerly called Mars Exploration Rover A, waited for a launch opportunity on Monday at Cape Canaveral Air Force Station. Opportunity, the second twin in what is still named the Mars Exploration Rover project, is being prepared for its first launch opportunity on June 25.

NASA's Jet Propulsion Laboratory, Pasadena, Calif., manages the Mars Exploration Rover project for the NASA Office of Space Science, Washington, D.C. JPL is a division of the California Institute of Technology, Pasadena.

Information about the rovers and the scientific instruments they carry is available online from JPL at http://mars.jpl.nasa.gov/mer and from Cornell University, Ithaca, N.Y., at http://athena.cornell.edu . Information about the naming contest is available at http://www.nametherovers.org .

Don Savage (202) 358-1727 NASA Headquarters, Washington D.C.
Guy Webster (321) 354-6278 Jet Propulsion Laboratory, Pasadena, Calif.
George Diller (321) 867-2468 Kennedy Space Center, Florida
Teresa Martini (646) 205-4508 LEGO Company

NASA's 'Spirit' Rises On Its Way To Mars

June 10, 2003 NEWS RELEASE 2003-081

Spirit lifts off from launch pad 17A

A NASA robotic geologist named Spirit began its seven-month journey to Mars at 1:58:47 p.m. Eastern Daylight Time (10:58:47 a.m. Pacific Daylight Time) today when its Delta II launch vehicle thundered aloft from Cape Canaveral Air Force Station, Fla.

The spacecraft, first of a twin pair in NASA's Mars Exploration Rover project, separated successfully from the Delta's third stage about 36 minutes after launch, while over the Indian Ocean. Flight controllers at NASA's Jet Propulsion Laboratory, Pasadena, Calif., received a signal from the spacecraft at 2:48 p.m. Eastern Daylight Time (11:48 a.m. Pacific Daylight Time) via the Canberra, Australia, antenna complex of NASA's Deep Space Network. All systems are operating as expected.

Spirit will roam a landing area on Mars that bears evidence of a wet history. The rover will examine rocks and soil for clues to whether the site may have been a hospitable place for life. Spirit's twin, Opportunity, which is being prepared for launch as early as 12:38 a.m. Eastern Daylight Time June 25 (9:38 p.m. Pacific Daylight Time on June 24) will be targeted to a separate site with different signs of a watery past.

"We have plenty of challenges ahead, but this launch went so well, we're delighted," said JPL's Pete Theisinger, project manager for the Mars Exploration Rover missions.

The spacecraft's cruise-phase schedule before arriving at Mars next Jan. 4, Universal Time (Jan. 3 in Eastern and Pacific time zones), includes a series of tests and calibrations, plus six opportunities for maneuvers to adjust its trajectory. JPL, a division of the California Institute of Technology, Pasadena, manages the Mars Exploration Rover project for the NASA Office of Space Science, Washington, D.C.

Information about the rovers and the scientific instruments they carry is available online from JPL at http://mars.jpl.nasa.gov/mer and from Cornell University, Ithaca, N.Y., at http://athena.cornell.edu .

Veronica McGregor (818) 354-9452
Jet Propulsion Laboratory, Pasadena, Calif.

Don Savage (202) 358-1727
NASA Headquarters, Washington, D.C.

Mars Rover Spirit Mission Status

June 12, 2003 NEWS RELEASE 2003-084

NASA's Spirit spacecraft, the first of twin Mars Exploration Rovers, has successfully reduced its spin rate as planned and switched to celestial navigation using a star scanner.

All systems on the spacecraft are in good health. As of 48 hours after the June 10 launch, Spirit had traveled 5,630,000 kilometers (3,500,000 miles) and was at a distance of 610,000 kilometers (380,000 miles) from Earth.

After separation from the third stage of its Delta II launch vehicle on Tuesday, Spirit was spinning 12.03 rotations per minute. Onboard thrusters were used Wednesday to reduce the spin rate to approximately 2 rotations per minute, the designed rate for the cruise to Mars. After the spinning slowed, Spirit's star scanner found stars that are being used as reference points for spacecraft attitude.

Navigators and other flight team members at NASA's Jet Propulsion Laboratory, Pasadena, Calif., will be deciding soon when to perform the first of several trajectory-correction maneuvers planned during the seven-month trip between Earth and Mars.

Spirit will arrive at Mars on Jan. 4, 2004, Universal Time (evening of Jan. 3, 2004, Eastern and Pacific times). The rover will examine its landing area in Mars' Gusev Crater for geological evidence about the history of water on Mars.

JPL, a division of the California Institute of Technology, manages the Mars Exploration Rover project for NASA's Office of Space Science, Washington, D.C. Additional information about the project is available from JPL at http://mars.jpl.nasa.gov/mer and from Cornell University, Ithaca, N.Y., at http://athena.cornell.edu .

Guy Webster (818) 354-0880 Jet Propulsion Laboratory, Pasadena, Calif.
Donald Savage (202) 358-1547 NASA Headquarters, Washington, D.C.

Mars Rover Spirit Mission Status

June 20, 2003 NEWS RELEASE: 2003-085

Artist's concept of Spirit

NASA's Spirit spacecraft, the first of twin Mars Exploration Rovers, performed its first trajectory correction maneuver today.

Following commands from the Mars Exploration Rover flight team at NASA's Jet Propulsion Laboratory, Pasadena, Calif., the spacecraft first performed a calibration and check of its eight thrusters, then fired the thrusters to fine-tune its flight path toward Mars.

The main burn had two components. Thrusters that accelerate the rotating spacecraft along the direction of the rotation axis burned steadily for about 28 minutes. Then, thrusters that accelerate the spacecraft in a direction perpendicular to the rotation axis fired in pulses timed to the spacecraft's rotation rate — with 264 pulses totaling about 22 minutes of burn time. The total maneuver increased Spirit's speed by 14.3 meters per second (32 miles per hour).

At the end of the trajectory correction, Spirit performed an attitude turn that adjusted its orientation in space to maintain the optimal combination of facing its solar array toward the Sun and pointing its low-gain antenna toward Earth. The spacecraft's next trajectory correction maneuver is scheduled for Aug. 1 and its next attitude turn for July 22.

All systems on the spacecraft are in good health. As of today at 6 a.m. Pacific Daylight Time, Spirit had traveled 27,390,000 kilometers (17,020,000 miles) since launch on June 10, and was at a distance of 2,660,000 kilometers (1,653,000 miles) from Earth. It was traveling at a speed of 32.22 kilometers per second (72,100 miles per hour) relative to the Sun. Spirit will arrive at Mars on Jan. 4, 2004, Universal Time (evening of Jan. 3, 2004, Eastern and Pacific times). The rover will examine its landing area in Mars' Gusev Crater for geological evidence about the history of water on Mars.

Spirit's twin, Opportunity, is being prepared at Cape Canaveral Air Force Station, Florida, for a first launch opportunity at 12:27:31 a.m. June 26, Eastern Daylight Time (9:27:31 p.m. June 25, PDT).

JPL, a division of the California Institute of Technology, manages the Mars Exploration Rover project for NASA's Office of Space Science, Washington, D.C. Additional information about the project is available from JPL at http://mars.jpl.nasa.gov/mer and from Cornell University, Ithaca, N.Y., at http://athena.cornell.edu .

Guy Webster (818) 354-6278
Jet Propulsion Laboratory, Pasadena, Calif.

Nancy Lovato (818) 354-9382
Jet Propulsion Laboratory, Pasadena, Calif.

Donald Savage (202) 358-1547
NASA Headquarters, Washington, D.C.
NEWS RELEASE: 2003-089

Newly Launched 'Opportunity' Follows Mars-Bound 'Spirit'

July 7, 2003

Opportunity Mars Exploration Rover lifts off

NASA launched its second Mars Exploration Rover, Opportunity, late Monday night aboard a Delta II launch vehicle whose bright glare briefly illuminated Florida Space Coast beaches.

Opportunity's dash to Mars began with liftoff at 11:18:15 p.m. Eastern Daylight Time (8:18:15 p.m. Pacific Daylight Time) from Cape Canaveral Air Force Station, Fla.

The spacecraft separated successfully from the Delta's third stage 83 minutes later, after it had been boosted out of Earth orbit and onto a course toward Mars. Flight controllers at NASA's Jet Propulsion Laboratory, Pasadena,

Calif., received a signal from Opportunity at 12:43 a.m. Tuesday EDT (9:43 p.m. Monday PDT) via the Goldstone, Calif., antenna complex of NASA's Deep Space Network.

All systems on the spacecraft are operating as expected, JPL's Richard Brace, Mars Exploration Rover deputy project manager, reported.

"We have a major step behind us now," said Pete Theisinger, project manager. "There are still high-risk parts of this mission ahead of us, but we have two spacecraft on the way to Mars, and that's wonderful."

NASA Associate Administrator for Space Science Dr. Ed Weiler said, "Opportunity joins Spirit and other Mars-bound missions from the European Space Agency, Japan and the United Kingdom, which together mark the most extensive exploration of another planet in history. This ambitious undertaking is an amazing feat for Planet Earth and the human spirit of exploration."

As of early Tuesday, Opportunity's twin, Spirit, has traveled 77 million kilometers (48 million miles) since its launch on June 10 and is operating in good health.

Opportunity is scheduled to arrive at a site on Mars called Meridiani Planum on Jan. 25, 2004, Universal Time (evening of Jan. 24, Eastern and Pacific times), three weeks after Spirit lands in a giant crater about halfway around the planet.

NASA's Mars Global Surveyor orbiter has identified deposits at Meridiani Planum of a type of mineral that usually forms in wet environments. Both rovers will function as robotic geologists, examining rocks and soil for clues about whether past environments at their landing sites may have been hospitable to life.

JPL is a division of the California Institute of Technology, Pasadena. It built the rovers and manages the Mars Exploration Rover project for the NASA Office of Space Science, Washington, D.C.

Information about the rovers and the scientific instruments they carry is available online from JPL at http://mars.jpl.nasa.gov/mer and from Cornell University, Ithaca, N.Y., at http://athena.cornell.edu.

Contact: JPL/Guy Webster (818) 354-6278
Donald Savage (202) 358-1727 NASA Headquarters, Washington, D.C. 2003-095

Mars Rover Opportunity Continues Healthy Journey

July 9, 2003

NASA's Opportunity spacecraft, the second of twin Mars Exploration Rovers, has successfully reduced its spin rate as planned and switched to celestial navigation using a star scanner.

Prior to today's maneuver, Opportunity was spinning 12.13 rotations per minute. Onboard thrusters were used to reduce the spin rate to approximately 2 rotations per minute, the designed rate for the cruise to Mars. After the spinning slowed, Opportunity's star scanner found stars that are being used as reference points for spacecraft attitude. One of the bright points in the star scanner's first field of view was Mars.

All systems on the spacecraft are in good health. As of 6 a.m. Pacific Daylight Time July 10, Opportunity will have traveled 6.6 million kilometers (4.1 million miles) since its July 7 launch. The Mars Exploration Rover flight team at NASA's Jet Propulsion Laboratory, Pasadena, Calif., is preparing to command Opportunity's first trajectory-correction maneuver, scheduled for July 18.

Opportunity will arrive at Mars on Jan. 25, 2004, Universal Time (evening of Jan. 24, 2004, Eastern and Pacific times). The rover will examine its landing area in Mars' Meridiani Planum area for geological evidence about the history of water on Mars.

Opportunity's twin, Spirit, also continues in good health on its cruise to Mars. As of 6 a.m. Pacific Daylight Time July 10, it will have traveled 82.6 million kilometers (51.3 million miles) since its June 10 launch.

JPL, a division of the California Institute of Technology, manages the Mars Exploration Rover project for NASA's Office of Space Science, Washington, D.C. Additional information about the project is available from JPL at http://mars.jpl.nasa.gov/mer or and from Cornell University, Ithaca, N.Y., at http://athena.cornell.edu .

Contact: JPL/Guy Webster (818) 354-6278
Donald Savage (202) 358-1727 NASA Headquarters, Washington, D.C.
2003-096

Students and Teachers to Explore Mars

July 18, 2003

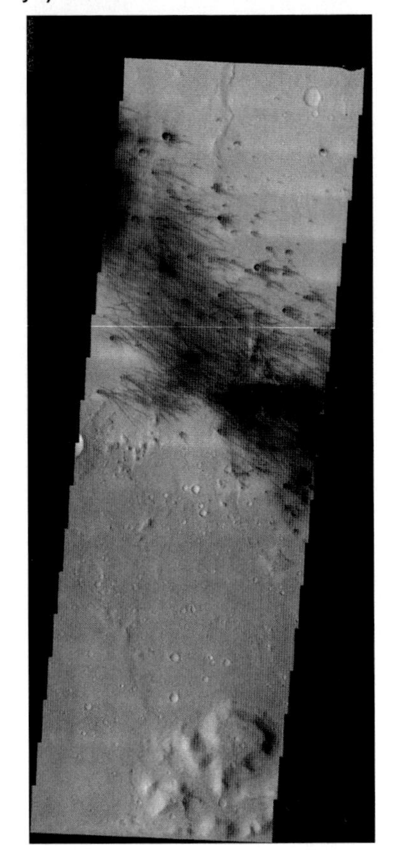

Gusev crater, destination of Mars Rover Spirit. The scene contains wispy dark streaks that probably arise from the removal by wind of a layer of bright dust.

While the ultimate field trip might someday be an actual journey to Mars, NASA is doing the next best thing – giving high school teams the opportunity to explore Mars by working on specific research projects during the Mars Exploration Rover missions, set to land on the red planet in January 2004.

Two programs designed to involve students in exploration and discovery enable high school teams to experience a space mission from launch through landing. Teams from 13 schools are participating in the Athena Student Interns program. The Mars Exploration Student Data Team has 51 participating schools. Advance studies will prepare the students for participating in the mission when the two rovers, Spirit and Opportunity, begin exploring Mars.

Participants in the Athena Student Interns program will work with mentors from the Mars science team and aid in data analysis. The students and teachers in the program will each spend a week at NASA's Jet Propulsion Laboratory, Pasadena, Calif., while the rovers are operating on the surface of Mars. Before arriving at JPL, the students will learn about the geology of Mars, the scientific and mechanical capabilities of the rovers, and the software needed to visualize the data that will be returned to Earth during the mission.

NASA is committed to developing programs to inspire students and give them hands-on experience to encourage the future scientists and engineers who will be crucial to space exploration. "More and more, we're trying to involve students directly in our missions, to give them real research opportunities," said Michelle Viotti, manager of NASA's Mars Public Engagement program at JPL. "They are our next generation of explorers."

Teachers will be part of each team and will help students in their investigations of Mars and its geologic history. Students will relay their experiences to other students in their schools and communities and to the public in order to share the excitement of exploring Mars.

Jaunine Fouché, a teacher from the Milton Hershey School in Hershey, Pa., said, "This program isn't simply about the student interns. It is about them learning and passing on their passion and fire to others. It is about inspiring, and it is about seeing and believing in the potential of children even before they see it and believe it for themselves."

The 51 teams participating in the Mars Exploration Student Data Team program will use data from Mars-orbiting spacecraft to help characterize aspects of Mars from the atmosphere to the surface that affect the rover missions. Two NASA orbiters, Mars Global Surveyor and Mars Odyssey, are actively examining the planet. The Mars Exploration Student Data Team will help compare orbital data to rover-collected data for "ground truthing," which means using ground-level observations to verify interpretations of remote observations.

"We can look around at our environment and surroundings and see the many similarities and differences we might share with other planets," said Joe Aragon, a teacher from Laguna-Acoma High School, New Laguna, N. M. "Learning about them will help us appreciate, respect and know this planet, and perhaps shed some light on our place in the solar system."

Future explorers in the Athena Student Interns Program were selected from around the country, including Alabama,

California, Colorado, Illinois, Nevada, New Mexico, New York, North Carolina, Pennsylvania, and Texas. Athena is the name of the main instrument payload on each rover — the toolkit the rovers will use to analyze rocks and other features on Mars.

The Mars Exploration Student Data Team teams are from 24 states plus the District of Columbia and an American school in Bolivia. The two programs will closely complement each other, just as both landed and orbital science teams work closely together in planetary missions.

Additional information about the Athena Student Interns Program, the Mars Exploration Student Data Team, and the Mars Exploration Rovers is available online at http://mars.jpl.nasa.gov and http://mars.jpl.nasa.gov/classroom/students/mer .

JPL, a division of the California Institute of Technology, Pasadena, Calif., manages the Mars Exploration Program's Mars Public Engagement efforts on behalf of NASA.

Contact: JPL/Nancy Lovato (818) 354-9382 2003-100

Mars Rover Opportunity Mission Status

July 18, 2003

NASA's Opportunity spacecraft made its first trajectory correction maneuver today, a scheduled operation to fine-tune its Mars-bound trajectory, or flight path.

The spacecraft and its twin, Spirit, in NASA's Mars Exploration Rover project are carrying field-geology robots for arrival at Mars in January.

For the trajectory adjustment, flight team members at NASA's Jet Propulsion Laboratory, Pasadena, Calif., commanded Opportunity to perform a prescribed sequence of thruster firings to adjust the spacecraft's flight path.

"It looks like a beautiful burn," said Jim Erickson, Mars Exploration Rover mission manager. "The thrusters fired correctly. We're on course for putting both spacecraft on Mars."

The thruster-firing sequence had three main components. First, the entire spacecraft, which is spinning at about 2 rotations per minute, turned to point its spin axis in the direction of the needed course correction. Next, thrusters that accelerate the spacecraft along the direction of that axis burned steadily for about 54 minutes. Afterwards, the spacecraft turned to its next standard cruise attitude. The attitude is changed periodically during the cruise from Earth to Mars to keep the spacecraft's antennas pointed toward Earth and its solar panels facing the Sun.

The total trajectory correction maneuver amounted to a velocity change of 16.2 meters per second (36 miles per hour) applied to Opportunity's flight path. This velocity change has two major effects. The first is to move the arrival time at Mars earlier by 1.48 days, to the intended landing date at Meridiani Planum on January 25, 2004, Universal Time (January 24, Pacific Standard Time). The second effect is to move the aimpoint at Mars from one that misses Mars by 340,000 kilometers (211,000 miles) to one that is targeted to enter the atmosphere. At launch, the spacecraft was intentionally targeted to miss Mars so that the upper stage of the Boeing Delta II launch vehicle, traveling on a nearly identical trajectory, would not hit Mars. A key purpose of today's maneuver was to adjust for that initial targeting.

As of 6 a.m. Pacific Daylight Time July 19, Opportunity will have traveled 31.5 million kilometers (19.6 million miles) since its July 7 launch. Spirit, launched on June 10, will have traveled 106.9 million kilometers (66.4 million miles). Spirit completed its first trajectory correction maneuver three weeks ago.

After arrival, the rovers will examine their landing areas for geological evidence about the history of water on Mars.

JPL, a division of the California Institute of Technology, manages the Mars Exploration Rover project for NASA's Office of Space Science, Washington, D.C. Additional information about the project is available from JPL at http://mars.jpl.nasa.gov/mer and from Cornell University, Ithaca, N.Y., at http://athena.cornell.edu .

Guy Webster (818) 354-6278 Jet Propulsion Laboratory, Pasadena, Calif.
Donald Savage (202) 358-1547 NASA Headquarters, Washington, D.C.
NEWS RELEASE: 2003-101

Mars Exploration Rover Mission Status

August 6, 2003

The first in-flight checkouts of the science instruments and engineering cameras on NASA's twin Spirit and Opportunity spacecraft on their way to Mars have provided an assessment of the instruments' condition after the stressful vibrations of launch.

The instrument tests run by the Mars Exploration Rover flight team at NASA's Jet Propulsion Laboratory, Pasadena, Calif., finished with performance data received Tuesday from two of the spectrometers on Opportunity.

Each rover's suite of science instruments includes a stereo panoramic camera pair, a microscope camera and three spectrometers. The tests also evaluated performance of each spacecraft's engineering cameras, which are a stereo navigation camera pair, stereo hazard-avoidance camera pairs on the front and back of the rover, and a downward-pointing descent camera on the lander to aid a system for reducing horizontal motion just before impact.

All 10 cameras on each spacecraft – three science cameras and seven engineering cameras on each – performed well. One of the three spectrometers on Spirit returned data that did not fit the expected pattern. The other two spectrometers on Spirit and all three on Opportunity worked properly. Teams have been busy since the tests began nearly three weeks ago analyzing about 200 megabits of instrument data generated from each spacecraft.

"All the engineering cameras are healthy," said JPL imaging scientist Dr. Justin Maki. "We took two pictures with each engineering camera — 14 pictures from each spacecraft. Even when the cameras are in the dark, the images give characteristic signatures that let us know whether the electronics are working correctly."

The science cameras on each rover – the Pancam color panoramic cameras and the Microscopic Imagers – all performed flawlessly. A spectrometer on each rover for identifying minerals from a distance, called the miniature thermal emission spectrometer, or mini-TES, also worked perfectly on each rover.

Two other spectrometers – an alpha particle X-ray spectrometer and a Mössbauer spectrometer – are mounted on an extendable arm for close-up examination of the composition of rocks and soil. Both instruments on Opportunity, as well as Spirit's alpha particle X-ray spectrometer worked properly. The Mössbauer spectrometer on Spirit is the one whose test data did not fit the pattern expected from normal operation.

"The Mössbauer results we just received from Opportunity are helping us interpret the data that we've been analyzing from Spirit," said Dr. Steve Squyres of Cornell University, Ithaca, N.Y., principal investigator for the suite of science tools on each rover. "Some of the theories we had developed for what might be causing the anomalous behavior of the Mössbauer instrument on Spirit have been eliminated by looking at the data from the one on Opportunity."

The remaining theories focus on an apparent problem in movement of a mechanism within the instrument that rapidly vibrates a gamma-ray source back and forth.

"The Mössbauer spectrometer on Spirit is working, and even if we don't come up with a way to improve its performance, we'll be able to get scientific information out of the data it sends us from Mars," Squyres said. "But it's a very flexible instrument, with lots of parameters we can change. We have high hopes that over the coming months we'll be able to understand exactly what's happened to it and make adjustments that will improve its performance. And if the Mössbauer spectrometer on Opportunity behaves on Mars the way it did today, we'll get beautiful data from that instrument."

The two types of spectrometers on the rovers' extendable arms complement each other. The alpha particle X-ray spectrometers provide information about what elements are in a rock. The Mössbauer spectrometers give information about the arrangement of iron atoms in the crystalline mineral structure within a rock.

As of 6 a.m. Pacific Daylight Time August 7, Spirit will have traveled 157.1 million kilometers (97.6 million miles) since its June 10 launch, and Opportunity will have traveled 82.7 million kilometers (51.4 million miles) since its July 7 launch. After arrival, the rovers will examine their landing areas for geological evidence about the history of water on Mars.

JPL, a division of the California Institute of Technology, manages the Mars Exploration Rover project for NASA's Office of Space Science, Washington, D.C. Additional information about the project is available from JPL at

http://mars.jpl.nasa.gov/mer and from Cornell University, Ithaca, N.Y., at http://athena.cornell.edu.
Contact: Guy Webster (818) 354-6278 JPL
Donald Savage (202) 358-1547 NASA Headquarters, Washington, D.C.
2003-109

Programs Will Share Inside Story of Mars-Bound Robots

August 19, 2003

Two free public programs in Pasadena this week will offer an introduction to the challenges and excitement of NASA's project to examine two areas of Mars with robotic rovers that are currently flying to Mars.

Peter Theisinger, Mars Exploration Rover project manager, will describe the project on Thursday evening, Aug. 21, at NASA's Jet Propulsion Laboratory, and on Friday evening, Aug. 22, at Pasadena City College.

"Three years of work by a great team got these spacecraft built and tested and launched, but the biggest hurdle is still in front of us," Theisinger said. "We have to get them safely onto the surface of Mars."

The two rovers, Spirit and Opportunity, will arrive three weeks apart in January at opposite sides of Mars. They will bounce and roll inside cocoons of inflated airbags. Unlike the much smaller Sojourner rover of the Mars Pathfinder mission in 1997, each Mars Exploration Rover will be independent of its stationary lander, capable of communicating directly with Earth and carrying a full set of cameras for scouting locations to explore. At selected rocks it will extend an arm bearing geological tools for close-up analysis. The landing sites were selected as places likely to hold geological clues about the history of water on Mars.

Theisinger, a La Crescenta resident, has worked on several interplanetary exploration missions since his 1967 graduation from the California Institute of Technology, including Voyager to the outer planets, Galileo to Jupiter and Mars Global Surveyor.

His two talks will be part of JPL's Theodore von Kármán Lecture Series. Both will begin at 7 p.m. Seating is first-come, first-served. The Thursday lecture will be in JPL's von Kármán Auditorium. JPL is at 4800 Oak Grove Dr., off the Oak Grove Drive exit of the 210 (Foothill) Freeway. The Friday lecture will be in Pasadena City College's Vosloh Forum, 1570 E. Colorado Blvd. For more information, call (818) 354-0112. Thursday's lecture will be webcast live and available afterwards at http://www.jpl.nasa.gov/events/lectures/aug03.html .

Guy Webster (818) 354-6278 Jet Propulsion Laboratory, Pasadena, Calif.
Donald Savage (202) 358-1547 NASA Headquarters, Washington, D.C.
NEWS RELEASE: 2003-109

Mars Rover Spirit Mission Status

November 04, 2003

A series of tests of one of the science instruments on NASA's Mars Exploration Rover Spirit has enabled engineers and scientists to identify how to work around an apparent problem detected in August.

Tests now indicate that all of the science instruments on both Spirit and its twin, Opportunity, are in suitable condition to provide full capabilities for examining the sites on Mars where they will land in January.

Spirit's Mössbauer spectrometer, a tool for identifying the types of iron-bearing minerals in rocks and soil, returned data that did not fit expectations during its first in-flight checkup three months ago. A drive system that rapidly vibrates a gamma-ray source back and forth inside the instrument appeared to show partial restriction in its motion.

"The drive system is adjustable. We can change its velocity. We can change its frequency," said Dr. Steve Squyres of Cornell University, Ithaca, N.Y., principal investigator for the rovers' science instruments. "We've found a set of parameters that will give us good Mössbauer science if the instrument behaves on Mars the way it is behaving now."

The corrective countermeasures include using a higher frequency of back-and-forth motion. "With these settings, whatever happened during launch will not decrease the quality of the data we get from the instrument," said Dr. Göstar Klingelhöfer, of Johannes Gutenberg University, Mainz, Germany, lead scientist for the Mössbauer spectrometers on both rovers. "The instrument was designed with enough margin in its performance that we can make this change with no significant science impact."

A possible explanation for the instrument's behavior since launch is that intense vibration of the spacecraft during launch shook something inside the spectrometer slightly out of position, he said.

Landings on Mars are risky. Most attempts over the years have failed. And even if the spacecraft survives the landing, there is the potential that individual components could be damaged. "One remaining issue with the Mössbauer Spectrometer on Spirit, as with all the instruments, is that we can't be one hundred percent sure it'll operate on Mars the way it's operating now," Squyres said. "We'll breathe easier once we've done all our post-landing health checks."

Another fact that has emerged from the in-flight checkouts of the Mössbauer spectrometers on both spacecraft is that the internal calibration channel of the Mössbauer spectrometer on Opportunity is not functioning properly. But because the instrument has the redundancy of a separate, completely independent external calibration method, this problem will not hamper use of that instrument, Squyres said.

Spirit is on course to arrive at Mars' Gusev Crater at 04:35 Jan. 4, 2004, Universal Time, which is 8:35 p.m. Jan. 3, Pacific Standard Time and 11:35 p.m. Jan. 3, Eastern Standard Time. (These are "Earth received times," meaning they reflect the delay necessary for a speed-of-light signal from Mars to reach Earth; on Mars, the landing will have happened nearly 10 minutes earlier.) Three weeks later, Opportunity will arrive at a level plain called Meridiani Planum on the opposite side of Mars from Gusev. Each rover will examine its landing area for geological evidence about the history of water there, key information for assessing whether the site ever could have been hospitable to life.

As of 13:00 Universal Time on Nov. 5 (5 a.m. PST; 8 a.m. EST), Spirit will have traveled 367.4 million kilometers (228.3 million miles) since its launch on June 10 and will still have 119.6 million kilometers (74.3 million miles) to go before reaching Mars. Opportunity will have traveled 296 million kilometers (184 million miles) since its launch on July 7 and will still have 160 million kilometers (99.2 million miles) to go to reach Mars.

The Jet Propulsion Laboratory, a division of the California Institute of Technology, manages the Mars Exploration Rover project for NASA's Office of Space Science, Washington, D.C. Additional information about the project is available from JPL at http://mars.jpl.nasa.gov/mer and from Cornell University, Ithaca, N.Y., at http://athena.cornell.edu .

Guy Webster (818) 354-6278 Jet Propulsion Laboratory, Pasadena, Calif.
NEWS RELEASE: 2003-144

Mars Rovers Head for Exciting Landings in January

December 02, 2003

NASA's robotic Mars geologist, Spirit, embodying America's enthusiasm for exploration, must run a grueling gantlet of challenges before it can start examining the red planet. Spirit's twin Mars Exploration Rover, Opportunity, also faces tough martian challenges.

"The risk is real, but so is the potential reward of using these advanced rovers to improve our understanding of how planets work," said Dr. Ed Weiler, associate administrator for space science at NASA Headquarters, Washington, D.C.

Spirit is the first of two golf-cart-sized rovers headed for Mars landings in January. The rovers will seek evidence about whether the environment in two regions might once have been capable of supporting life. Engineers at NASA's Jet Propulsion Laboratory, Pasadena, Calif., have navigated Spirit to arrive during the evening of Jan. 3, 2004, in the Eastern time zone.

Spirit will land near the center of Gusev Crater, which may have once held a lake. Three weeks later, Opportunity will reach the Meridiani Planum, a region containing exposed deposits of a mineral that usually forms under watery conditions.

"We've cleared two of the big hurdles, building both spacecraft and launching them," said JPL's Peter Theisinger, project manager for the Mars Exploration Rover Project. "Now we're coming up on a third, getting them safely onto the ground."

Since their launches on June 10 and July 7 respectively, each rover has been flying tucked inside a folded-up lander.

The lander is wrapped in deflated airbags, cocooned within a protective aeroshell and attached to a cruise stage that provides solar panels, antennas and steering for the approximately seven month journey. Spirit will cast off its cruise stage 15 minutes before hitting the top of the martian atmosphere at 5,400 meters per second (12,000 miles per hour). Atmospheric friction during the next four minutes will heat part of the aeroshell to about 1,400 C (2,600 F) and slow the descent to about 430 meters per second (960 mph). Less than two minutes before landing, the spacecraft will open its parachute.

Twenty seconds later, it will jettison the bottom half of its aeroshell, exposing the lander. The top half of the shell, still riding the parachute, will lower the lander on a tether. In the final six seconds, airbags will inflate, retro rockets on the upper shell will fire, and the tether will be cut about 15 meters (49 feet) above the ground.

Several bounces and rolls could take the airbag-cushioned lander about a kilometer (0.6 mile) from where it initially lands. If any of the initial few bounces hits a big rock that's too sharp, or if the spacecraft doesn't complete each task at just the right point during the descent, the mission could be over. More than half of all the missions launched to Mars have failed.

JPL Director Dr. Charles Elachi said, "We have done everything we know that could be humanly done to ensure success. We have conducted more testing and external reviews for the Mars Exploration Rovers than for any previous interplanetary mission."

Landing safely is the first step for three months of Mars exploration by each rover. Before rolling off its lander, each rover will spend a week or more unfolding itself, rising to full height, and scanning surroundings. Spirit and Opportunity each weigh about 17 times as much as the Sojourner rover of the 1997 Mars Pathfinder mission. They are big enough to roll right over obstacles nearly as tall as Sojourner.

"Think of Spirit and Opportunity as robotic field geologists," said Dr. Steve Squyres of Cornell University, Ithaca, N.Y., principal investigator for the rovers' identical sets of science instruments. "They look around with a stereo, color camera and with an infrared instrument that can classify rock types from a distance. They go to the rocks that seem most interesting. When they get to one, they reach out with a robotic arm that has a handful of tools, a microscope, two instruments for identifying what the rock is made of, and a grinder for getting to a fresh, unweathered surface inside the rock."

JPL, a division of the California Institute of Technology in Pasadena, manages the Mars Exploration Rover project for NASA's Office of Space Science, Washington. For information about the Mars Exploration Rover project on the Internet, visit http://mars.jpl.nasa.gov/mer . For Cornell University's Web site about the science payload, visit http://athena.cornell.edu

Guy Webster (818) 354-6278 Jet Propulsion Laboratory, Pasadena, Calif.
NEWS RELEASE: 2003-158

Mars Exploration Rover Mission Status

December 29, 2003

NASA's Spirit rover spacecraft fired its thrusters for 3.4 seconds on Friday, Dec. 26, to make a slight and possibly final correction in its flight path about one week before landing on Mars.

Radio tracking of the spacecraft during the 24 hours after the maneuver showed it to be right on course for its landing inside Mars' Gusev Crater at 04:35 Jan. 4, 2004, Universal Time (8:35 p.m. Jan. 3, Pacific Standard Time.) Spirit's twin, Opportunity, will reach Mars three weeks later.

"The maneuver went flawlessly," said Dr. Mark Adler, Spirit mission manager at NASA's Jet Propulsion Laboratory, Pasadena, Calif.

This was Spirit's fourth trajectory correction maneuver since launch on June 10. Two more are on the schedule for the flight's final three days, if needed. Adler said, "It seems unlikely we'll have to do a fifth trajectory correction maneuver, but we'll make the final call Thursday morning after we have a few more days of tracking data. Right now, it looks as though we hit the bull's-eye."

The adjustment was a quick nudge approximately perpendicular to the spacecraft's spin axis, said JPL's Chris Potts, deputy navigation team chief for the NASA Mars Exploration Rover project. "It moved the arrival time later by 2 seconds and moved the landing point on the surface northeast by about 54 kilometers" (33 miles), Potts said. The engine firing changed the velocity of the spacecraft by only 25 millimeters per second (about one-twentieth of one mile per hour).

For both NASA rovers approaching Mars, the most daunting challenges will be descending through Mars' atmosphere, landing on the surface, and opening up properly from the enclosed and folded configuration in which the rovers arrive. Most previous Mars landing attempts, by various nations, have failed.

Each rover, if it arrives successfully, will then spend more than a week in a careful sequence of steps before rolling off its lander platform. The rovers' mission is to examine their landing areas for geological evidence about past environmental conditions. In particular, they will seek evidence about the local history of liquid water, which is key information for assessing whether the sites ever could have been hospitable to life. Opportunity will land halfway around Mars from Spirit.

As of 13:00 Universal Time (6 a.m. PST) on New Year's Day, Spirit will have traveled 481.9 million kilometers (299.4 million miles) since launch and have will have 5.1 million kilometers (3.2 million miles) left to go. Opportunity will have traveled 411 million kilometers (255 million miles) since its July 7 launch and will have 45 million kilometers (27.9 million miles) to go, with three remaining scheduled opportunities for trajectory correction maneuvers.

JPL, a division of the California Institute of Technology, manages the Mars Exploration Rover project for NASA's Office of Space Science, Washington. Additional information about the project is available from JPL at http://marsrovers.jpl.nasa.gov and from Cornell University, Ithaca, N.Y., at http://athena.cornell.edu .

Guy Webster (818) 354-6278Jet Propulsion Laboratory, Pasadena, Calif.
NEWS RELEASE: 2003-173

Mars Exploration Rover Mission Status

January 3, 2004

Navigators for NASA's Spirit Mars Exploration Rover put the spacecraft so close to a bull's-eye with earlier maneuvers that mission managers chose to skip the final two optional maneuvers for adjusting course before arrival at Mars.

With less than four hours of flight time remaining, Spirit was on course to land within a targeted ellipse 62 kilometers long by 3 kilometers wide (39 miles by 2 miles) within Mars' Gusev Crater. A trajectory correction maneuver scheduled for four hours before landing was cancelled.

"The navigation status is truly excellent," said Dr. Lou D'Amario, the mission's navigation team chief at NASA's Jet Propulsion Laboratory, Pasadena, Calif. A slight trajectory adjustment on Dec. 26 was the fourth and final for the flight.

Preparations in the past two days for arrival at Mars have included an adjustment that will open Spirit's parachute about two seconds earlier than it would have been without the change, in order to compensate for recent weather on Mars. "A dust storm seen on the other side of the planet has caused global heating and thinning of the atmosphere at high altitudes" said JPL's Dr. Mark Adler, Spirit mission manager.

Also, engineers sent commands today to alter the timing when several pyro devices (explosive bolts) will be put into an enabled condition prior to firing. Enabling will begin 40 minutes earlier than it would have under previous commands. These pyro devices will be fired to carry out necessary steps of descent and landing, such as deploying the parachute and jettisoning the heat shield.

Mars is 170 million kilometers (106 million miles) away from Earth today, a distance that takes nearly 10 minutes for radio signals to cross at the speed of light. Counting that communication delay, Spirit will hit the top of Mars' atmosphere at about 04:29 Jan. 4, Universal Time (8:29 p.m. Jan. 3, Pacific Standard Time), and reach the surface six minutes later.

JPL, a division of the California Institute of Technology in Pasadena, manages the Mars Exploration Rover project for NASA's Office of Space Science, Washington, D.C. Additional information about the project is available at http://marsrovers.jpl.nasa.gov http://www.nasa.gov and from Cornell University, Ithaca, N.Y., at http://athena.cornell.edu .

JPL Newsroom (818) 354-5011 Jet Propulsion Laboratory, Pasadena, Calif.
NEWS RELEASE: 2004-002

Spirit Lands On Mars and Sends Postcards

January 4, 2004

First Look at Spirit on Mars

This mosaic image taken by the navigation camera on the Mars Exploration Rover Spirit shows a 360 degree panoramic view of the rover on the surface of Mars.

Image credit: NASA/JPL

View in Front of Spirit

This image taken by the hazard avoidance camera on the Mars Exploration Rover Spirit shows the rover's front wheels in stowed configuration.

Image credit: NASA/JPL

First Look Behind Spirit

This image taken by the hazard avoidance camera on the Mars Exploration Rover Spirit shows the rover's rear lander petal and, in the background, the Martian horizon. Spirit took the picture right after successfully landing on the surface of Mars.

Image credit: NASA/JPL

First Look at Spirit at Landing Site

This is one of the first images beamed back to Earth shortly after the Mars Exploration Rover Spirit landed on the red planet.

Image credit: NASA/JPL

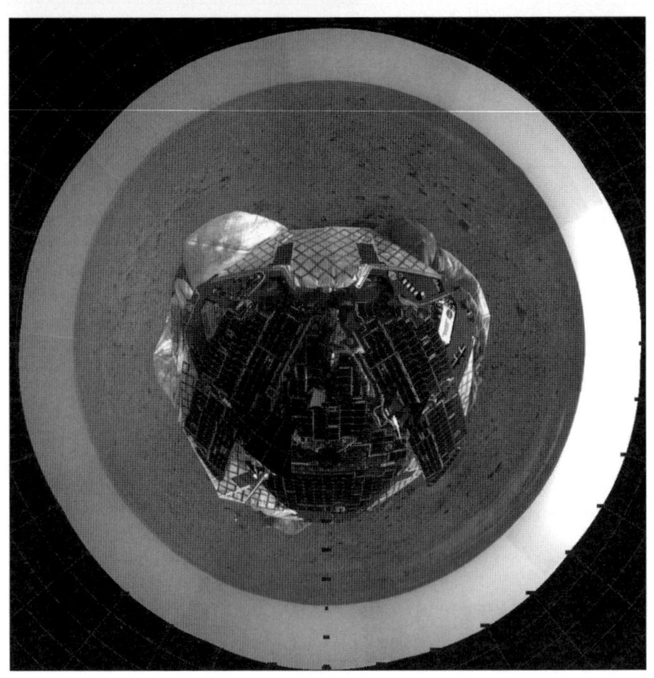

View From Above Spirit on Mars

This mosaic image taken by the navigation camera on the Mars Exploration Rover Spirit represents an overhead view of the rover on the surface of Mars.

Image credit: NASA/JPL

A traveling robotic geologist from NASA has landed on Mars and returned stunning images of the area around its landing site in Gusev Crater.

Mars Exploration Rover Spirit successfully sent a radio signal after the spacecraft had bounced and rolled for several minutes following its initial impact at 11:35 p.m. EST (8:35 p.m. Pacific Standard Time) on January 3.

"This is a big night for NASA," said NASA Administrator Sean O'Keefe. "We're back. I am very, very proud of this team, and we're on Mars."

Members of the mission's flight team at NASA's Jet Propulsion Laboratory, Pasadena, Calif., cheered and clapped when they learned that NASA's Deep Space Network had received a post-landing signal from Spirit. The cheering resumed about three hours later when the rover transmitted its first images to Earth, relaying them through NASA's Mars Odyssey orbiter.

"We've got many steps to go before this mission is over, but we've retired a lot of risk with this landing," said JPL's Pete Theisinger, project manager for the Mars Exploration Rover Project.

Deputy project manager for the rovers, JPL's Richard Cook, said, "We're certainly looking forward to Opportunity landing three weeks from now." Opportunity is Spirit's twin rover, headed for the opposite side of Mars.

Dr. Charles Elachi, JPL director, said, "To achieve this mission, we have assembled the best team of young women and men this country can put together. Essential work was done by other NASA centers and by our industrial and academic partners.

Spirit stopped rolling with its base petal down, though that favorable position could change as airbags deflate, said JPL's Rob Manning, development manager for the rover's descent through Mars' atmosphere and landing on the surface.

NASA chose Spirit's landing site, within Gusev Crater, based on evidence from Mars orbiters that this crater may have held a lake long ago.A long, deep valley, apparently carved by ancient flows of water, leads into Gusev.The crater itself is basin the size of Connecticut created by an asteroid or comet impact early in Mars' history. Spirit's task is to spend the next three months exploring for clues in rocks and soil about whether the past environment at this part of Mars was ever watery and suitable to sustain life.

Spirit traveled 487 million kilometers (302.6 million) miles to reach Mars after its launch from Cape Canaveral Air Force Station, Fla., on June 10, 2003. Its twin, Mars Exploration Rover Opportunity, was launched July 7, 2003, and is on course for a landing on the opposite side of Mars on Jan. 25 (Universal Time and EST; 9:05 p.m. on Jan. 24, PST).

The flight team expects to spend more than a week directing Spirit through a series of steps in unfolding, standing up and other preparations necessary before the rover rolls off of its lander platform to get its wheels onto the ground. Meanwhile, Spirit's cameras and a mineral-identifying infrared instrument will begin examining the surrounding terrain. That information will help engineers and scientists decide which direction to send the rover first.

JPL, a division of the California Institute of Technology, manages the Mars Exploration Rover project for NASA's Office of Space Science, Washington. Additional information about the project is available from JPL at: http://marsrovers.jpl.nasa.gov and from Cornell University, Ithaca, N.Y., at: http://athena.cornell.edu .

Guy Webster (818) 354-6278 Jet Propulsion Laboratory, Pasadena, Calif.
JPL Newsroom (818) 354-5011 Jet Propulsion Laboratory, Pasadena, Calif.
NEWS RELEASE: 2004-003

Healthy Rover Shows Its New Neighborhood on Mars

January 4, 2004

Spirit's Descent to Mars-1690m

This image, taken by the descent image motion estimation system camera located on the bottom of the Mars Exploration Rover Spirit's lander, shows a view of Gusev Crater as the lander descends to Mars. The picture is taken at an altitude of 1690 meters. Numerous small impact craters can be seen on the surface of the planet. These images help the onboard software to minimize the lander's horizontal velocity before its bridal is cut, and it falls freely to the surface of Mars.

Image credit: NASA/JPL

Spirit's Descent to Mars-1985m

This image, taken by the descent image motion estimation system camera located on the bottom of the Mars Exploration Rover Spirit's lander, shows a view of Gusev Crater as the lander descends to Mars. The picture is taken at an altitude of 1985 meters. Numerous small impact craters can be seen on the surface of the planet. These images help the onboard software to minimize the lander's horizontal velocity before its bridal is cut, and it falls freely to the surface of Mars.

Image credit: NASA/JPL

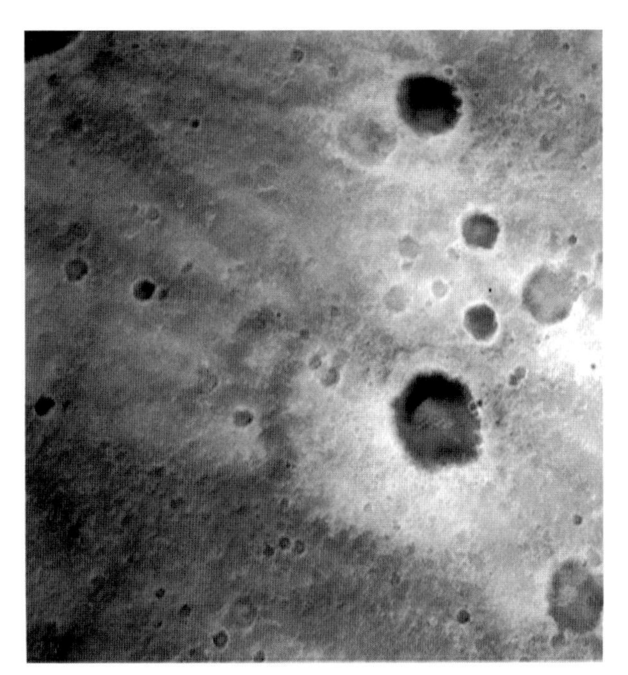

Spirit's Descent to Mars-1400m

This image, taken by the descent image motion estimation system camera located on the bottom of the Mars Exploration Rover Spirit's lander, shows a view of Gusev Crater as the lander descends to Mars. The picture is taken at an altitude of 1400 meters. Numerous small impact craters can be seen on the surface of the planet. These images help the onboard software to minimize the lander's horizontal velocity before its bridal is cut, and it falls freely to the surface of Mars.

Image credit: NASA/JPL

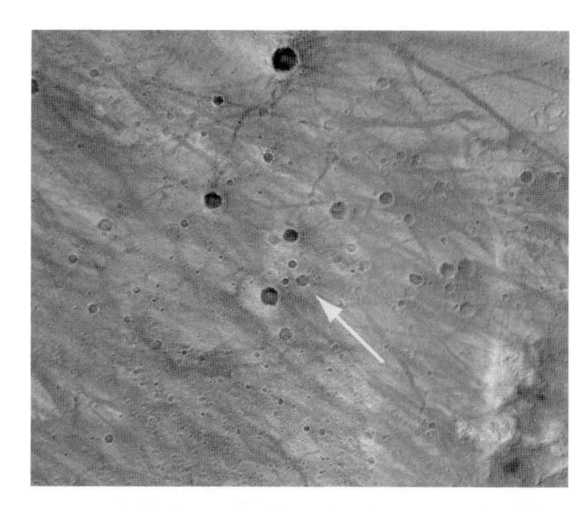

Approximate Location of Spirit

This image, taken previously by the thermal emission spectrometer onboard Mars Global Surveyor, highlights the same cluster of craters captured by the Mars Exploration Rover Spirit as it descends to Mars.

Image credit: NASA/JPL

View From Above Spirit on Mars-2

This mosaic image taken by the navigation camera on the Mars Exploration Rover Spirit has been reprocessed to project a clear overhead view of the rover on the surface of Mars.

Image credit: NASA/JPL

First Look at Spirit on Mars-2

This mosaic image taken by the navigation camera on the Mars Exploration Rover Spirit has been further processed, resulting in a significantly improved 360 degree panoramic view of the rover on the surface of Mars.

Image credit: NASA/JPL

NASA's Spirit Rover is starting to examine its new surroundings, revealing a vast flatland well suited to the robot's unprecedented mobility and scientific toolkit.

"Spirit has told us that it is healthy," Jennifer Trosper of NASA's Jet Propulsion Laboratory, Pasadena, Calif., said today. Trosper is Spirit mission manager for operations on Mars' surface. The rover remains perched on its lander platform, and the next nine days or more will be spent preparing for egress, or rolling off, onto the martian surface.

With only two degrees of tilt, with the deck toward the front an average of only about 37 centimeters (15 inches) off the ground, and with apparently no large rocks blocking the way, the lander is in good position for egress. "The egress path we're working toward is straight ahead," Trosper said.

The rover's initial images excited scientists about the prospects of exploring the region after the roll-off.

"My hat is off to the navigation team because they did a fantastic job of getting us right where we wanted to be," said Dr. Steve Squyres of Cornell University, Ithaca, N.Y., principal investigator for the science payload. By correlating images taken by Spirit with earlier images from spacecraft orbiting Mars, the mission team has determined that the rover appears to be in a region marked with numerous swaths where dust devils have removed brighter dust and left darker gravel behind.

"This is our new neighborhood," Squyres said. "We hit the sweet spot. We wanted someplace where the wind had cleared off the rocks for us. We've landed in a place that's so thick with dust devil tracks that a lot of the dust has been blown away."

The terrain looks different from any of the sites examined by NASA's three previous successful landers — the two Vikings in 1976 and Mars Pathfinder in 1987.

"What we're seeing is a section of surface that is remarkably devoid of big boulders, at least in our immediate vicinity, and that's good news because big boulders are something we would have trouble driving over," Squyres said. "We see a rock population that is different from anything we've seen elsewhere on Mars, and it comes out very much in our favor."

Spirit arrived at Mars Jan. 3 (EST and PST; Jan. 4 Universal Time) after a seven month journey. Its task is to spend the next three months exploring for clues in rocks and soil about whether the past environment at this part of Mars was ever watery and suitable to sustain life.

Spirit's twin Mars Exploration Rover, Opportunity, will reach its landing site on the opposite side of Mars on Jan. 25 (EST and Universal Time; Jan. 24 PST) to begin a similar examination of a site on the opposite side of the planet from Gusev Crater.

JPL, a division of the California Institute of Technology, manages the Mars Exploration Rover project for NASA's Office of Space Science, Washington. Additional information about the project is available from JPL at: http://marsrovers.jpl.nasa.gov and from Cornell University at: http://athena.cornell.edu.

Donald Savage (202) 358-1547 NASA Headquarters, Washington, D.C.
Guy Webster (818) 354-6278 Jet Propulsion Laboratory, Pasadena, Calif.
NEWS RELEASE: 2004-004

Mars Team Energized About "Sleepy Hollow" Near Rover

January 5, 2004

"Sleepy Hollow," a shallow depression in the Mars ground near NASA's Spirit rover, may become an early destination when the rover drives off its lander platform in a week or so.

That possible crater and other features delighted engineers and scientists examining pictures from the Mars Exploration Rover Spirit's first look around.

First 3-D Panorama of Spirit's Landing Site (See Fold-out at back of book)

This sprawling look at the martian landscape surrounding the Mars Exploration Rover Spirit is the first 3-D stereo image from the rover's navigation camera. A surface depression nicknamed "Sleepy Hollow" can be seen to center left of the image. Scientists theorize that this topographic feature, measuring about 10 meters (30 feet) in diameter and located approximately 10 to 20 meters (30 to 60 feet) away from Spirit, is either an impact crater or a product of wind erosion.

Image credit: NASA/JPL

"Reality has surpassed fantasy. We're like kids in a candy store," said Art Thompson, rover tactical activity lead at NASA's Jet Propulsion Laboratory, Pasadena, Calif. "We can hardly wait until we get off the lander and start doing fun stuff on the surface."

A clean bill of health from a checkout of all three science instruments on Spirit's robotic arm fortified scientists' anticipation of beginning to use those tools after the rover gets its six wheels onto the ground.

Also, Spirit succeeded Sunday in finding the Sun with its panoramic camera and calculating how to point its main antenna toward Earth by knowing the Sun's position.

"Just as the ancient mariners used sextants for 'shooting the Sun,' as they called it, we were successfully able to shoot the Sun with our panorama camera, then use that information to point the antenna," said JPL's Matt Wallace, mission manger.

Within sight of Spirit are several wide, shallow bowls that may be impact craters, said Dr. Steve Squyres of Cornell University, Ithaca, New York, principal investigator for the spacecraft's science payload. "It's clear that while we have a generally flat surface, it is pockmarked with these things.

The mission's scientists, who are getting little rest as they examine the pictures from Spirit, chose the name "Sleepy Hollow" for one of these circular depressions. This one is about 9 meters (30 feet) across and about 12 meters (40 feet) north of the lander, Squyres said.

"It's a hole in the ground," he said. "It's a window into the interior of Mars."

One of the next steps in preparing Spirit for rolling onto the soil is to extend the front wheels, which are tucked in for fitting inside a tight space during the flight from Earth.

Spirit arrived at Mars Jan. 3 (EST and PST; Jan. 4 Universal Time) after a seven month journey. Its task is to spend the next three months exploring for clues in rocks and soil about whether the past environment at this part of Mars was ever watery and possibly suitable to sustain life.

Spirit's twin Mars Exploration Rover, Opportunity, will reach its landing site on the opposite side of Mars on Jan. 25 (EST and Universal Time; Jan. 24 PST) to begin a similar examination of a site on the opposite side of the planet from Gusev Crater.

JPL, a division of the California Institute of Technology, manages the Mars Exploration Rover project for NASA's Office of Space Science, Washington. Additional information about the project is available from JPL at http://marsrovers.jpl.nasa.gov and and from Cornell University at http://athena.cornell.edu .

Donald Savage (202) 358-1547 NASA Headquarters, Washington, D.C.
Guy Webster (818) 354-6278 Jet Propulsion Laboratory, Pasadena, Calif.
NEWS RELEASE: 2004-005

Space Shuttle Columbia Crew Memorialized On Mars

January 6, 2004

Plaque on Spirit Honors Columbia Astronauts

NASA Administrator Sean O'Keefe today announced plans to name the landing site of the Mars Spirit rover in honor of the astronauts who died in the tragic accident of the Space Shuttle Columbia in February. The area in the vast flatland of the Gusev Crater where Spirit landed this weekend will be called the Columbia Memorial Station.

Since its historic landing, Spirit has been sending extraordinary images of its new surroundings on the red planet over the past few days. Among them, an image of a memorial plaque placed on the spacecraft to Columbia's astronauts and the STS-107 mission.

The plaque is mounted on the back of Spirit's high-gain antenna, a disc-shaped tool used for communicating directly with Earth. The plaque is aluminum and approximately six inches in diameter. The memorial plaque was attached March 28, 2003, at the Payload Hazardous Servicing Facility at NASA's Kennedy Space Center, Fla. Chris Voorhees and Peter Illsley, Mars Exploration Rover engineers at NASA's Jet Propulsion Laboratory, Pasadena, Calif., designed the plaque.

"During this time of great joy for NASA, the Mars Exploration Rover team and the entire NASA family paused to remember our lost colleagues from the Columbia mission. To venture into space, into the unknown, is a calling heard by the bravest, most dedicated individuals," said NASA Administrator Sean O'Keefe." As team members gazed at Mars through Spirit's eyes, the Columbia memorial appeared in images returned to Earth, a fitting tribute to their own spirit and dedication. Spirit carries the dream of exploration the brave astronauts of Columbia held in their hearts."

Spirit successfully landed on Mars Jan. 3. It will spend the next three months exploring the barren landscape to determine if Mars was ever watery and suitable to sustain life. Spirit's twin, Opportunity, will reach Mars on Jan. 25 to begin a similar examination of a site on the opposite side of the planet.

A copy of the image is available on the Internet at: http://www.nasa.gov

Glenn Mahone (202) 358-1898 NASA Headquarters, Washington, D.C.
Bob Jacobs (202) 358-1600 NASA Headquarters, Washington
NEWS RELEASE: 2004-006

Mars Mania Lands Online

January 6, 2004

As the spacecraft flies, Mars is millions of miles away. Thanks to the Internet, NASA can bring it into your living room, to a local Internet cafe, or anywhere else with access to the World Wide Web.

Between 12 noon Pacific Standard Time (3 p.m. Eastern Standard Time) Saturday and 6:30 a.m. PST (9:30 a.m. EST) Tuesday, NASA's Web portal, which includes the agency's home page, the Mars program Web and the Spaceflight Web, received 916 million hits, and users downloaded 154 million Web pages. The site's one-billionth hit was expected at about 12 noon PST (3 p.m. EST) Tuesday. In comparison, the portal received 2.8 billion hits for all of 2003. A hit is counted each time a Web site visitor downloads a picture, graphic element or the text on a Web page.

Internet users began tuning in to the webcast of NASA Television on Saturday, Jan. 4, and kept coming back. By Tuesday, more than 250,000 people had watched some of the mission coverage. More than 48,000 people tuned into mission control for the landing at 8:30 a.m. PST (11:30 p.m. EST) on Saturday.

"The wonders of space are now a mouse click away," said Dr. Charles Elachi, director of NASA's Jet Propulsion Laboratory, Pasadena, Calif., which manages the Mars Exploration Rover program. "Who knows how many kids will be inspired to study science or engineering because of the martian journey they're experiencing on our Web sites." The JPL site at http://www.jpl.nasa.gov , which features the latest news and images from the Mars rover Spirit, has received 107 million hits since Saturday. The NASA Portal site includes Mars information at http://marsrovers.nasa.gov .

By early Tuesday, users downloaded nearly 15 terabytes of information from the portal (a terabyte is a million megabytes. A terabyte of data would fill about one million standard floppy disks or more than 1,300 data CDs. It would take more about 20,000 CDs to store 15 terabytes. That's a stack of CDs, without cases, more than 100 feet high.

"Since 1994, when Comet Shoemaker-Levy collided with Jupiter, NASA has been using the Internet to bring the excitement of exploration directly to the public," said Brian Dunbar, NASA's Internet services manager. "Most of the time we host these sites on the NASA network, but events of this magnitude require more bandwidth than we can provide ourselves. So when we were defining requirements for the portal, a scalable, secure, offsite hosting environment was a requirement." For comparison, 24-hour traffic figures for major NASA events in the Internet era:

Pathfinder, July 9, 1997, hits: 47 million; Mars Polar Lander, Dec. 3, 1999, hits: 69 million; Columbia loss, Feb. 1, 2003, hits: 75, 539, 052; sessions: 1,060,887; page views: 10,042,668; terabytes: 0.41; Stardust, Jan. 2, 2004, hits: 12,011,502; sessions: 120,339; page views: 1,651,898; terabytes: 0.12; Spirit landing, Jan. 3-4, 2004, hits: 109,172,900; terabytes: 2.2.

Brought online less than a year ago, the NASA Web portal uses a commercial hosting infrastructure with capacity that can be readily increased to accommodate short-term, high-visibility events. Content is replicated and stored on 1,300 computers worldwide to shorten download times for users.

In 1997, the Mars Pathfinder team built a volunteer network of reflector sites and served one of the biggest Internet events to that time, if not the biggest. For the Mars Exploration Rovers, the existing portal infrastructure was available, so the Mars Web content was incorporated into the environment.

The portal prime contractor is eTouch Systems Corp. of Fremont, Calif. Speedera Networks, Inc., of Santa Clara, Calif. is delivering the NASA Web content over its globally distributed on-demand computer network. Content is replicated and stored on thousands of computer servers around the world to shorten download times for users.

This infrastructure enables NASA to provide access to the latest images from Mars, which will automatically be added to the Mars Exploration Rover site as they are received on Earth. The network also allows NASA's museum partners to access high- resolution images and video for big-screen, highly immersive experiences in local communities. Students and teachers will also find weekly classroom activities so that they can be a part of discovery on Mars.

"The portal was designed technically and graphically to enable NASA to communicate directly with members of the public, especially young people," said Dunbar. "It's a key element of NASA's mission to inspire the next generation of explorers as only NASA can."

For more information about NASA programs on the Internet, visit http://www.nasa.gov

Jane Platt (818) 354-0880 Jet Propulsion Laboratory, Pasadena, Calif.
Bob Jacobs (202) 358-1600 NASA Headquarters, Washington
NEWS RELEASE: 2004-009

First Color Image from Spirit

This is the first color image of Mars taken by the panoramic camera on the Mars Exploration Rover Spirit. It is the highest resolution image ever taken on the surface of another planet.

Image credit: NASA/JPL/Cornell

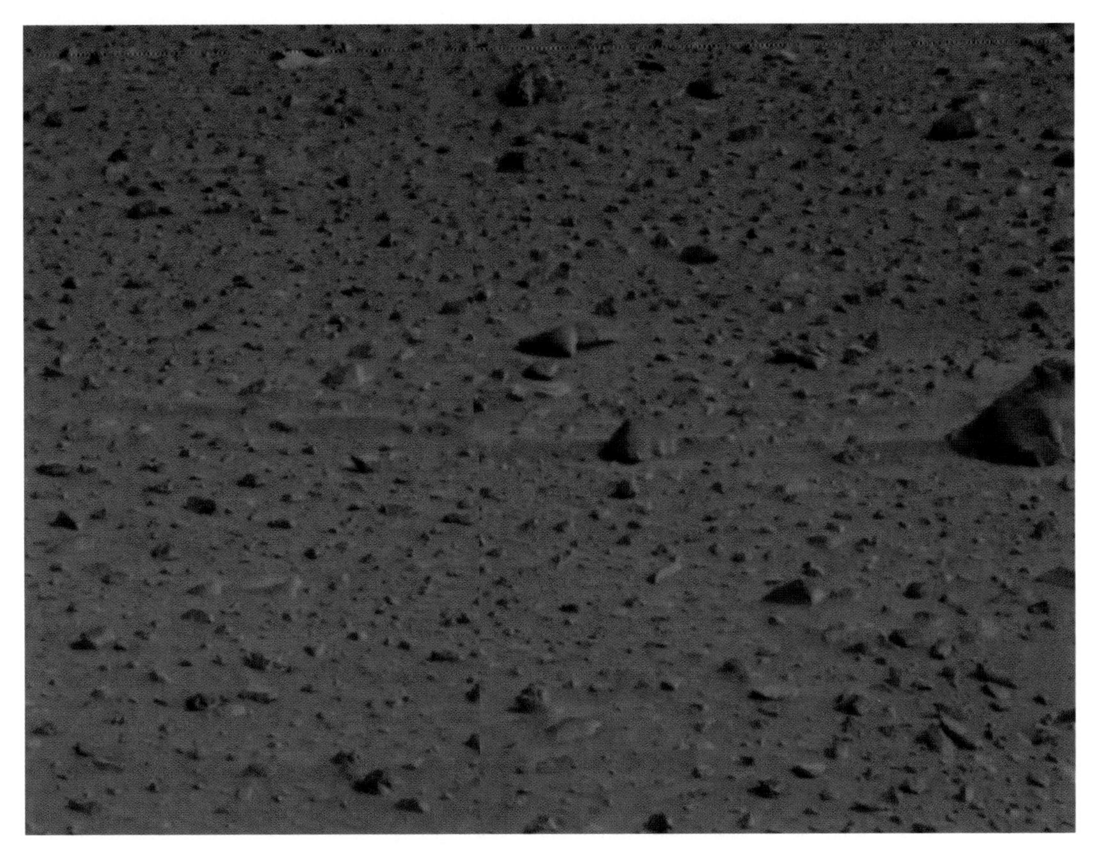

Windtails Show Direction of Martian Winds

This image highlights streaks or tails of loose debris in the martian soil, which reveal the direction of prevailing winds. The picture was taken by the panoramic camera on Mars Exploration Rover Spirit.
Image credit: NASA/JPL/Cornell

Wind-polished rocks

The smooth surfaces of angular and rounded rocks seen in this image of the martian terrain may have been polished by wind-blown debris. The picture was taken by the panoramic camera on the Mars Exploration Rover Spirit.
Image credit: NASA/JPL/Cornell

Spirit's Airbags Leave Trail

This image shows marks in the martian soil (upper right) made by the Mars Exploration Rover Spirit's airbags during their final deflation and retraction. The picture was taken by the panoramic camera on the rover.
Image credit: NASA/JPL/Cornell

In Full Bloom

The airbags are fully inflated in this photograph taken at the JPL In-Situ Instrument Laboratory or "Testbed," where engineers simulated the orientation of the airbags during the deflation process. The airbags had to be inflated seconds before landing on Mars' surface and deflated once safely on the ground.
Image credit: NASA/JPL

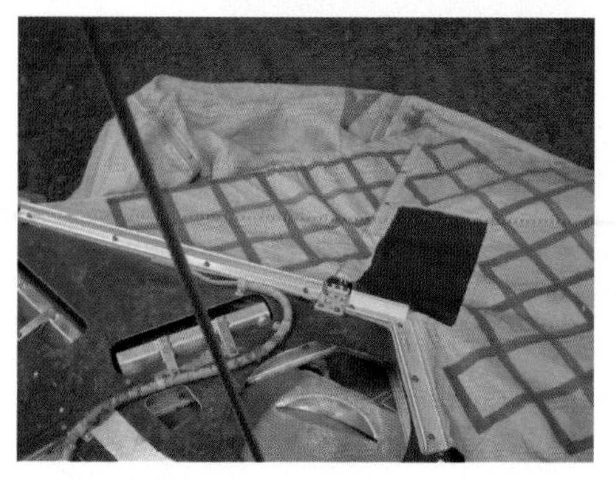

Out of Gas

This image shows the airbags in deflated position at the JPL In-Situ Instrument Laboratory, where engineers did some tests on the airbags to ensure a safe landing on Mars.

Image credit: NASA/JPL

Airbag Retraction

This image shows the deflated airbags retracted underneath the lander petal at the JPL In-Situ Instrument Laboratory. Retracting the airbags helps clear the path for the rover to roll off the lander and onto the martian surface.

Image credit: NASA/JPL

Airbag Deflates on Mars

This image, taken by the navigation camera onboard the Mars Exploration Rover Spirit, shows the airbags used to protect the rover during landing. One bright, dust-covered bag is slightly puffed up against the lander.

Image credit: NASA/JPL

Martian Horizon

This is a portion of the first color image captured by the panoramic camera on the Mars Exploration Rover Spirit.

Image credit: NASA/JPL/Cornell

Color Picture from Spirit is Most Detailed View of Mars Ever Seen

January 6, 2004

The people operating NASA's Spirit have received the first color pictures from the rover and a congratulatory call from the president.

President George W. Bush called today to congratulate the rover flight team for reconfirming the American spirit of exploration, said Dr. Charles Elachi, director of NASA's Jet Propulsion Laboratory, Pasadena, Calif., where the mission is managed. Later in the day, the Spirit team awakened the rover with the Mormon Tabernacle Choir's rendition of "Hail to the Chief."

Color images in a mosaic released today are the highest-resolution pictures ever sent from Mars, more than three times as detailed as images from Mars Pathfinder in 1997. Spirit's panoramic camera took 12 contiguous frames that the camera team combined into the mosaic.

"This is the day we've been waiting for," said Dr. Jim Bell of Cornell University, Ithaca, N.Y., leader of the panoramic camera team.

The scene rises from near the edge of Spirit's lander platform to the sky. Scientists are examining every detail to learn about the landing area within Gusev Crater. In one section of particular interest, retraction of the spacecraft's deflated airbags has disturbed the surface.

"There are places where rocks were dragged through the soil and the soil was stripped off and folded into bizarre textures," Bell said. Other areas show tails of debris to one side of rocks, possibly shaped by martian winds. "There's a wonderful mix of both smooth and angular rocks near the landing site, and this is something we'll be trying to puzzle out in the next few weeks," he said.

Scientists and the public may soon have even more to look at. The panoramic camera mosaic released today shows about one-eighth of a full-circle panorama of the landing region. The camera team plans to have the camera finish taking a full panorama this week. The pictures will share priority with other data during communication sessions either directly from the rover to Earth or relayed via NASA's Mars Global Surveyor and Mars Odyssey orbiters.

Engineers are conducting test movements of Spirit's high-gain, direct-to-Earth antenna today to learn more about spikes in the amount of electricity drawn by one of the antenna's motors when the antenna was first used Jan. 4, said JPL's Jennifer Trosper, Spirit mission manager. Meanwhile, the spacecraft will continue using the orbiter relays and its low-gain, direct-to-Earth antenna.

The flight team is also finding ways to prevent overheating of electronics inside Spirit. "Our robot geologist was dressed a little warm for the weather on Mars," Trosper said. The atmosphere and surface at the landing site this week are not as cold as anticipated. However, the rover's temperatures are expected to drop when it rolls off its lander platform and gets its wheels onto the ground.

Roll-off is now planned no sooner than Jan. 12. One of the next steps in preparing for that event will be to further retract a deflated airbag protruding from under the lander, said JPL's Jessica Collisson, flight director. The team tried out the planned retraction steps on a test rover at JPL. "We're hoping we'll have similar results to what we had in the test bed and we can get that airbag out of the way," Collisson said.

Seeing real panoramic camera pictures from Mars, instead of just from tests of the camera inside laboratories or spacecraft assembly areas, put the camera into new perspective for Bell. "Until now, it's been like having an animal in a cage, but now this beast is out, taking incredible pictures in the native habitat it was designed to work in," he said. He praised "the talented and heroic teamwork of people at Cornell and around the country who helped develop this camera — its optics, filters, electronics."

Spirit's twin Mars Exploration Rover, Opportunity, will reach its landing site on the opposite side of Mars on Jan. 25 (EST and Universal Time; Jan. 24 PST). The rovers' task is to explore for clues in rocks and soil about whether the past environments in their landing areas were ever watery and suitable to sustain life.

JPL, a division of the California Institute of Technology, manages the Mars Exploration Rover project for NASA's Office of Space Science, Washington. Images from Spirit and additional information about the project are available from JPL at http://marsrovers.jpl.nasa.gov and from Cornell University, Ithaca, N.Y., at http://athena.cornell.edu .

Guy Webster (818) 354-5011 JPL
Donald Savage (202) 358-1547 NASA Headquarters, Washington
NEWS RELEASE: 2004-010

Rover Airbag to Get Another Tug

January 7, 2004

The engineers and scientists for NASA's Spirit are eager to get the rover off its lander and out exploring the terrain that Spirit's pictures are revealing, but caution comes first.

An added "lift and tuck" to get deflated airbag material out of the way extends the number of activities Spirit needs to finish before it can get its wheels onto martian ground.

"We'll lift up the left petal of the lander, retract the airbag, then let the petal back down," said Art Thompson, rover tactical uplink lead at NASA's Jet Propulsion Laboratory, Pasadena, Calif. This and other added activities have pushed the earliest scenario for roll-off to Jan. 14, and it could be later.

The first stereo image mosaic from Spirit's panoramic camera provided new details of the landscape's shapes, including hills about 2 kilometers (1.2 miles) away that scientists are discussing as a possible drive target for the rover. The rover's infrared sensing instrument, called the miniature thermal emission spectrometer, has begun returning data about the surroundings, too, indicating that it is in good health. Now, positive health reports are in for all of Spirit's science instruments.

The rover carried out commands late Tuesday to pull in the cords to its base-petal airbags with three turns of the airbag retraction motor. "We got about a 5 centimeter (2 inch) lowering of the airbag to the left of the front of the lander, which is the one we're most concerned about," said JPL's Arthur Amador, mission manager. "That airbag is still a little too high, and we're concerned that we might hit it with our solar panel on the way down."

The rover could also turn to roll off in a different direction, but the maneuver to lift a petal and pull airbags further under it is designed to improve conditions for exiting to the front.

"We have experienced a couple of hiccups, so we're being very cautious about how we deal with them," Thompson said. One concern from Sunday and Monday was resolved late Tuesday, when results of testing a motor that moves the high-gain antenna showed no sign of a problem.

"We're chomping at the bit to get this puppy off the lander," Thompson said.

Dr. Ray Arvidson of Washington University in St. Louis, Mo., deputy principal investigator for the rover's science instruments, said the science team gathered in Pasadena has been offering diverse theories for how the landscape surrounding Spirit was shaped, and anticipating ways to test the theories with the rover's instruments.

"A lake bed is typically flat, with very fine-grain sediments," Arvidson said. "That's not what we're looking at. If these are lake sediments, then they've been chewed up by impacts and rocks have been brought in."

Besides looking forward to exploring away from the lander, the rover teams are looking forward to getting Spirit's twin Mars Exploration Rover, Opportunity, safely landed on Mars. Atmospheric conditions in the region of Opportunity's landing site are being monitored from orbit, said Dr. Joy Crisp, project scientist for both rovers. Information about the actual conditions Spirit experienced on its descent through Mars' atmosphere are being compared with the conditions predicted ahead of time in order to refine the predictions for what Opportunity will experience.

Spirit arrived at Mars Jan. 3 (EST and PST; Jan. 4 Universal Time) after a seven-month journey. Its task is to spend the next three months exploring for clues in rocks and soil about whether the past environment at this part of Mars was ever watery and suitable to sustain life.

Spirit's twin Mars Exploration Rover, Opportunity, will reach its landing site on the opposite side of Mars on Jan. 25 (EST and Universal Time; Jan. 24 PST) to begin a similar examination of a site on the opposite side of the planet from Gusev Crater.

JPL, a division of the California Institute of Technology in Pasadena, manages the Mars Exploration Rover project for

NASA's Office of Space Science, Washington, D.C. Additional information about the project is available from JPL at http://marsrovers.jpl.nasa.gov and from Cornell University, Ithaca, N.Y., at http://athena.cornell.edu .
Guy Webster (818) 354-5011 JPL
Donald Savage (202) 358-1547 NASA Headquarters, Washington

Mars in Stereo

This image shows the martian terrain in 3-D. The Mars Exploration Rover Spirit captured the image with its two high-resolution stereo panoramic cameras.
Image credit: NASA/JPL/Cornell

Shrouded in Dust

Dust-covered rocks can be seen in this portion of the 3-D image taken by the panoramic camera on the Mars Exploration Rover Spirit. Scientists plan to use the rover's rock abrasion tool to grind away dusty and weathered rock, exposing fresh rock underneath.

Image credit: NASA/JPL/Cornell

Airbag Trails

This segment of the first color image from the panoramic camera on the Mars Exploration Rover Spirit shows the rover's airbag trails. These depressions in the soil were made when the airbags were deflated and retracted after landing.

Image credit: NASA/JPL/Cornell

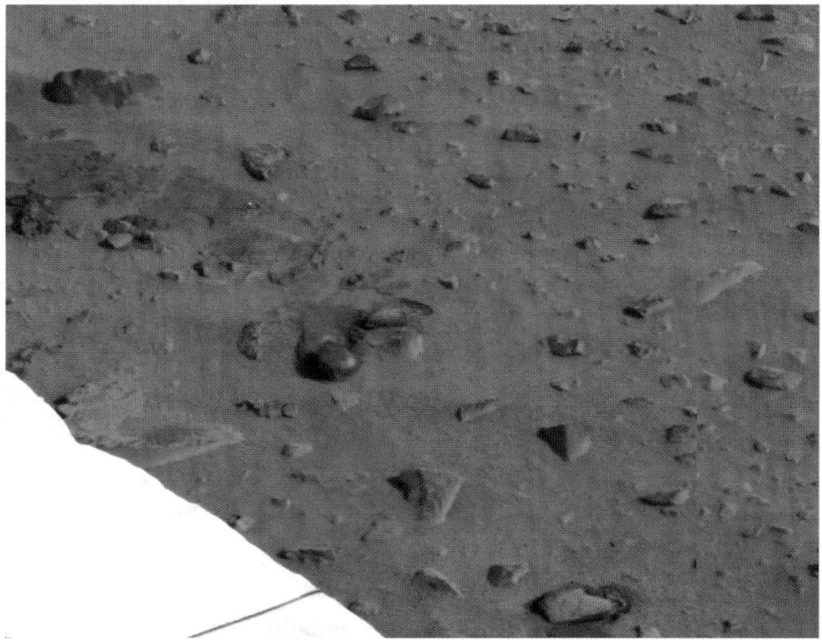

Airbag Trails-2

This segment of the first color image from the panoramic camera on the Mars Exploration Rover Spirit shows the rover's airbag trails (upper left). These depressions in the soil were made when the airbags were deflated and retracted after landing.
Image credit: NASA/JPL/Cornell

Plaque on Spirit Honors Columbia Astronauts

A plaque commemorating the astronauts who died in the tragic accident of the Space Shuttle Columbia is mounted on the back of the Mars Exploration Rover Spirit's high-gain antenna. The plaque was designed by Mars Exploration Rover engineers. The astronauts are also honored by the new name of the rover landing site, the Columbia Memorial Station. This image was taken on Mars by Spirit's navigation camera.

Image credit: NASA/JPL

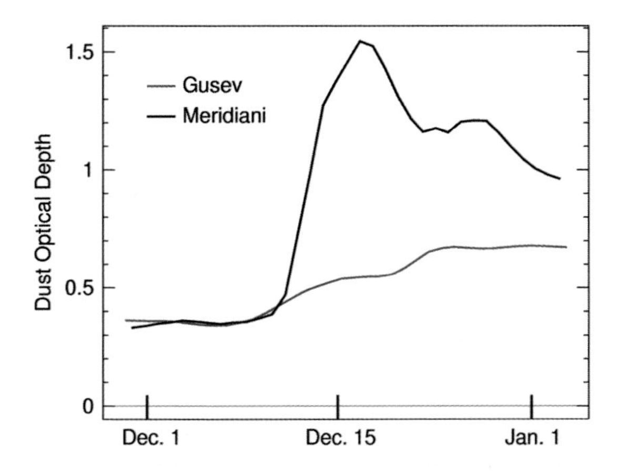

Dust in the Wind

This plot shows the estimated change in dust levels from December 2003 to early January 2004 at Gusev Crater (red curve) and Meridiani Planum (black curve), the two Mars Exploration Rover landings sites. The measurements, retrieved from Mars Global Surveyor Thermal Emission Spectrometer, indicate that a large regional dust storm beginning in mid-December raised significant dust near Meridiani. Smaller amounts of dust were spread globally by winds, the effects of which were seen at Gusev Crater. For comparison, a dust optical depth value of 1.0 would correspond to a very smoggy day in Los Angeles or Houston, and a value of 0.1 to a relatively clear day in Los Angeles.

Image credit: NASA/Goddard/Arizona State University

High School Students Land on Mars

January 8, 2004

While their peers sweat out their next geometry quiz, high school students Courtney Dressing and Rafael Morozowski are sweating out the commencement of surface activities with the rest of the Mars Exploration Rover team.

Dressing, a sophomore from Virginia, and Morozowski, a senior from Brazil, are members of an international team of students working directly with scientists and engineers overseeing the science payload on the Mars Exploration Rovers. The two 16-year-olds are the first of 16 "Student Astronauts" aged 13 to 17 who, all told, call twelve different nations home. They won their places aboard the Mars Exploration Rover team through an essay contest run by the Planetary Society followed by oral interviews. This is the first time that an international group of young people has been selected through open competition to participate in an active planetary spacecraft mission.

"We were there in the control room when Spirit landed," said Morozowski. "I was totally speechless. Then, we were there when the first pictures from Mars came back. We were seeing great things that were over 170 million kilometers (about 100 million miles) away. It was fascinating."

But Dressing and Morozowski, and the 14 Student Astronauts to follow, are there to do more than observe. They are involved in one of three programs where high school students become official Mars Rover participants. As Student Astronauts, Dressing and Morozowski have their own offices, attend team meetings, serve as ambassadors communicating to the world about life inside a Mars mission team, and are tasked to process images taken of the rover's "MarsDial".

"The MarsDial is a sundial but this one is different and cooler because it is on Mars," said Dressing. "It is used by the rover's cameras for calibration purposes. We are going to process these calibration images, add hour marks electronically and provide them online. This should gets kids excited about space exploration and help them and their teachers learn about time and celestial motion."

Also receiving an education from the Mars imagery on a daily - if not hourly - basis, are the full-time scientists and engineers working on Spirit. During this morning's briefing, Cornell University's Dr. Jim Bell, the mission's payload element lead for the Panoramic Camera, revealed some more remarkable imagery of Columbia Memorial Station. This latest "Postcard From Mars," downloaded using the rover's high-speed, high gain antenna, depicts the view north — behind the rover. The image has an apparent slope with relation to the horizon due to a tilt of the lander deck. There is a dune-like object to the right side of the image that has piqued the interest of the science team and may become a future target of the rover's onboard instruments. On the left of the image, the circular topographic feature dubbed Sleepy Hollow can be seen along with dark markings that scientists think may very well be surface disturbances caused by the airbag-encased lander as it bounced and rolled to rest.

"Originally, we thought that prominence right in front of the rover was a big rock," said Matt Wallace, a mission manager at the Jet Propulsion Laboratory in Pasadena, Calif. "But after we downloaded the more high-resolution images, we realized it is actually just a very, very dirty, dust-coated airbag."

The airbags themselves have become a topic of conversation among the team. An attempt during Sol 5 to fully deflate and retract an airbag at the front-left side of the rover was not as successful as the team had hoped. Commands will soon be beamed up to Spirit to again retract the uncooperative airbag and then lower the lander's pedestal.

"This rover is a thoroughbred, and I have a great deal of confidence we could drive right over that airbag," added Wallace. "But we are 100 million miles away and have several other options for egress, so we will take our time and get it right. We are going to be brave but we're not going to be stupid."

While Spirit's exploration is just getting warmed and Opportunity is waiting for its moment in the martian sun, JPL's first duo of Student Astronauts is wrapping up a one-week stint living and breathing the red planet. But while their tour of duty ends Saturday, their interest in the mission will continue, apparently far beyond the rover's lifespan.

"After I finish my education I think I want to work right here at JPL," added Dressing. "I love space and I think this is a great place for that. After all, where else can you go to Mars?"

Spirit's twin Mars Exploration Rover, Opportunity, will reach its landing site on the opposite side of Mars on Jan. 25 (EST and Universal Time; Jan. 24 PST). The rovers' task is to explore for clues in rocks and soil about whether the past environments in their landing areas were ever watery and suitable to sustain life.

JPL, a division of the California Institute of Technology in Pasadena, manages the Mars Exploration Rover project for NASA's Office of Space Science, Washington. Images from Spirit and additional information about the project are available from JPL at http://marsrovers.jpl.nasa.gov and from Cornell University, Ithaca, N.Y., at http://athena.cornell.edu .

Guy Webster (818) 354-5011JPL
Donald Savage (202) 358-1547 NASA Headquarters, Washington

Martian Surface at an Angle

This latest color "postcard from Mars," taken on Sol 5 by the panoramic camera on the Mars Exploration Rover Spirit, looks to the north. The apparent slope of the horizon is due to the several-degree tilt of the lander deck. On the left, the circular topographic feature dubbed Sleepy Hollow can be seen along with dark markings that may be surface disturbances caused by the airbag-encased lander as it bounced and rolled to rest. A dust-coated airbag is prominent in the foreground, and a dune-like object that has piqued the interest of the science team with its dark, possibly armored top coating, can be seen on the right.

Image credit: NASA/JPL/Cornell

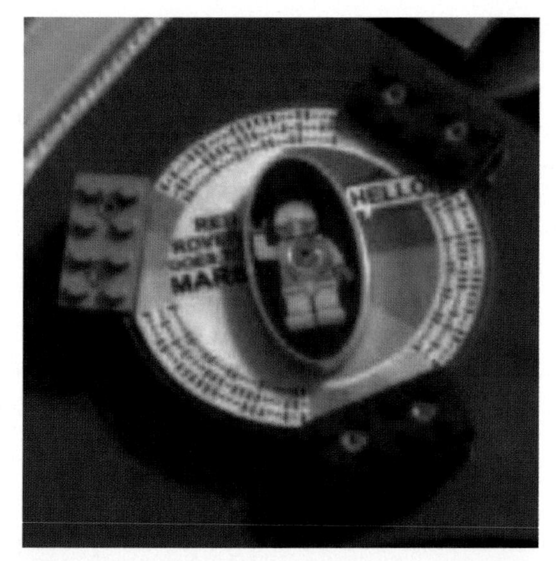

Students, Public Connect with Mars

This DVD carries nearly 4 million names collected by NASA in the "Send Your Name to Mars" project as well as various student activities. At the center of the DVD is a Lego "astrobot" minifigure that allows children to follow the mission via the astrobot diaries of Biff Starling and Sandy Moondust. Magnets on the outer edge of the DVD will collect dust for student analysis, and children can also decode the hidden message in the black dashes around the edges of the DVD. The DVD was provided and supported by the Planetary Society, the LEGO Company, Visionary Products, Inc., Plasmon OMS and the Danish magnet team.

Image credit: NASA/JPL/Cornell

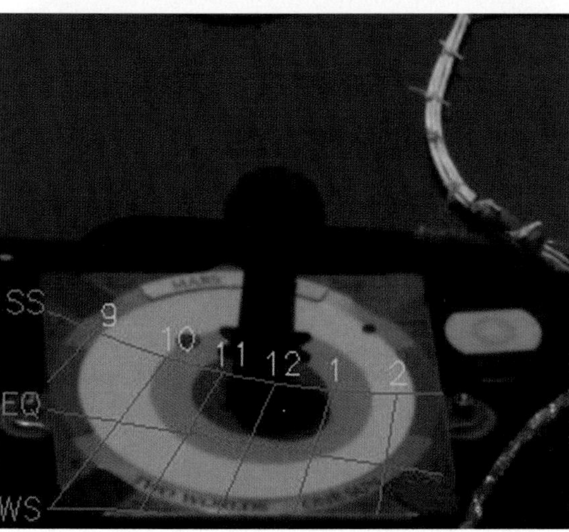

What time is it on Mars?

This image of the martian sundial onboard the Mars Exploration Rover Spirit was processed by the Student Astronauts to impose hour markings on the face of the dial. The position of the shadow of the sundial's post within the markings indicates the time of day and the season, which in this image is 12:17 p.m. local solar time, late summer. The Student Astronauts are a team of 16 students from 12 countries selected by the Planetary Society to participate in the Mars Exploration Rover program. This image was taken on Mars by the rover's panoramic camera.

Image credit: NASA/JPL/Cornell University

Pancam Calibration Target

High Sun

Low Sun

Sundial Lands on Mars

Two views of a sundial called the MarsDial can be seen in this image taken on Mars by the Mars Exploration Rover Spirit's panoramic camera. These calibration instruments, positioned on the solar panels of both Spirit and the Mars Exploration Rover Opportunity, are tools for both scientists and educators. Scientists use the sundial to adjust the rovers' panoramic cameras, while students participating in NASA's Red Rover Goes to Mars program will monitor the dial to track time on Mars. Students worldwide will also have the opportunity to build their own Earth sundial and compare it to that on Mars.

The left image was captured near martian noon when the Sun was very high in the sky. The right image was acquired later in the afternoon when the Sun was lower in sky, casting longer shadows. The colored blocks in the corners of the sundial are used to fine-tune the panoramic camera's sense of color. Shadows cast on the sundial help scientists adjust the brightness of images.

The sundial is embellished with artwork from children, and displays the word Mars in 17 different languages.

Image credit: NASA/JPL/Cornell University

Spirit Lowers Front Wheels, Looks Around in Infrared

January 09, 2004

NASA's Spirit, the first of two Mars Exploration Rovers on the martian surface, has stood up and extended its front wheels while continuing to delight its human partners with new information about its neighborhood within Mars' Gusev Crater.

Traces of carbonate minerals showed up in the rover's first survey of the site with its infrared sensing instrument, called the miniature thermal emission spectrometer or Mini-TES. Carbonates form in the presence of water, but it's too early to tell whether the amounts detected come from interaction with water vapor in Mars' atmosphere or are evidence of a watery local environment in the past, scientists emphasized.

"We came looking for carbonates. We have them. We're going to chase them," said Dr. Phil Christensen of Arizona State University, Tempe, leader of the Mini-TES team. Previous infrared readings from Mars orbit have revealed a low concentration of carbonates distributed globally. Christensen has interpreted that as the result of dust interaction with atmospheric water. First indications are that the carbonate concentration near Spirit may be higher than the Mars global average.

After the rover drives off its lander platform, infrared measurements it takes as it explores the area may allow scientists to judge whether the water indicated by the nearby carbonates was in the air or in a suspected ancient lake.

"The beauty is we know how to find out," said Dr. Steve Squyres of Cornell University, Ithaca, N.Y., principal investigator for the mission. "Is the carbonate concentrated in fluffy dust? That might favor the atmospheric hypothesis. Is it concentrated in coarser material? That might favor the water hypothesis."

Spirit accomplished a key step late Thursday in preparing for rolling off the lander. In anticipation, the flight team at NASA Jet Propulsion Laboratory in Pasadena, Calif., played Bob Marley's "Get Up, Stand Up" as wake-up music for the sixth morning on Mars, said JPL's Matt Wallace, mission manager. In the following hours, the rover was raised by a lift mechanism under its belly, and its front wheels were fully extended. Then the rover was set back down, raised again and set down again to check whether suspension mechanisms had latched properly.

Pictures returned from the rover's navigation camera and front hazard-identification camera, plus other data, confirmed success.

"We are very, very, very pleased to see the rover complete the most critical part of the stand-up process," Wallace said. Next steps include retracting the lift mechanism and extending the rear wheels.

A tug on airbag tendons by the airbag retraction motor Thursday evening did not lower puffed up portions of airbag material that are a potential obstacle to driving the rover straight forward to exit the lander. The most likely path for driving off will be to turn 120 degrees to the right before rolling off. "This is something we have practiced many times. We are very comfortable doing it," Wallace said.

The earliest scenario for getting the rover off the lander, if all goes smoothly, is Spirit's 13th or 14th day on Mars, Jan. 16 or 17.

"We're proceeding in a measured, temperate way," said JPL's Peter Theisinger, project manager for the Mars Exploration Rover project. "This is a priceless asset. It is fully functioning. It is sitting in a beautiful scientific target. We're not going to take any inappropriate risks."

While preparing to learn more about what Mars rocks are made of, Christensen announced an educational project to involve school children and other people in getting rocks from all over Earth for comparison. "Send me your rocks and we'll see if there are rocks in your back yard that are similar to what we're seeing on Mars," he said. Information about how to send rocks to Arizona State University is on the rovers' Web site at http://marsrovers.jpl.nasa.gov.

Spirit's twin Mars Exploration Rover, Opportunity, will reach Mars on Jan. 25 (Universal Time and EST; Jan. 24 PST). The rovers' main task is to spend three months exploring for clues in rocks and soil about whether past environments near the landing sites were ever watery and possibly suitable to sustain life.

JPL, a division of the California Institute of Technology, manages the Mars Exploration Rover project for NASA's Office of Space Science, Washington.
Guy Webster (818) 354-5011 Jet Propulsion Laboratory, Pasadena, California
Donald Savage (202) 358-1547 NASA Headquarters, Washington
NEWS RELEASE: 2004-13

Spirit Rises to the Occasion

JPL engineers played Bob Marley's "Get Up, Stand Up" in the control room as they watched new images confirming that the Mars Exploration Rover Spirit successfully stood up on its lander late Thursday night Pacific time, a major step in preparing for egress. This image from the rover's front hazard avoidance camera shows the rover in the final stage of its stand-up process. The two wheels on the bottom right and left are locked into position, along with the suspension system. The martian landscape is in the background. With a deflated airbag partially blocking one exit route, engineers will decide whether Spirit should use a different route to roll off the lander.

Image credit: NASA/JPL

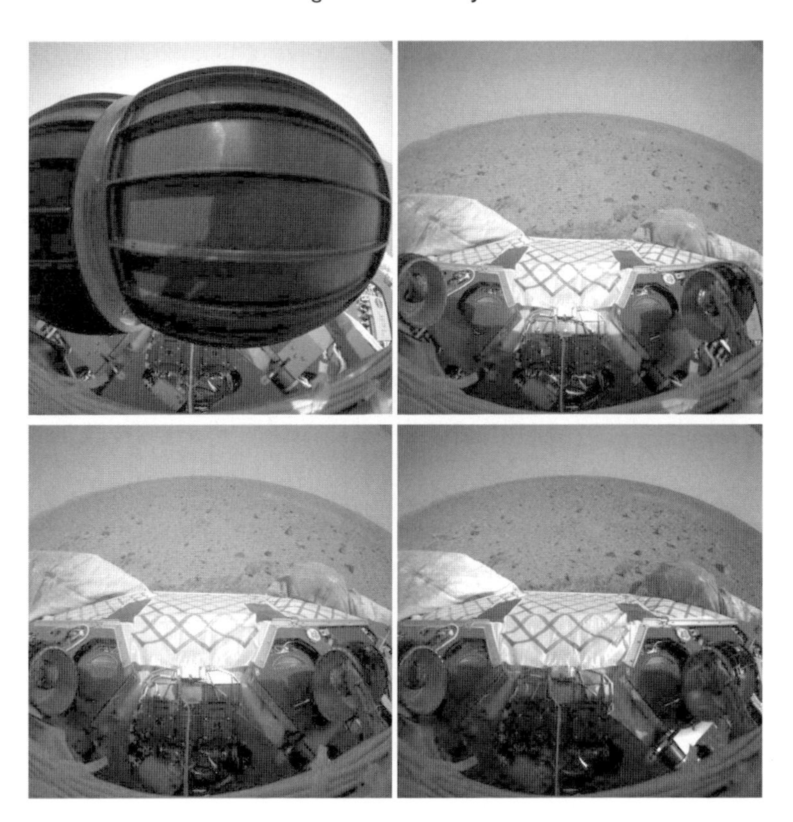

Spirit Rises to the Occasion (Animation sequence)

This animation strings together images from the rover's front hazard avoidance camera taken during the stand-up process of the Mars Exploration Rover Spirit. The first frame shows the rover's wheels tucked under in pre-stand-up position. The following frames show the stages of the stand-up process. The rover first elevates itself and unfolds the wheels. It then lowers, lifts and lowers again into its final position. Note the changing camera perspectives of the martian landscape, indicating the rover's heightened and lowered positions.

Image credit: NASA/JPL

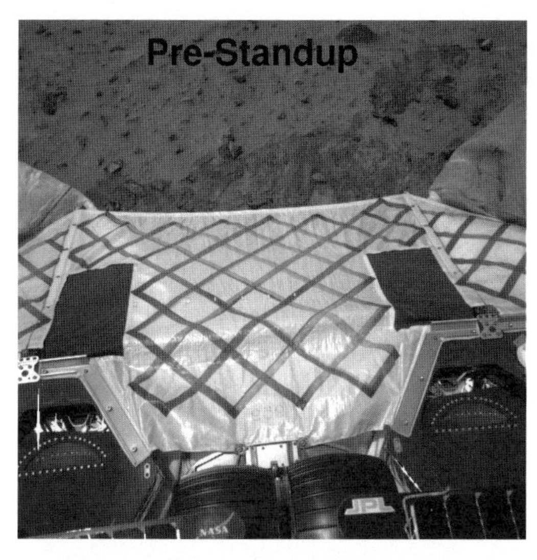

From Spirit's Perspective

This is a perspective from the navigation camera on the Mars Exploration Rover Spirit prior to beginning the stand-up process.

Image credit: NASA/JPL

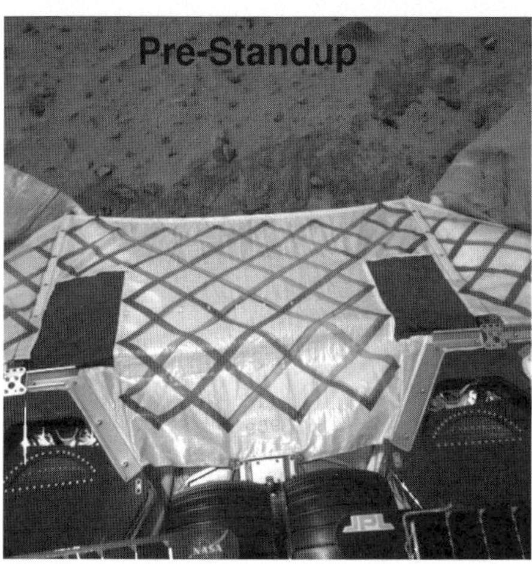

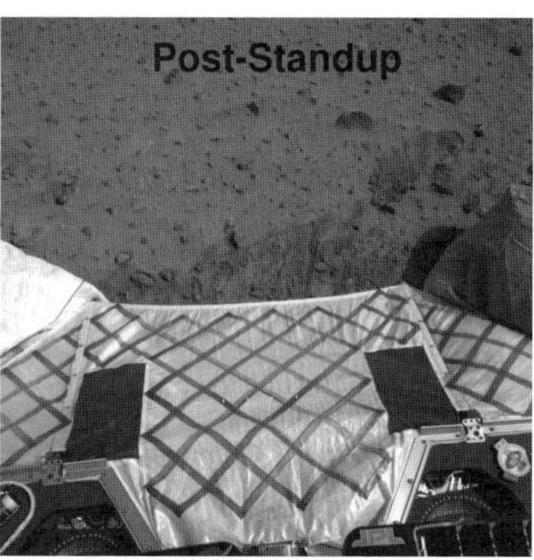

From Spirit's Perspective (Animation sequence)

This animation shows the perspective from the navigation camera on the Mars Exploration Rover Spirit before and after its automated stand-up process. After standing up, the rover is approximately 12 inches higher off of the lander, resulting in a better view of the surrounding terrain.

Image credit: NASA/JPL

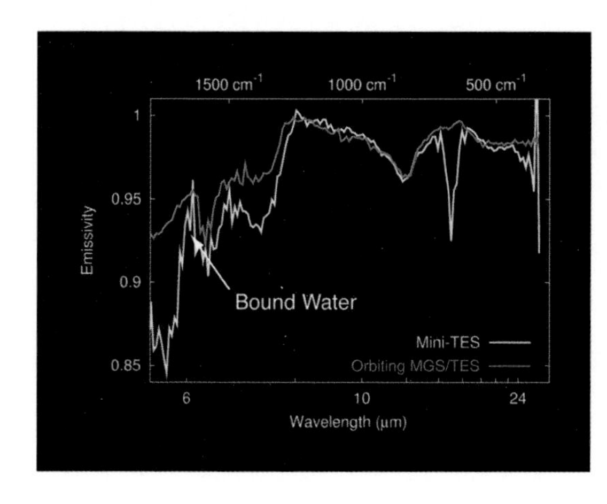

Rover Senses Bound Water

This graph, consisting of data acquired on Mars from the Mars Exploration Rover Spirit's mini-thermal emission spectrometer, shows the light, or spectral, signature of an as-of-yet unidentified mineral that contains bound water in its crystal structure. Minerals such as gypsum and zeolites are possible candidates.

Image credit: NASA/JPL/Arizona State University

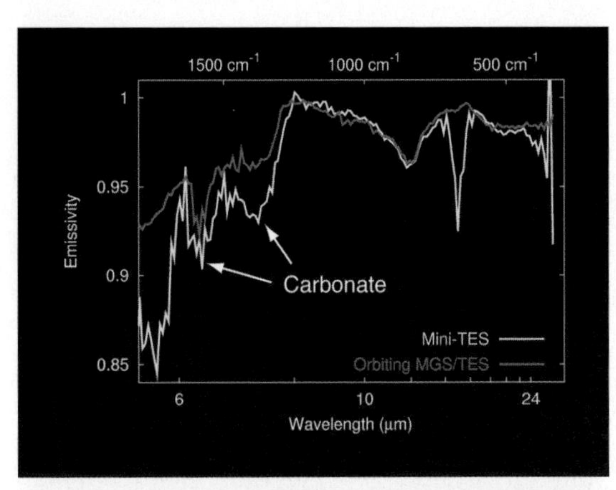

Rover Senses Carbonates

This graph, consisting of data from the Mars Exploration Rover Spirit's mini-thermal emission spectrometer, shows the light, or spectral, signatures of carbonates - organic molecules common to Earth that form only in water. The detection of trace amounts of carbonates on Mars may be due to an interaction between the water vapor in the atmosphere and minerals on the surface.

Image credit: NASA/JPL/Arizona State University

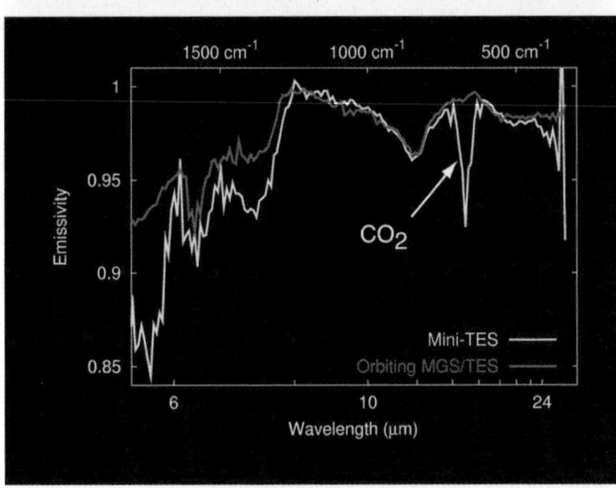

Rover Senses Carbon Dioxide

This graph, consisting of data acquired on Mars from the Mars Exploration Rover Spirit's mini-thermal emission spectrometer, shows the light, or spectral, signature of carbon dioxide. Carbon dioxide makes up the bulk of the thin martian atmosphere.

Image credit: NASA/JPL/Arizona State University

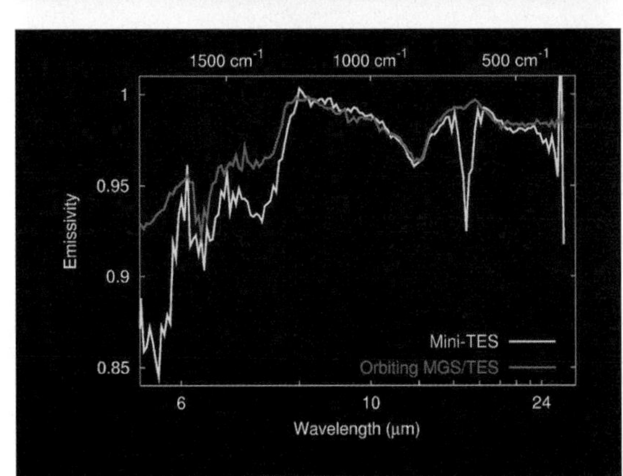

Dust Spectra from Above and Below

Spectra of martian dust taken by the Mars Exploration Rover Spirit's mini-thermal emission spectrometer are compared to that of the oribital Mars Global Surveyor's thermal emission spectrometer. The graph shows that the two instruments are in excellent agreement.

Image credit: NASA/JPL/Arizona State University

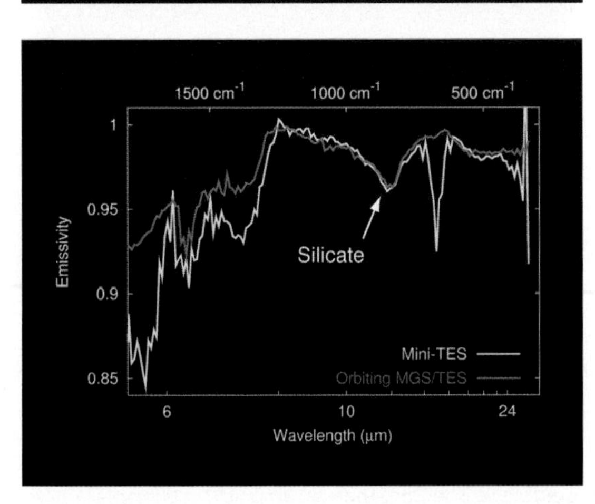

Rover Senses Silicates

This graph, consisting of data acquired on Mars by the Mars Exploration Rover Spirit's mini-thermal emission spectrometer, shows the light, or spectral, signature of silicates - a group of minerals that form the majority of Earth's crust. Minerals called feldspars and zeolites are likely candidates responsible for this feature.

NASA/JPL/Arizona State University

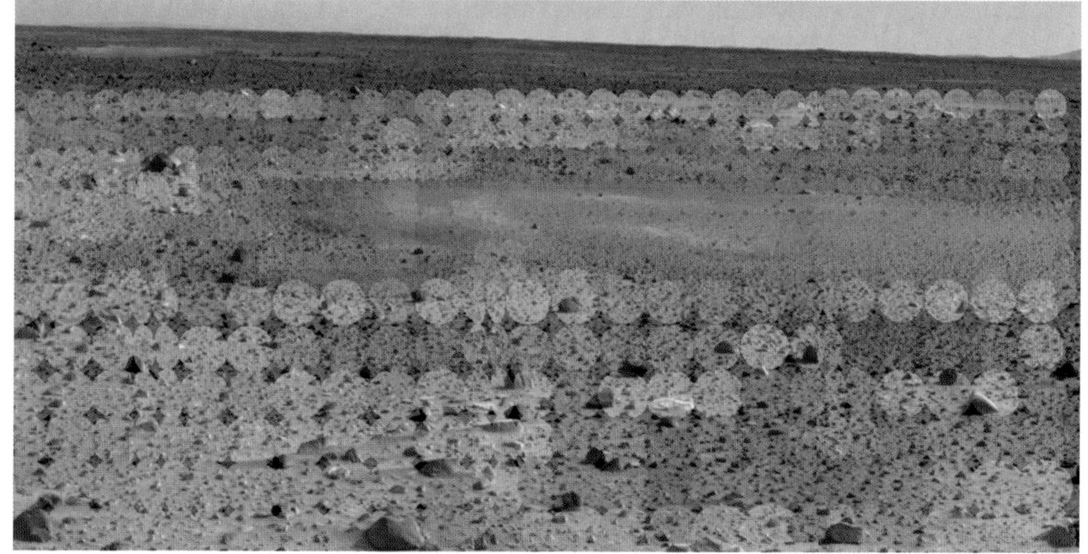

Mars Through Infrared Eyes of Spirit

This image shows the martian terrain through the eyes of the Mars Exploration Rover Spirit's mini-thermal emission spectrometer, an instrument that detects the infrared light, or heat, emitted by objects. The different colored circles show a spectrum of soil and rock temperatures, with red representing warmer regions and blue, cooler. A warm and dusty depression similar to the one dubbed Sleepy Hollow stands out to the upper right. Scientists and engineers will use this data to pinpoint features of interest, and to plot a safe course for the rover free of loose dust. The mini-thermal emission spectrometer data are superimposed on an image taken by the rover's panoramic camera.
Image credit: NASA/JPL/Arizona State University/Cornell University

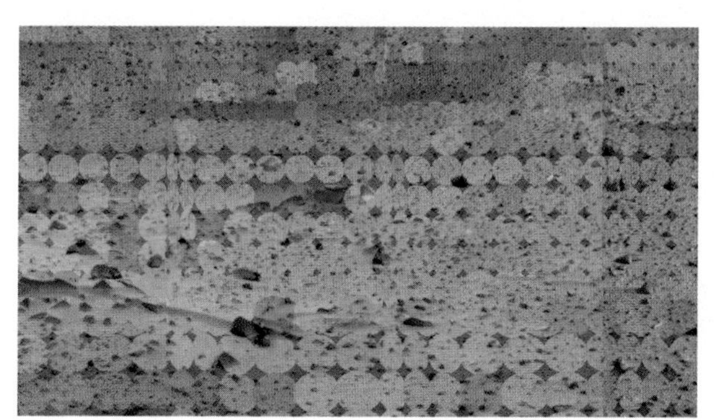

Mars Through Infrared Eyes of Spirit-2

This image shows the martian terrain through the eyes of the Mars Exploration Rover Spirit's mini-thermal emission spectrometer, an instrument that detects the infrared light, or heat, emitted by objects. The different colored circles show a spectrum of soil and rock temperatures, with red representing warmer regions and blue, cooler. Clusters of cool rocks can be seen to the left and center. Scientists and engineers will use this data to pinpoint features of interest, and to plot a safe course for the rover free of loose dust. The mini-thermal emission spectrometer data are superimposed on an image taken by the rover's panoramic camera. Image credit: NASA/JPL/Arizona State University/Cornell University

Mars Through Infrared Eyes of Spirit-3

This image shows the martian terrain through the eyes of the Mars Exploration Rover Spirit's mini-thermal emission spectrometer, an instrument that detects the infrared light, or heat, emitted by objects. The different colored circles show a spectrum of soil and rock temperatures, with red representing warmer regions and blue, cooler. Clusters of cool rocks stand out to the left, and a warm, dusty depression similar to the one dubbed Sleepy Hollow can be seen to the upper right. Scientists and engineers will use this data to pinpoint features of interest, and to plot a safe course for the rover free of loose dust. The mini-thermal emission spectrometer data are superimposed on an image taken by the rover's panoramic camera.
Image credit: NASA/JPL/Arizona State University/Cornell University

Mars Through Infrared Eyes of Spirit-3 - Zoomed

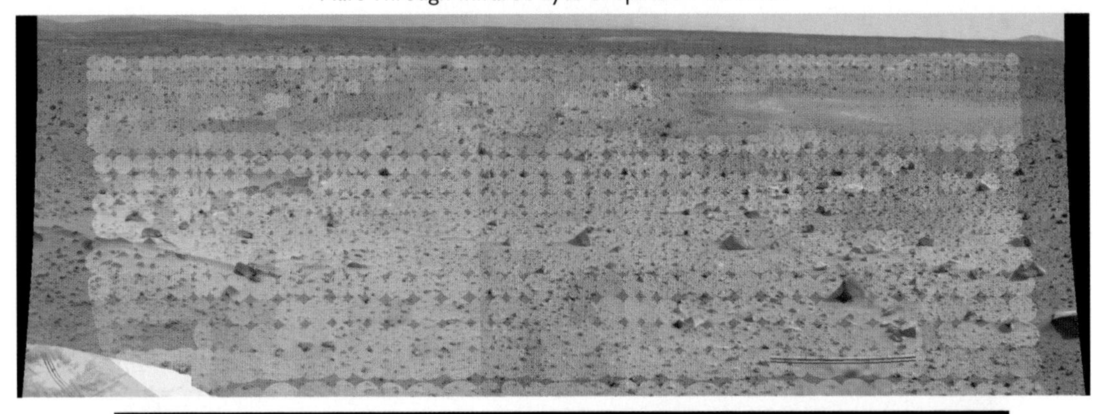

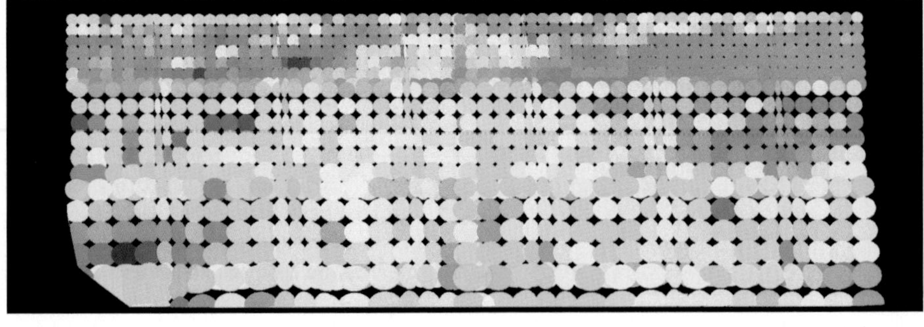

America On Board

Nestled atop an electronics module on a lander side petal of the Mars Exploration Rover Spirit, a miniature American flag accompanies a patch bearing the names and signatures of U.S. officials including President George W. Bush and NASA Administrator Sean O'Keefe. This image was taken on Mars by the rover's panoramic camera.

Image credit: NASA/JPL/Cornell University

Spirit Rises at JPL

This image taken at JPL shows engineers testing the rover stand-up motions. The rover is in an elevated pose, preparing to fold down its wheels, descend and stand up.

Image credit: NASA/JPL/Arizona State University/Cornell University

NASA's Spirit Stages Martian Stand-up Performance

January 10, 2004

NASA's Mars Exploration Rover Spirit has successfully completed its stand-up activities by extending the rear wheels. This puts the rover into a fully opened configuration for the first time since pre-launch testing in Florida last spring.

Meanwhile, the rover is sending home sections of a 360-degree color panorama it has taken and stored onboard, plus other information about the terrain around its landing site, Columbia Memorial Station in Mars' Gusev Crater.

Mission managers at NASA's Jet Propulsion Laboratory, Pasadena, Calif., have decided that changing the tilt of the lander platform will not be necessary before the rover drives off, possibly allowing drive-off to occur late Tuesday night or early Wednesday, Pacific Standard Time.

JPL's Chris Voorhees, who led the engineering team that planned the unfolding sequences for Spirit and its sister rover, Opportunity, said "Spirit has spent most of the last seven months scrunched up inside of a tetrahedral-shaped lander, and that is not the shape a rover wants to be. Over the last several days, Spirit has performed a sort of reverse robotic origami."

"The rover now stands at its full height and all six wheels are in position for driving on the surface of Mars," said Jennifer Trosper, mission manager at JPL.

The rover is still attached to the lander. The next step planned for Saturday evening (Pacific Standard Time) is to command the rover to release connections between the middle wheels and the lander. Under best-case conditions, severing the final cable connection is planned for Sunday night, followed by clockwise turns totaling 120 degrees on Monday night into Tuesday, then drive-off toward the northwest on the following martian day.

Pictures from Spirit's panoramic camera continue to provide details about the martian ground and sky. The rover transmitted home about 180 megabits of science data in the past martian day, nearly 10 times the maximum daily capability of Mars Pathfinder in 1997.

JPL geologist Dr. Matt Golombek, co-chair of the steering committee that evaluated potential landing sites for Spirit and Opportunity, said the pictures are confirming some predictions about the Gusev site. Rocks cover less of the ground than at the three previous Mars landing sites — about three percent of ground area around Spirit compared with about 20 percent of the ground around each of Mars Pathfinder, Viking 1 and Viking 2.

Presenting the latest high-resolution color mosaic from Spirit, Golombek said, "This is without question the smoothest, flattest place we've ever landed on Mars, with the possible exception of Viking 2."

Dr. Mark Lemmon a member of the rover science team from Texas A & M University, College Station, said the atmosphere at Spirit's site is dustier than at previous landing sites, except during dust storms observed by the Viking landers. The dust colors the sky and affects the appearance of objects on the ground.

Higher above the ground, atmospheric densities predicted for Spirit's descent closely matched the true conditions measured from the spacecraft's deceleration, said JPL's Dr. Joy Crisp. That is a good sign for Opportunity's descent two weeks from now, though risks remain high for any landing on Mars.

Spirit arrived at Mars Jan. 3 (EST and PST; Jan. 4 Universal Time) after a seven-month journey. Its task is to spend the next three months exploring for clues in rocks and soil about whether the past environment in Gusev Crater was ever watery and suitable to sustain life.

Spirit's twin Mars Exploration Rover, Opportunity, will reach its landing site on the opposite side of Mars on Jan. 25 (EST and Universal Time; 9:05 p.m., Jan. 24, PST) to begin a similar examination of a site on the opposite side of the planet from Gusev Crater. As of Sunday morning, Opportunity will have flown 428 million kilometers (266 million miles) since launch and will still have 28 million kilometers (17 million miles) to go before landing.

JPL, a division of the California Institute of Technology in Pasadena, manages the Mars Exploration Rover project for NASA's Office of Space Science, Washington, D.C. Images and additional information about the project are available from JPL at http://marsrovers.jpl.nasa.gov and from Cornell University, Ithaca, N.Y., at http://athena.cornell.edu .

Guy Webster (818) 354-6278 Jet Propulsion Laboratory, Pasadena, Calif.
Donald Savage (202) 358-1547 NASA Headquarters, Washington, D.C.
NEWS RELEASE: 2004-xxx

Martian Landscape in 3-D

This 3-D stereo image taken by the Mars Exploration Rover Spirit's navigation camera shows the rover's lander and, in the background, the surrounding martian terrain.

Image credit: NASA/JPL

New Real Estate on Mars

This image mosaic taken by the Mars Exploration Rover Spirit's panoramic camera shows a new slice of martian real estate southwest of the rover's landing site. The landscape shows little variation in local topography, though a narrow peak only seven to eight kilometers away is visible on the horizon. A circular depression, similar to the one dubbed Sleepy Hollow, can be seen in the foreground. Compared to the Viking and Pathfinder landing sites (PIA02405, PIA00563, PIA00393, PIA00568), the terrain at Gusev Crater, Spirit's landing site, is flat and speckled with a sparse array of rocks.

Image credit: NASA/JPL/Cornell University

Hazy Martian Skies

This image mosaic taken by the Mars Exploration Rover Spirit's panoramic camera shows the hills southeast of Spirit's landing site. Like a smoggy day in Los Angeles, dusty martian skies limit how much detail can be seen. This lack in visibility is demonstrated by comparing hills on the left to those on the right, located nearly two times farther away. The left panel of this image was captured in the late morning martian hours, looking toward the Sun. The right image was taken in the early afternoon, when the Sun was higher and the skies appeared darker.

Image credit: NASA/JPL/Cornell University

Spirit Stretches Out (Animation sequence)

This animation flips back and forth between images taken before and after deployment of the Mars Exploration Rover Spirit's bogie, a part of the rover's suspension system that extends the wheel base. These images were taken by Spirit's rear hazard avoidance camera.

Image credit: NASA/JPL

Spirit Rises to the Occasion (Animation sequence)

This animation strings together images from the rover's front hazard avoidance camera taken during the stand-up process of the Mars Exploration Rover Spirit. The first frame shows the rover's wheels tucked under in pre-stand-up position. The following frames show the stages of the stand-up process. The rover first elevates itself and unfolds the wheels. It then lowers, lifts and lowers again into its final position. Note the changing camera perspectives of the martian landscape, indicating the rover's heightened and lowered positions.

Image credit: NASA/JPL

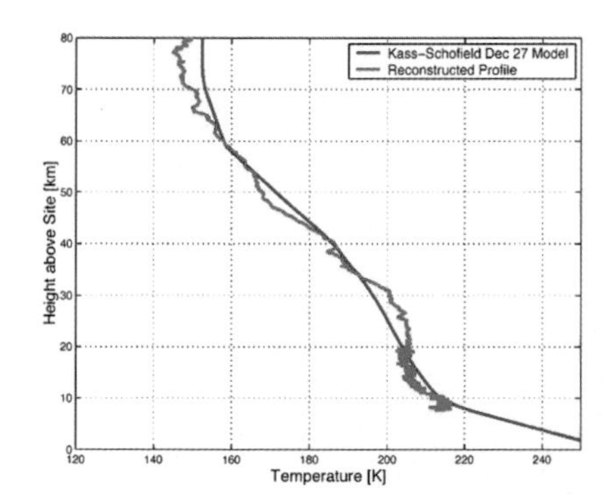

Lessons from Spirit's Landing

This graph illustrates that models used to predict the atmospheric entry details of the Mars Exploration Rover Spirit were right on track. The red curve shows the estimated change in temperature as Spirit descended through the martian atmosphere, from 80 kilometers (50 miles) to the altitude just above parachute deployment. The estimated profile was reconstructed from accelerometer and gyro readings taken by the spacecraft during its descent. This data roughly matches that of the blue curve, which represents the temperature profile predicted before landing. The predicted profile was generated from observations made at Gusev Crater on December 27, 2003, by Mars Global Surveyor Thermal Emission Spectrometer.

Image credit: NASA/JPL/George Washington University/Langley/ASU

Overhead View

This mosaic image taken by the navigation camera on the Mars Exploration Rover Spirit has been reprocessed to project a clear overhead view of the rover on the surface of Mars.

Image credit: NASA/JPL

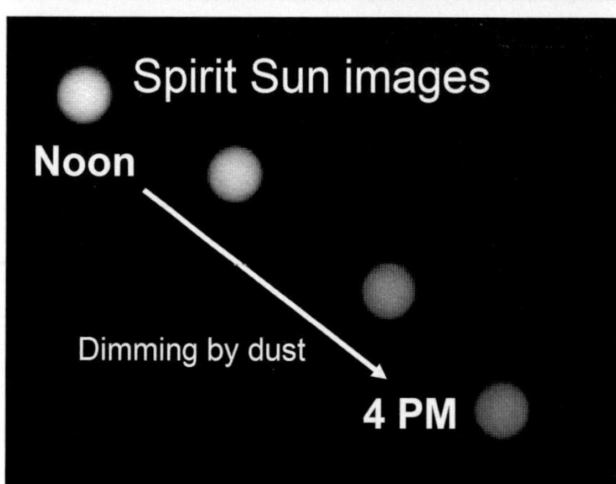

Martian Sunsets More Than Just Pretty

This image shows the Sun as it appears on Mars throughout the day. Scientists monitor the dimming of the setting Sun to assess how much dust is in the martian atmosphere. The pictures were taken by the Mars Exploration Rover Spirit's panoramic camera.

Image credit: NASA/JPL/Cornell University

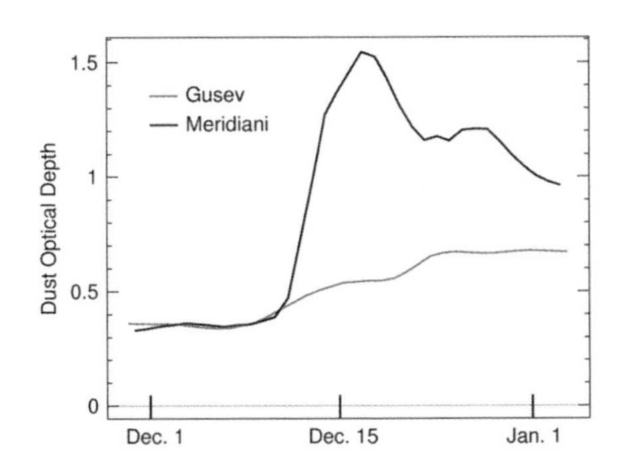

Dust in the Wind

This plot shows the estimated change in dust levels from December 2003 to early January 2004 at Gusev Crater (red curve) and Meridiani Planum (black curve), the two Mars Exploration Rover landings sites. The measurements, retrieved from Mars Global Surveyor Thermal Emission Spectrometer, indicate that a large regional dust storm beginning in mid-December raised significant dust near Meridiani. Smaller amounts of dust were spread globally by winds, the effects of which were seen at Gusev Crater. For comparison, a dust optical depth value of 1.0 would correspond to a very smoggy day in Los Angeles or Houston, and a value of 0.1 to a relatively clear day in Los Angeles.

Image credit: NASA/Goddard/Arizona State University

Spirit color calibration target

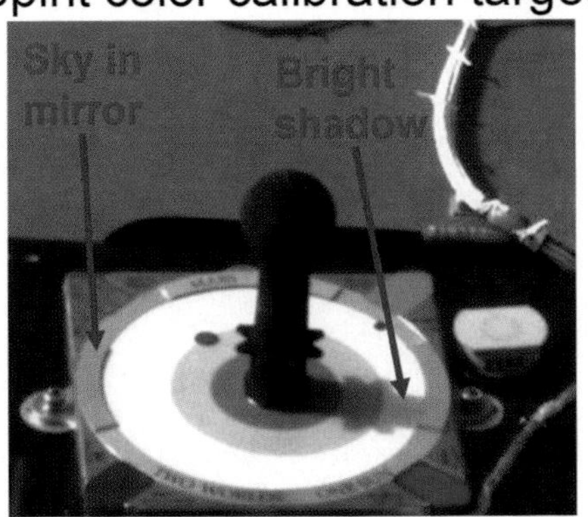

True Colors of Mars

This image taken on Mars by the panoramic camera on the Mars Exploration Rover Spirit shows the rover's color calibration target, also known as the MarsDial. The target's mirror and the shadows cast on it by the Sun help scientists determine the degree to which dusty martian skies alter the panoramic camera's perception of color. By adjusting for this effect, Mars can be seen in all its true colors.

Image credit: NASA/JPL/Cornell University

Pathfinder dust devils

Color image Enhanced image

Dust devils

Pathfinder Spies Dust Devils

This set of images from NASA's 1997 Pathfinder mission highlight the dust devils that gust across the surface of Mars. The left image shows the dusty martian sky as our eye would see it. The right image has been enhanced to expose the dust devils that lurk in the hazy sky.

Image credit: NASA/JPL/University of Arizona

Images from press conference used to compare landing sites of Viking and Pathfinder

Spirit Rover Nearly Ready to Roll

January 11, 2004

NASA's Spirit rover now has its arm and all six of its wheels free, and only a single cable must be cut before it can turn and roll off its lander onto the soil of Mars. As that milestone is completed, scientists are taking opportunities to take extra pictures and other data.

During the past 24 hours — the rover's 8th martian day on the planet, or "sol 8" — pyro devices were fired slicing cables to free the rover's middle wheels and releasing pins that held in place its instrumented arm. The arm was then locked onto a hook where it will be stowed when the rover is driving.

Because one airbag remains adjacent to the lander's forward ramp, the rover will turn about 120 degrees to its right and exit the lander from the side facing west-northwest on the planet — also the direction of an intriguing depression that scientists have dubbed Sleepy Hollow.

Current plans call for the rover to complete that turn in three steps, said Arthur Amador, one of the mission managers at NASA's Jet Propulsion Laboratory, Pasadena, Calif. As currently envisioned, during the coming martian day engineers will complete ground tests and execute dress rehearsals of the drive-off, or "egress."

On sol 10 — the night of Monday-Tuesday, Jan. 12-13, California time — engineers expect to sever the umbilical cord that connects the rover to its lander by firing a pyro device, the last of 126 pyro firings since Spirit separated from its cruise stage shortly before landing on Jan. 4 (Jan. 3 in U. S. time zones). Also on that day, the rover will execute the first of three parts of its turn when it moves clockwise (as viewed from above) about 45 degrees.

After taking and analyzing pictures to verify the first part of the turn, engineers anticipate completing it on sol 11 (night of Tuesday-Wednesday, Jan. 13-14). First, the rover will turn an additional 50 degrees and stop to take pictures. Then, if all is well, it will turn a final 20 to 25 degrees to position it precisely in front of one of its three exit ramps.

If no issues crop up as those steps are completed, the rover could drive off onto the martian soil no earlier than sol 12 (night of Wednesday-Thursday, Jan. 14-15). "But we adjust our schedule every day, based on flight events, so this remains an estimate," said Amador.

The rover's status overall is "pretty darn perfect," said Amador. He described the communication link from Mars to Earth as excellent, allowing the team to receive 170 megabits of data during the past day. All science data stored on the rover has been sent to Earth. The rover is generating 900 watt-hours of power per day and using 750 watt-hours, and its thermal condition is good, he added.

While engineers are completing and testing commands to execute the rover's turn and egress, the science team is enjoying an "unexpected dividend" of time to collect data, said Dr. John Callas, Mars Exploration Rover science manager at JPL.

Until now, all science observations have been planned far in advance, but the unfolding schedule of rover activities gave the team the opportunity to do their first on-the-fly planning for observations driven by previous results, Callas explained. In doing so they segued to a working style that they will practice on a day to day basis as the rover rolls across the surface of its landing site in Gusev Crater, named the Columbia Memorial Station.

In the next 24 hours, the team will collect 270 megabits of science data, considerably more than on any previous martian day. This will include a high-quality, 14-color mosaic taken by the panoramic camera of a third of the horizon toward Sleepy Hollow, the direction in which the rover will leave its lander.

In addition, they plan to complete two remaining "octants" (each a pie slice showing an eighth of the horizon) with the rover's miniature thermal emission spectrometer. These areas will also be rephotographed with the rover's panoramic camera in order to allow the camera and spectrometer data to be co-registered. Plans also call for the spectrometer to "stare" at three selected sites to collect very low-noise data, as well as calibration of another science instrument, the alpha particle X-ray spectrometer.

Spirit's twin Mars Exploration Rover, Opportunity, will reach Mars on Jan. 25 (Universal Time and EST; Jan. 24 PST). The rovers' main task is to spend three months exploring for clues in rocks and soil about whether the landing sites may have had abundant water for long enough in the past for life to appear. Pictures and detailed information from the mission is available at the project's Web site: http://marsrovers.jpl.nasa.gov .

JPL, a division of the California Institute of Technology, manages the Mars Exploration Rover project for NASA's Office of Space Science, Washington.

Franklin O'Donnell (818) 354-5011 Jet Propulsion Laboratory, Pasadena, Calif.
Donald Savage (202) 358-1547 NASA Headquarters, Washington, D.C.
NEWS RELEASE: 2004-15

In and Out (Animation sequence)

This animation links two images taken by the front hazard avoidance camera on the Mars Exploration Rover Spirit. The rover is stowing and unstowing its robotic arm, or instrument deployment device. The device is designed to hold and maneuver the various instruments on board that will help scientists get up-close and personal with martian rocks and soil. Image credit: NASA/JPL

"Bird's Eye" View of Egress

This mosaic image taken by the navigation camera on the Mars Exploration Rover Spirit represents an overhead view of the rover as it prepares to roll from its current position on the lander down to the martian surface.

Image credit: NASA/JPL

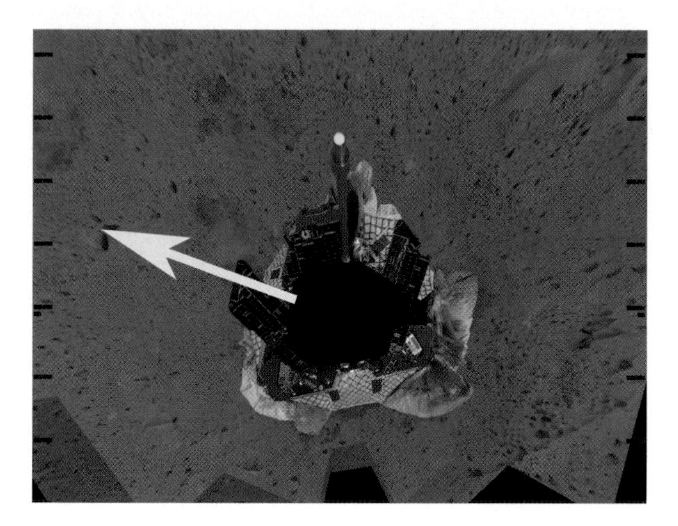

Plan B

This mosaic image taken by the navigation camera on the Mars Exploration Rover Spirit represents an overhead view of the rover as it prepares to roll off the lander and onto the martian surface. The yellow arrow illustrates the direction the rover may take to roll safely off the lander. The rover was originally positioned to roll straight forward off the lander (south side of image). However, an airbag is blocking its path. To take this northwestern route, the rover must back up and perform what is likened to a 3-point turn in a cramped parking lot.

Image credit: NASA/JPL

Roll-Off Test at JPL

This still image illustrates what the Mars Exploration Rover Spirit will look like as it rolls off the northeastern side of the lander on Mars. The image was taken from footage of rover testing at JPL's In-Situ Instruments Laboratory, or "Testbed."

Image credit: NASA/JPL

Spirit's Surroundings Beckon in Color Panorama

January 12, 2004

The first 360-degree color view from NASA's Spirit Mars Exploration Rover presents a range of tempting targets from nearby rocks to hills on the horizon.

"The whole panorama is there before us," said rover science-team member Dr. Michael Malin of Malin Space Science Systems, San Diego. "It's a great opening to the next stage of our mission."

Spirit's flight team at NASA's Jet Propulsion Laboratory, Pasadena, Calif., continues making progress toward getting the rover off its lander platform, but expected no sooner than early Thursday morning. "We're about to kick the baby bird out of its nest," said JPL's Kevin Burke, lead mechanical engineer for the rover's egress off the lander.

The color panorama is a mosaic stitched from 225 frames taken by Spirit's panoramic camera. It spans 75 frames across, three frames tall, with color information from shots through three different filters. The images were calibrated at Cornell University, Ithaca, N.Y., home institution for Dr. Jim Bell, panoramic camera team leader.

Malin said, "Seeing the panorama totally assembled instead of in individual pieces gives a much greater appreciation for the position of things and helps in developing a sense of direction. I find it easier to visualize where I am on Mars when I can look at different directions in one view. For a field geologist, it's exactly the kind of thing you want to look at to understand where you are."

Another new image product from Spirit shows a patch of intriguing soil near the lander in greater detail than an earlier view of the same area. Scientists have dubbed the patch "Magic Carpet" for how some soil behaved when scraped by a retracting airbag.

"It has been detached and folded like a piece of carpet sliding across the floor," said science-team member Dr. John Grotzinger of Massachusetts Institute of Technology, Cambridge.

Spirit's next step in preparing to drive onto the surface of Mars is to sever its final connection with the lander platform by firing a cable cutter, which Burke described as "an explosive guillotine." The planned sequence after that is a turn in place of 115 degrees clockwise, completed in three steps over the next two days. If no obstacles are seen from images taken partway through that turn, drive-off is planned toward the northwestern compass point of 286 degrees.

Spirit landed on Mars Jan. 3 after a seven-month journey. Its task is to spend the next three months exploring rocks and soil for clues about whether the past environment in Gusev Crater was ever watery and suitable to sustain life. Spirit's twin Mars Exploration Rover, Opportunity, will reach Mars Jan. 24 PST (Jan. 25 Univeral Time and EST) to begin a similar examination of a site on a broad plain called Meridiani Planum, on the opposite side of the planet from Gusev Crater.

NASA JPL, a division of the California Institute of Technology, Pasadena, manages the Mars Exploration Rover project for NASA's Office of Space Science, Washington. For information about NASA and the Mars mission on the Internet, visit: http://www.nasa.gov. Additional information about the project is available on the Internet at: http://marsrovers.jpl.nasa.gov. Mission information is also available from Cornell University, at: http://athena.cornell.edu.

Guy Webster (818) 354-5011 Jet Propulsion Laboratory, Pasadena, Calif.
Donald Savage (202) 358-1547 NASA Headquarters, Washington
NEWS RELEASE: 2004-016

Mars in Full View (See fold-out at rear of book)

This is a medium-resolution version of the first 360-degree panoramic view of the martian surface, taken on Mars by the Mars Exploration Rover Spirit's panoramic camera. Part of the spacecraft can be seen in the lower corner regions.

Image credit: NASA/JPL/Cornell

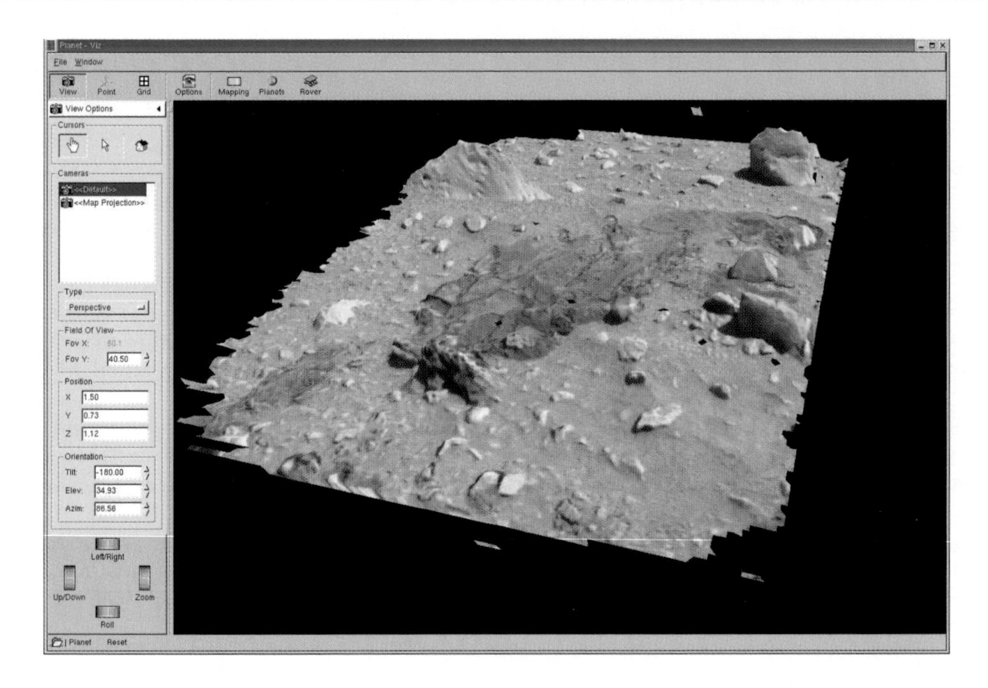

Landing Trail in 3-D

A three-dimensional color model created using data from the Mars Exploration Rover's panoramic camera shows images of airbag drag marks on the martian surface. The triangular rock in the upper left corner is approximately 20 centimeters (8 inches) tall. The meatball-shaped rock in the upper right corner is approximately 10 centimeters (4 inches) tall. The dark portion of the surface, or "trough" is approximately 1 centimeter (0.4 inches) deep at its deepest point. This model is displayed using software developed by NASA's Ames Research Center.

Image credit: NASA/JPL/Cornell

Airbag Trail dubbed "Magic Carpet"

This section of the first color image from the Mars Exploration Rover Spirit has been further processed to produce a sharper look at a trail left by the one of rover's airbags. The drag mark was made after the rover landed and its airbags were deflated and retracted. Scientists have dubbed the region the "Magic Carpet" after a crumpled portion of the soil that appears to have been peeled away (lower left side of the drag mark). Rocks were also dragged by the airbags, leaving impressions and "bow waves" in the soil. The mission team plans to drive the rover over to this site to look for additional clues about the composition of the martian soil. This image was taken by Spirit's panoramic camera.

Image credit: NASA/JPL/Cornell

Airbag Trail Dubbed "Magic Carpet" (Zoom)

This section of the first color image from the Mars Exploration Rover Spirit has been further processed to produce a sharper look at a trail left by the one of rover's airbags. The drag mark was made after the rover landed and its airbags were deflated and retracted. Scientists have dubbed the region the "Magic Carpet" after a crumpled portion of the soil that appears to have been peeled away (lower left side of the drag mark). Rocks were also dragged by the airbags, leaving impressions and "bow waves" in the soil. The mission team plans to drive the rover over to this site to look for additional clues about the composition of the martian soil. This image was taken by Spirit's panoramic camera.

Image credit: NASA/JPL/Cornell

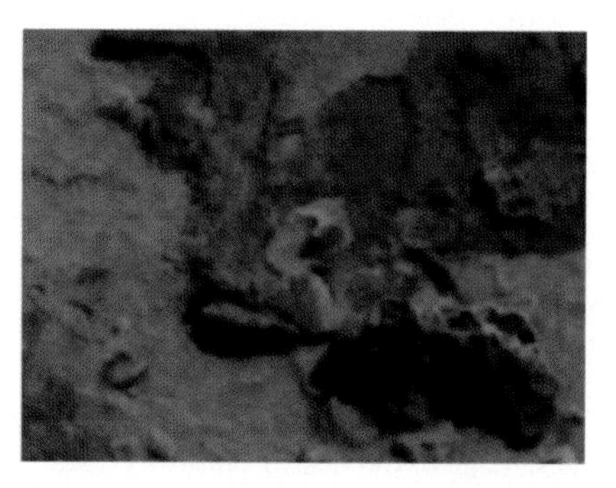

Airbag Trail Dubbed "Magic Carpet" (Zoom2)

This extreme close-up image highlights the martian feature that scientists have named "Magic Carpet" because of its resemblance to a crumpled carpet fold. Scientists think the soil here may have detached from its underlying layer, possibly due to interaction with the Mars Exploration Rover Spirit's airbag after landing. This image was taken on Mars by the rover's panoramic camera.

Image credit: NASA/JPL/Cornell

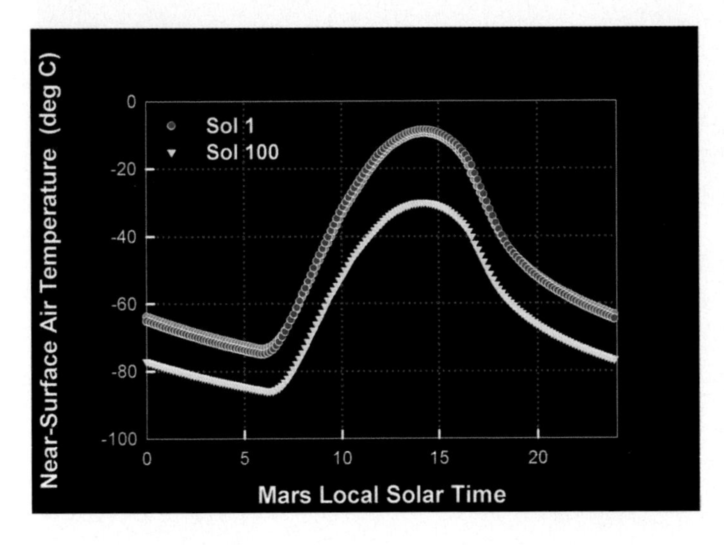

How Warm is Mars?

This graph shows the predicted daily change in the atmospheric temperature one meter above the surface of Mars at Gusev Crater, the Mars Exploration Rover Spirit's landing site. The blue curve denotes predicted values for sol 1 (the first day of Spirit's mission) and the yellow for sol 100 (100 days into the mission). The light blue symbols represent temperatures for a total atmospheric dust abundance of 0.7 visible optical depth units, and the darker blue symbols for a total atmospheric dust abundance of 1.0 visible optical depth units. Scientists use this data to ensure that Spirit stays within the right temperature range.

Image credit: NASA/JPL/ARC/New Mexico State University

Go To That Crater And Turn Right: Spirit Gets A Travel Itinerary

January 13, 2004

NASA's Spirit has begun pivoting atop its lander platform on Mars, and the robot's human partners have announced plans to send it toward a crater, then toward some hills, during the mission.

Determining exactly where the spacecraft landed, in the context of images taken from orbit, has given planners a useful map of the vicinity. After Spirit drives off its lander and examines nearby soil and rocks, the scientists and engineers managing it from NASA's Jet Propulsion Laboratory, Pasadena, Calif., intend to tell it to head for a crater that is about 250 meters (about 270 yards) northeast of the lander.

"We'll be careful as we approach. No one has ever driven up to a martian crater before," said Dr. Steve Squyres of Cornell University, Ithaca, N.Y., principal investigator for the science instruments on Spirit and on its twin Mars Exploration Rover, Opportunity.

The impact that dug the crater about 200 meters (about 220 yards)wide probably flung rocks from as deep as 20 to 30 meters (22 to 33 yards) onto the surrounding surface, where Spirit may find them and examine them. "It will provide a window into the subsurface of Mars," Squyres said.

Craters come in all sizes. The main scientific goal for Spirit is to determine whether the Connecticut-sized Gusev Crater ever contained a lake. Taking advantage of the nearby unnamed crater for access to buried deposits will add to what Spirit can learn from surface materials near the lander. After that, if all goes well, the rover will head toward a range of hills about 3 kilometers (2 miles) away for a look at rocks that sit higher than the landing neighborhood's surface. That distance is about five times as far as NASA's mission-success criteria for how far either rover would drive. The highest hills in the group rise about 100 meters (110 yards) above the plain.

"I cannot tell you we're going to reach those hills," Squyres said. "We're going to go toward them." Getting closer would improve the detail resolved by Spirit's panoramic camera and by the infrared instrument used for identifying minerals from a distance.

First, though, comes drive-off. Overnight Monday to Tuesday, Spirit began rolling. It backed up 25 centimeters (10 inches), turned its wheels and pivoted 45 degrees.

"The engineering team is just elated that we're driving," said JPL's Chris Lewicki, flight director. "We've cut loose our ties and we're ready to rove." After two more pivots, for a total clockwise turn of 115 degrees, Spirit will be ready for driving onto the martian surface very early Thursday morning, according to latest plans.

Engineers and scientists have determined where on the martian surface the lander came to rest. NASA's Mars Odyssey orbiter was used in a technique similar to satellite-based global positioning systems on Earth to estimate the location of the landing site, said JPL's Joe Guinn of the rover mission's navigation team. Other researchers correlated features seen on the horizon in Spirit's panoramic views with hills and craters identifiable in images taken by Mars Global Surveyor and Odyssey. "We've got a tremendous vista here with all kinds of features on the horizon," said JPL's Dr.Tim Parker, landing site-mapping geologist.

The spacecraft came to rest only about 250 to 300 meters (270 to 330 yards) southeast of its first impact. Transverse rockets successful slowed horizontal motion seconds before impact, said JPL's Rob Manning, who headed development of the entry, descent and landing system.The spacecraft, encased in airbags, was just 8.5 meters (27.9 feet) off the ground when its bridle was cut for the finalfreefall to the surface. It first bounced about 8.4 meters (27.6 feet) high, then bounced 27 more times before stopping.

Analysis of Spirit's landing may aid in minor adjustments for Opportunity, on track for landing on the opposite side of Mars on Jan. 25 (Universal Time and EST; 9:05 p.m. Jan. 24, PST).

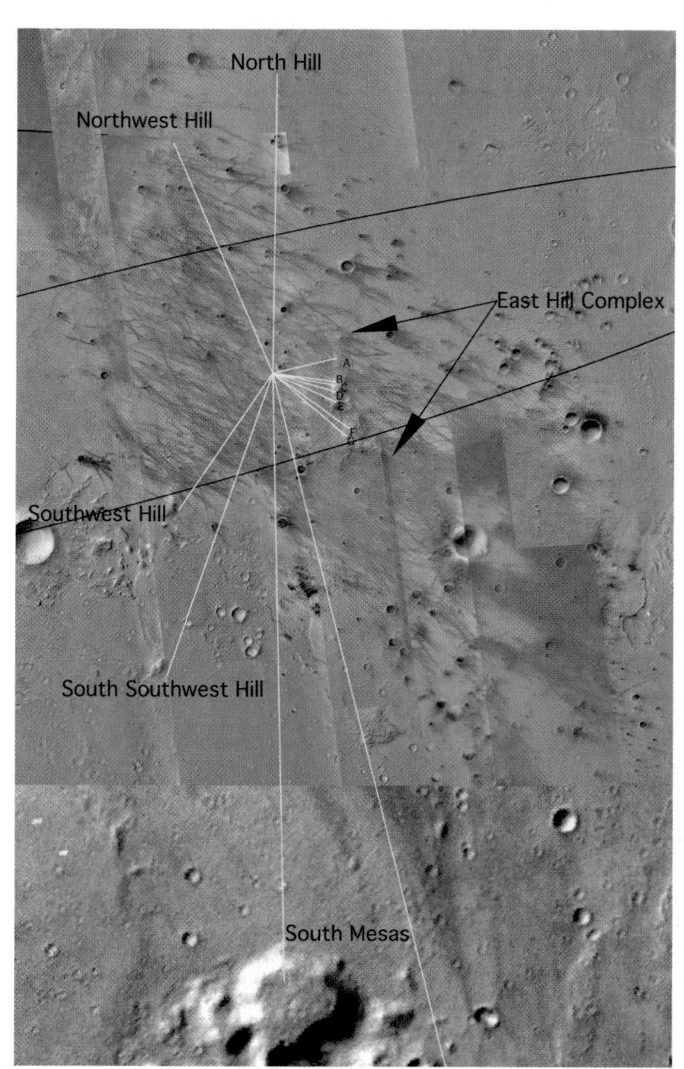

JPL, a division of the California Institute of Technology in Pasadena, manages the Mars Exploration Rover project for NASA's Office of Space Science, Washington. For more information about NASA and the Mars mission on the Internet, visit http://www.nasa.gov. Additional information about the rover project is available from NASA's JPL at http://marsrovers.jpl.nasa.gov and from Cornell University at http://athena.cornell.edu.

Guy Webster (818) 354-5011 Jet Propulsion Laboratory, Pasadena, Calif.
Donald Savage (202) 358-1547 NASA Headquarters, Washington, D.C.
NEWS RELEASE: 2004-018

Map of Hills on the Horizon

This overhead view maps the Mars Exploration Rover Spirit's approximate location in relation to nearby craters and hills. By combining images from both the camera on Mars Global Surveyor and the descent image motion estimation system camera located on the bottom of the rover's lander, scientists and engineers can tell how far away the hills are from the rover. This information would be more difficult to obtain from the panoramic images.The hills and hill ranges are marked by yellow lines, and the rover is located where the yellow lines intersect. Black arrows locate the east hill complex, a potential rover destination.

Image credit: NASA/JPL/MSSS

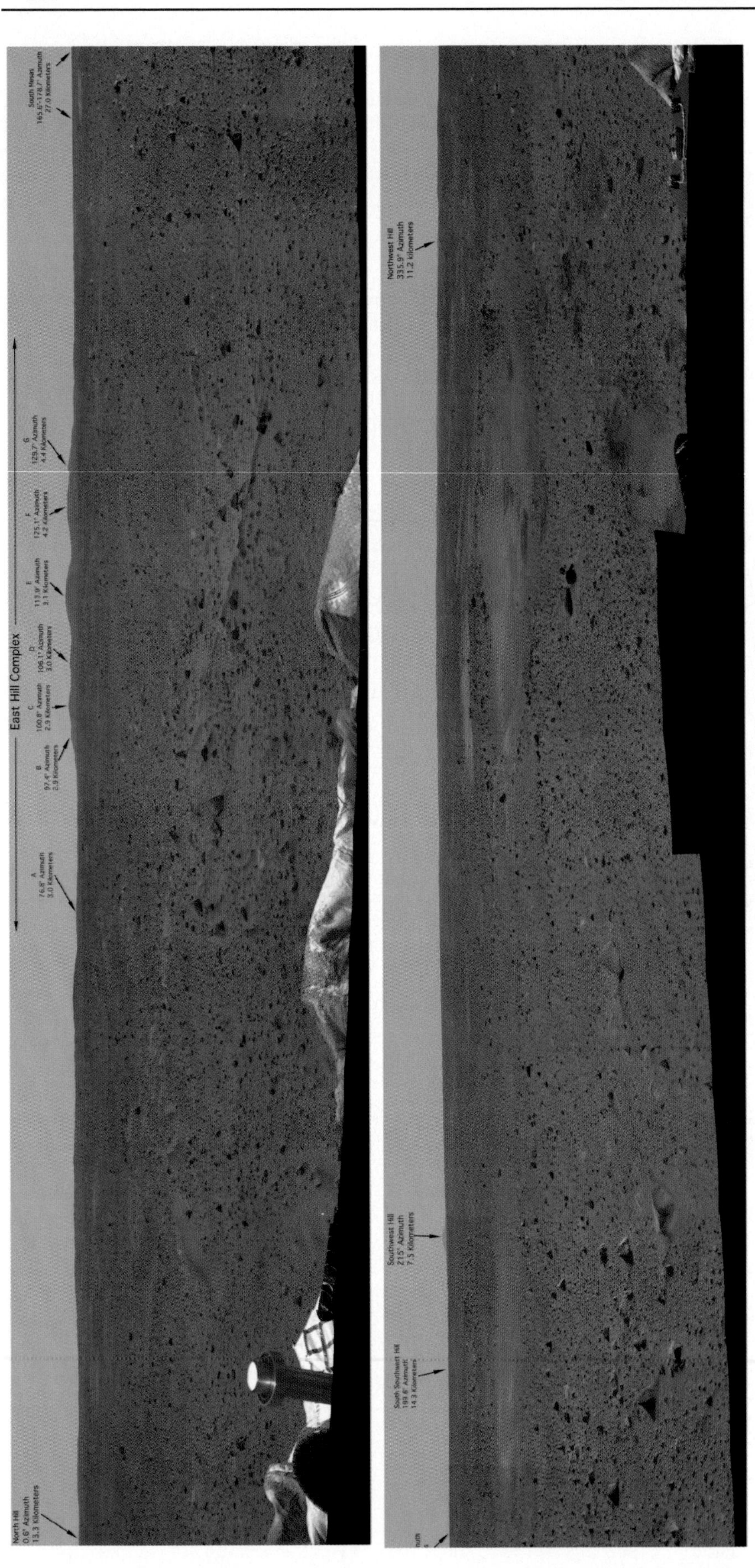

Hills Over Yonder

The arrows in this 360-degree panoramic view (left) of the martian surface identify hills and craters on the martian horizon that scientists can easily find with orbiters Mars Global Surveyor and Mars Odyssey. The image was taken on Mars by the panoramic camera on the Mars Exploration Rover Spirit.

Image credit:
NASA/JPL/MSSS

In the Far East

In the distance stand the east hills, which are closest to the Mars Exploration Rover Spirit in comparison to other hill ranges seen on the martian horizon. The top of the east hills are approximately 2 to 3 kilometers (1 to 2 miles) away from the rover's approximate location. This image was taken on Mars by the rover's panoramic camera.

Image credit: NASA/JPL/Cornell

Virtual Rover Takes its First Turn

This image shows a screenshot from the software used by engineers to drive the Mars Exploration Rover Spirit. The software simulates the rover's movements across the martian terrain, helping to plot a safe course for the rover. The virtual 3-D world around the rover is built from images taken by Spirit's stereo navigation cameras. Regions for which the rover has not yet acquired 3-D data are represented in beige. This image depicts the state of the rover before it backed up and turned 45 degrees on Sol 11 (01-13-04).

Image credit: NASA/JPL

Belly Dancing on Mars

This image shows a screenshot from the software used by engineers to drive the Mars Exploration Rover Spirit. The software simulates the rover's movements across the martian terrain, helping to plot a safe course for the rover. The virtual 3-D world around the rover is built from images taken by Spirit's stereo navigation cameras. Regions for which the rover has not yet acquired 3-D data are represented in beige. The red dart to the left shows a target destination for the rover. Red lines indicate the path the rover's wheels will follow to reach the target, and the blue line denotes the path of the rover's "belly button," as engineers like to call it.

Image credit: NASA/JPL

Cutting the Cord (Animation sequence)

This animation shows the view from the front hazard avoidance cameras on the Mars Exploration Rover Spirit as the rover turns 45 degrees clockwise. This maneuver is the first step in a 3-point turn that will rotate the rover 115 degrees to face west. The rover must make this turn before rolling off the lander because airbags are blocking it from exiting off the front lander petal. Before this crucial turn could take place, engineers instructed the rover to cut the final cord linking it to the lander. The turn took around 30 minutes to complete.

Image credit: NASA/JPL

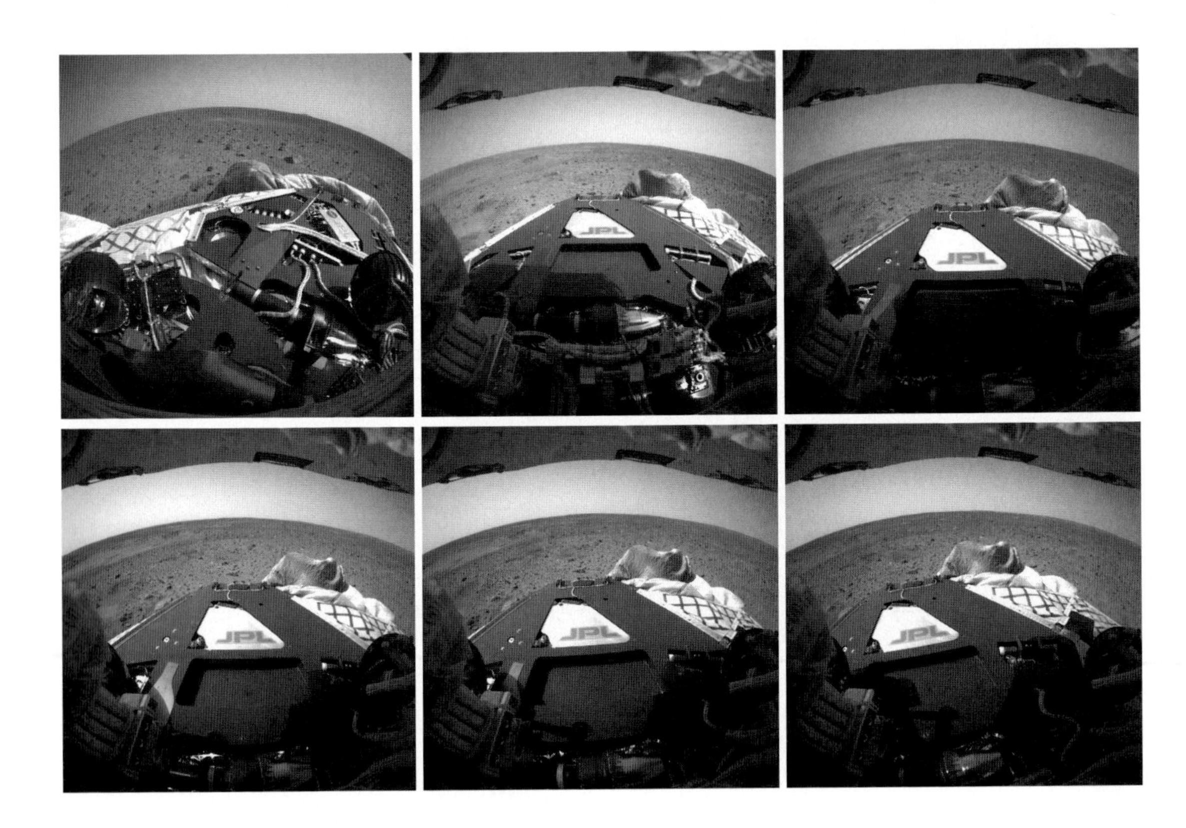

Cutting the Cord-2 (animation sequence)

This animation shows the view from the rear hazard avoidance cameras on the Mars Exploration Rover Spirit as the rover turns 45 degrees clockwise. This maneuver is the first step in a 3-point turn that will rotate the rover 115 degrees to face west. The rover must make this turn before rolling off the lander because airbags are blocking it from exiting from the front lander pedal. Before this crucial turn took place, engineers instructed the rover to cut the final cord linking it to the lander. The turn took around 30 minutes to complete.

Image credit: NASA/JPL

Rover Pre-Turn

This image shows the view from the front hazard avoidance cameras on the Mars Exploration Rover Spirit before the rover begins a crucial 3-point turn to face in a west direction and roll off the lander.
Image credit: NASA/JPL

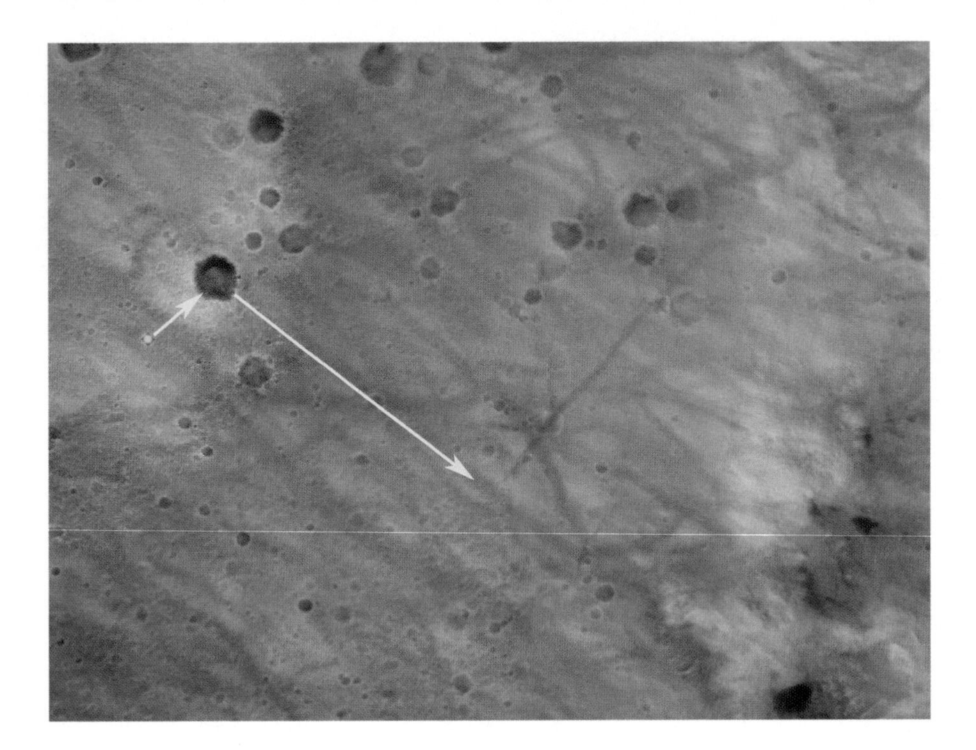

Spirit's Travel Plan

This zoomed-in overhead view of the Mars Exploration Rover Spirit's estimated landing site and surrounding area shows the rover's potential "itinerary." Scientists and engineers plan to drive the rover approximately 250 meters (820 feet) from the green point to the rim of a nearby crater measuring 192 meters (630 feet) in diameter. They then plan to drive toward the east hills, the tops of which measure 2-3 kilometers (1-2 miles) away from the rover's estimated landing site. This image is a composite of images taken by the camera on Mars Global Surveyor and the descent image motion estimation system camera located on the bottom of the rover's lander.

Image credit: NASA/JPL/MSSS

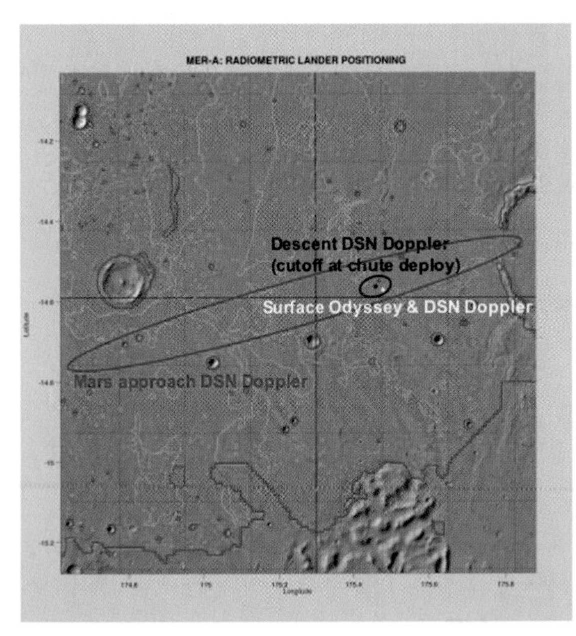

Right on Target

This map shows the estimated location of the Mars Exploration Rover Spirit within Gusev Crater, Mars. Engineers targeted Spirit for the center of the blue ellipse. Measurements taken during the rover's descent by the Deep Space Network predicted its landing site to be the spot marked with a black dot. Later measurements taken on the ground by both the Deep Space Network and the orbiter Mars Odyssey narrowed the predicted landing site to a spot marked with a white dot. When initially choosing a landing site for the rover, engineers avoided hazardous terrain outlined here in yellow and red. This map consists of data from Mars Odyssey and Mars Global Surveyor.

Image credit: NASA/JPL/Arizona State University

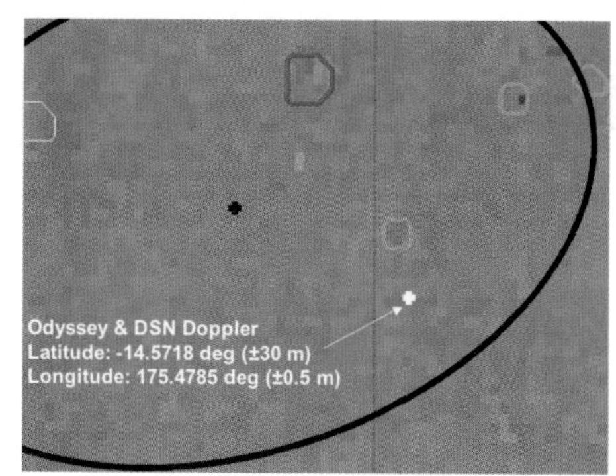

Right on Target-2

This map shows a close-up look at the estimated location of the Mars Exploration Rover Spirit within Gusev Crater, Mars. Measurements taken during the rover's descent by the Deep Space Network predicted its landing site to be the spot marked with a black dot. Later measurements taken on the ground by both the Deep Space Network and the orbiter Mars Odyssey narrowed the predicted landing site to a spot marked with a white dot. When initially choosing a landing site for the rover, engineers avoided hazardous craters outlined here in yellow and red. This map consists of data from Mars Odyssey and Mars Global Surveyor.

Image credit: NASA/JPL/Arizona State University

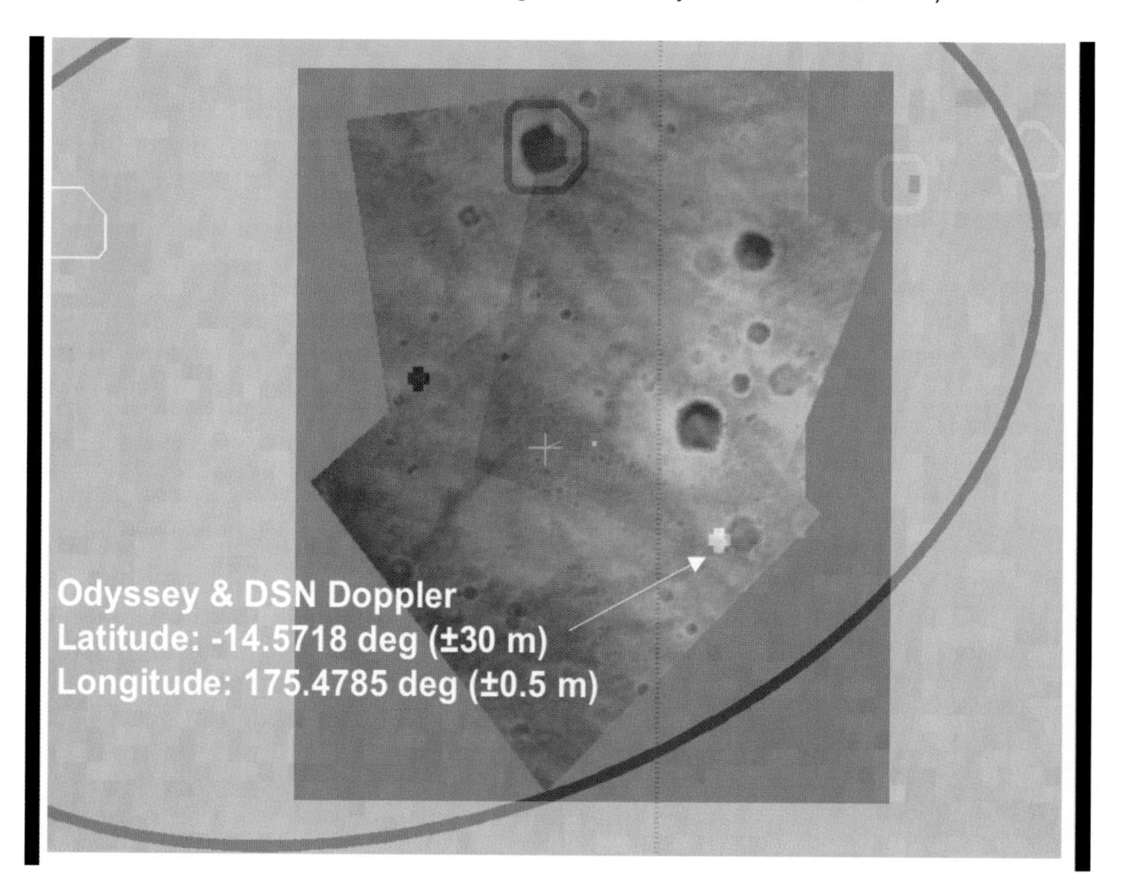

Right on Target-3

This map shows the estimated location of the Mars Exploration Rover Spirit within Gusev Crater, Mars. Measurements taken during the rover's descent by the Deep Space Network predicted its landing site to be the spot marked with a black cross. Later measurements taken on the ground by both the Deep Space Network and the orbiter Mars Odyssey narrowed the predicted landing site to a spot marked with a white cross. When initially choosing a landing site for the rover, engineers avoided hazardous craters outlined here in yellow and red. This map consists of data taken during Spirit's descent by the descent image motion estimation system located at the bottom of the rover.

Image credit: NASA/JPL/Cornell

Turning in the Testbed

This image, taken in the JPL In-Situ Instruments Laboratory or "Testbed," shows the view from the front hazard avoidance cameras on the Mars Exploration Rover Spirit after the rover has backed up and turned 45 degrees counterclockwise. Engineers rehearsed this maneuver at JPL before performing it on Mars. This maneuver is the first step in a 3-point turn that will rotate the rover 115 degrees to face the rear direction and drive off a rear side lander petal.

Image credit: NASA/JPL

Turning on Mars

This image, taken on Mars, shows the view from the front hazard avoidance cameras on the Mars Exploration Rover Spirit after the rover has backed up 25 centimeters (10 inches) and turned 45 degrees clockwise. This maneuver is the first step in a 3-point turn that will rotate the rover 115 degrees to face the rear direction and drive off a rear side lander petal. Note that the view in this image matches that of the image taken during rehearsal of this maneuver in the JPL testbed.

Image credit: NASA/JPL

Spirit Ready to Drive onto Mars Surface

January 14, 2004

NASA's Spirit completed a three-stage turn early today, the last step before a drive planned early Thursday to take the rover off its lander platform and onto martian soil for the first time.

"We are very excited about where we are today. We've just completed the exploration of our lander and we're ready to explore Mars," said Kevin Burke of NASA's Jet Propulsion Laboratory, Pasadena, Calif., leader of the engineering team that planned the rover's egress from the lander. "We are headed in a north-northwest direction. That is our exit path, and we're sitting just where we want to be."

Late tonight, mission managers at JPL plan to send the command for Spirit to drive forward 3 meters (10 feet), enough to get all six wheels onto the soil.

After the move, one of the rover's first jobs will be to locate the Sun with its panoramic camera and calculate from the Sun's position how to point its main antenna at Earth, JPL's Jennifer Trosper, mission manager, explained.

On Friday, Spirit's science team will take advantage of special possibilities presented by the European Space Agency's Mars Express orbiter flying almost directly overhead, about 300 kilometers (186 miles) high. Mars Express successfully entered orbit around Mars last month. Spirit will be looking up while Mars Express uses three instruments to look down.

"This is an historic opportunity," said Dr. Ray Arvidson of Washington University in St. Louis, deputy principal investigator for the science instruments on Spirit and on its twin Mars Exploration Rover, Opportunity. "The intent is to get observations from above and to get observations from below at the same time to do the best possible job of determining the dynamics of the atmosphere." The Mars Express observations are also expected to supplement earlier information from two NASA Mars orbiters about the surface minerals and landforms in Spirit's neighborhood within Gusev Crater.

Mars Express will be looking down with a high-resolution stereo camera, a spectrometer for identifying minerals in infrared and visible wavelengths, and another spectrometer for studying atmospheric circulation and composition. Spirit will be looking up with its panoramic camera and its infrared spectrometer.

Dr. Michael Smith of NASA's Goddard Space Flight Center, Greenbelt, Md., reported how Spirit's miniature thermal emission spectrometer can be used to assess the temperatures in Mars' atmosphere from near the planet's surface to several kilometers or miles high. Spirit'smeasurements are most sensitive for the lower portion of the atmosphere, while Mars Express' measurements will be most sensitive for the upper atmosphere, he said.

Spirit arrived at Mars Jan. 3 (EST and PST; Jan. 4 Universal Time) after a seven-month journey. In coming weeks and months, according to plans, it will be exploring for clues in rocks and soil to decipher whether the past environment in Gusev Crater was ever watery and possibly suitable to sustain life.

Opportunity will reach Mars on Jan. 25 (EST and Universal Time; 9:05 p.m., Jan. 24, PST) to begin a similar examination of a site on the opposite side of the planet from Gusev Crater. As of Thursday morning, Opportunity will have flown 438 million kilometers (272 million miles) since launch and will still have 18 million kilometers (11 million miles) to go before landing.

JPL, a division of the California Institute of Technology inPasadena, manages the Mars Exploration Rover project for NASA's Office of Space Science, Washington, D.C. Images and additional information about the project are available from JPL at http://marsrovers.jpl.nasa.gov and from Cornell University, Ithaca, N.Y., at http://athena.cornell.edu.

Guy Webster (818) 354-5011 Jet Propulsion Laboratory, Pasadena, Calif.
Donald Savage (202) 358-1547 NASA Headquarters, Washington, D.C.
NEWS RELEASE: 2004-19

This image shows the view from the Mars Exploration Rover Spirit after it successfully completed a 115 degree turn to face northwest, the direction it will roll off the lander. The image was taken by the rover's front hazard avoidance camera.

Image credit: NASA/JPL

Ready to Roll-2

This image shows the view from the Mars Exploration Rover Spirit after it successfully completed a 115 degree turn to face northwest, the direction it will roll off the lander. The image was taken by the rover's navigation camera.

Image credit: NASA/JPL

95-degree Position on Mars

This image from the hazard avoidance camera on the Mars Exploration Rover Spirit shows the rover in its near-final turned position on the lander at Gusev Crater. At this point, the rover has turned 95 degrees, with 115 degrees being its goal position. This picture looks remarkably similar to the image taken during a "dress rehearsal" at the JPL In-Situ Laboratory, or "testbed," prior to the maneuver on Mars.

Image credit: NASA/JPL

95-degree Position at JPL Testbed

This image shows a test rover in a near-final turned position on the lander in the JPL In-Situ Instruments Laboratory, or "testbed." This is where engineers tested the rover's three-point turn before completing the manuever with the Mars Exploration Rover Spirit on Mars. At this point, the test rover has turned 95 degrees, with 115 degrees being its goal position. This picture looks remarkably similar to the image taken by the rover's hazard avoidance camera while in the same position on Mars.

Image credit: NASA/JPL

Location of Spirit's Homeland

This image shows where Earth would set on the martian horizon from the perspective of the Mars Exploration Rover Spirit if it were facing northwest atop its lander at Gusev Crater. Earth cannot be seen in this image, but engineers have mapped its location. This image mosaic was taken by the hazard-identification camera onboard Spirit.

Image credit: NASA/JPL

Panoramic View of Lander During Turn

This 360-degree panoramic mosaic image composed of data from the hazard avoidance camera on the Mars Exploration Rover Spirit shows a view of the lander from under the rover deck. The images were taken as the rover turned from its landing position 95 degrees toward the northwest side of the lander.

Image credit: NASA/JPL

Roll-Off Dress Rehearsal at JPL

This image shows a test rover as it attempts a complete 115-degree turn on the lander in the JPL In-Situ Instruments Laboratory, or "testbed." This is where engineers tested the rover's three-point turn before completing the manuever with the Mars Exploration Rover Spirit at Gusev Crater on Mars. At this point, the test rover has turned 95 degrees, with 115 degrees being its goal position.

Image credit: NASA/JPL

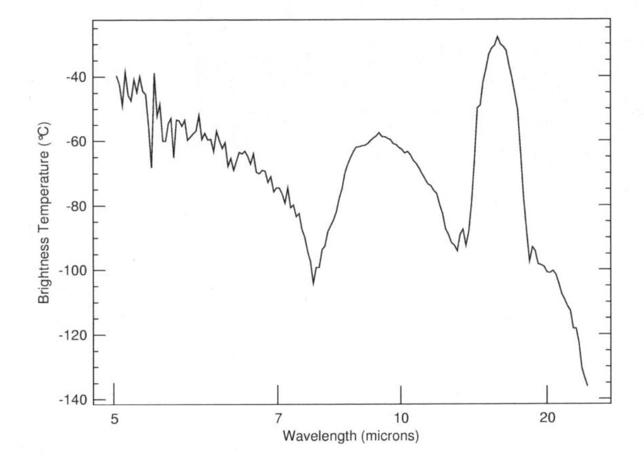

What Makes Up a Martian Sky?

A view of the sky as observed by the mini-thermal emission spectrometer onboard the Mars Exploration Rover Spirit. This instrument detects the different wavelengths of infrared light emitted by an object, in this case the sky, producing a graph called a spectrum that reveals the presence of specific chemicals. This spectrum, taken on Sol 7 in the early afternoon (night of January 9/10, 2004), contains the signatures of carbon dioxide (15 microns), atmospheric dust (9 microns) and water vapor (6 microns). Scientists also expect to see water ice clouds in the martian atmosphere, but did not observe them at the time of this observation. The thermal brightness of carbon dioxide allows the atmospheric temperature as a function of height to be determined. Carbon dioxide makes up 95 percent of the martian atmosphere.

Image credit: NASA/JPL/Arizona State University

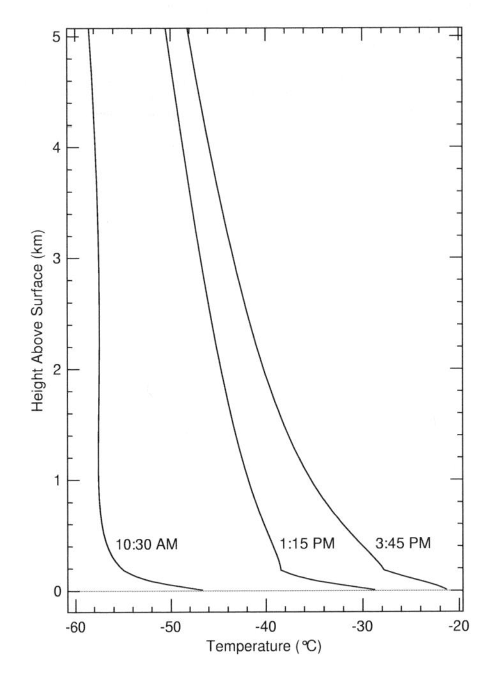

The Unpredictable Weather on Mars

This graph shows the temperature above the the surface of Mars at three different points in time: Sol 5, late afternoon; Sol 6, mid-morning; and Sol 7, early afternoon (Sol 5 occurred on the night of January 7/8, 2004). These temperature profiles were derived from data taken by the mini-thermal emission spectrometer onboard the Mars Exploration Rover Spirit. By measuring the brightness of the carbon dioxide gas that makes up the martian atmosphere, scientists can deduce the surface temperature above Mars between 20 meters (65 feet) and 2 kilometers (1.2 miles). The observations show large changes in atmospheric temperature both as a function of time of day, and as a function of height near the surface.

Image credit: NASA/JPL/Arizona State University

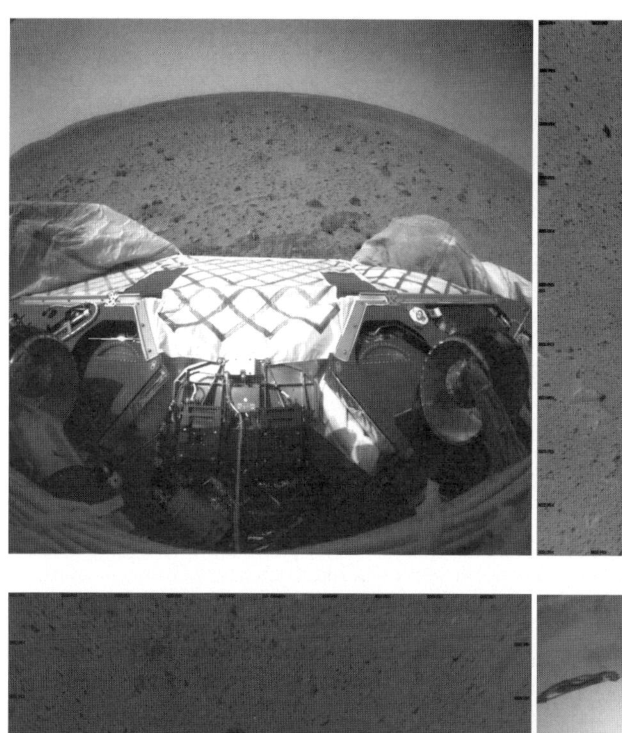

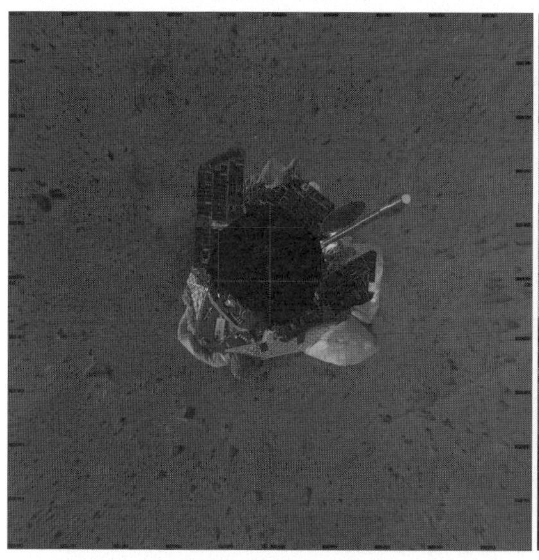

Spirit Rolls All Six Wheels Onto Martian Soil

January 15, 2004

NASA's Mars Exploration Rover Spirit successfully drove off its lander platform and onto the soil of Mars early today.

The robot's first picture looking back at the now-empty lander and showing wheel tracks in the soil set off cheers from the robot's flight team at NASA's Jet Propulsion Laboratory, Pasadena, Calif.

"Spirit is now ready to start its mission of exploration and discovery. We have six wheels in the dirt," said JPL Director Dr. Charles Elachi.

Since Spirit landed inside Mars' Gusev Crater on Jan. 3

(PST and EST; Jan. 4 Universal Time), JPL engineers have put it through a careful sequence of unfolding, standing up, checking its surroundings and other steps leading up to today's drive-off.

"It has taken an incredible effort by an incredible group of people," said Mars Exploration Rover Project Manager Peter Theisinger of JPL.

The drive moved Spirit 3 meters (10 feet) in 78 seconds, ending with the back of the rover about 80 centimeters (2.6 feet) away from the foot of the egress ramp, said JPL's Joel Krajewski, leader of the team that developed the sequence of events from landing to drive-off. The flight time sent the command for the drive-off at 12:21 a.m. PST today and received data confirming the event at 1:53 a.m. PST. The data showed that the rover completed the drive-off at 08:41 Universal Time (12:41 a.m. PST).

"There was a great sigh of relief from me," said JPL's Kevin Burke, lead mechanical engineer for the drive-off. "We are now on the surface of Mars."

With the rover on the ground, an international team of scientists assembled at JPL will be making daily decisions about how to use the rover for examining rocks, soils and atmosphere with a suite of scientific instruments onboard.

"Now, we are the mission that we all envisioned three-and-a-half years ago, and that's tremendously exciting," said JPL's Jennifer Trosper, mission manager.

JPL engineer Chris Lewicki, flight director, said "It's as if we get to drive a nice sports car, but in the end we're just the valets who bring it around to the front and give the keys to the science team."

Spirit was launched from Cape Canaveral Air Force Station, Fla., on June 10, 2003. Now that it is on Mars, its task is to spend the rest of its mission exploring for clues in rocks and soil about whether the past environment in Gusev Crater was ever watery and suitable to sustain life. Spirit's twin Mars Exploration Rover, Opportunity, will reach Mars on Jan. 25 (EST and Universal Time; 9:05 p.m., Jan. 24, PST) to begin a similar examination of a site on the opposite side of the planet.

JPL, a division of the California Institute of Technology in Pasadena, manages the Mars Exploration Rover project for NASA's Office of Space Science, Washington, D.C. Images and additional information about the project are available from JPL at http://marsrovers.jpl.nasa.gov and from Cornell University, Ithaca, N.Y., at http://athena.cornell.edu .

Guy Webster (818) 354-5011 Jet Propulsion Laboratory, Pasadena, California
Donald Savage (202) 358-1547 NASA Headquarters, Washington, D.C.
NEWS RELEASE: 2004-020.

JPL engineers played Baha Men's "Who Let the Dogs Out" in the control room as they watched new images confirming that the Mars Exploration Rover Spirit successfully rolled off its lander platform early Thursday morning. This image from the rover's front hazard identification camera shows the rover's view of the martian landscape from its new position 1 meter (3 feet) northwest of the lander. One of the rover's next tasks will be to locate the Sun with its panoramic camera and calculate from the Sun's position how to point its main antenna toward Earth.

Image credit: NASA/JPL

Spirit Looks Back

This image from the Mars Exploration Rover Spirit's rear hazard identification camera shows the rover's hind view of the lander platform, its nest for the past 12 sols, or martian days. The rover is approximately 1 meter (3 feet) in front of the airbag-cushioned lander, facing northwest. Note the tracks left in the martian soil by the rovers' wheels, all six of which have rolled off the lander. This is the first time the rover has touched martian soil.

Image credit: NASA/JPL

Spirit Flexes Its Arm to Use Microscope on Mars' Soil

January 16, 2004

This image taken by the front hazard-identification camera on the Mars Exploration Rover Spirit, shows the rover's robotic arm, or instrument deployment device. The arm was deployed from its stowed position beneath the "front porch" of the rover body early Friday morning. This is the first use of the arm to deploy the microscopic imager, one of four geological instruments located on the arm. The instrument will help scientists analyze and understand martian rocks and soils by taking very high resolution, close-up images.

NASA's Spirit rover reached out with its versatile robotic arm early today and examined a patch of fine-grained martian soil with a microscope at the end of the arm.

"We made our first use of the arm and took the first microscopic image of the surface of another planet," said Dr. Mark Adler, Spirit mission manager at NASA's Jet Propulsion Laboratory, Pasadena, Calif.

The rover's microscopic imager, one of four tools on a turret at the end of the arm, serves as the functional equivalent of a field geologist's hand lens for examining structural details of rocks and soils.

"I'm elated and relieved at how well things are going. We got some great images in our first day of using the microscopic imager on Mars," said Dr. Ken Herkenhoff of the U.S. Geological Survey Astrogeology Team, Flagstaff, Ariz. Herkenhoff is the lead scientist for the microscopic imagers on Spirit and on Spirit's twin Mars Exploration Rover, Opportunity.

The microscope can show features as small as the width of a human hair. While analysis of today's images from the instrument has barely begun, Herkenhoff said his first impression is that some of the tiny particles appear to be stuck together.

Before driving to a selected rock early next week, Spirit will rotate the turret of tools to use two spectrometer instruments this weekend on the same patch of soil examined by the microsope, said Jessica Collisson, mission flight director. The Mössbauer Spectrometer identifies types of iron-bearing minerals. The Alpha Particle X-ray Spectrometer identifies the elements in rocks and soils.

The rover's arm is about the same size as a human arm, with comparable shoulder, elbow and wrist joints. It is "one of the most dextrous and capable robotic devices ever flown in space," said JPL's Dr. Eric Baumgartner, lead engineer for the robotic arm, which also goes by the name "instrument deployment device."

"Best of all," Baumgartner said, "this robotic arm sits on a rover, and a rover is meant to rove. Spirit will take this arm and the tremendous science package along with it, and reach out to investigate the surface."

The wheels Spirit travels on provide other ways to examine Mars' soil. Details visible in images of the wheel tracks from the rover's first drive onto the soil give information about the soil's physical properties.

"Rover tracks are great," said Dr. Rob Sullivan of Cornell University, Ithaca, N.Y., a member of the science team for Spirit and Opportunity. "For one thing, they mean we're on the surface of Mars! We look at them for engineering reasons and for science reasons." The first tracks show that the wheels did not sink too deep for driving and that the soil has very small particles that provide a finely detailed imprint of the wheels, he said.

Opportunity, equipped identically to Spirit, will arrive at Mars Jan. 25 (Universal Time and EST; 9:05 p.m. Jan. 24, PST). The amount of dust in the atmosphere over Opportunity's planned landing site has been declining in recent days, said JPL's Dr. Joy Crisp, project scientist for the Mars Exploration Rover Project.

Today, Spirit completes its 13th martian day, or "sol", at its landing site in Gusev Crater. Each sol lasts 39 minutes and 35 seconds longer than an Earth day. The rover project's goal is for Spirit and Opportunity to explore the areas around their landing sites for clues in the rocks and the soil about whether the past environments there were ever watery and possibly suitable for sustaining life.

JPL, a division of the California Institute of Technology in Pasadena, manages the Mars Exploration Rover Project for NASA's Office of Space Science, Washington, D.C. Pictures and additional information about the project are available from JPL at http://marsrovers.jpl.nasa.gov and from Cornell University, Ithaca, N.Y., at http://athena.cornell.edu .

Guy Webster (818) 354-5011 Jet Propulsion Laboratory, Pasadena, Calif.
Donald Savage (202) 358-1547 NASA Headquarters, Washington, D.C.

NEWS RELEASE: 2004-022

This image taken by the front hazard-identification camera on the Mars Exploration Rover Spirit, shows the rover's robotic arm, or instrument deployment device. The arm was deployed from its stowed position beneath the "front porch" of the rover body early Friday morning. This is the first use of the arm to deploy the microscopic imager, one of four geological instruments located on the arm. The instrument will help scientists analyze and understand martian rocks and soils by taking very high resolution, close-up images.

Mars Rover Opportunity Mission Status

January 16, 2004

With barely a week before reaching Mars, NASA's Opportunity spacecraft adjusted its trajectory, or flight path, today for the first time in four months.

The spacecraft carries a twin to the Spirit rover, which is now exploring Mars' Gusev Crater. It will land halfway around Mars, in a region called Meridiani Planum, on Jan. 25 (Universal Time and EST; Jan. 24 at 9:05 p.m., PST). For today's trajectory correction maneuver, engineers at NASA's Jet Propulsion Laboratory, Pasadena, Calif., commanded Opportunity at 6 p.m. PST to fire thrusters in a sequence carefully calculated by the mission's navigators. The spacecraft is spinning at two rotations per minute. The maneuver began with a 20-second burn in the direction of the axis of rotation, then included two 5-second pulses perpendicular to that axis.

"Looks like we got a nice burn out of Opportunity," said JPL's Jim Erickson, mission manager. "We're on target for our date on the plains of Meridiani next Saturday with a healthy spacecraft."

Before the thruster firings, Opportunity was headed for a landing about 384 kilometers (239 miles) west and south of the intended landing site, said JPL's Christopher Potts, deputy navigation team chief for the Mars Exploration Rover Project. The maneuver was designed to put it on course for the target.

Opportunity's schedule still includes two more possible trajectory correction maneuvers, on Jan. 22 and Jan. 24, but the maneuvers will only be commanded if needed.

As of 5 a.m. Sunday, PST, Opportunity will have traveled 444 million kilometers (276 million miles) since its July 7 launch, and will have 12.5 million kilometers (7.8 million miles) left to go.

JPL, a division of the California Institute of Technology, manages the Mars Exploration Rover project for NASA's Office of Space Science, Washington, D.C. Additional information about the project is available from JPL at http://marsrovers.jpl.nasa.gov and from Cornell University, Ithaca, N.Y., at http://athena.cornell.edu .

Guy Webster (818) 354-5011 Jet Propulsion Laboratory, Pasadena, Calif.
Donald Savage (202) 358-1547 NASA Headquarters, Washington, D.C.
NEWS RELEASE: 2004-023

Spirit Drives to a Rock Called 'Adirondack' for Close Inspection

January 19, 2004

NASA's Spirit rover has successfully driven to its first target on Mars, a football-sized rock that scientists have dubbed Adirondack.

The Mars Exploration Rover flight team at NASA's Jet Propulsion Laboratory, Pasadena, Calif., plans to send commands to Spirit early Tuesday to examine Adirondack with a microscope and two instruments that reveal the composition of rocks, said JPL's Dr. Mark Adler, Spirit mission manager. The instruments are the Mössbauer spectrometer and the alpha particle X-ray spectrometer.

Spirit successfully rolled off the lander and onto the martian surface last Thursday. To make the drive to Adirondack, the rover turned 40 degrees in short arcs totaling 95 centimeters (3.1 feet). It then turned in place to face the target rock and drove four short moves straightforward totaling 1.9 meters (6.2 feet). The moves covered a span of 30 minutes on Sunday, though most of that was sitting still and taking pictures between moves. The total amount of time when Spirit was actually moving was about two minutes.

"These are the sorts of baby steps we're taking," said JPL's Dr. Eddie Tunstel, rover mobility engineer.

"The drive was designed for two purposes, one of which was to get to the rock," Tunstel said. "From the mobility engineers' standpoint, this drive was geared to testing out how we do drives on this new surface." Gathering new information such as how much the wheels slip in the martian soil will give the team confidence for more ambitious drives in future weeks and months.

"Adirondack is now about one foot (30 centimeters) in front of the front wheels," he said.

Scientists chose Adirondack to be Spirit's first target rock rather than another rock, called Sashimi, that would have been a shorter, straight-ahead drive. Rocks are time capsules containing evidence of the environmental conditions of the past, said Dr. Dave Des Marais, a rover science-team member from NASA Ames Research Center, Moffett Field, Calif. "We needed to decide which of these time capsules to open."

Sashimi appears dustier than Adirondack. The dust layer could obscure good observations of the rock's surface, which may give information about chemical changes and other weathering from environmental conditions affecting the rock since its surface was fresh. Also, Sashimi is more pitted than Adirondack. That makes it a poorer candidate for the rover's rock abrasion tool, which scrapes away a rock's surface for a view of the interior evidence about environmental conditions when the rock first formed. Adirondack has a "nice, flat surface" well suited to trying out the rover's tools on their first martian rock, Des Marais said.

"The hypothesis is that this is a volcanic rock, but we'll test that hypothesis," he said.

Spirit arrived at Mars Jan. 3 (EST and PST; Jan. 4 Universal Time) after a seven-month journey. In coming weeks and months, according to plans, it will be exploring for clues in rocks and soil to decipher whether the past environment in Gusev Crater was ever watery and possibly suitable to sustain life.

Spirit's twin Mars Exploration Rover, Opportunity, will reach Mars on Jan. 25 (EST and Universal Time; 9:05 p.m., Jan. 24, PST) to begin a similar examination of a site on the opposite side of the planet from Gusev Crater.

JPL, a division of the California Institute of Technology in Pasadena, manages the Mars Exploration Rover project for NASA's Office of Space Science, Washington, D.C. Images and additional information about the project are available from JPL at http://marsrovers.jpl.nasa.gov and from Cornell University, Ithaca, N.Y., at http://athena.cornell.edu .

Guy Webster (818) 354-5011
Jet Propulsion Laboratory, Pasadena, Calif.

Donald Savage (202) 358-1547
NASA Headquarters, Washington, D.C.

NEWS RELEASE: 2004-024

"They of the Great Rocks"

This approximate true color image taken by the panoramic camera onboard the Mars Exploration Rover Spirit shows "Adirondack," the rover's first target rock. Spirit traversed the sandy martian terrain at Gusev Crater to arrive in front of the football-sized rock on Sunday, Jan. 18, 2004, just three days after it successfully rolled off the lander. The rock was selected as Spirit's first target because its dust-free, flat surface is ideally suited for grinding. Clean surfaces also are better for examining a rock's top coating. Scientists named the angular rock after the Adirondack mountain range in New York. The word Adirondack is Native American and is interpreted by some to mean "They of the great rocks."

Image credit: NASA/JPL/Cornell

"They of the Great Rocks"-2

This approximate true color image taken by the panoramic camera onboard the Mars Exploration Rover Spirit shows "Adirondack," the rover's first target rock. Spirit traversed the sandy martian terrain at Gusev Crater to arrive in front of the football-sized rock on Sunday, Jan. 18, 2004, just three days after it successfully rolled off the lander. The rock was selected as Spirit's first target because its dust-free, flat surface is ideally suited for grinding. Clean surfaces also are better for examining a rock's top coating. Scientists named the angular rock after the Adirondack mountain range in New York. The word Adirondack is Native American and is interpreted by some to mean "They of the great rocks."

Image credit: NASA/JPL/Cornell

"They of the Great Rocks"-3

This 3-D perspective image taken by the panoramic camera onboard the Mars Exploration Rover Spirit shows "Adirondack," the rover's first target rock. Spirit traversed the sandy martian terrain at Gusev Crater to arrive in front of the football-sized rock on Sunday, Jan. 18, 2004, just three days after it successfully rolled off the lander. The rock was selected as Spirit's first target because it has a flat surface and is relatively free of dust - ideal conditions for grinding into the rock to expose fresh rock underneath. Clean surfaces also are better for examining a rock's top coating. Scientists named the angular rock after the Adirondack mountain range in New York. The word Adirondack is Native American and is interpreted by some to mean "They of the great rocks." Data from the panoramic camera's red, green and blue filters were combined to create this approximate true color image. Image credit: NASA/JPL/Cornell

Hungry for Rocks

This image taken by the panoramic camera onboard the Mars Exploration Rover Spirit shows "Sashimi" and "Sushi" - two rocks that scientists considered investigating first. Ultimately, these rocks were not chosen because their rough and dusty surfaces are ill-suited for grinding. Dusty surfaces also can obscure observations of a rock's top coating. Data from the panoramic camera's red, blue and green filters were combined to create this approximate true color picture.

Image credit: NASA/JPL

Approaching Rock Target No. 1

This 3-D stereo anaglyph image was taken by the Mars Exploration Rover Spirit front hazard-identification camera after the rover's first post-egress drive on Mars Sunday. Engineers drove the rover approximately 3 meters (10 feet) from the Columbia Memorial Station toward the first rock target, seen in the foreground. The football-sized rock was dubbed Adirondack because of its mountain-shaped appearance. Scientists plan to use instruments at the end of the rover's robotic arm to examine the rock and understand how it formed.

Image credit: NASA/JPL

Rover Takes a Sunday Drive (Animation sequence)

This animation, made with images from the Mars Exploration Rover Spirit hazard-identification camera, shows the rover's perspective of its first post-egress drive on Mars Sunday. Engineers drove Spirit approximately 3 meters (10 feet) toward its first rock target, a football-sized, mountain-shaped rock called Adirondack. The drive took approximately 30 minutes to complete, including time stopped to take images. Spirit first made a series of arcing turns totaling approximately 1 meter (3 feet). It then turned in place and made a series of short, straightforward movements totaling approximately 2 meters (6.5 feet). Image credit: NASA/JPL

Ready to Rock and Roll

This image from the Mars Exploration Rover Spirit hazard-identification camera shows the rover's perspective just before its first post-egress drive on Mars. On Sunday, the 15th martian day, or sol, of Spirit's journey, engineers drove Spirit approximately 3 meters (10 feet) toward its first rock target, a football-sized, mountain-shaped rock called Adirondack (not pictured). In the foreground of this image are "Sashimi" and "Sushi" - two rocks that scientists considered investigating first. Ultimately, these rocks were not chosen because their rough and dusty surfaces are ill-suited for grinding.

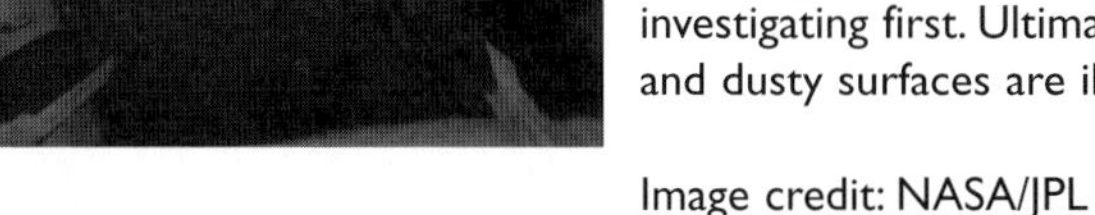

Image credit: NASA/JPL

Spirit Takes a Turn for Adirondack

This rear hazard-identification camera image looks back at the circular tracks made in the martian soil when the Mars Exploration Rover Spirit drove about 3 meters (10 feet) toward the mountain-shaped rock called Adirondack, Spirit's first rock target. Spirit made a series of arcing turns totaling approximately 1 meter (3 feet). It then turned in place and made a series of short, straightforward movements totaling approximately 2 meters (6.5 feet). The drive took about 30 minutes to complete, including time stopped to take images. The two rocks in the upper left corner of the image are called "Sashimi" and "Sushi." In the upper right corner is a portion of the lander, now known as the Columbia Memorial Station. Image credit: NASA/JPL

Rocks: Windows to History of Mars

This full-resolution image taken by the panoramic camera onboard the Mars Exploration Rover Spirit before it rolled off the lander shows the rocky surface of Mars. Scientists are eager to begin examining the rocks because, unlike soil, these "little time capsules" hold memories of the ancient processes that formed them. Data from the camera's red, green and blue filters were combined to create this approximate true color picture.
Image credit: NASA/JPL/Cornell

Rocks: Windows to History of Mars-2

This full-resolution image taken by the panoramic camera onboard the Mars Exploration Rover Spirit before it rolled off the lander shows the rocky surface of Mars. Scientists are eager to begin examining the rocks because, unlike soil, these "little time capsules" hold memories of the ancient processes that formed them. The lander's deflated airbags can be seen in the foreground. Data from the camera's red, green and blue filters were combined to create this approximate true color picture.
Image credit: NASA/JPL/Cornell

Virtual Rover Drives Toward Rock

This image shows a screenshot from the software used by engineers to test and drive the Mars Exploration Rover Spirit. The software simulates the rover's movements across the martian terrain, helping to plot a safe course. Here, engineers simulated Spirit's first post-egress drive on Mars Sunday. The 3-meter (10-foot) drive totaled approximately 30 minutes, including time to stop and take images. The rover drove toward its first rock target, a mountain-shaped rock called Adirondack. The blue line denotes the path of the rover's "belly button," as engineers like to call it, as the rover drove toward Adirondack. The virtual 3-D world around the rover was built from images taken by Spirit's stereo navigation cameras. Regions for which the rover has not yet acquired 3-D data are represented in beige.

Image credit: NASA/JPL

NASA Mars Rover's First Soil Analysis Yields Surprises

January 20, 2004

The first use of the tools on the arm of NASA's Mars Exploration Rover Spirit reveals puzzles about the soil it examined and raises anticipation about what the tool will find during its studies of a martian rock.

Today and overnight tonight, Spirit is using its microscope and two up-close spectrometers on a football-sized rock called Adirondack, said Jennifer Trosper, mission manager at NASA's Jet Propulsion Laboratory, Pasadena, Calif.

"We're really happy with the way the spacecraft continues to work for us," Trosper said. The large amount of data — nearly 100 megabits — transmitted from Spirit in a single relay session through NASA's Mars Odyssey spacecraft today "is like getting an upgrade to our Internet connection."

Scientists today reported initial impressions from using Spirit's alpha particle X-ray spectrometer, Moessbauer spectrometer and microscopic imager on a patch of soil that was directly in front of the rover after Spirit drove off its lander Jan. 15.

"We're starting to put together a picture of what the soil at this particular place in Gusev Crater is like. There are some puzzles and there are surprises," said Dr. Steve Squyres of Cornell University, Ithaca, N.Y., principal investigator for the suite of instruments on Spirit and on Spirit's twin, Opportunity.

One unexpected finding was the Mössbauer spectrometer's detection of a mineral called olivine, which does not survive weathering well. This spectrometer identifies different types of iron-containing minerals; scientists believe many of the minerals on Mars contain iron. "This soil contains a mixture of minerals, and each mineral has its own distinctive Mössbauer pattern, like a fingerprint," said Dr. Goestar Klingelhoefer of Johannes Gutenberg University, Mainz, Germany, lead scientist for this instrument.

The lack of weathering suggested by the presence of olivine might be evidence that the soil particles are finely ground volcanic material, Squyres said. Another possible explanation is that the soil layer where the measurements were taken is extremely thin, and the olivine is actually in a rock under the soil.

Scientists were also surprised by how little the soil was disturbed when Spirit's robotic arm pressed the Moessbauer spectrometer's contact plate directly onto the patch being examined. Microscopic images from before and after that pressing showed almost no change. "I thought it would scrunch down the soil particles," Squyres said. "Nothing collapsed. What is holding these grains together?"

Information from another instrument on the arm, an alpha particle X-ray spectrometer, may point to an answer. This instrument "measures X-ray radiation emitted by Mars samples, and from this data we can derive the elemental composition of martian soils and rocks," said Dr. Johannes Brueckner, rover science team member from the Max Planck Institute for Chemistry, Mainz, Germany. The instrument found the most prevalent elements in the soil patch were silicon and iron. It also found significant levels of chlorine and sulfur, characteristic of soils at previous martian landing sites but unlike soil composition on Earth.

Squyres said, "There may be sulfates and chlorides binding the little particles together." Those types of salts could be left behind by evaporating water, or could come from volcanic eruptions, he said. The soil may not have even originated anywhere near Spirit's landing site, because Mars has dust storms that redistribute fine particles around the planet. The next target for use of the rover's full set of instruments is a rock, which is more likely to have originated nearby.

Spirit landed in the Connecticut-sized Gusev Crater on Jan. 3 (EST and PST; Jan. 4 Universal Time). In coming weeks and months, according to plans, it will examine rocks and soil for clues about whether the past environment there was ever watery and possibly suitable to sustaining life. Spirit's twin Mars Exploration Rover, Opportunity, will reach Mars on Jan. 25 (EST and Universal Time; 9:05 p.m., Jan. 24, PST) to begin a similar examination of a site on the opposite side of the planet.

JPL, a division of the California Institute of Technology in Pasadena, manages the Mars Exploration Rover project for NASA's Office of Space Science, Washington, D.C. Images and additional information about the project are available from JPL at http://marsrovers.jpl.nasa.gov and from Cornell University, Ithaca, N.Y., at http://athena.cornell.edu .

Guy Webster (818) 354-5011 Jet Propulsion Laboratory, Pasadena, Calif.
Donald Savage (202) 358-1547 NASA Headquarters, Washington, D.C.
NEWS RELEASE: 2004-025

Spirit Switches on Its X-ray Vision

This image shows the Mars Exploration Rover Spirit probing its first target rock, Adirondack. At the time this picture was snapped, the rover had begun analyzing the rock with the alpha particle X-ray spectrometer located on its robotic arm. This instrument uses alpha particles and X-rays to determine the elemental composition of martian rocks and soil. The image was taken by the rover's hazard-identification camera.

Image credit: NASA/JPL

The Mystery Soil

This high-resolution image from the panoramic camera on the Mars Exploration Rover Spirit shows the region containing the patch of soil scientists examined at Gusev Crater just after Spirit rolled off the Columbia Memorial Station. Scientists examined this patch on the 13th and 15th martian days, or sols, of Spirit's journey. Using nearly all the science instruments located on the rover's instrument deployment device or "arm," scientists yielded some puzzling results including the detection of a mineral called olivine and the appearance that the soil is stronger and more cohesive than they expected. Like detectives searching for clues, the science team will continue to peruse the landscape for explanations of their findings. Data taken from the camera's red, green and blue filters were combined to create this approximate true color picture, acquired on the 12th martian day, or sol, of Spirit's journey.
Image credit: NASA/JPL/Cornell

A Puzzling Patch

The yellow box in this high-resolution image from the panoramic camera on the Mars Exploration Rover Spirit outlines the patch of soil scientists examined at Gusev Crater just after Spirit rolled off the Columbia Memorial Station. Scientists examined this patch on the 13th and 15th Martian days, or sols, of Spirit's journey. Using nearly all the science instruments located on the rover's instrument deployment device or "arm," scientists yielded some puzzling results including the detection of a mineral called olivine and the appearance that the soil is stronger and more cohesive than they expected. Like detectives searching for clues, the science team will continue to peruse the landscape for explanations of their findings.

Data taken from the camera's red, green and blue filters were combined to create this approximate true color picture, acquired on the 12th Martian day, or sol, of Spirit's journey.
Image credit: NASA/JPL/Cornell

Super Soil?

This animation made of images from the microscopic imager instrument on the Mars Exploration Rover Spirit shows the patch of soil scientists examined at Gusev Crater just after Spirit rolled off the Columbia Memorial Station. The upper left corner of the soil patch in part of this animation is illuminated by direct sunlight and thus appears brighter. The actual size of the patch is about 3 centimeters (1.2 inches) across. Scientists initially thought that the soil was dust-like and therefore would collapse as the instrument pressed down on it with approximately 4 ounces (113 grams) of force. But they were surprised when, as the rotating frames show, the soil barely moved under the instrument's weight. Scientists are still determining why this happened.

Image credit: NASA/JPL/US Geological Survey

Super Rover's X-Ray Vision

Located on the arm of the Mars Exploration Rover Spirit, the alpha particle X-ray spectrometer uses alpha particles and X-rays to determine the chemical make up of martian rocks and soils. This type of information helps scientists understand how the planet's crust was weathered and formed. Mars Exploration Rover team members used this palm-sized instrument on a small patch of martian soil just after Spirit rolled off the Columbia Memorial Station. They found that although the soil was very similar to what they had seen previously on Mars, the instrument's improved sensitivity allowed them to see new elements and subtle differences not detected before.

Image credit: NASA/JPL/Max-Planck-Institute for Chemistry

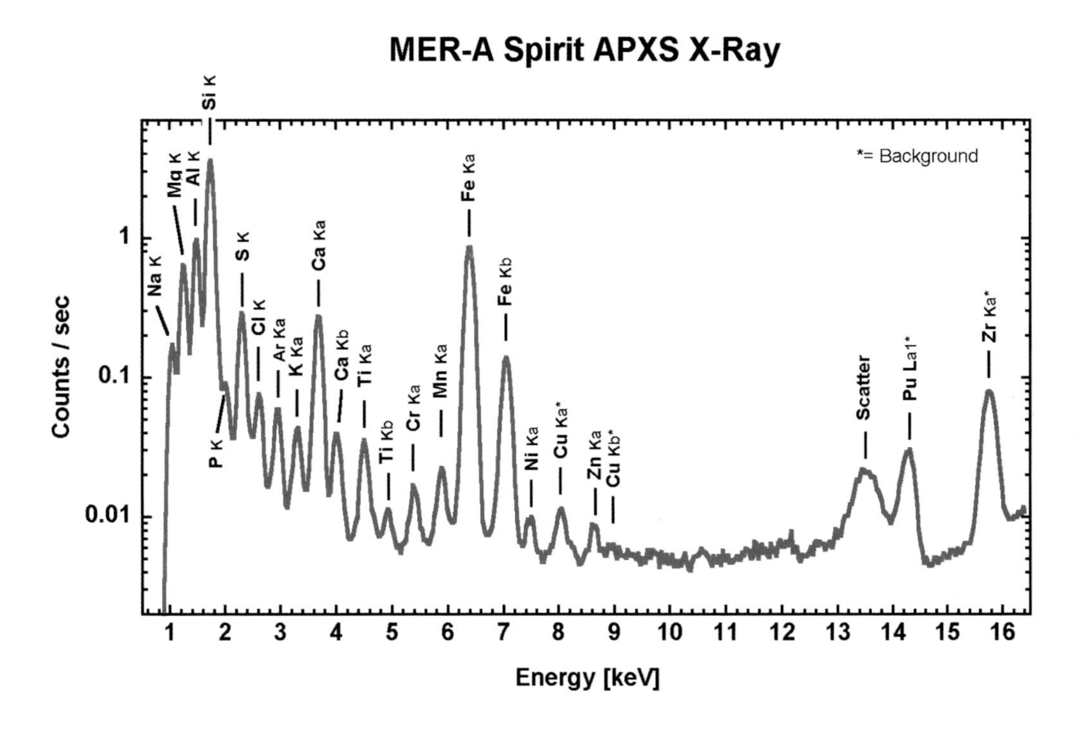

A Rainbow of Martian Elements

This graph or spectrum taken by the alpha particle X-ray spectrometer onboard the Mars Exploration Rover Spirit shows the variety of elements present in the soil at the rover's landing site. In agreement with past missions to Mars, iron and silicon make up the majority of the martian soil. Sulfur and chlorine were also observed as expected. Trace elements detected for the first time include zinc and nickel. These latter observations demonstrate the power of the alpha particle X-ray spectrometer to pick up the signatures of elements too faint to be seen before. The alpha particle X-ray spectrometer uses alpha particles and X-rays to measure the presence and abundance of all major rock-forming elements except hydrogen.

Image credit: NASA/JPL/Max-Planck-Institute for Chemistry

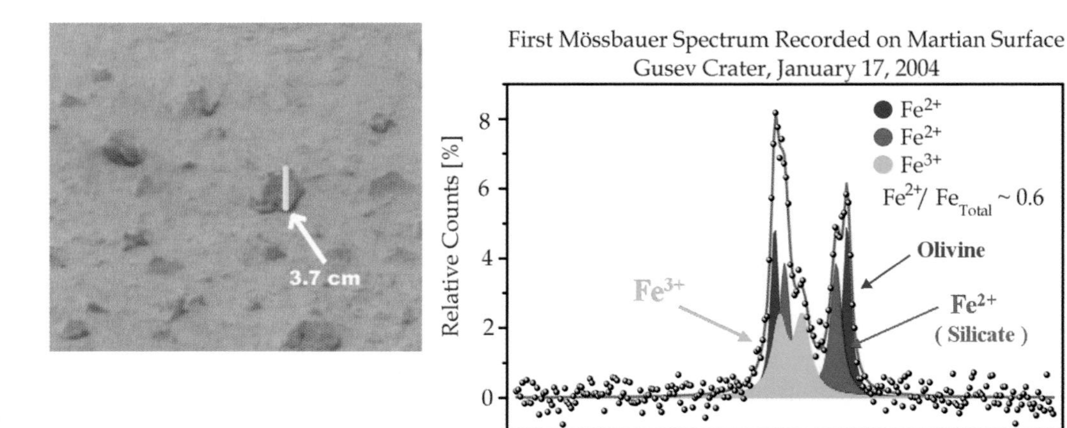

Mysterious Lava Mineral on Mars

This graph or spectrum captured by the Mössbauer spectrometer onboard the Mars Exploration Rover Spirit shows the presence of three different iron-bearing minerals in the soil at the rover's landing site. One of these minerals has been identified as olivine, a shiny green rock commonly found in lava on Earth. The other two have yet to be pinned down. Scientists were puzzled by the discovery of olivine because it implies the soil consists at least partially of ground up rocks that have not been weathered or chemically altered. The black line in this graph represents the original data; the three colored regions denote individual minerals and add up to equal the black line.

The Mössbauer spectrometer uses two pieces of radioactive cobalt-57, each about the size of pencil erasers, to determine with a high degree of accuracy the composition and abundance of iron-bearing minerals in martian rocks and soil. It is located on the rover's instrument deployment device, or "arm."

Image credit: NASA/JPL/University of Mainz

A Powerful Yet Tiny Machine

This image taken at JPL shows the Mössbauer spectrometer, an instrument on the Mars Exploration Rover Spirit that detects iron-bearing minerals in martian rocks and soil. Located on the rover's instrument deployment device, or "arm," this machine uses two pieces of radioactive cobalt-57, each about the size of pencil erasers, to determine with a high degree of accuracy the composition and abundance of iron-bearing minerals too difficult to detect by other means.

Image credit: NASA/JPL/University of Mainz

Mars Exploration Rover Mission Status

January 21, 2004

Ground controllers were able to send commands to the Mars Exploration Rover Spirit early Wednesday and received a simple signal acknowledging that the rover heard them, but they did not receive expected scientific and engineering data during scheduled communication passes during the rest of that martian day.

Project managers have not yet determined the cause, but similar events occurred several times during the Mars Pathfinder mission. The team is examining a number of different scenarios, some of which would be resolved when the rover wakes up after powering down at the end of the martian day (around midday Pacific time Wednesday).

The next opportunity to hear from the vehicle is when the rover may attempt to communicate with the Mars Global Surveyor orbiter at about 8:30 p.m. Pacific time tonight. A second communication opportunity may occur about two hours later during a relay pass via the Mars Odyssey orbiter. If necessary, the flight team will take additional recovery steps early Thursday morning (the morning of sol 19 on Mars) when the rover wakes up and can communicate directly with Earth.

Full details on the rover's status will be described in the next daily news conference Thursday at 9 a.m. Pacific time at the Jet Propulsion Laboratory, which will be broadcast live on NASA Television.

NEWS RELEASE: 2004-027

Mars Exploration Rover Mission Status

January 22, 2004

Flight-team engineers for NASA's Mars Exploration Rover Mission were encouraged this morning when Spirit sent a simple radio signal acknowledging that the rover had received a transmission from Earth.

However, the team is still trying to diagnose the cause of earlier communications difficulties that have prevented any data being returned from Spirit since early Wednesday.

"We have a very serious situation," said Pete Theisinger of NASA's Jet Propulsion Laboratory, project manager for Spirit and its twin, Opportunity.

Spirit did send a radio signal via NASA's Mars Global Surveyor orbiter Wednesday evening, but the transmission did not carry any data. Spirit did not make radio contact with NASA's Mars Odyssey during a scheduled session two hours later or during another one Thursday morning. It also did not respond to the first two attempts Thursday to elicit an acknowledgment signal with direct communications between Earth and the rover, and it did not send a signal at a time pre-set for doing so when its computer recognizes certain communication problems. The successful attempt to get a response signal came shortly before 9 a.m. Pacific Standard Time.

No single explanation considered so far fits all of the events observed, Theisinger said. When the team tried to replicate the situation in its testing facility at JPL, the testbed rover did not have any trouble communicating. Two of the possibilities under consideration are a corruption of flight software or corruption of computer memory, either of which could leave Spirit's power supply healthy and allow adequate time for recovering control of the rover.

Engineers will continue efforts to understand the situation in preparation for scheduled communication relay sessions using Mars Global Surveyor at 7:10 p.m. PST and Mars Odyssey at 10:35 PST. Efforts to resume direct communications between Spirit and antennas of NASA's Deep Space Network will resume after the rover's expected wake-up at about 3 a.m. PST Friday.

Meanwhile, mission leaders decided to skip an optional trajectory correction maneuver today for Opportunity, the other Mars Exploration Rover. Opportunity is on course to land halfway around Mars from Spirit, in a region called Meridiani Planum, on Jan. 25 (Universal Time and EST; Jan. 24 at 9:05 p.m. PST).

JPL, a division of the California Institute of Technology in Pasadena, manages the Mars Exploration Rover project for NASA's Office of Space Science, Washington, D.C. Additional information about the project is available from JPL at http://marsrovers.jpl.nasa.gov and from Cornell University, Ithaca, N.Y., at http://athena.cornell.edu .

Guy Webster (818) 354-5011 Jet Propulsion Laboratory, Pasadena, Calif.
Donald Savage (202) 358-1547 NASA Headquarters, Washington, D.C.
NEWS RELEASE: 2004-028

Mars Exploration Rover Mission Status

January 23, 2004

NASA's Spirit rover communicated with Earth in a signal detected by NASA's Deep Space Network antenna complex near Madrid, Spain, at 12:34 Universal Time (4:34 a.m. PST) this morning.

The transmissions came during a communication window about 90 minutes after Spirit woke up for the morning on Mars. The signal lasted for 10 minutes at a data rate of 10 bits per second.

Mission controllers at NASA's Jet Propulsion Laboratory, Pasadena, Calif., plan to send commands to Spirit seeking additional data from the spacecraft during the subsequent few hours.

JPL, a division of the California Institute of Technology, Pasadena, manages the Mars Exploration Rover project for NASA's Office of Space Science, Washington, D.C. Additional information about the project is available from JPL at http://marsrovers.jpl.nasa.gov and from Cornell University, Ithaca, N.Y., at http://athena.cornell.edu .

Guy Webster (818) 354-5011 Jet Propulsion Laboratory, Pasadena, Calif.
Donald Savage (202) 358-1547 NASA Headquarters, Washington, D.C.
NEWS RELEASE: 2004-29

Mars Exploration Rover Updated Mission Status

January 23, 2004

The flight team for NASA's Spirit received data from the rover in a communication session that began at 13:26 Universal Time (5:26 a.m. PST) and lasted 20 minutes at a data rate of 120 bits per second. "The spacecraft sent limited data in a proper response to a ground command, and we're planning for commanding further communication sessions later today," said Mars Exploration Rover Project Manager Pete Theisinger at NASA's Jet Propulsion

Laboratory, Pasadena, Calif.

The flight team at JPL had sent a command to Spirit at 13:02 Universal Time (5:02 PST) via the NASA Deep Space Network antenna complex near Madrid, Spain, telling Spirit to begin transmitting.

Meanwhile, the other Mars Exploration Rover, Opportunity is on course to land halfway around Mars from Spirit, in a region called Meridiani Planum, on Jan. 25 (Universal Time and EST; Jan. 24 at 9:05 p.m. PST).

JPL, a division of the California Institute of Technology, Pasadena, manages the Mars Exploration Rover project for NASA's Office of Space Science, Washington, D.C. Additional information about the project is available from JPL at http://marsrovers.jpl.nasa.gov and from Cornell University, Ithaca, N.Y., at http://athena.cornell.edu .

Guy Webster (818) 354-5011 Jet Propulsion Laboratory, Pasadena, Calif.
Donald Savage (202) 358-1547 NASA Headquarters, Washington, D.C.
NEWS RELEASE: 2004-30

Rover Team Readies for Second Landing While Trying to Mend Spirit

January 23, 2004

Some members of the flight team for NASA's Mars Exploration Rovers are preparing for this weekend's landing of the second rover, Opportunity, while others are focused on trying to restore the first rover, Spirit, to working order.

"We should expect we will not be restoring functionality to Spirit for a significant amount of time — many days, perhaps two weeks — even in the best of circumstances," said Peter Theisinger, rover project manager at NASA's Jet Propulsion Laboratory, Pasadena, Calif.

Spirit transmitted data to Earth today for the first time since early Wednesday. The information about the rover's status arrived during three sessions lasting 10 minutes, 20 minutes and 15 minutes. Engineers will be examining it overnight and developing a plan for obtaining more on Saturday morning.

Spirit's flight software is not functioning normally. It appears to have rebooted the rover's computer more than 60 times in the past three days. A motor that moves a mirror for the rover's infrared spectrometer was partway through an operation when the problem arose, so the possibility of a mechanical problem with that hardware will be one theory investigated.

"We believe, based on everything we know now, we can sustain the current state of the spacecraft from a health standpoint for an indefinite amount of time," Theisinger said. That will give the team time to work on the problem.

Meanwhile, Spirit's twin, Opportunity, will reach Mars at 05:05 Universal Time on Jan. 25 (12:05 a.m. Sunday EST or 9:05 p.m. Saturday PST) at a landing site on the opposite side of the planet from Spirit. Opportunity's landing site is on plains called Meridiani Planum within an Oklahoma-sized outcropping of gray hematite, a mineral that usually forms in the presence of water. Scientists plan to use the research instruments on Opportunity to determine whether the gray hematite layer comes from sediments of a long-gone ocean, from volcanic deposits altered by hot water, or from other ancient environmental conditions.

Analysis of Spirit's descent through Mars' atmosphere for its landing at Gusev has contributed to a decision by flight controllers to program Opportunity to open its parachute higher than had been planned earlier, said JPL's Dr. Wayne Lee, chief engineer for development of the rover's descent and landing systems.

The Mars Orbiter Camera on NASA's Mars Global Surveyor orbiter has taken an image of Spirit's landing region that shows the spacecraft's lander platform on the ground. The jettisoned parachute, backshell and heat shield are also visible, noted Dr. Michael Malin of Malin Space Science Systems, San Diego, lead investigator for the orbiter's camera and a member of the rover science team.

JPL, a division of the California Institute of Technology in Pasadena, manages the Mars Exploration Rover project for NASA's Office of Space Science, Washington, D.C. Images and additional information about the project are available from JPL at http://marsrovers.jpl.nasa.gov and from Cornell University, Ithaca, N.Y., at http://athena.cornell.edu .

Guy Webster (818) 354-5011 Jet Propulsion Laboratory, Pasadena, Calif.
Donald Savage (202) 358-1547 NASA Headquarters, Washington, D.C.
NEWS RELEASE: 2004-031

Behold Spirit

This high-resolution image shows a computer-generated model of Spirit's lander at Gusev Crater as engineers and scientists would have expected to see it from a perfect overhead view. The background is a reprojected image taken by the Spirit panoramic camera on Sol 19 (Jan. 21-22, 2004). The top of the image faces north.

Image credit: NASA/JPL/Cornell

View from above Landing Site

The image on the left is a computer-generated model of Spirit's lander at Gusev Crater as engineers and scientists would have expected to see it from a perfect overhead view. The background is a reprojected image taken by the Spirit panoramic camera on Sol 19 (Jan. 21-22, 2004). The picture on the right is an actual image of the lander on Mars taken Jan. 19, 2004, by the camera on board Mars Global Surveyor. The tops of both images face north.

Image credit: NASA/JPL/MSSS/Cornell

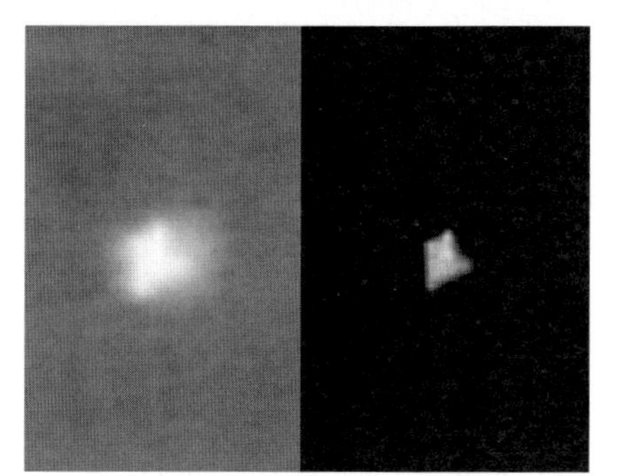

Spirit's Successful Landing

The bright triangle seen in these images is Spirit's lander resting at the Gusev Crater landing site on Mars after a nerve-wracking entry, descent and landing process on Jan. 3, 2004. The left image was taken by the camera on board the orbiting Mars Global Surveyor on Jan. 19, 2004. The right image is the same image enhanced to show the contrast between the lander and the martian surface. The rover is not visible in this image due to the bright glare of the lander.

Image credit: NASA/JPL/MSSS

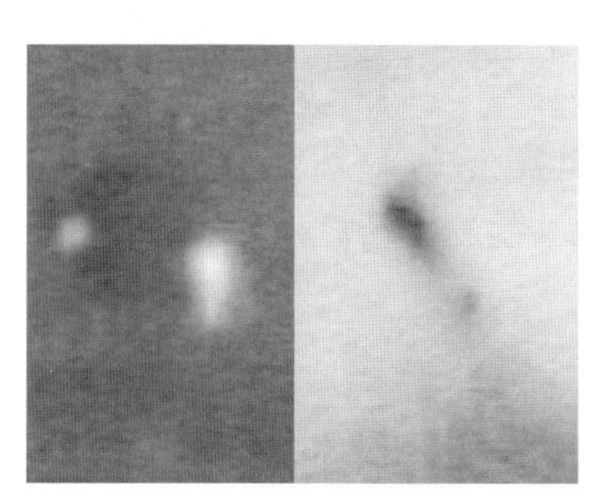

Spirit's Hardware Up Close on Mars

This image shows, on the left, a close-up of the backshell and parachute Spirit dropped while landing at Gusev Crater on Mars on Jan. 3, 2004. The backshell is the smaller white mark to the left of the parachute. On the right is a close-up of the location believed to be where the heat shield impacted, leaving a visible dark streak. Both images were taken on Jan. 19, 2004, by the camera on board Mars Global Surveyor.

Image credit: NASA/JPL/MSSS

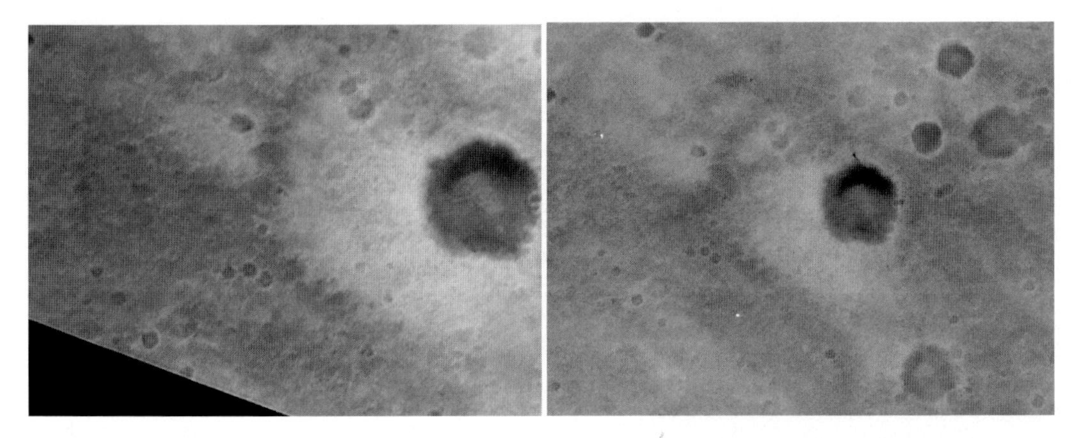

Spirit Lightens the Load

The history of Spirit's descent and landing on the surface of Mars is recorded in this image taken more than two weeks later on Jan. 19, 2004, by the camera on the orbiting Mars Global Surveyor. Spirit landed on Jan. 3, 2004. The two dots in the upper left are the spacecraft's backshell and parachute, which were shed as Spirit's bridle was cut, allowing the lander to bounce to a rest while safely encased in airbags. The white dot near the bottom of the image is the lander, also known as the Columbia Memorial Station, at the Gusev Crater landing site. The lander appears white to the camera because sunlight is hitting the lander's highly reflective surface, causing contrast between the lander and the surrounding martian terrain. This image was taken in the early martian afternoon.

Image credit: NASA/JPL/MSSS

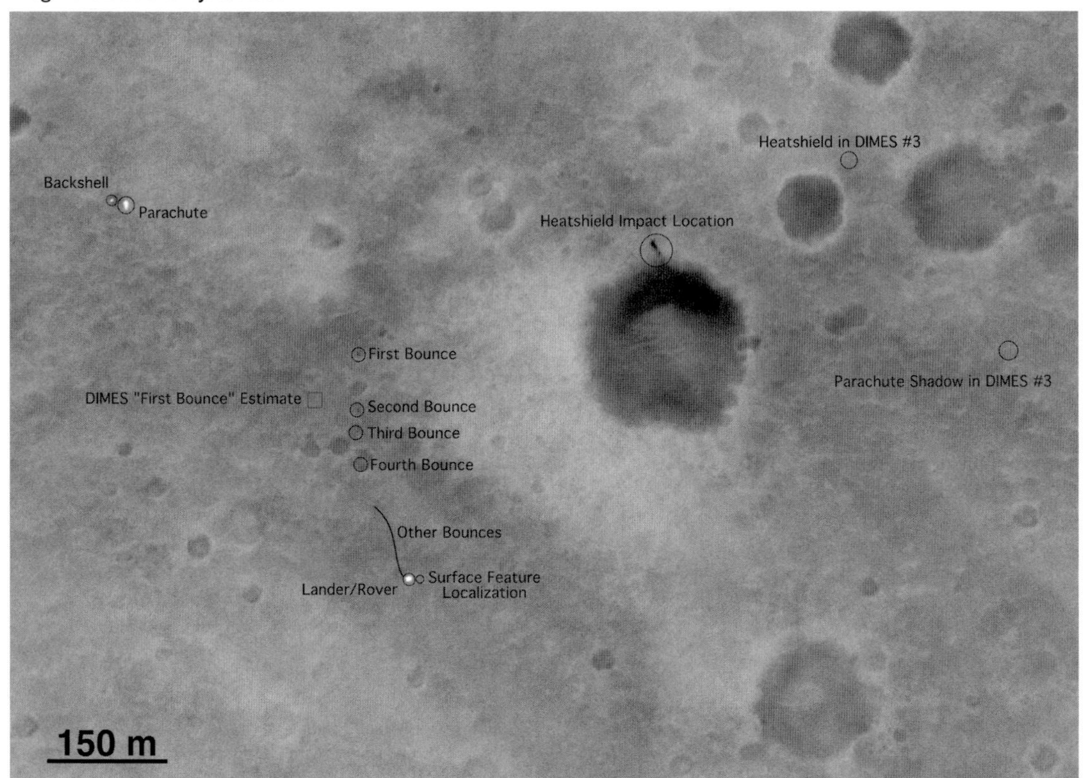

Following Spirit's Tracks

The history of Spirit's descent and landing on the surface of Mars is recorded in this image taken more than two weeks later on Jan. 19, 2004, by the camera on the orbiting Mars Global Surveyor. Spirit landed on Jan. 3, 2004. The two dots in the upper left are the spacecraft's backshell and parachute, which were shed as Spirit's bridle was cut, allowing the lander to bounce to a rest while safely encased in airbags. To the far right of the image, a dark streak above a large crater is believed to be the location where the heat shield impacted. The heat shield had protected the spacecraft during its descent through the martian atmosphere and was jettisoned several kilometers above the surface. A trail of bounce marks made by the airbags as Spirit bounced to a stop can be seen in the middle of the image. To the left of the second bounce mark is a square showing the location where engineers had calculated Spirit's airbags first hit the martian surface, based on data from the descent image motion estimation system located on the bottom of the rover's lander. The white dot near the bottom of the image is the lander, also known as the Columbia Memorial Station, at the Gusev Crater landing site. Beside it is a dot marked "surface feature location," showing the location of the lander estimated by the Spirit team using sight lines to landmarks in the lander's panoramic images. This image was taken in the early martian afternoon.

Image credit: NASA/JPL/MSSS

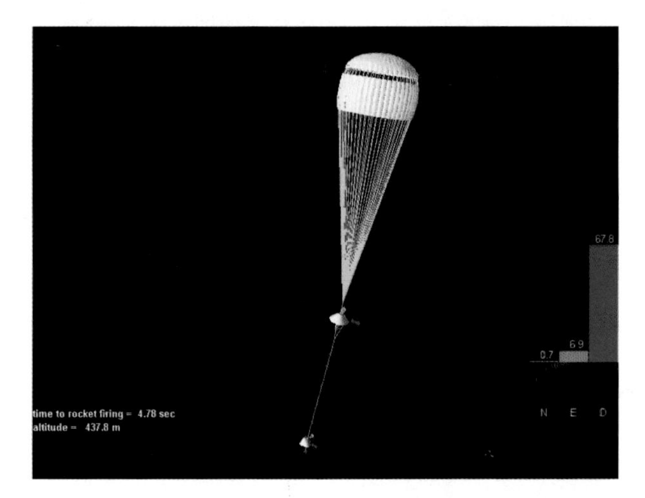

A Flyby Tour of Spirit's Descent

Telemetry sent down to Earth from the Mars Exploration Rover Spirit has been reconstructed to create this reenactment of the rover's final 30 seconds before landing at Gusev Crater, Mars. Just seconds before the rover touched down and its airbags were inflated, a gust of wind threatened to significantly increase the rover's horizontal speed. But the firing of a lateral rocket, called the Tranverse Impulse Rocket System (blue), kept the rover on course, orienting the main retrorockets (white) to the their correct upright position. Subsequent igniting of these rockets reduced the rover's speed to near zero, 23 feet (7 meters) above the martian surface. The colored bars to the right indicate Spirit's north, east and downward velocities. The telemetry was acquired through the Mars Global Surveyor.

Image credit: NASA/JPL/Langley

A Flyby Tour of Spirit's Descent-2

Telemetry sent down to Earth from the Mars Exploration Rover Spirit has been reconstructed to create this computer-generated movie of the rover's final 30 seconds before landing at Gusev Crater, Mars. Just seconds before the rover touched down and its airbags were inflated, a gust of wind threatened to significantly increase the rover's horizontal speed. But the firing of a lateral rocket, called the Tranverse Impulse Rocket System (blue), kept the rover on course, orienting the main retrorockets (white) to the their correct upright position. Subsequent igniting of these rockets reduced the rover's speed to near zero, 23 feet (7 meters) above the martian surface. The telemetry was acquired through the Mars Global Surveyor.

Image credit: NASA/JPL/Analytical Mechanics Associates

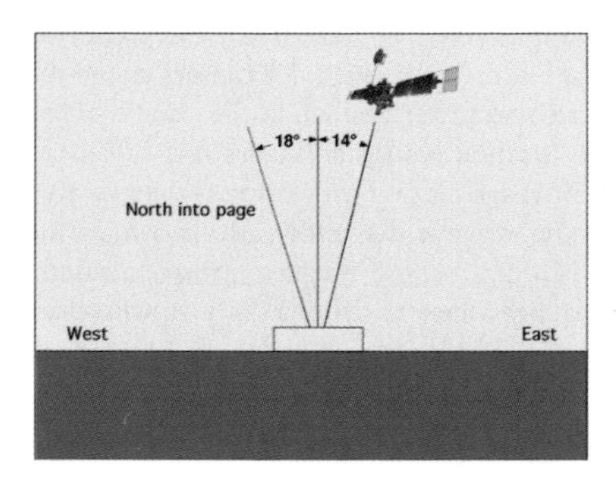

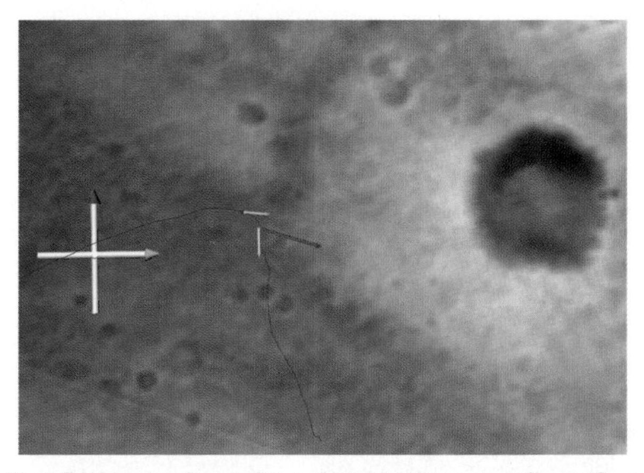

Wind Gusts: No Longer a Rover's Achilles Heel

This image shows the path (blue line) taken by the Mars Exploration Rover Spirit during its descent to Gusev Crater, Mars. Just seconds before landing, the rover fired its lateral rocket, called the Tranverse Impulse Rocket System, to protect against a horizontal gust of wind. The turquoise and yellow arrows show the actual speed and direction of Spirit; the purple arrow indicates what the rover's speed and direction would have been without the corrective maneuver. The red dot indicates where the parachute bridle was cut. North is denoted by the red-tipped arrow in the white cross. This picture consists of reconstructed telemetry mapped on top of surface images captured by the descent image motion estimation system camera located on the bottom of the rover.

Image credit: NASA/JPL

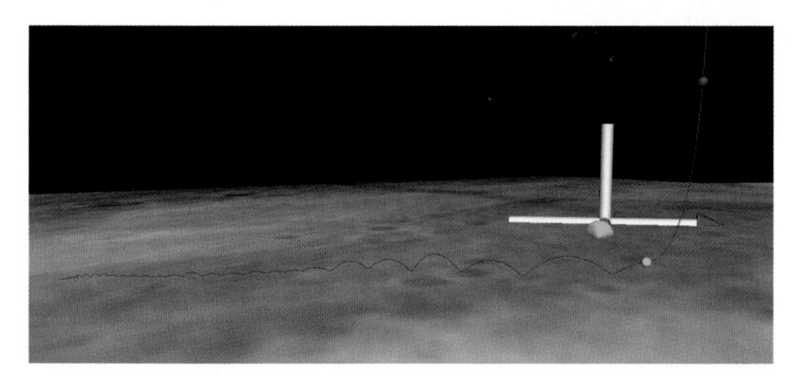

Bouncing Down to Mars

This image shows the path (blue line) taken by the Mars Exploration Rover Spirit as it bounced down to its final resting spot in Gusev Crater, Mars. Data taken by Spirit during descent indicates that the rover bounced 28 times, including one dip into a crater. The green dot shows where the parachute bridle was cut, and the red dot indicates where the main retrorockets were fired. North is denoted by the red-tipped arrow in the white cross. This picture consists of reconstructed telemetry data mapped on top of surface images captured by the descent image motion estimation system camera located on the bottom of the rover.

Image credit: NASA/JPL

On Its Own

This 3-D image combines computer-generated models of the Mars Exploration Rover Spirit and its lander with real surface data from the rover's panoramic camera. It shows Spirit's position just after it rolled off the lander on Jan. 15, 2004.

Image credit: NASA/JPL

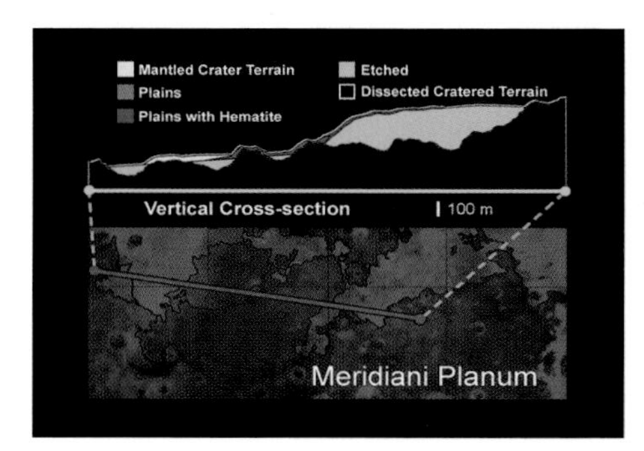

Hematite Deposits at Opportunity Landing Site

This vertical cross-section of the Meridiani Planum region shows that the hematite-bearing plains are part of an extensive set of deposits on top of the ancient, heavily cratered terrain. The Mars Exploration Rover Opportunity is targeted to land here on January 24, 2004 Pacific Standard Time. The background surface image of Meridiani Planum was acquired by the Mars Orbital Camera on NASA's Mars Global Surveyor. On Earth, grey hematite is an iron oxide mineral that typically forms in the presence of liquid water. The rover Opportunity will study the martian terrain and examine the hematite deposits to determine whether liquid water was present in the past when rocks were being formed.

Image credit: NASA/JPL/ASU

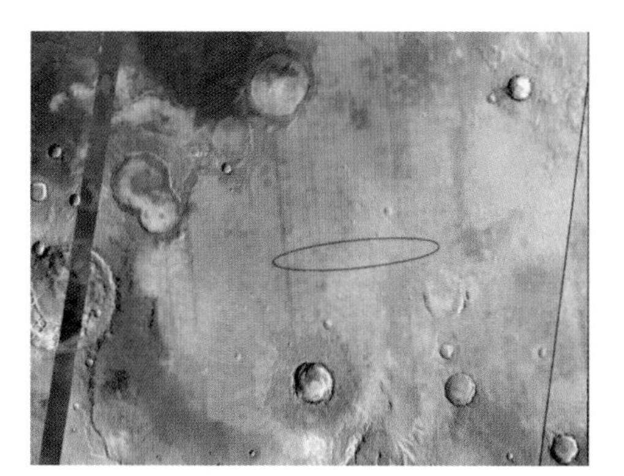

Targeting a Hematite-rich Terrain

This image shows the abundance and location of the mineral grey hematite at the Mars Exploration Rover Opportunity's landing site, Meridiani Planum, Mars. Opportunity is targeted to land somewhere inside the oval, approximately 71 kilometers (45 miles) long, on January 24, 2004 Pacific Standard Time. The background surface image of Meridiani Planum is a mosaic of daytime infrared images acquired by the thermal emission imaging system instrument on NASA's Mars Odyssey orbiter. Superimposed on this image mosaic is a rainbow-colored map showing the abundance and location of grey hematite, as mapped by the thermal emission spectrometer on NASA's Mars Global Surveyor orbiter. Red and yellow indicates higher concentrations, whereas green and blue areas denote lower levels. On Earth, grey hematite is an iron oxide mineral that typically forms in the presence of liquid water. The rover Opportunity will study the martian terrain to determine whether liquid water was present in the past when rocks were being formed, and ultimately will address whether that past environment was favorable for life.

Image credit: NASA/JPL/ASU

Mars Exploration Rover Mission Status

January 23, 2004

NASA's Spirit rover did not go to sleep today even after ground controllers sent commands twice for it to do so.

Shortly before noon, controllers were surprised to receive a relay of data from Spirit via the Mars Odyssey orbiter. Spirit sent 73 megabits at a rate of 128 kilobits per second. The transmission included power subsystem engineering data, no science data, and several frames of "fill data." Fill data are sets of intentionally random numbers that do not provide information.

Spirit had not communicated successfully through Odyssey since the rover's communications difficulties began on Wednesday.

Spirit's twin, Opportunity, will reach Mars at 05:05 Universal Time on Jan. 25 (12:05 a.m. Sunday EST or 9:05 p.m. Saturday PST) at a landing site on the opposite side of the planet from Spirit.

JPL, a division of the California Institute of Technology, Pasadena, manages the Mars Exploration Rover project for NASA's Office of Space Science, Washington, D.C. Additional information about the project is available from JPL at http://marsrovers.jpl.nasa.gov and from Cornell University, Ithaca, N.Y., at http://athena.cornell.edu .

Guy Webster (818) 354-5011 Jet Propulsion Laboratory, Pasadena, California
Donald Savage (202) 358-1547 NASA Headquarters, Washington, D.C.
NEWS RELEASE: 2004-033

Spirit Condition Upgraded As Twin Rover Nears Mars

January 24, 2004

Hours before NASA's Opportunity rover will reach Mars, engineers have found a way to communicate reliably with its twin, Spirit, and to get Spirit's computer out of a cycle of rebooting many times a day.

Spirit's responses to commands sent this morning confirm a theory developed overnight that the problem is related to the rover's two "flash" memories or software controlling those memories.

"The rover has been upgraded from critical to serious," said Mars Exploration Rover Project Manager Peter Theisinger at NASA's Jet Propulsion Laboratory, Pasadena, Calif. Significant work is still ahead for restoring Spirit, he predicted.

Opportunity is on course for landing in the Meridiani Planum region of Mars. The center of an ellipse covering the area where the spacecraft has a 99 percent chance of landing is just 11 kilometers (7 miles) from the target point. That point was selected months ago. Mission managers chose not to use an option for making a final adjustment to the flight path. Previously, the third and fifth out of five scheduled maneuvers were skipped as unnecessary. " We managed to target Opportunity to the desired atmospheric entry point, which will bring us to the target landing site, in only three maneuvers," said JPL's Dr. Louis D'Amario, navigation team chief for the rovers.

Opportunity will reach Mars at 05:05 Sunday, Universal Time (12:05 a.m. Sunday EST or 9:05 p.m. Saturday PST).

From the time Opportunity hits the top of Mars' atmosphere at about 5.4 kilometers per second (12,000 miles per hour) to the time it hits the surface 6 minutes later, then bounces, the rover will be going through the riskiest part of its mission. Based on analysis of Spirit's descent and on weather reports about the atmosphere above Meridiani Planum, mission controllers have decided to program Opportunity to open its parachute slightly earlier than Spirit did.

Mars is more than 10 percent farther from Earth than it was when Spirit landed. That means radio signals from Opportunity during its descent and after rolling to a stop have a lower chance of being detected on Earth. About four hours after the landing, news from the spacecraft may arrive by relay from NASA's Mars Odyssey orbiter. However, that will depend on Opportunity finishing critical activities, such as opening the lander petals and unfolding the rover's solar panels, before Odyssey flies overhead.

Spirit has 256 megabytes of flash memory, a type commonly used on gear such as digital cameras for holding data even when the power is off. Engineers confirmed this morning that Spirit's recent symptoms are related to the flash memory when they commanded the rover to boot up and utilize its random-access memory instead of flash memory. The rover then obeyed commands about communicating and going into sleep mode. Spirit communicated successfully at 120 bits per second for nearly an hour.

"We have a vehicle that is stable in power and thermal, and we have a working hypothesis we have confirmed," Theisinger said. By commanding Spirit each morning into a mode that avoids using flash memory, engineers plan to get it to communicate at a higher data rate, to diagnose the root cause of the problem and develop ways to restore as much functioning as possible.

The work on restoring Spirit is not expected to slow the steps in getting Opportunity ready to roll off its lander platform if Opportunity lands safely. For Spirit, those steps took 12 days. The rovers' main task is to explore their landing sites for evidence in the rocks and soil about whether the sites' past environments were ever watery and possibly suitable for sustaining life.

JPL, a division of the California Institute of Technology in Pasadena, manages the Mars Exploration Rover project for NASA's Office of Space Science, Washington. Images and additional information about the project are available from JPL at http://marsrovers.jpl.nasa.gov and from Cornell University, Ithaca, N.Y., at http://athena.cornell.edu .

Guy Webster (818) 354-5011 Jet Propulsion Laboratory, Pasadena, California
Donald Savage (202) 358-1547 NASA Headquarters, Washington, D.C.
NEWS RELEASE: 2004-033

NASA Hears From Opportunity Rover On Mars

January 25, 2004

Mission members celebrate Opportunity's arrival on Mars

NASA's second Mars Exploration Rover successfully sent signals to Earth during its bouncy landing and after it came to rest on one of the three side petals of its four-sided lander.

Mission engineers at NASA's Jet Propulsion Laboratory, Pasadena, Calif., received the first signal from Opportunity on the ground at 9:05 p.m. Pacific Standard Time Saturday via the NASA Deep Space Network, which was listening with antennas in California and Australia.

"We're on Mars, everybody!" JPL's Rob Manning, manager for development of the landing system, announced to the cheering flight team.

NASA Administrator Sean O'Keefe said at a subsequent press briefing, "This was a tremendous testament to how NASA, when really focused on an objective, can put every ounce of effort, energy, emotion and talent to an important task. This team is the best in the world, no doubt about it."

Opportunity landed in a region called Meridiani Planum, halfway around the planet from the Gusev Crater site where its twin rover, Spirit, landed three weeks ago. Earlier today, mission managers reported progress in understanding and dealing with communications and computer problems on Spirit.

"In the last 48 hours, we've been on a roller coaster," said Dr. Ed Weiler, NASA associate administrator for space science. "We resurrected one rover and saw the birth of another." JPL's Pete Theisinger, project manager for the rovers, said, "We are two for two. Here we are tonight with Spirit on a path to recovery and with Opportunity on Mars."

By initial estimates, Opportunity landed about 24 kilometers (15 miles) down range from the center of the target landing area. That is well within an outcropping of a mineral called gray hematite, which usually forms in the presence of water. "We're going to have a good place to do science," said JPL's Richard Cook, deputy project manager for the rovers.

Once it pushed itself upright by opening the petals of the lander, Opportunity was expected to be facing east.

The main task for both rovers in coming months is to explore the areas around their landing sites for evidence in rocks and soils about whether those areas ever had environments that were watery and possibly suitable for sustaining life.

JPL, a division of the California Institute of Technology in Pasadena, manages the Mars Exploration Rover project for NASA's Office of Space Science, Washington. Images and additional information about the project are available from JPL at http://marsrovers.jpl.nasa.gov and from Cornell University, Ithaca, N.Y., at http://athena.cornell.edu .

Guy Webster (818) 354-5011 Jet Propulsion Laboratory, Pasadena, California
Donald Savage (202) 358-1547 NASA Headquarters, Washington, D.C.
NEWS RELEASE: 2004-035

First Images Of Opportunity Site Show Bizarre Landscape

January 25, 2004

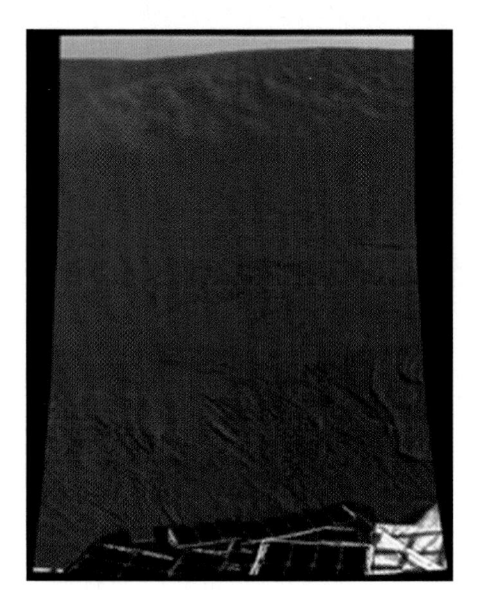

First color image from Opportunity

NASA's Opportunity rover returned the first pictures of its landing site early today, revealing a surreal, dark landscape unlike any ever seen before on Mars.

Opportunity relayed the images and other data via NASA's Mars Odyssey orbiter. The data showed that the spacecraft is healthy, said Matt Wallace, mission manager at NASA's Jet Propulsion Laboratory.

"Opportunity has touched down in a bizarre, alien landscape," said Dr. Steve Squyres of Cornell University, Ithaca, N.Y., principal investigator for the science instruments on Opportunity and its twin, Spirit. "I'm flabbergasted. I'm astonished. I'm blown away."

The terrain is darker than at any previous Mars landing site and has the first accessible bedrock outcropping ever seen on Mars. The outcropping immediately became a candidate target for the rover to visit and examine up close.

Wallace noted that the straight-ahead path looks clear for the rover to roll off its lander platform. The rover is facing north-northeast.

JPL Administrator Dr. Charles Elachi said, "This team succeeded the old fashioned way. They were excellent, they were determined, and they worked very hard."

JPL, a division of the California Institute of Technology in Pasadena, manages the Mars Exploration Rover project for NASA's Office of Space Science, Washington. Images and additional information about the project are available from JPL at http://marsrovers.jpl.nasa.gov and from Cornell University, Ithaca, N.Y., at http://athena.cornell.edu .

Guy Webster (818) 354-5011 Jet Propulsion Laboratory, Pasadena, California

Donald Savage (202) 358-1547 NASA Headquarters, Washington, D.C.
NEWS RELEASE: 2004-036

Opportunity Sits In A Small Crater, Near A Bigger One

January 25, 2004

A small impact crater on Mars is the new home for NASA's Opportunity rover, and a larger crater lies nearby. Scientists value such crater locations as a way to see what's beneath the surface without needing to dig.

Encouraging developments continued for Opportunity's twin, Spirit, too. Engineers have determined that Spirit's flash memory hardware is functional, strengthening a theory that Spirit's main problem is in software that controls file management of the memory. "I think we've got a patient that's well on the way to recovery," said Mars Exploration Rover Project Manager Pete Theisinger at NASA's Jet Propulsion Laboratory, Pasadena, Calif.

Opportunity returned the first pictures of its landing site early today, about four hours after reaching Mars. The pictures indicate that the spacecraft sits in a shallow crater about 20 meters (66 feet) across.

"We have scored a 300-million mile interplanetary hole in one," said Dr. Steve Squyres of Cornell University, Ithaca, N.Y., principal investigator for the science instruments on both rovers.

NASA selected Opportunity's general landing area within a region called Meridiani Planum because of extensive deposits of a mineral called crystalline hematite, which usually forms in the presence of liquid water. Scientists had hoped for a specific landing site where they could examine both the surface layer that's rich in hematite and an underlying geological feature of light-colored layered rock. The small crater appears to have exposures of both, with soil that could be the hematite unit and an exposed outcropping of the lighter rock layer.

"If it got any better, I couldn't stand it," said Dr. Doug Ming, rover science team member from NASA Johnson Space Center, Houston. With the instruments on the rover and just the rocks and soil within the small crater, Opportunity should allow scientists to determine which of several theories about the region's past environment is right, he said. Those theories include that the hematite may have formed in a long-lasting lake or in a volcanic environment.

An even bigger crater, which could provide access to deeper layers for more clues to the past, lies nearby. Images taken by a camera on the bottom of the lander during Opportunity's final descent show a crater about 150 meters (about 500 feet) across likely to be within about one kilometer or half mile of the landing site, said Dr. Andrew Johnson of JPL. He is an engineer for the descent imaging system that calculated the spacecraft's horizontal motion during its final seconds of flight. The system determined that sideways motion was small, so Opportunity's computer decided not to fire the lateral rockets carried specifically for slowing that motion.

Squyres presented an outline for Opportunity's potential activities in coming weeks and months. After driving off the lander, the rover will first examine the soil right next to the lander, then drive to the outcrop of layered-looking rocks and spend considerable time examining it. Then the rover may climb out of the small crater, take a look around, and head for the bigger crater.

But first, Opportunity will spend more than a week — perhaps two — getting ready to drive off the lander, if all goes well. Engineering data from Opportunity returned in relays via NASA's Mars Odyssey orbiter early this morning and at midday indicate the spacecraft is in excellent health, said JPL's Arthur Amador, mission manager. The rover will try its first direct-to-Earth communications this evening.

The main task for both rovers in coming months is to explore the areas around their landing sites for evidence in rocks and soils about whether those areas ever had environments that were watery and possibly suitable for sustaining life.

JPL, a division of the California Institute of Technology in Pasadena, manages the Mars Exploration Rover project for NASA's Office of Space Science, Washington. Images and additional information about the project are available from JPL at http://marsrovers.jpl.nasa.gov and from Cornell University, Ithaca, N.Y., at http://athena.cornell.edu .

Guy Webster (818) 354-5011 Jet Propulsion Laboratory, Pasadena, California
Donald Savage (202) 358-1547 NASA Headquarters, Washington, D.C.
NEWS RELEASE: 2004-037

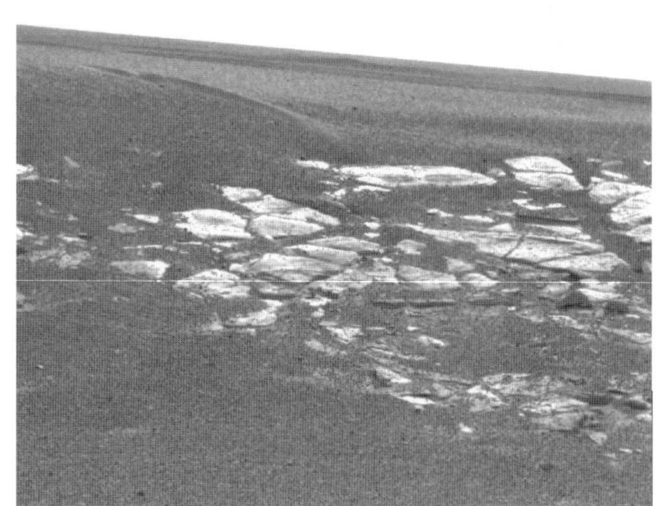

This "postcard" from the panoramic camera on the Mars Exploration Rover Opportunity shows the view of the martian landscape southwest of the rover. The image was taken in the late martian afternoon at Meridiani Planum on Mars, where Opportunity landed at approximately 9:05 p.m. PST on Saturday, Jan. 24.

Image credit: NASA/JPL/Cornell

Crater Down Below

Scientists believe the circular feature in this image to be a crater near the Mars Exploration Rover Opportunity. The rover landed at Meridiani Planum on Mars at approximately 9:05 p.m. PST on Saturday, Jan. 24. This image was taken at an altitude of 1,986 meters (6,516 feet) by the descent image motion estimation system camera located on the bottom of the rover. The image spans approximately 1.6 kilometers (1 mile) across the surface of Mars.

Image credit: NASA/JPL

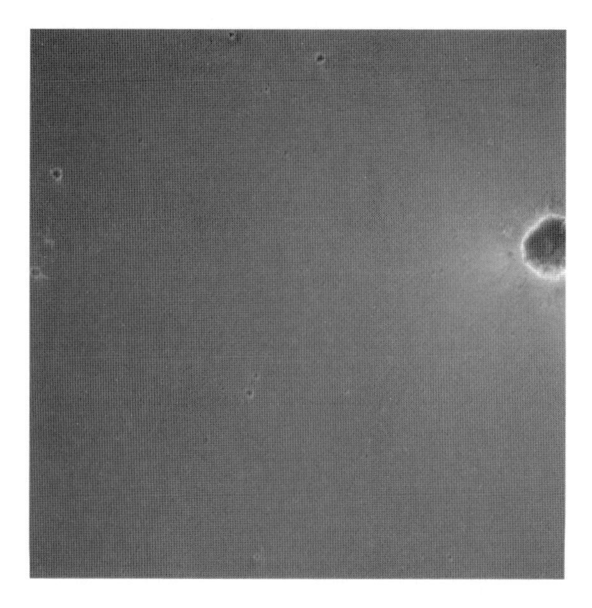

Crater Down Below-2

Scientists believe the circular feature in this image to be a crater near the Mars Exploration Rover Opportunity. The rover landed at Meridiani Planum on Mars at approximately 9:05 p.m. PST on Saturday, Jan. 24. This image was taken at an altitude of 1,690 meters (5,545 feet) by the descent image motion estimation system camera located on the bottom of the rover. The image spans approximately 1.4 kilometers (7/8 of a mile) across the surface of Mars.

Image credit: NASA/JPL

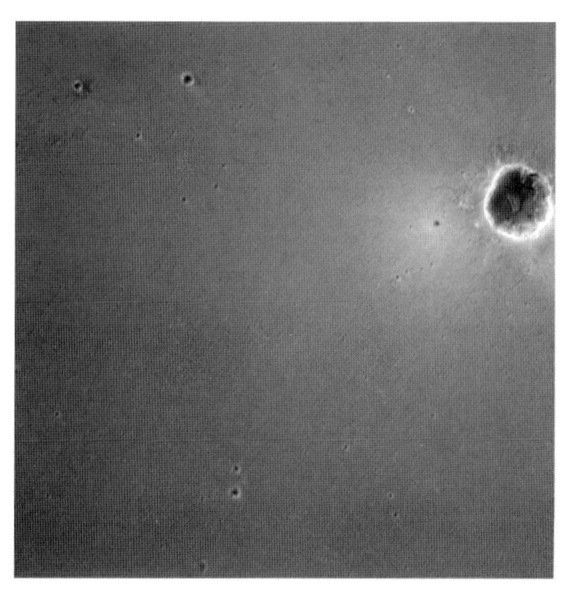

Crater Down Below-3

Scientists believe the circular feature in this image to be a crater near the Mars Exploration Rover Opportunity. The rover landed at Meridiani Planum on Mars at approximately 9:05 p.m. PST on Saturday, Jan. 24. This image was taken at an altitude of 1,404 meters (4,606 feet) by the descent image motion estimation system camera located on the bottom of the rover. The image spans approximately 1.2 kilometers (3/4 of a mile) across the surface of Mars.

Image credit: NASA/JPL

Gear On Opportunity Rover Passes Martian Health Check

January 26, 2004

During the second day on Mars for NASA's Opportunity rover, key science instruments passed health tests and the rover made important steps in communicating directly with Earth.

Halfway around the planet, during its 22nd day on Mars, NASA's Spirit obeyed commands for transmitting information that is helping engineers set a strategy for fixing problems with the rover's computer memory.

On Earth this morning, scientists marveled at a high-resolution color "postcard" of Opportunity's surroundings. The mosaic of 24 frames from the panoramic camera shows details from the edge of the lander to the distant horizon beyond the rim of the rover's small home crater.

"We're looking out across a pretty spectacular landscape," said Dr. Jim Bell of Cornell University, Ithaca, N.Y., lead scientist for the panoramic cameras on Spirit and Opportunity. "It's going to be a wonderful area for geologists to explore with the rover."

The color view shows dark soil that brightened where it was compacted by the rolling spacecraft, and an outcropping of bedrock on the inside slope of the 20-meter (66-foot) crater in which the rover sits. Opportunity will be commanded to finish taking a 360-degree color panorama of the site during its third Mars day, which began at 12:01 p.m. PST today.

Another major step planned for Opportunity's third day is to begin using its high-gain antenna for communicating directly with Earth at a high data rate, said Jackie Lyra of NASA's Jet Propulsion Laboratory, Pasadena, Calif., activity lead for this rover event. In preparation for this transition, Opportunity found the Sun with its panoramic camera yesterday. Once oriented by knowing the position of the Sun, it can calculate how to point its high-gain antenna toward Earth.

"We're making steady progress in our effort to get the wheels of the rover dirty," said Mission Manager Jim Erickson of JPL. Still the earliest scenario for the rover to drive off its lander platform is more than a week away.

Opportunity has tested the three scientific sensing instruments on its robotic arm that will be used for up-close examination of rocks and soil: the microscopic imager, the alpha particle X-ray spectrometer for determining what elements are present, and a Moessbauer spectrometer for identifying iron-containing minerals. "I'm pleased to report that all are in perfect health," said Dr. Steve Squyres of Cornell University, Ithaca, N.Y., principal investigator for the science instruments on the rovers.

Squyres had been especially concerned about the Moessbauer spectrometer because tests conducted while the spacecraft was on its way to Mars showed that an internal calibration system was not working as intended. However, after the rover landed on Mars, the instrument is functioning normally again. The Moessbauer spectrometer's function for identifying iron-bearing minerals will be important in the scientific goal of determining the origin of iron-bearing hematite deposits in the Meridiani Planum region selected as Opportunity's landing site.

"We have a perfectly functioning Moessbauer spectrometer, and given that we are now perched atop the hematite capital of the Solar System, that's a good thing," Squyres said.

Restoration efforts continue making progress on Spirit. "We have a patient in rehab, and we're nursing her back to health," said JPL's Jennifer Trosper, mission manager.

Engineers found a way to stop Spirit's computer from resetting itself about once an hour by putting the spacecraft into a mode that avoids use of flash memory. Flash memory is a type common in many electronic products, such as digital cameras, for storing information even when the power is off. The rover also has random-access memory, which cannot hold information during the rover's overnight sleep. One of the next steps planned is to erase from flash memory the files stored there from the spacecraft's cruise to Mars from Earth. That is intended to lessen the task of managing the flash memory files.

The rovers' main task is to explore their landing sites during coming months for evidence in the rocks and soil about whether the sites' past environments were ever watery and possibly suitable for sustaining life.

JPL, a division of the California Institute of Technology in Pasadena, manages the Mars Exploration Rover project for NASA's Office of Space Science, Washington, D.C. Images and additional information about the project are available from JPL at http://marsrovers.jpl.nasa.gov and from Cornell University, Ithaca, N.Y., at http://athena.cornell.edu .

Guy Webster (818) 354-5011 Jet Propulsion Laboratory, Pasadena, Calif.
Donald Savage (202) 358-1547 NASA Headquarters, Washington, D.C.
NEWS RELEASE: 2004-036

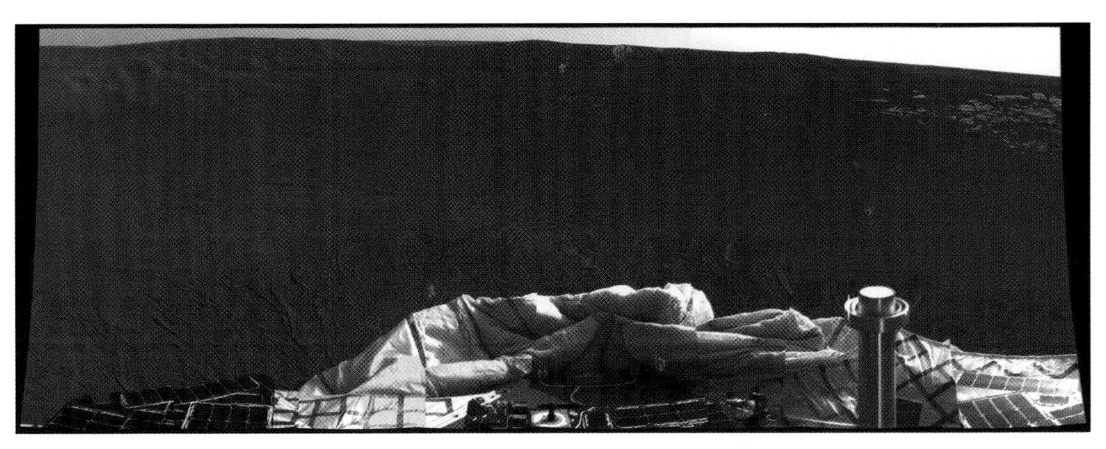

A Hole in One

The interior of a crater surrounding the Mars Exploration Rover Opportunity at Meridiani Planum on Mars can be seen in this color image from the rover's panoramic camera. This is the darkest landing site ever visited by a spacecraft on Mars. The rim of the crater is approximately 10 meters (32 feet) from the rover. The crater is estimated to be 20 meters (65 feet) in diameter. Scientists are intrigued by the abundance of rock outcrops dispersed throughout the crater, as well as the crater's soil, which appears to be a mixture of coarse gray grains and fine reddish grains.

Data taken from the camera's near-infrared, green and blue filters were combined to create this approximate true color picture, taken on the first day of Opportunity's journey. The view is to the west-southwest of the rover.

Image credit: NASA/JPL/Cornell

Over the Rover

This image from the navigation camera on the Mars Exploration Rover Opportunity has been projected to show an overhead perspective of the rover. This image was taken shortly after the rover touched down at Meridiani Planum, Mars, at approximately 9:05 p.m. PST on Saturday, Jan. 24.

Image credit: NASA/JPL

Ring Around the Rover

This polar projection of an image from the navigation camera on the Mars Exploration Rover Opportunity shows an overhead perspective of the rover. Opportunity's view of the martian horizon can also be seen in this image, taken shortly after the rover touched down at Meridiani Planum, Mars at 9:05 p.m. PST on Saturday, Jan. 24.

Image credit: NASA/JPL

Scientists Thrilled To See Layers in Mars Rocks Near Opportunity

January 27, 2004

New pictures from NASA's Mars Exploration Rover Opportunity reveal thin layers in rocks just a stone's throw from the lander platform where the rover temporarily sits.

Geologists said that the layers — some no thicker than a finger — indicate the rocks likely originated either from sediments carried by water or wind, or from falling volcanic ash. "We should be able to distinguish between those two hypotheses," said Dr. Andrew Knoll of Harvard University, Cambridge, a member of the science team for Opportunity and its twin, Spirit. If the rocks are sedimentary, water is a more likely source than wind, he said.

The prime goal for both rovers is to explore their landing areas for clues in the rocks and soil about whether those areas ever had watery environments that could possibly have sustained life.

Controllers at NASA's Jet Propulsion Laboratory, Pasadena, Calif., plan to tell Opportunity tonight to start standing up from the crouched and folded posture in which it traveled to Mars.

"We're going to lift the entire rover, then the front wheels will be turned out," said Mission Manager Jim Erickson of JPL. Several more days of activities are still ahead before the rover will be ready to drive off the lander.

"We're about to embark on what could be the coolest geological field trip in history," said Dr. Steve Sqyures of Cornell University, Ithaca, N.Y., principal investigator for the rovers' science payload.

The layered rocks are in a bedrock outcrop about 30 to 45 centimeters (12 to 18 inches) tall, and only about eight meters (26 feet) away from where Opportunity came to rest after bouncing to a landing three days ago. Examination of their texture and composition with the cameras and spectrometers on the rover may soon reveal whether they are sedimentary, Knoll predicted.

Scientists also hope to determine the relationship between those light-colored rocks and the dark soil that covers most of the surrounding terrain. The soil may contain the mineral hematite, which was identified from orbit and motivated the choice of Opportunity's landing area, Squyres said.

Opportunity successfully used its high-gain antenna for the first time yesterday. The rover is losing some if its battery charge each night, apparently due to an electric heater at the shoulder joint of the rover's robotic arm. A thermostat turns on the heater whenever the air temperature falls to levels that Opportunity is experiencing every night. The heater is not really needed when the arm is not in use, but ground control has not been able to activate a switch designed to override the thermostat, Erickson said. Mission engineers are working to confirm the diagnosis, determine the ramifications of the power drain, and propose workarounds or fixes.

Meanwhile, engineers working on Spirit have determined that the high-gain antenna on that rover is likely in working order despite earlier indications of a possible problem. They are continuing to take information out of Spirit's flash memory. Results from a testbed simulator of the rover's electronics supported the diagnosis of a problem with management of the flash memory, reported JPL's Jennifer Trosper, mission manager.

JPL, a division of the California Institute of Technology in Pasadena, manages the Mars Exploration Rover project for NASA's Office of Space Science, Washington, D.C. Images and additional information about the project are available from JPL at http://marsrovers.jpl.nasa.gov and from Cornell University at http://athena.cornell.edu .

Guy Webster (818) 354-5011 Jet Propulsion Laboratory, Pasadena, Calif.
Donald Savage (202) 358-1547 NASA Headquarters, Washington, D.C.
NEWS RELEASE: 2004-039

Not of this Earth (3-D) (See fold-out)

This sweeping 3-D look at the unusual rock outcropping near the Mars Exploration Rover Opportunity was captured by the rover's panoramic camera. Scientists believe the layered rocks are either volcanic ash deposits, or sediments laid down by wind or water. Opportunity landed at Meridiani Planum, Mars on January 24 at 9:05 p.m. PST.

Image credit: NASA/JPL/Cornell

Not of this Earth

This sweeping look at the unusual rock outcropping near the Mars Exploration Rover Opportunity was captured by the rover's right panoramic camera. Scientists believe the layered rocks are either volcanic ash deposits, or sediments laid down by wind or water. Opportunity landed at Meridiani Planum, Mars on January 24 at 9:05 p.m. PST.

Image credit: NASA/JPL/Cornell

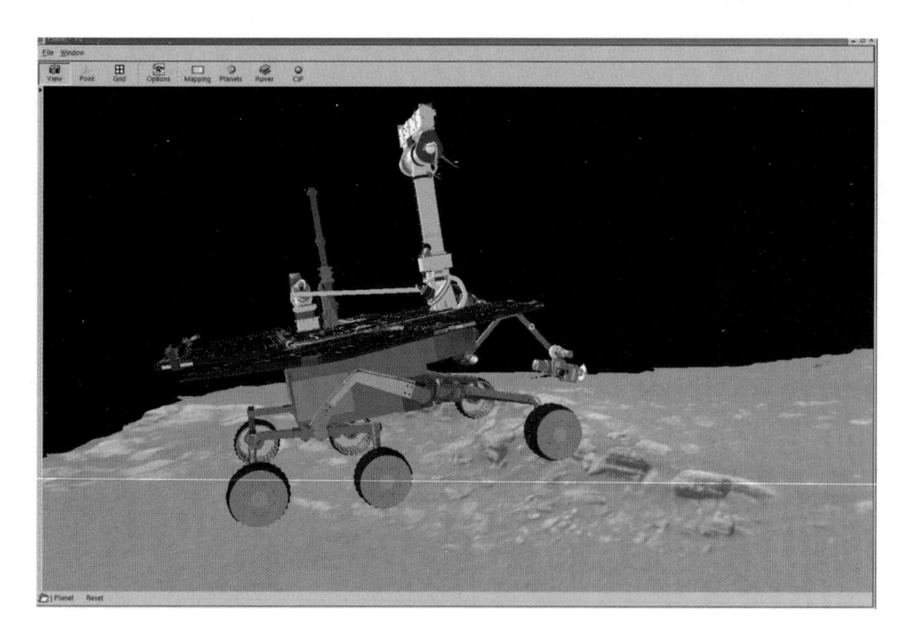

A Precious Opportunity

This three-dimensional model superimposes the Mars Exploration Rover Opportunity on one of its potential targets, a scientific treasure chest of martian rocks contained within the landing site, a crater on Meridiani Planum, Mars. The rover is placed on the rock outcrop for scale. Opportunity has not yet visited these rocks; it is currently still on its lander. Scientists plan to use the tools on the rover's instrument deployment device, or robotic "arm," to examine these rocks, which are about 10 centimeters (4 inches) high and approximately 8 meters (26 feet) away from the rover. The image of the terrain was acquired on Sol, or martian day, 2 of Opportunity's journey. This model was created using data from the rover's panoramic camera and is displayed using software developed by NASA's Ames Research Center.

Image credit: NASA/JPL/Cornell/Ames/Maas Digital LLC

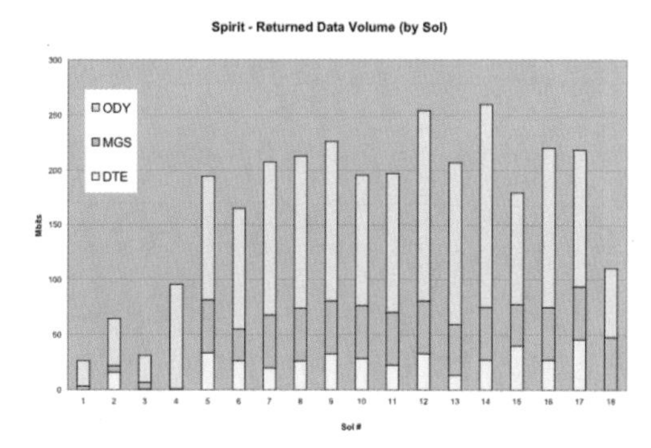

Martian Landmarks Dedicated to Apollo 1 Crew

January 27, 2004

NASA memorialized the Apollo 1 crew — Gus Grissom, Ed White and Roger Chaffee — by dedicating the hills surrounding the Mars Exploration Rover Spirit's landing site to the astronauts. The crew of Apollo 1 perished in flash fire during a launch pad test of their Apollo spacecraft at Kennedy Space Center, Fla., 37 years ago today.

"Through recorded history explorers have had both the honor and responsibility of naming significant landmarks," said NASA administrator Sean O'Keefe. "Gus, Ed and Roger's contributions, as much as their sacrifice, helped make our giant leap for mankind possible. Today, as America strides towards our next giant leap, NASA and the Mars Exploration Rover team created a fitting tribute to these brave explorers and their legacy."

Newly christened "Grissom Hill" is located 7.5 kilometers (4.7 miles) to the southwest of Spirit's position. "White Hill" is 11.2 kilometers (7 miles) northwest of its position and "Chaffee Hill" is 14.3 kilometers (8.9 miles) south-southwest of rover's position.

Lt. Colonel Virgil I. "Gus" Grissom was a U.S. Air Force test pilot when he was selected in 1959 as one of NASA's Original Seven Mercury Astronauts. On July 21, 1961, Grissom became the second American and third human in space when he piloted Liberty Bell 7 on a 15 minute sub-orbital flight. On March 23, 1965 he became the first human to make the voyage to space twice when he commanded the first manned flight of the Gemini space program, Gemini 3. Selected as commander of the first manned Apollo mission, Grissom perished along with White and Chaffee in the Apollo 1 fire. He is buried at Arlington National Cemetery, Va.

Captain Edward White was a US Air Force test pilot when selected in 1962 as a member of the "Next Nine," NASA's second astronaut selection. On June 3, 1965, White became the first American to walk in space during the flight of Gemini 4. Selected as senior pilot for the first manned Apollo mission, White perished along with Grissom and Chaffee in the Apollo 1 fire. He is buried at his alma mater, the United States Military Academy, West Point, N.Y.

Selected in 1963 as a member of NASA's third astronaut class, U.S. Navy Lieutenant Commander Roger Chaffee worked as a Gemini capsule communicator. He also researched flight control communications systems, instrumentation systems, and attitude and translation control systems for the Apollo Branch of the Astronaut office. On March 21, 1966, he was selected as pilot for the first 3-man Apollo flight. He is buried at Arlington National Cemetery, Va.

Images of the Grissom, White and Chaffee Hills can be found at: http://www.jpl.nasa.gov/mer2004/rover-images/jan-27-2004/captions/image-1.html

The Jet Propulsion Laboratory, Pasadena, Calif., manages the Mars Exploration Rover project for NASA's Office of Space Science, Washington, D.C. JPL is a division of the California Institute of Technology, also in Pasadena. Additional information about the project is available from JPL at http://marsrovers.jpl.nasa.gov and from Cornell University, Ithaca, N.Y., at http://athena.cornell.edu .

DC Agle (818) 393-9011 Jet Propulsion Laboratory
Donald Savage (202) 358-1547 NASA Headquarters, Washington, D.C.
NEWS RELEASE: 2004-40

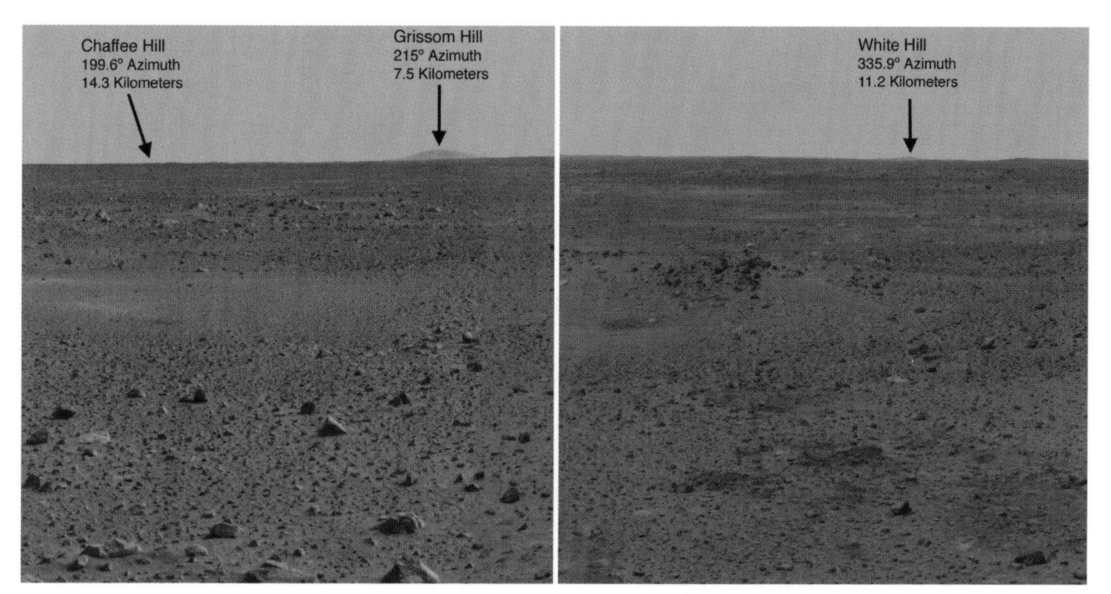

NASA Dedicates Martian Landmarks To Apollo 1 Crew

An image taken from Spirit's PanCam looking west depicts the nearby hills named after the astronauts of the Apollo 1. The crew of Apollo 1 perished in flash fire during a launch pad test of their Apollo spacecraft at Kennedy Space Center, Fl. on January 27, 1967.

Image credit: NASA/JPL/Cornell

An image taken by the Mars Global Surveyor's Mars Orbiter Camera of the Columbia Memorial Station and the nearby hills named after the Apollo I crew. "Grissom Hill" is located 7.5 kilometers (4.7 miles) to the Southwest of the rover Spirit's landing site. "White Hill" is 11.2 kilometers (7 miles) Northwest of its position and "Chaffee Hill" is 14.3 kilometers (8.9 miles) south-Southwest of Spirit.

Image credit: NASA/JPL/Cornell

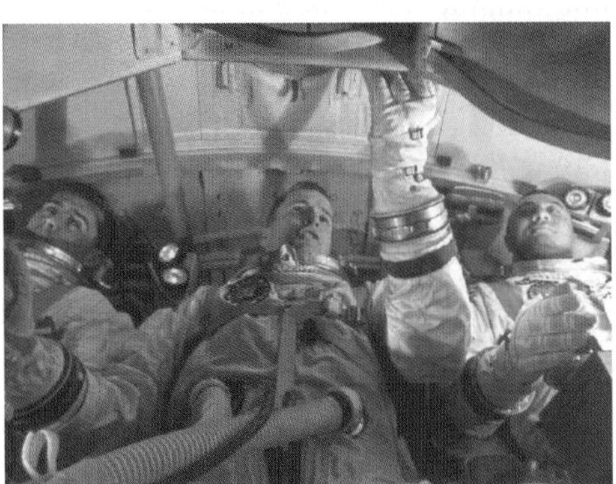

The Apollo I crew, from left to right, Roger Chaffee, Ed White and Gus Grissom.

Space Shuttle Challenger Crew Memorialized on Mars

January 28, 2004

NASA announced plans to name the landing site of the Mars Opportunity rover in honor of the Space Shuttle Challenger's final crew. The area in the vast flatland called Meridiani Planum, where Opportunity landed this weekend, will be called the Challenger Memorial Station.

The seven-member crew of Space Shuttle Challenger was lost when the orbiter suffered an in-flight breakup during launch Jan. 28, 1986, 18 years ago today.

NASA selected Meridiani Planum as a landing site because of extensive deposits of a mineral called crystalline hematite, which usually forms in the presence of liquid water. Scientists had hoped for a specific landing site where they could examine both the surface layer that's rich in hematite and an underlying geological feature of light-colored layered rock. The small crater in which Opportunity alighted appears to have exposures of both, with soil that could be the hematite unit and an exposed outcropping of the lighter rock layer.

Challenger's 10th flight was to have been a six-day mission dedicated to research and education, as well as the deployment of the Tracking and Data Relay Satellite-B communications satellite.

Challenger's commander was Francis R. Scobee and the mission pilot was Michael J. Smith. Mission specialists included Judith A. Resnik, Ellison S. Onizuka and Ronald E. McNair. The mission also carried two payload specialists, Gregory B. Jarvis and Sharon Christa McAuliffe, who was the agency's first teacher in space.

Opportunity successfully landed on Mars January 25 (Eastern and Universal Time; January 24 Pacific Time). It will spend the next three months exploring the region surrounding what is now known as Challenger Memorial Station to determine if Mars was ever watery and suitable to sustain life.

Opportunity's twin, Spirit, is trailblazing a similar path on the other side of the planet, in a Connecticut-sized feature called Gusev Crater.

A composite image depicting the location of the Challenger Memorial Station can be found on the Web at:
http://www.jpl.nasa.gov/mer2004/rover-images/jan-28-2004/captions/image-1.html

NASA's Jet Propulsion Laboratory, Pasadena, Calif., is a division of the California Institute of Technology, also in Pasadena. JPL manages the Mars Exploration Rover mission for NASA's Office of Space Science in Washington, D.C.

Additional information about the project is available from NASA, JPL and Cornell University, Ithaca, N.Y., on the Internet at: http://www.nasa.gov/ , http://marsrovers.jpl.nasa.gov and http://athena.cornell.edu .

Glenn Mahone/Donald Savage (202) 358-1898/1547 NASA Headquarters, Washington, D.C.
DC Agle (818) 393-9011 Jet Propulsion Laboratory, Pasadena, Calif.
NEWS RELEASE: 2004-042

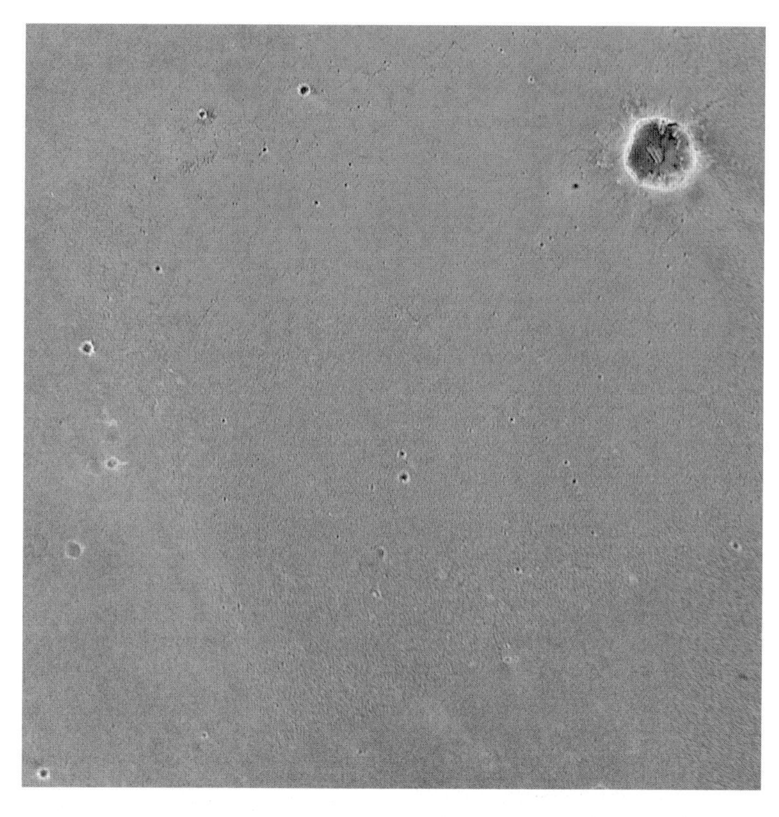

Challenger Memorial Station, Meridiani Planum, Mars

A composite image reveals the local region surrounding the Challenger Memorial Station. The image is actually an amalgamation of a Mars Global Surveyor's Mars Orbiter Camera image and the third and final picture taken by Opportunity's DIMES camera (Descent Image Motion Estimation System) during descent. The location of the site is a 20-meter (65.6 foot) wide, 2-meter (6.6 foot) deep crater somewhere in this composite image. The final crew of the Space Shuttle Challenger was lost when the shuttle suffered an in-flight breakup during launch on Jan. 28, 1986.

Image credit: NASA/JPL/MSSS

The Challenger crew.

Opportunity Rover Begins Standing Up

January 28, 2004

NASA's Opportunity rover has untucked its front wheels and latched its suspension system in place, key steps in preparing to drive off its lander and onto martian soil.

Overnight tonight, mission controllers at NASA's Jet Propulsion Laboratory, Pasadena, Calif., plan to try tilting the lander platform down in the front by pressing the rear petal downward to raise the back.

"What we want to do is lower the front edge by about 5 degrees," said JPL's Dr. Rick Welch, activity lead for preparing the rover for roll-off. Plans call for driving off straight ahead, possibly as early as overnight Sunday-Monday, if all goes well.

Meanwhile, halfway around Mars, Opportunity's twin, Spirit, continues on the mend from a computer memory problem that struck it a week ago. "Right now we're working to get complete control of the vehicle, and we're still not quite there," said JPL's Jennifer Trosper, mission manager. "If we're on the right track, we hope to be back doing some science by early next week. If we're not on the right track, it could take longer than that."

Opportunity's infrared sensing instrument, the miniature thermal emission spectrometer, passed a health check last night. Scientists plan to begin using it tonight. The instrument detects the composition of rocks and soils from a distance. That information will help scientists decide what targets to approach after Opportunity drives off the lander.

Scientists and rover engineers are already discussing which specific rocks within an outcropping near the lander will make the best targets, said Dr. Jim Bell of Cornell University, Ithaca, N.Y., lead scientist for the panoramic cameras on Opportunity and Spirit. Details of the outcrop can be seen in a new a color-picture mosaic Bell presented, the first portion of a full-circle panorama that has been taken and partially transmitted.

Other new images show how Opportunity's airbags left detailed impressions in the fine-textured soil as the spacecraft was rolling to a stop in the small crater where it now sits. "These marks are telling us about the physical properties of the material," Bell said.

Some scientists believe that dark colored granules covering most of the crater's surface were pressed down into an underlying layer of powdery, lighter red material when the airbags hit. Others hold to a theory that the dark granules are agglomerations that crumble into the finer, lighter material when disturbed. After roll-off, soil near the lander will be the rover's first target for close-up examination with a microscope and two tools for detecting the composition of the target. The soil at Opportunity's landing site appears to have different properties than the soil at Spirit's landing site, Bell said.

Opportunity has already validated predictions about the landing site made on the basis of images and measurements taken by spacecraft orbiting Mars, said JPL's Dr. Matt Golombek, a member of the rover science team and co-chair of a steering committee that evaluated potential landing sites for the rovers. The predictions included that the region of Meridiani Planum where Opportunity landed would be safe for landing, would be safe for rover driving, would have very few rocks and would look unlike any place previously seen on Mars.

"This bodes well for our ability to use remote sensing data in the future for picking landing sites," Golombek said.

Engineers have been able to confirm a diagnosis that an unplanned drawdown of battery power each night on

Opportunity is due to a heater on the rover's robotic arm. A switch designed to overrule the heater's thermostatic control has not been working. "In the near term, it's not providing any operational constraints," Welch said.

JPL, a division of the California Institute of Technology in Pasadena, manages the Mars Exploration Rover project for NASA's Office of Space Science, Washington, D.C. Images and additional information about the project are available from JPL at http://marsrovers.jpl.nasa.gov and from Cornell University at http://athena.cornell.edu .

Guy Webster (818) 354-5011 Jet Propulsion Laboratory, Pasadena, California
Donald Savage (202) 358-1547 NASA Headquarters, Washington, D.C.
NEWS RELEASE: 2004-43

An Opportunity to Rise

This image shows the Mars Exploration Rover Opportunity's wheels in their stowed configuration. As of 9:00 a.m. January 28, 2004, the rover had deployed its wheels and completed the first half of the stand-up process. This image was taken at Opportunity's landing site, Meridiani Planum, by the hazard-identification camera.

Image credit: NASA/JPL

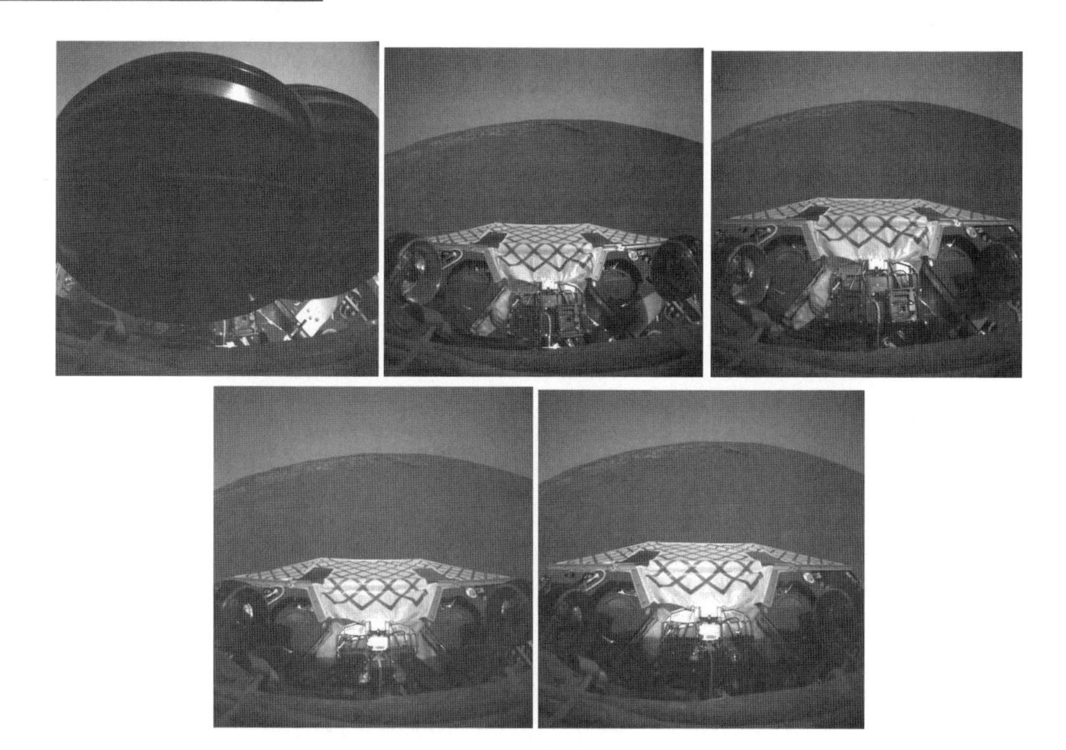

A Higher Opportunity (Animation sequence)

This animation strings together images from the rover's front hazard-identification camera taken during the first half of the stand-up process of the Mars Exploration Rover Opportunity at Meridiani Planum, Mars. The first frame shows the rover's wheels tucked under in pre-stand-up position. The following frames show the first two stages of the stand-up process, in which the rover elevates itself and unfolds the wheels.

Image credit: NASA/JPL

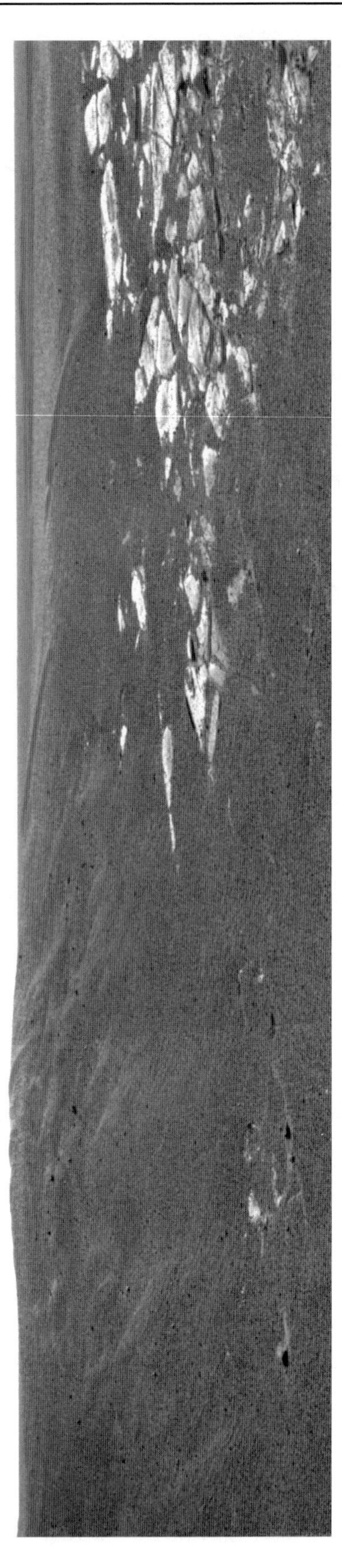

Shades and Shapes of Mars

This image captured by the Mars Exploration Rover Opportunity's panoramic camera highlights the flat and dark terrain of its landing site at Meridiani Planum. The landscape is in contrast to that of past landing sites on Mars, which show variations in color and topography. For example, the Viking 1 and Viking 2 missions observed rocky, dust-covered surfaces (PIA00393, PIA00568) much like those observed at Pathfinder's landing site (PIA02405). Gusev Crater, the landing site of the Mars Exploration Rover Spirit, is slightly darker in color but flat and speckled with a sparse array of rocks (PIA05102). Meridiani Planum has even fewer rocks than Gusev Crater and is darkest and cleanest of all the landing sites. This assortment of martian shades and shapes are further revealed in an image of the red planet taken by the Hubble Space Telescope (PIA03154).

Image credit: NASA/JPL/Cornell

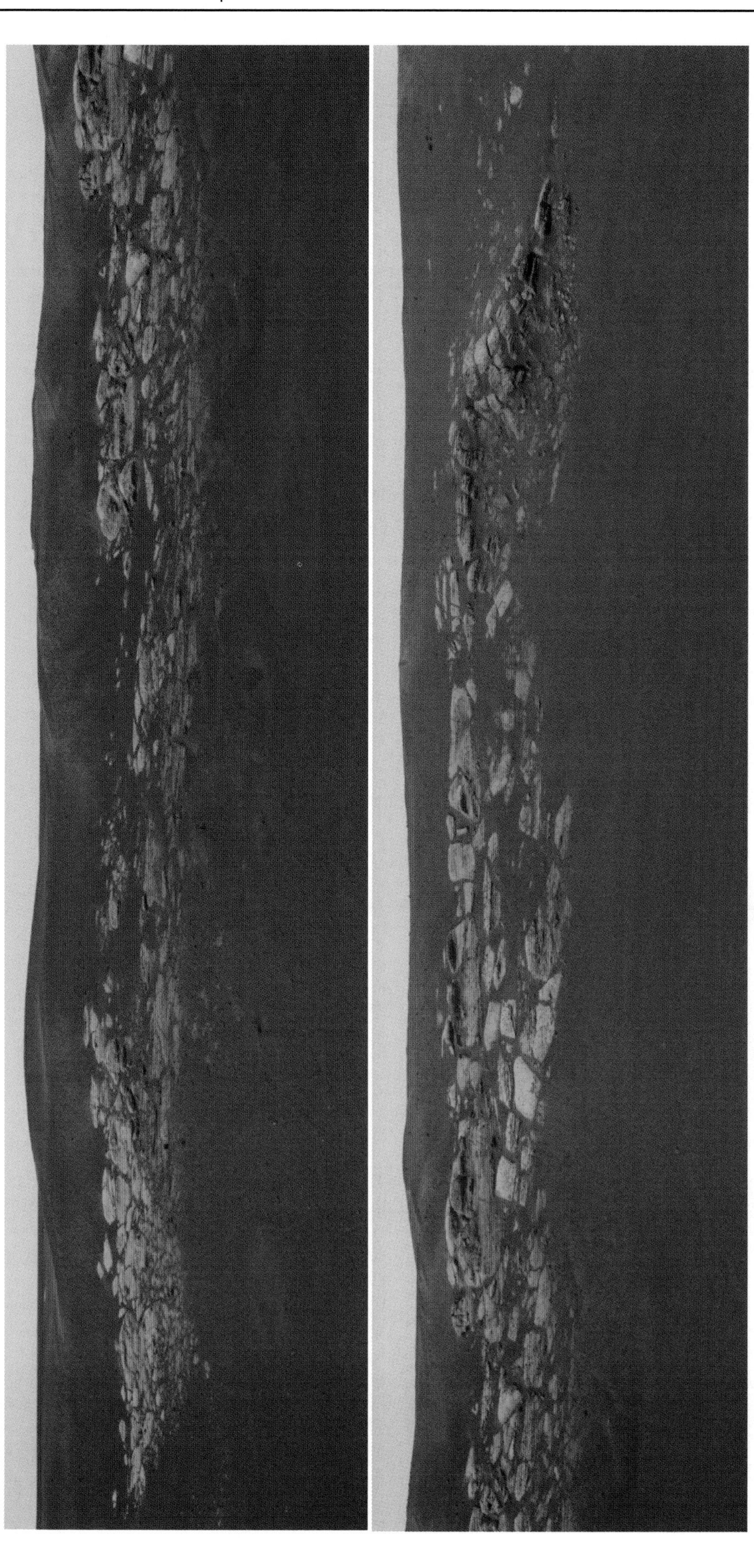

A Geologist's Treasure Trove

This high-resolution image captured by the Mars Exploration Rover Opportunity's panoramic camera highlights the puzzling rock outcropping that scientists are eagerly planning to investigate. Presently, Opportunity is on its lander facing northeast; the outcropping lies to the northwest. These layered rocks measure only 10 centimeters (4 inches) tall and are thought to be either volcanic ash deposits or sediments carried by water or wind. Data from the panoramic camera's near-infrared, blue and green filters were combined to create this approximate, true-color image.

Image credit: NASA/JPL/Cornell

Opportunity Rocks!

This high-resolution image captured by the Mars Exploration Rover Opportunity's panoramic camera shows in superb detail a portion of the puzzling rock outcropping that scientists are eagerly planning to investigate. Presently, Opportunity is on its lander facing northeast; the outcropping lies to the northwest. These layered rocks measure only 10 centimeters (4 inches) tall and are thought to be either volcanic ash deposits or sediments carried by water or wind. The small rock in the center is about the size of a golf ball.

Image credit: NASA/JPL/Cornell

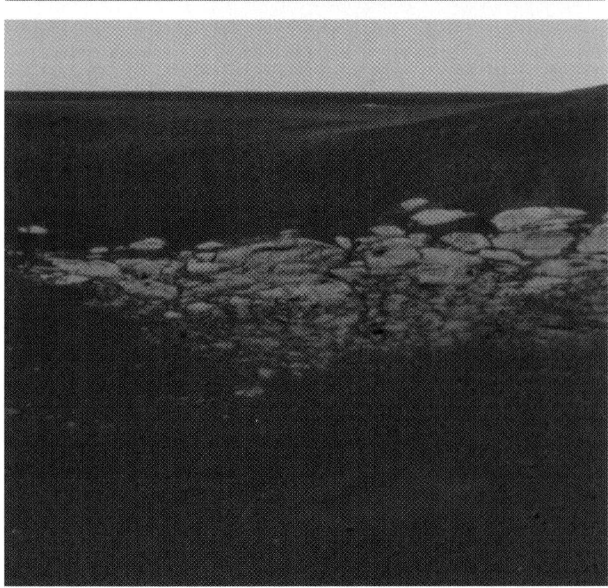

Opportunity Rocks Again!

This high-resolution image captured by the Mars Exploration Rover Opportunity's panoramic camera highlights a portion of the puzzling rock outcropping that scientists eagerly wait to investigate. Presently, Opportunity is on its lander facing northeast; the outcropping lies to the northwest. These layered rocks measure only 10 centimeters (4 inches) tall and are thought to be either volcanic ash deposits or sediments carried by water or wind. Data from the panoramic camera's near-infrared, blue and green filters were combined to create this approximate true color image.

Image credit: NASA/JPL/Cornell

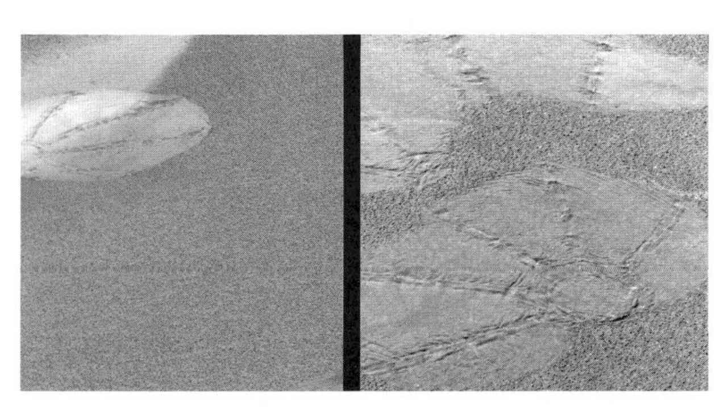

Airbag Tracks on Mars

The circular shapes seen on the martian surface in these images are "footprints" left by the Mars Exploration Rover Opportunity's airbags during landing as the spacecraft gently rolled to a stop. Opportunity landed at approximately 9:05 p.m. PST on Saturday, Jan. 24, 2004, Earth-received time. The circular region of the flower-like feature on the right is about the size of a basketball. Scientists are studying the prints for more clues about the makeup of martian soil. The images were taken at Meridiani Planum, Mars, by the panoramic camera on the Mars Exploration Rover Opportunity.

Image credit: NASA/JPL

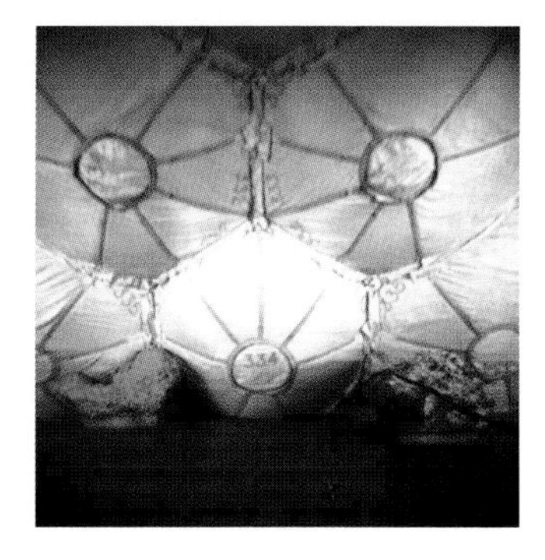

Tracks 'Seam' Like Airbags

Bearing a striking resemblance to a cluster of paper lanterns, these inflated airbags show a pattern of seams exactly like those left in the martian soil by the Mars Exploration Rover Opportunity during landing at Meridiani Planum, Mars. This image was taken during airbag testing at NASA's Plum Brook Station, located about 50 miles west of Cleveland in Sandusky, Ohio and operated by NASA's Glenn Research Center.

Image credit: NASA/JPL/Cornell

Healthier Spirit Gets Back to Work While Opportunity Prepares to Roll

January 29, 2004

An image from Spirit's front hazard identification camera shows the robotic arm extended to the rock called Adirondack.

NASA's Spirit rover on Mars has resumed taking pictures as engineers continue work on restoring its health. Meanwhile, Spirit's twin, Opportunity, extended its rear wheels backward to driving position last night as part of preparations to roll off its lander, possibly as early as overnight Saturday-to-Sunday.

Spirit shot and transmitted a picture yesterday to show the position of its robotic arm. "The arm is exactly where we expected," said Jennifer Trosper, mission manager at NASA's Jet Propulsion Laboratory, Pasadena, Calif. It is still extended in the same position as when the rover developed communication and computer problems on Jan. 22. A mineral-identifying instrument called a Moessbauer spectrometer, at the tip of the arm, is positioned at a rock nicknamed Adirondack.

Engineers have been carefully nursing Spirit back toward full operations for the past week. They are sending commands today for the rover to begin making new scientific observations again, starting with panoramic camera images of nearby rocks. Today's commands also tell the rover to send data stored by two instruments since they took readings on Adirondack last week — the Moessbauer spectrometer and the alpha particle X-ray spectrometer, which identifies the chemical elements in a target.

"We know we still have some engineering work to do, but we think we understand the problem well enough to do science in parallel with that work," Trosper said. Several attempts to get a full trace of data related to the rover's problem have only partially succeeded. The engineers might choose to reformat the rover's flash memory in the next few days.

A health check of Spirit's camera mast is on the agenda for today. Another health check, of an actuator motor for a periscope mirror of the miniature thermal emission spectrometer, is planned for Friday.

Halfway around Mars from Spirit, Opportunity's lander platform successfully tilted itself forward by pulling airbag material under the rear portion of the lander then flexing its rear petal downward. "What this did is drive our front edge lower," said JPL's Matt Wallace, mission manager. "The tips of the egress aid (a reinforced fabric ramp) are now in the soil. That makes egress look perfect. It's going to be an easy ride." The rover also retracted a lift mechanism underneath the rover, to get it out of the way for the egress, or drive-off.

During Opportunity's sol 6, the martian day that started today at 10:26 a.m. PST, the rover will be commanded to lower the middle pair of its six wheels and to release its robotic arm from the latch that has held it since before launch.

Yesterday, Opportunity used its minature thermal emission spectrometer on a portion of the landing neighborhood that includes a rock outcrop. The instrument identifies the composition of rocks and soils from a distance. Opportunity did not return the data from those observations before going to sleep for the martian night, but may later today.

The rovers' main task in coming weeks and months is to explore their landing sites for evidence in the rocks and soil about whether the sites' past environments were ever watery and possibly suitable for sustaining life.

JPL, a division of the California Institute of Technology in Pasadena, manages the Mars Exploration Rover project for NASA's Office of Space Science, Washington, D.C. Images and additional information about the project are available from JPL at http://marsrovers.jpl.nasa.gov and from Cornell University, Ithaca, N.Y., at http://athena.cornell.edu .

Guy Webster (818) 354-5011 Jet Propulsion Laboratory, Pasadena, Calif.
Donald Savage (202) 358-1547 NASA Headquarters, Washington, D.C.
NEWS RELEASE: 2004-044

Two Working Rovers on Martian Soil Expected by Saturday Morning

January 30, 2004

Ground controllers plan to tell Opportunity to drive off its lander early Saturday, and with Spirit now back in working order, NASA should soon have two healthy rovers loose on Mars.

Early today, the controllers at NASA's Jet Propulsion Laboratory, Pasadena, Calif., decided to move up the time for Opportunity's roll-off by nearly 24 hours, to the rover's seventh martian day since landing last weekend. "We're ahead of schedule and taking advantage of the fact that Opportunity treats us well," said JPL's Daniel Limonadi, rover systems engineer. "We feel it's good to egress today and get ready to do science earlier with six wheels on the ground in Meridiani Planum."

Dr. Ray Arvidson of Washington University in St. Louis, deputy principal investigator for the rover science instruments, said, "We're totally ecstatic that we're going to be on the surface."

If a final check finds conditions OK for sending the egress commands at about 12:30 a.m. Saturday, Pacific Standard Time, confirmation of the roll-off would be expected between 3 a.m. and 4 a.m. PST.

Opportunity's twin Mars Exploration Rover, Spirit, has sent back its first new science data in more than a week. On Thursday, it took and transmitted panoramic camera images including views of two light-colored rocks, nicknamed Cake and Blanco. Scientists are considering those rocks as possible targets for up-close examination after Spirit finishes inspection of the rock called Adirondack over the next few days.

Spirit has also returned microscopic images and Mössbauer spectrometer readings of Adirondack taken the day before the rover developed computer and communication problems on Jan. 22. Both are unprecedented investigations of any rock on another planet.

The microscopic images indicate Adirondack is a hard, crystalline rock. "If you had a hammer and whacked that rock, it would ring," Arvidson said.

Mössbauer readings allow scientists to determine what types of iron-bearing minerals are in a rock. "What made us extremely happy when we saw the graph for the first time were the small peaks," said Dr. Bodo Bernhardt, a member of the rover science team from the University of Mainz, Germany, which provided the instrument. The peaks large and small in the spectrum reveal that the minerals in Adirondack include olivine, pyroxene and magnetite. That composition is common in volcanic basalt rocks on Earth, said science-team member Dr. Dick Morris of NASA's Johnson Space Center, Houston.

In coming days, scientists plan to use Spirit's rock abrasion tool to grind the weathered surface off of a small area on Adirondack to inspect its interior. Later plans include examining a nearby whitish rock, then driving toward a crater nicknamed Bonneville that's about 250 meters (820 feet) away. Researchers will use the rover to search for rocks that may have been excavated from below the surface and tossed outward by the impact that dug the crater. If Spirit can reach the rim, scientists hope to see outcrops in the crater walls.

Engineers are continuing to restore Spirit to full health as the rover makes scientific observations, said JPL's Dr. Mark Adler, mission manager. They plan to delete from the rover's flash memory a large amount of information stored before landing, then resume operating Spirit in a normal mode that uses flash memory.

Halfway around the planet, Opportunity's main task in the days after roll-off will be to take microscopic images and spectrometer readings of the soil close to the lander. Within about a week, controllers anticipate sending the rover to an outcrop of bedrock about 8 meters (26 feet) northwest of the lander.

Opportunity currently sits near the center of a crater 22 meters (72 feet) across and 3 meters (10 feet) deep. A new three-dimensional model of the crater, created from information in stereo images, will provide a reference for rover driving within the crater and later for choosing a route out onto the surrounding plains, said Dr. Ron Li, a rover science team member from Ohio State University, Columbus. This is the first time a crater on another planet has been mapped from inside the crater.

JPL, a division of the California Institute of Technology in Pasadena, manages the Mars Exploration Rover project for NASA's Office of Space Science, Washington, D.C. Images and additional information about the project are available from JPL at http://marsrovers.jpl.nasa.gov and from Cornell University, Ithaca, N.Y., at http://athena.cornell.edu .

Guy Webster (818) 354-5011 Jet Propulsion Laboratory, Pasadena, Calif.
Donald Savage (202) 358-1547 NASA Headquarters, Washington, D.C.
NEWS RELEASE: 2004-046

Adirondack Under the Microscope

This image was taken by the Mars Exploration Rover Spirit front hazard-identification camera after the rover's first post-egress drive on Mars Sunday, Jan. 15, 2004. Engineers drove the rover approximately 3 meters (10 feet) from the Columbia Memorial Station toward the first rock target, seen in the foreground. The football-sized rock was dubbed Adirondack because of its mountain-shaped appearance. Scientists have begun using the microscopic imager instrument at the end of the rover's robotic arm to examine the rock and understand how it formed.

Image credit: NASA/JPL

Adirondack Under the Microscope-2

This overhead look at the martian rock dubbed Adirondack was captured by the Mars Exploration Rover Spirit's panoramic camera. It shows the approximate region where the rover's microscopic imager began its first close-up inspection.

Image credit: NASA/JPL/Cornell

Adirondack's Finer Side

This close-up look at the martian rock dubbed Adirondack was captured by the Mars Exploration Rover Spirit's microscopic imager before Spirit stopped communicating with Earth on the 18th martian day, or sol, of its mission. The rock's smooth and pitted surface is revealed in this first-ever microscopic image of a rock on another planet. The examined patch of rock is 3 centimeters (1.2 inches) across; features within the rock as small as 1/10 of a millimeter (.04 inch) can be detected. The rover's shadow appears at the bottom of the image.

Image credit: NASA/JPL/US Geological Survey

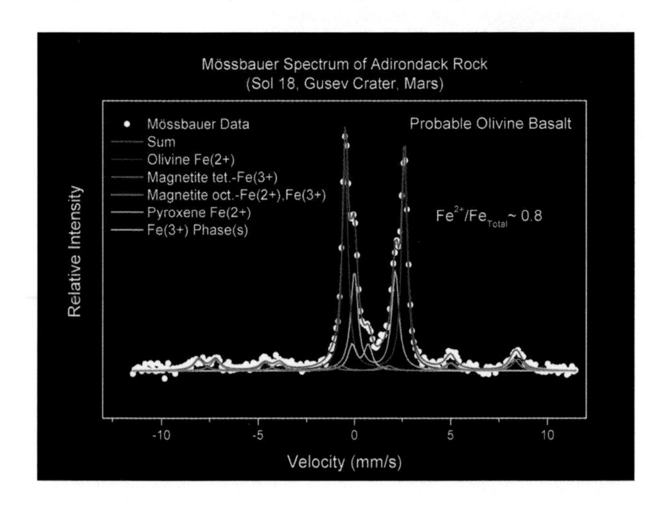

Adirondack's Inner Self

This spectrum - the first taken of a rock on another planet - reveals the different iron-containing minerals that makeup the martian rock dubbed Adirondack. It shows that Adirondack is a type of volcanic rock known as basalt. Specifically, the rock is what is called olivine basalt because in addition to magnetite and pyroxene, two key ingredients of basalt, it contains a mineral called olivine. This data was acquired by Spirit's Mössbauer spectrometer before the rover developed communication problems with Earth on the 18th martian day, or sol, of its mission.

Image credit: NASA/JPL/University of Mainz

Recovering Spirit Sets Sight on Cake

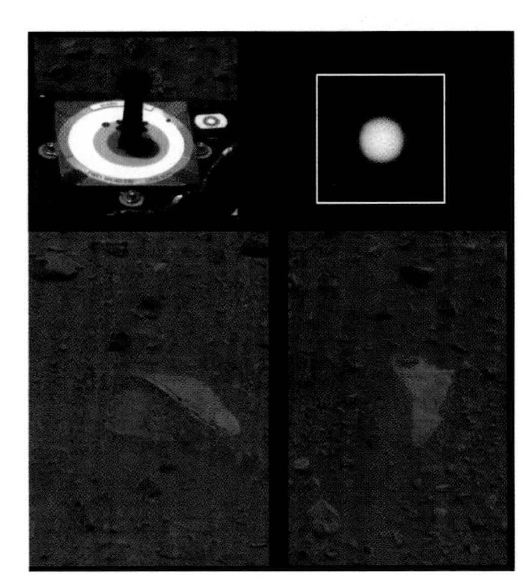

These are the first images sent back from the panoramic camera on the Mars Exploration Rover Spirit since the rover experienced communications problems on the 18th sol, or martian day, of its mission. They were acquired at Gusev Crater, Mars, on Sol 26 (Jan. 29, 2004), showing that the camera's health remained excellent during Spirit's recovery. Two of Spirit's potential target rocks, which are near the rock called Adirondack, can be seen on the lower left and right. The rock on the left has been named "Cake," and the white rock on the right has been named "Blanco."

In the upper left is a color image of the panoramic camera calibration target, also known as the martian sundial. The color panel of the calibration target looks almost exactly like it did on Earth, indicating that the color shown of Mars, though approximated, is close to true color.

The monochrome image in the upper right shows the sun, magnified five times. This image was acquired by the panoramic camera as part of a routine sequence of images designed to monitor the dust abundance in the martian atmosphere. The dust abundance appears to be decreasing slowly with time, consistent with the atmosphere continuing to clear after the large dust storm of last December.

Image credit: NASA/JPL/Cornell

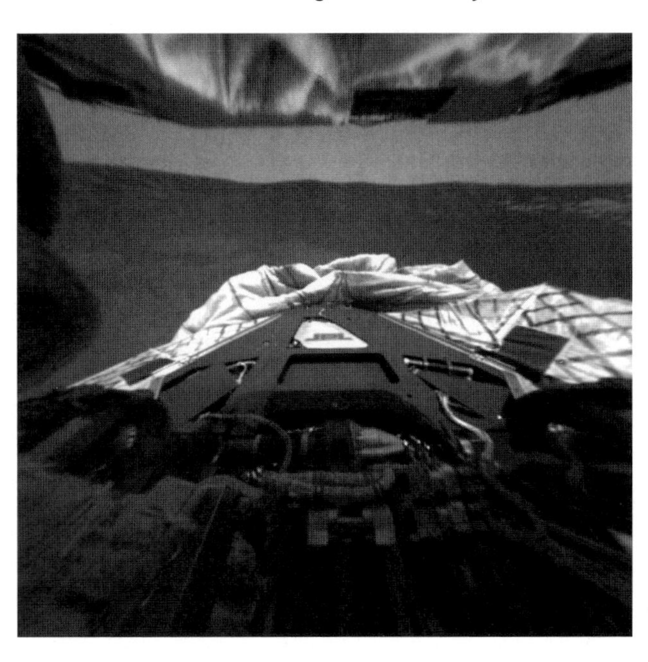

Opportunity Prepares for Egress

This image from the rear hazard-identification camera on the Mars Exploration Rover Opportunity shows the spacecraft's rear point of view before the rear lander petal was hyperextended. This was one of the steps taken to successfully tilt the lander forward in preparation for egress, or rolling off the lander, at Meridiani Planum, Mars. The rover will roll north off the lander, opposite this viewpoint.

Image credit: NASA/JPL

Opportunity Lowers for Egress (animation sequence)

This animation strings together three images from the rear hazard-identification camera on the Mars Exploration Rover Opportunity. The "movie" shows the lander before and after it successfully tilted itself forward by hyperextending its rear lander petal downward. This manuever was performed in preparation for egress, or rolling off the lander at Meridiani Planum, Mars. Opportunity will roll north off the lander, opposite this viewpoint.

Image credit: NASA/JPL

Opportunity Egress Aid Contacts Soil

This image from the navigation camera on the Mars Exploration Rover Opportunity shows the rover's egress aid touching the martian soil at Meridiani Planum, Mars. The image was taken after the rear lander petal hyperextended in a manuever to tilt the lander forward. The maneuver pushed the front edge lower, placing the tips of the egress aids in the soil. The rover will drive straight ahead to exit the lander.

Image credit: NASA/JPL

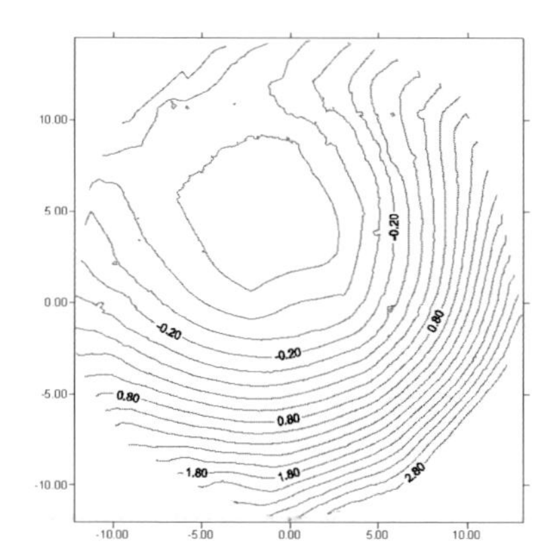

An Intimate Look at a Martian Crater

This 3-D contour map shows the martian crater currently cradling the Mars Exploration Rover Opportunity. It is the first look at the shape of a crater on another planet from the unique vantage point of inside the crater itself. Engineers and scientists will use this data to plot an exit route for Opportunity once it is ready to roll out of the crater; to characterize geological features of the crater; and to help pinpoint the rover's location on the surface of Mars. The crater is estimated to be 3 meters (9.8 feet) deep and 22 (72.2) meters across. The map consists of data from the rover's panoramic camera.

Image credit: NASA/JPL/Ohio State University

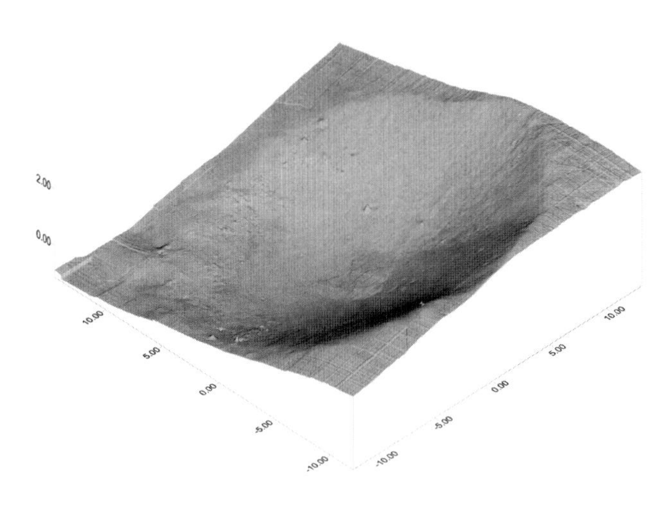

A Unique Opportunity

This 3-D topographic map shows the martian crater currently cradling the Mars Exploration Rover Opportunity. It is the first look at the shape of a crater on another planet from the unique vantage point of inside the crater itself. Engineers and scientists will use this data to plot an exit route for Opportunity once it is ready to roll out of the crater; to characterize geological features of the crater; and to help pinpoint the rover's location on the surface of Mars. The crater is estimated to be 3 meters (9.8 feet) deep and 22 (72.2) meters across. The map consists of data from the rover's panoramic camera.

Image credit: NASA/JPL/Ohio State University

Opportunity Rolls Onto Martian Ground

January 31, 2004

NASA's Mars Exploration Rover Opportunity drove down a reinforced fabric ramp at the front of its lander platform and onto the soil of Mars' Meridiani Planum this morning.

Also, new science results from the rover indicate that the site does indeed have a type of mineral, crystalline hematite, that was the principal reason the site was selected for exploration.

Controllers at NASA's Jet Propulsion Laboratory received confirmation of the successful drive at 3:01 a.m. Pacific Standard Time via a relay from the Mars Odyssey orbiter and Earth reception by the Deep Space Network. Cheers erupted a minute later when Opportunity sent a picture looking back at the now-empty lander and showing wheel tracks in the martian soil.

For the first time in history, two mobile robots are exploring the surface of another planet at the same time. Opportunity's twin, Spirit, started making wheel tracks halfway around Mars from Meridiani on Jan. 15.

"We're two for two! One dozen wheels on the soil." JPL's Chris Lewicki, flight director, announced to the control room.

Matt Wallace, mission manager at JPL, told a subsequent news briefing, "We knew it was going to be a good day. The rover woke up fit and healthy to Bruce Springsteen's 'Born to Run,' and it turned out to be a good choice."

The flight team needed only seven days since Opportunity's landing to get the rover off its lander, compared with 12 days for Spirit earlier this month. "We're getting practice at it," said JPL's Joel Krajewski, activity lead for the procedure. Also, the configuration of the deflated airbags and lander presented no trouble for Opportunity, while some of the extra time needed for Spirit was due to airbags at the front of the lander presenting a potential obstacle.

Looking at a photo from Opportunity showing wheel tracks between the empty lander and the rear of the rover about one meter or three feet away, JPL's Kevin Burke, lead mechanical engineer for getting the rover off the lander, said "We're glad to be seeing soil behind our rover."

JPL's Chris Salvo, flight director, reported that Opportunity will be preparing over the next couple days to reach out with it robotic arm for a close inspection of the soil.

Gray granules covering most of the crater floor surrounding Opportunity contain hematite, said Dr. Phil

Christensen, lead scientist for both rovers' miniature thermal emission spectrometers, which are infrared-sensing instruments used for identifying rock types from a distance. Crystalline hematite is of special interest because, on Earth, it usually forms under wet environmental conditions. The main task for both Mars Exploration Rovers in coming weeks and months is to read clues in the rocks and soil to learn about past environmental conditions at their landing sites, particularly about whether the areas were ever watery and possibly suitable for sustaining life.

The concentration of hematite appears strongest in a layer of dark material above a light-covered outcrop in the wall of the crater where Opportunity sits, Christensen said. "As we get out of the bowl we're in, I think we'll get onto a surface that is rich in hematite," he said.

JPL, a division of the California Institute of Technology in Pasadena, manages the Mars Exploration Rover project for NASA's Office of Space Science, Washington, D.C. Images and additional information about the project are available from JPL at http://marsrovers.jpl.nasa.gov and from Cornell University, Ithaca, N.Y., at http://athena.cornell.edu .

Guy Webster (818) 354-5011 Jet Propulsion Laboratory, Pasadena, California
Donald Savage (202) 358-1547 NASA Headquarters, Washington, D.C.
NEWS RELEASE: 2004-047

This image shows the Mars Exploration Rover Opportunity's view of the martian horizon from its new position on the surface of Mars. Engineers received confirmation that Opportunity's six wheels rolled off the lander and onto martian soil at 3:01 a.m. PST, January 31, 2004, on the seventh martian day, or sol, of the mission. The rover is approximately 1 meter (3 feet) in front of the lander, facing north. The image was taken at Meridiani Planum by the rover's front hazard-identification camera.

Image credit: NASA/JPL

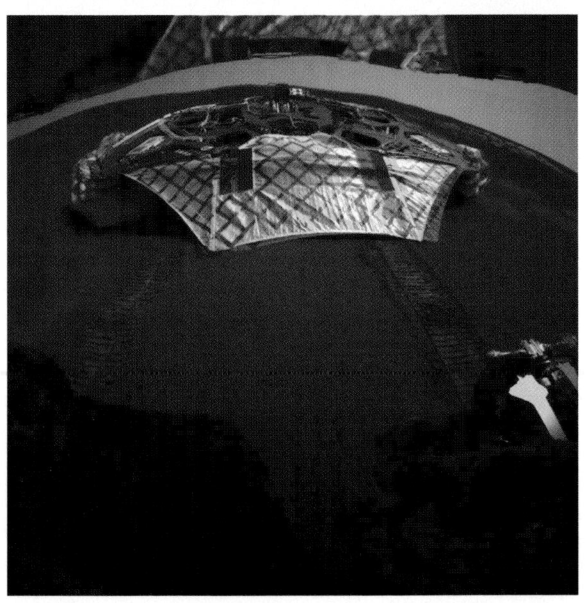

Vacant Lander in 3-D

This 3-D image captured by the Mars Exploration Rover Opportunity's rear hazard-identification camera shows the now-empty lander that carried the rover 283 million miles to Meridiani Planum, Mars. Engineers received confirmation that Opportunity's six wheels successfully rolled off the lander and onto martian soil at 3:01 a.m. PST, January 31, 2004, on the seventh martian day, or sol, of the mission. The rover is approximately 1 meter (3 feet) in front of the lander, facing north.

Image credit: NASA/JPL

Opportunity on Its Own

This image captured by the Mars Exploration Rover Opportunity's rear hazard-identification camera shows the now-empty lander that carried the rover 283 million miles to Meridiani Planum, Mars. Engineers received confirmation that Opportunity's six wheels successfully rolled off the lander and onto martian soil at 3:01 a.m. PST, January 31, 2004, on the seventh martian day, or sol, of the mission. The rover is approximately 1 meter (3 feet) in front of the lander, facing north.

Image credit: NASA/JPL

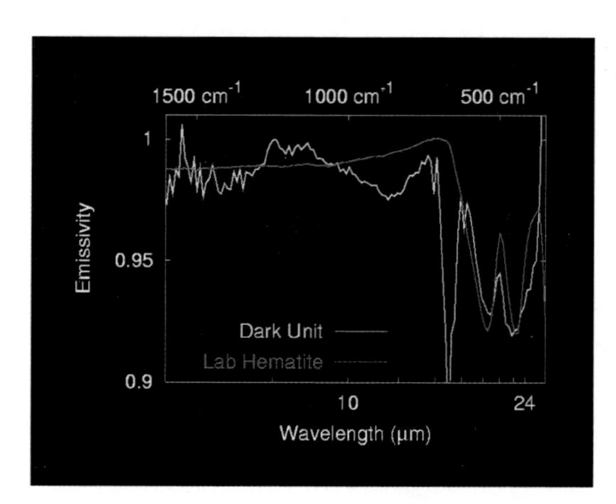

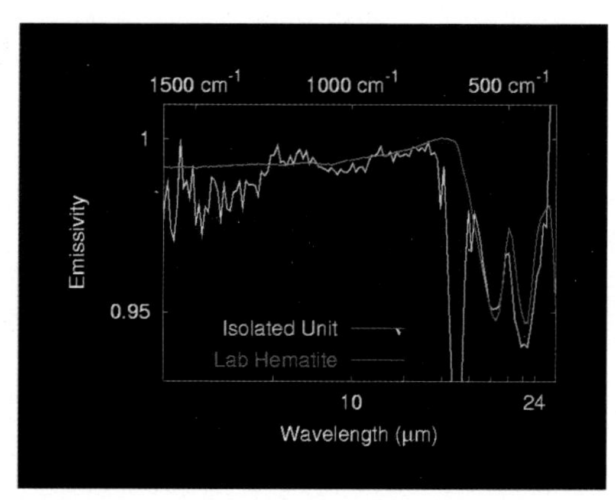

Hematite on Mars!

This spectrum captured by the Mars Exploration Rover Opportunity's mini-thermal emission spectrometer shows the presence of grey hematite in the martian soil at Meridiani Planum, Mars. On Earth, hematite forms in the presence of water, at the bottom of lakes, springs and other bodies of standing water. But it can also arise without water in volcanic regions. Scientists hope to discover the origins of martian hematite with the help of Opportunity's robotic set of geological tools. The yellow line represents the spectrum, or light signature, of the martian soil, while the red line shows the spectrum of pure hematite.

Image credit: NASA/JPL/Arizona State University

Virtual Rover on Its Own

This image shows a screenshot from the software used by engineers to roll the Mars Exploration Rover Opportunity off its lander and onto martian soil. Engineers received confirmation that Opportunity's six wheels had touched ground at 3:01 a.m. PST, January 31, 2004, on the seventh martian day, or sol, of the mission. The software simulates the rover's movements, helping to plot a safe course. The virtual 3-D world around the rover is built from images taken by Opportunity's stereo navigation cameras. Regions for which the rover has not yet acquired 3-D data are represented in beige. The rover is approximately 1 meter (3 feet) in front of the lander, facing north.

Image credit: NASA/JPL

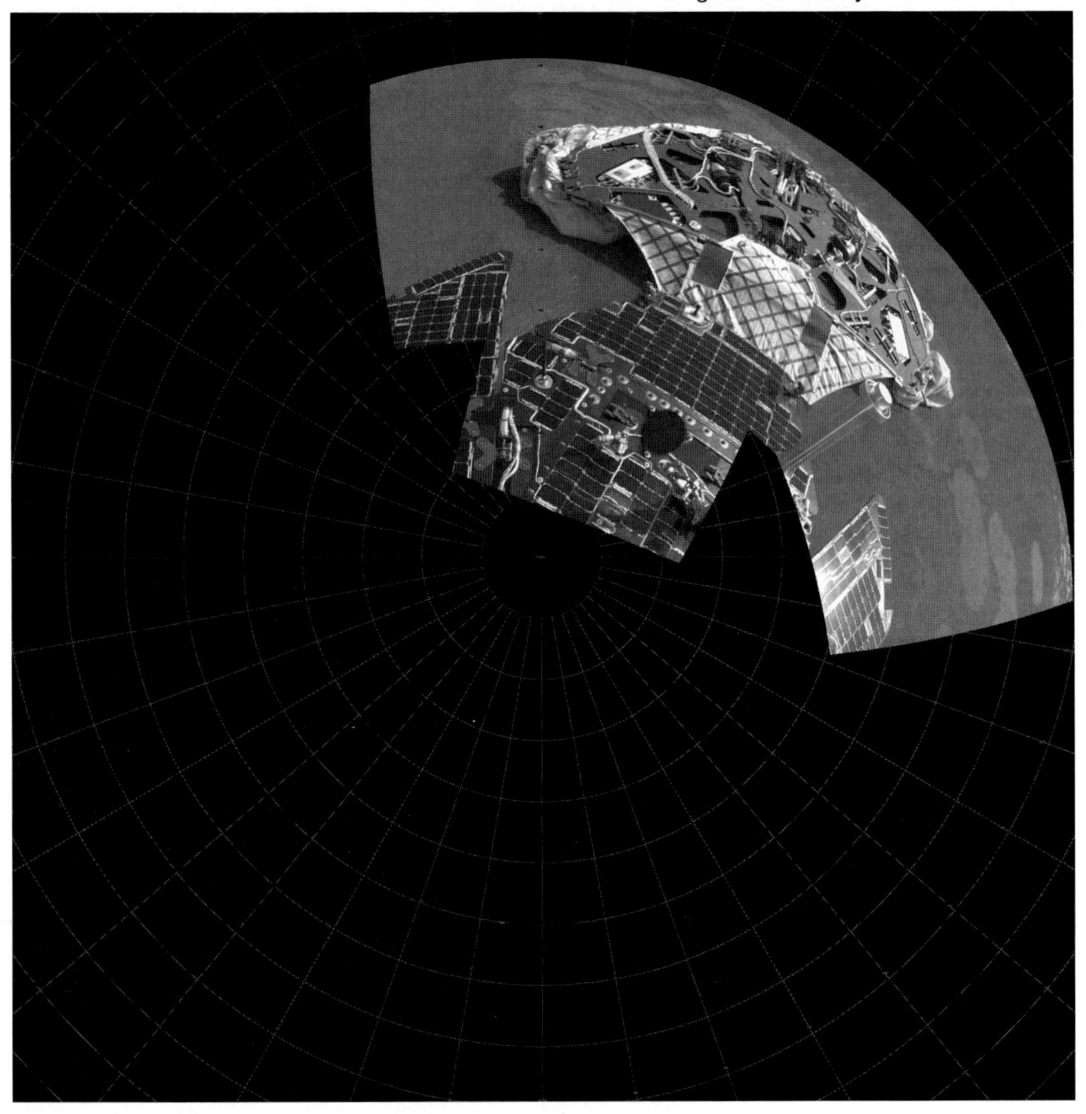

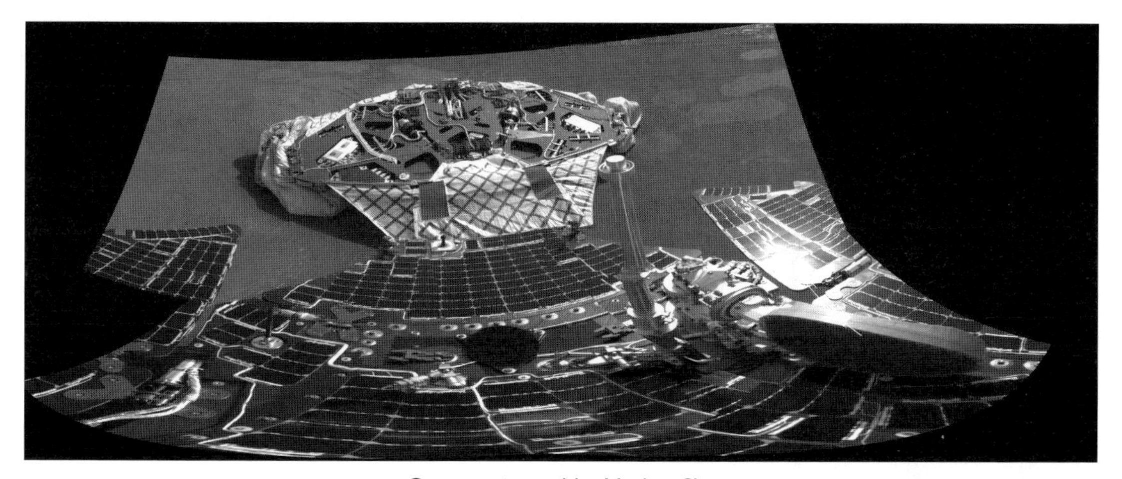

Opportunity and Its Mother Ship

This image mosaic captured by the Mars Exploration Rover Opportunity's navigation camera shows the rover and the now-empty lander that carried it 283 million miles to Meridiani Planum, Mars. Engineers received confirmation that Opportunity's six wheels rolled off the lander and onto martian soil at 3:02 a.m. PST, January 31, 2004, on the seventh martian day, or sol, of the mission. The rover, seen at the bottom of the image, is approximately 1 meter (3 feet) in front of the lander, facing north.

Image credit: NASA/JPL

Mars Rover Spirit Restored To Health

February 01, 2004

NASA's Mars Exploration Rover Spirit is healthy again, the result of recovery work by mission engineers since the robot developed computer-memory and communications problems 10 days ago.

"We have confirmed that Spirit is booting up normally. Tomorrow we'll be doing some preventive maintenance," Dr. Mark Adler, mission manager at NASA's Jet Propulsion Laboratory, Pasadena, Calif., said Sunday morning.

Spirit's twin, Opportunity, which drove off its lander platform early Saturday, will be commanded tonight to reach out with its robot arm early Monday, said JPL's Matt Wallace, mission manager. Opportunity will examine the soil in front of it over the next few days with a microscope and with a pair of spectrometer instruments for determining what elements and minerals are present.

This image from Spirit's front hazard identification camera shows the robotic arm extended to the rock called Adirondack.

For Spirit, part of the cure has been deleting thousands of files from the rover's flash memory — a type of rewritable electronic memory that retains information even when power is off. Many of the deleted files were left over from the seven-month flight from Florida to Mars. Onboard software was having difficulty managing the flash memory, triggering Spirit's computer to reset itself about once an hour.

Two days after the problem arose, engineers began using a temporary workaround of sending commands every day to put Spirit into an operations mode that avoided use of flash memory. Now, however, the computer is stable even when operating in the normal mode, which uses the flash memory.

"To be safe, we want to reformat the flash and start again with a clean slate," Adler said. That reformatting is planned for Monday. It will erase everything stored in the flash file system and install a clean version of the flight software.

Today, Spirit is being told to transmit priority data remaining in the flash memory. The information includes data from atmospheric observations made Jan. 16 in coordination with downward-looking observations by the European Space Agency's Mars Express orbiter. Also today, Spirit will make new observations coordinated with another Mars Express overflight and will run a check of the rover's miniature thermal emission spectrometer.

Spirit will resume examination of a rock nicknamed Adirondack later this week and possibly move on to a lighter-colored rock by week's end.

Each martian day, or "sol" lasts about 40 minutes longer than an Earth day. Spirit begins its 30th sol on Mars at 12:44 a.m. Monday, Pacific Standard Time. Opportunity begins its 10th sol on Mars at 1:05 p.m. Monday, PST. The two rovers are halfway around Mars from each other.

The main task for both Spirit and Opportunity in coming weeks and months is to find geological clues about past environmental conditions at their landing sites, particularly about whether the areas were ever watery and possibly suitable for sustaining life.

JPL, a division of the California Institute of Technology in Pasadena, manages the Mars Exploration Rover project for NASA's Office of Space Science, Washington, D.C. Images and additional information about the project are available from JPL at http://marsrovers.jpl.nasa.gov and from Cornell University, Ithaca, N.Y., at http://athena.cornell.edu .

Guy Webster (818) 354-5011 Jet Propulsion Laboratory, Pasadena, California
Donald Savage (202) 358-1547 NASA Headquarters, Washington, D.C.
NEWS RELEASE: 2004-048

Opportunity And Spirit Reach Out

February 02, 2004

Each of NASA's two Mars Exploration Rovers is using its versatile robotic arm for positioning tools at selected targets on the red planet.

Also, a newly completed 360-degree color panorama from Opportunity shows a trail of bounce marks coming down the inner slope of the small crater where the spacecraft came to rest when it landed on Mars nine days ago.

Opportunity extended its arm early today for the first time since pre-launch testing. "This was a great confirmation for the team," said Joe Melko of NASA's Jet Propulsion Laboratory, Pasadena, Calif. Melko is mechanical systems engineer for the arm, which is also called the instrument deployment device.

Mission controllers at JPL are telling Opportunity to use two of the instruments on the arm overnight tonight to examine a patch of soil in front of the rover. A microscope on the arm will reveal structures as thin as a human hair and a Mössbauer Spectrometer will collect information to identify minerals in the soil, according to plans. Tomorrow, the rover will be told to turn the turret at the end of the arm in order to examine the same patch of soil with another instrument, the alpha particle X-ray spectrometer, which reveals the chemical elements in a target.

Spirit is now in good working order after more than a week of computer-memory problems. It is brushing dust off of a rock today with the rock abrasion tool on its robotic arm. After the brushing, Spirit will use the microscope and two spectrometers on the arm to examine the rock.

"We're moving forward with our science on the rock Adirondack," said JPL's Jennifer Trosper, Spirit mission manager. Reformatting of Spirit's flash memory was postponed from today to tomorrow. The reformatting is a precautionary measure against recurrence of the problem that prevented Spirit from doing much science last week.

Later in the week, Spirit will grind the surface off of a sample area on Adirondack with the rock abrasion tool to inspect the rock's interior. After observations of Adirondack are completed, the rover will begin rolling again. "We are already strategizing how to drive far and fast," Trosper said.

Observations by each rover's panoramic camera help scientists choose where to drive and what to examine with the instruments on each rover's arm. Dr. Jeff Johnson, a rover science team member from the U.S. Geological Survey's Astrogeology Team, Flagstaff, Ariz., said that 14 filters available on each rover's panoramic camera allow the instrument to provide much more information for identifying different types of rocks than can be gleaned from color images such as the new panoramic view.

"By looking at the brightness values in each of these wavelengths, we can start to get an idea of the things we're interested in, especially to unravel the geological history of these landing sites," Johnson said.

The main task for both rovers in coming weeks and months is to find clues in rocks and soil about past environmental conditions, particularly about whether the landing areas were ever watery and possibly suitable for sustaining life.

Each martian day, or "sol" lasts about 40 minutes longer than an Earth day. Spirit begins its 31st sol on Mars at 1:23 a.m. Tuesday, Pacific Standard Time. Opportunity begins its 11th sol on Mars at 1:44 p.m. Tuesday, PST. The two rovers are halfway around Mars from each other.

JPL, a division of the California Institute of Technology in Pasadena, manages the Mars Exploration Rover project for NASA's Office of Space Science, Washington, D.C. Images and additional information about the project are available from JPL at http://marsrovers.jpl.nasa.gov and from Cornell University, Ithaca, N.Y., at http://athena.cornell.edu .

Guy Webster (818) 354-5011 Jet Propulsion Laboratory, Pasadena, California Donald Savage (202) 358-1547 NASA Headquarters, Washington, D.C.
NEWS RELEASE: 2004-049

Opportunity Stretches Out

This image taken by the front hazard-identification camera onboard the Mars Exploration Rover Opportunity shows the rover's arm in its extended position. The arm, or instrument deployment device, was deployed on the ninth martian day, or sol, of the mission. The rover, now sitting 1 meter (3 feet) away from the lander, can be seen in the foreground. Image credit: NASA/JPL

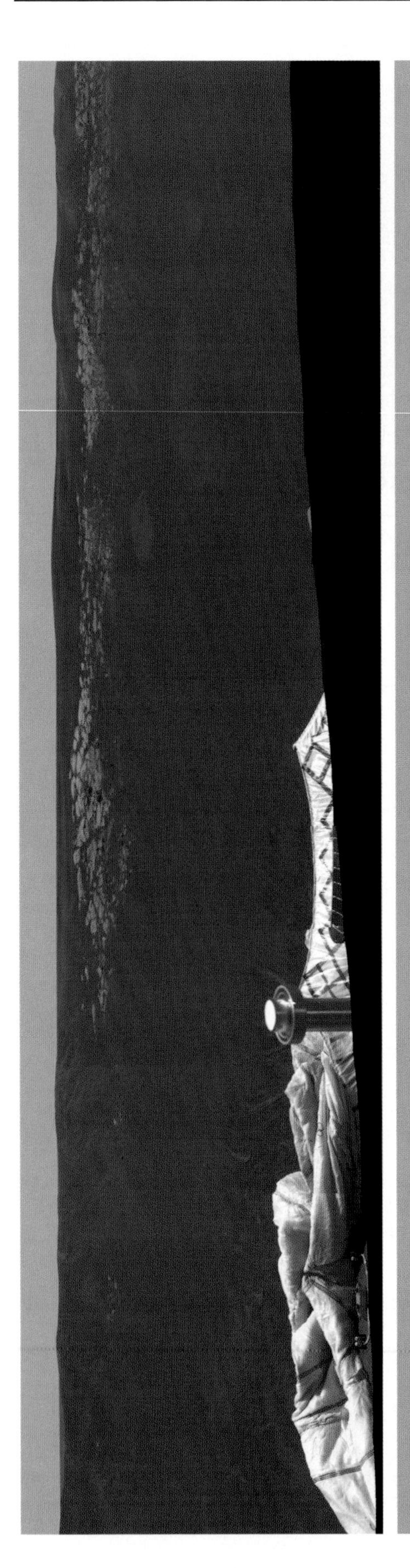

As Far as Opportunity's Eye Can See

This expansive view of the martian real estate surrounding the Mars Exploration Rover Opportunity is the first 360 degree, high-resolution color image taken by the rover's panoramic camera. The airbag marks, or footprints, seen in the soil trace the route by which Opportunity rolled to its final resting spot inside a small crater at Meridiani Planum, Mars. The exposed rock outcropping is a future target for further examination. This image mosaic consists of 225 individual frames.

Image credit: NASA/JPL/Cornell

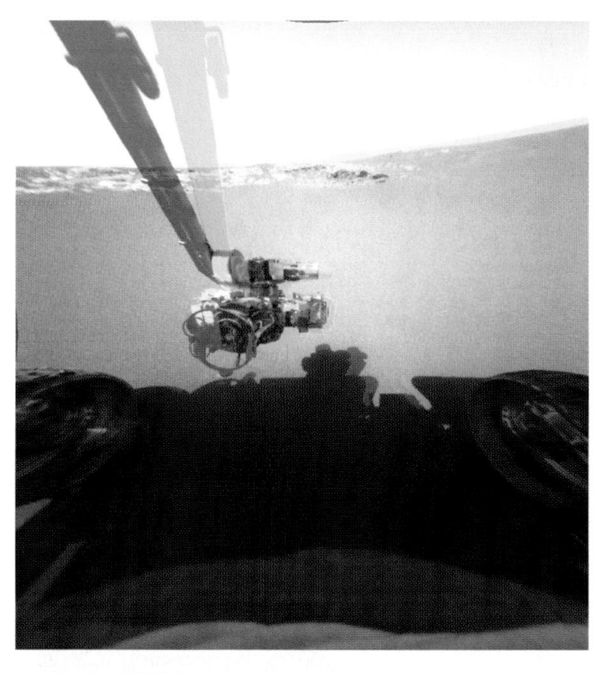

Opportunity Stretches Out (3-D)

This is a three-dimensional stereo anaglyph of an image taken by the front hazard-identification camera onboard the Mars Exploration Rover Opportunity, showing the rover's arm in its extended position. The arm, or instrument deployment device, was deployed on the ninth martian day, or sol, of the mission. The rover, now sitting 1 meter (3 feet) away from the lander, can be seen in the foreground.

Image credit: NASA/JPL

Rat on Mars

This image taken on Mars by the panoramic camera on the Mars Exploration Rover Opportunity shows the rover's rock abrasion tool, also known as "rat" (circular device in center), located on its instrument deployment device, or "arm." The image was acquired on the ninth martian day or sol of the rover's mission.

Image credit: NASA/JPL

Microscope on Mars

This image taken at Meridiani Planum, Mars by the panoramic camera on the Mars Exploration Rover Opportunity shows the rover's microscopic imager (circular device in center), located on its instrument deployment device, or "arm." The image was acquired on the ninth martian day or sol of the rover's mission.

Image credit: NASA/JPL

X-ray Machine on Mars

This image taken at Meridiani Planum, Mars by the panoramic camera on the Mars Exploration Rover Opportunity shows the rover's alpha particle X-ray spectrometer (circular device in center), located on its instrument deployment device, or "arm." The image was acquired on the ninth martian day or sol of the rover's mission.

Image credit: NASA/JPL

Moessbauer on Mars

This image taken at Meridiani Planum, Mars, by the panoramic camera on the Mars Exploration Rover Opportunity shows the rover's Moessbauer spectrometer (circular device in center), located on its instrument deployment device, or "arm." The image was acquired on the ninth martian day or sol of the rover's mission.

Image credit: NASA/JPL

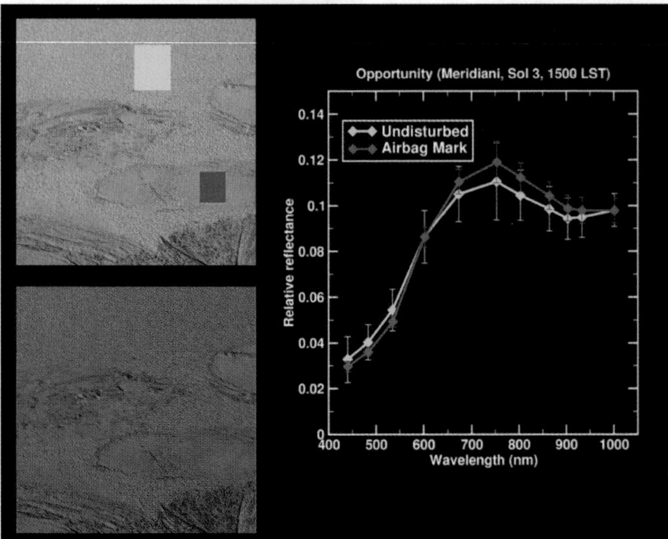

Lily Pad Spectra

The color image on the lower left from the panoramic camera on the Mars Exploration Rover Opportunity shows the "Lily Pad" bounce-mark area at Meridiani Planum, Mars. This image was acquired on the 3rd sol, or martian day, of Opportunity's mission (Jan.26, 2004). The upper left image is a monochrome (single filter) image from the rover's panoramic camera, showing regions from which spectra were extracted from the "Lily Pad" area. As noted by the line graph on the right, the green spectra is from the undisturbed surface and the red spectra is from the airbag bounce mark.

Image credit: NASA/JPL/Cornell

Rock Outcrop Spectra

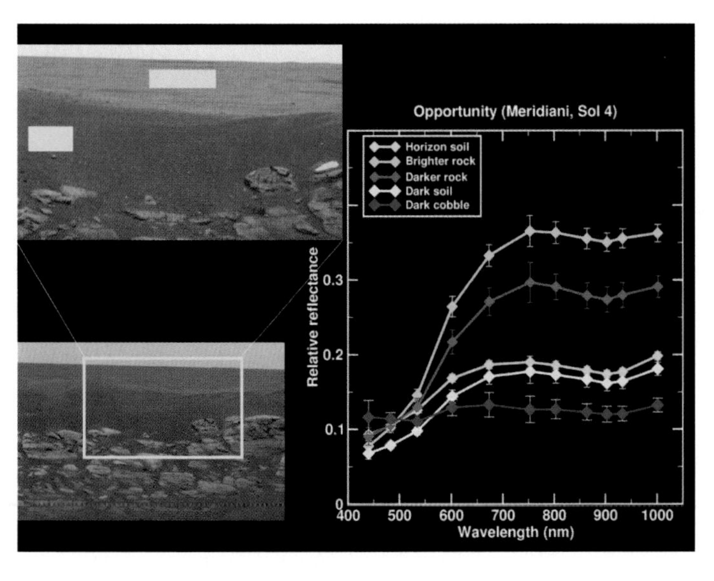

The color image on the lower left shows a rock outcrop at Meridiani Planum, Mars. This image was taken by the panoramic camera on the Mars Exploration Rover Opportunity, looking north, and was acquired on the 4th sol, or martian day, of the rover's mission (Jan. 27, 2004). The yellow box outlines an area detailed in the top left image, which is a monochrome (single filter) image from the rover's panoramic camera. The top image uses solid colors to show several regions on or near the rock outcrop from which spectra were extracted: the dark soil above the outcrop (yellow), the distant horizon surface (aqua), a bright rock in the outcrop (green), a darker rock in the outcrop (red), and a small dark cobblestone (blue). Spectra from these regions are shown in the plot to the right.

Image credit: NASA/JPL/Cornell

Multihued Mars

This image taken at JPL shows the panoramic camera used onboard both Mars Exploration Rovers. The panel to the lower right highlights the multicolored filter wheel that allows the camera to see a rainbow of colors, in addition to infrared bands of light. By seeing Mars in all its colors, scientists can gain insight into the different minerals that constitute its rocks and soil.

Image credit: NASA/JPL

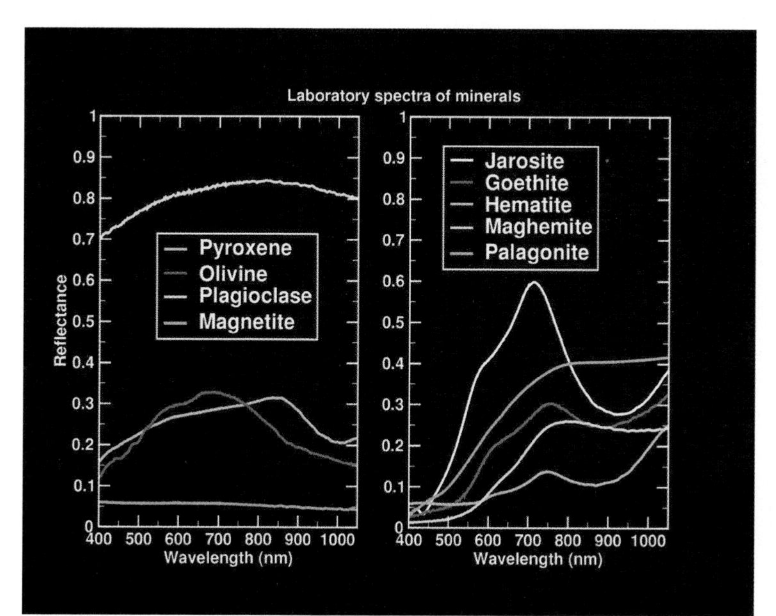

Different Strokes

In these line graphs of laboratory spectra, it is evident that different minerals have different spectra. The graph on the left shows the typical minerals found in igneous rocks, which are rocks related to magma or volcanic activity. The graph on the right shows iron-bearing candidates for further study and comparison to spectra from the Mars Exploration Rover panoramic cameras on Mars.

Image credit: NASA/JPL/U.S. Geological Survey

NASA Dedicates Mars Landmarks To Columbia Crew

February 02, 2004

NASA Administrator Sean O'Keefe today announced the martian hills, located east of the Spirit Mars Exploration Rover's landing site, would be dedicated to the Space Shuttle Columbia STS-107 crew.

"These seven hills on Mars are named for those seven brave souls, the final crew of the Space Shuttle Columbia. The Columbia crew faced the challenge of space and made the supreme sacrifice in the name of exploration," Administrator O'Keefe said.

The Shuttle Columbia was commanded by Rick Husband and piloted by William McCool. The mission specialists were Michael Anderson, Kalpana Chawla, David Brown, Laurel Clark; and the payload specialist was Israeli astronaut Ilan Ramon. On February 1, 2003, the Columbia and its crew were lost over the western United States during re-entry into Earth's atmosphere. The 28th and final flight of Columbia was a 16-day mission dedicated to research in physical, life and space sciences. The Columbia crew successfully conducted approximately 80 separate experiments during their mission.

NASA will submit the names of the Mars features to the International Astronomical Union for official designation. The organization serves as the internationally recognized authority for assigning designations to celestial bodies and their surface features.

An image taken by the Mars Global Surveyor Mars Orbiter Camera of the Columbia Memorial Station and Columbia Hills is available on the Internet at: http://www.jpl.nasa.gov/mer2004/rover-images/feb-02-2004/captions/image-10.html .

For information about NASA and the Mars mission on the Internet, visit: http://www.nasa.gov .
The Jet Propulsion Laboratory, a division of the California Institute of Technology, Pasadena, manages the Mars Exploration Rover project for NASA's Office of Space Science, Washington, D.C. Additional information about the project is available on the Internet at: http://marsrovers.jpl.nasa.gov .

DC Agle (818) 393-9011Jet Propulsion Laboratory, Pasadena, California
Donald Savage (202) 358-1547NASA Headquarters, Washington, D.C.
NEWS RELEASE: 2004-050

NASA Dedicates Mars Landmarks To Columbia Crew

An image taken from Spirit's PanCam looking east depicts the nearby hills dedicated to the final crew of Space Shuttle Columbia. Arranged alphabetically from left to right - "Anderson Hill" is the most northeast of Spirit's landing site and 3 kilometers away. Next are "Brown Hill" and "Chawla Hill", both 2.9 kilometers distant. Next is "Clark Hill" at 3 kilometers. "Husband Hill" and "McCool Hill", named for Columbia's commander and pilot respectively, are 3.1 and 4.2 kilometers distant. "Ramon Hill" is furthest southeast of Spirit's landing site and 4.4 kilometers away.

Image Credit: NASA/JPL/Cornell

The Columbia Crew

Columbia Landmarks From Orbit

An image taken by the Mars Global Surveyor's Mars Orbiter Camera of the Columbia Memorial Station and the nearby hills named after the Columbia crew. The 28th and final flight of Columbia (STS-107) was a 16-day mission dedicated to research in physical, life and space sciences. The Columbia crew worked 24 hours a day in two alternating shifts, successfully conducting approximately 80 separate experiments. On February 1, 2003, the Columbia and its crew were lost over the southern United States during the spacecraft's re-entry into Earth's atmosphere.

Image credit: NASA/JPL

Opportunity Sees Tiny Spheres In Martian Soil

February 04, 2004

NASA's Opportunity has examined its first patch of soil in the small crater where the rover landed on Mars and found strikingly spherical pebbles among the mix of particles there.

"There are features in this soil unlike anything ever seen on Mars before," said Dr. Steve Squyres of Cornell University, Ithaca, N.Y., principal investigator for the science instruments on the two Mars Exploration Rovers.

For better understanding of the soil, mission controllers at NASA's Jet Propulsion Laboratory, Pasadena, Calif., plan to use Opportunity's wheels later this week to scoop a trench to expose deeper material. One front wheel will rotate to dig the hole while the other five wheels hold still.

The spherical particles appear in new pictures from Opportunity's microscopic imager, the last of 20 cameras to be used on the two rover missions. Other particles in the image have jagged shapes. "The variety of shapes and colors indicates we're having particles brought in from a variety of sources," said Dr. Ken Herkenhoff of the U.S. Geological Survey's Astrogeology Team, Flagstaff, Ariz.

The shapes by themselves don't reveal the particles' origin with certainty. "A number of straightforward geological processes can yield round shapes," said Dr. Hap McSween, a rover science team member from the University of Tennessee, Knoxville. They include accretion under water, but apparent pores in the particles make alternative possibilities of meteor impacts or volcanic eruptions more likely origins, he said.

A new mineral map of Opportunity's surroundings, the first ever done from the surface of another planet, shows that concentrations of coarse-grained hematite vary in different parts of the crater. The soil patch in the new microscopic images is in an area low in hematite. The map shows higher hematite concentrations inside the crater in a layer above an outcrop of bedrock and on the slope just under the outcrop.

Hematite usually forms in association with liquid water, so it holds special interest for the scientists trying to determine whether the rover landing sites ever had watery environments possibly suitable for sustaining life. The map uses data from Opportunity's miniature thermal emission spectrometer, which identifies rock types from a distance.

"We're seeing little bits and pieces of this mystery, but we haven't pieced all the clues together yet," Squyres said.

Opportunity's Mössbauer spectrometer, an instrument on the rover's robotic arm designed to identify the types of iron-bearing minerals in a target, found a strong signal in the soil patch for olivine. Olivine is a common ingredient in volcanic rocks. A few days of analysis may be needed to discern whether any fainter signals are from hematite, said Dr. Franz Renz, science team member from the University of Mainz, Germany.

To get a better look at the hematite closer to the outcrop, Opportunity will go there. It will begin by driving about 3 meters (10 feet) tomorrow, taking it about halfway to the outcrop. On Friday it will dig a trench with one of its front wheels, said JPL's Dr. Mark Adler, mission manager.

Opportunity's twin, Spirit, today is reformatting its flash memory, a preventive measure that had been planned for earlier in the week. "We spent the last four days in the testbed testing this," Adler said. "It's not an operation we do lightly. We've got to be sure it works right." Tomorrow, Spirit will resume examining a rock called Adirondack after a two-week interruption by computer memory problems. Controllers plan to tell Spirit to brush dust off of a rock and examine the cleaned surface tomorrow.

Each martian day, or "sol," lasts about 40 minutes longer than an Earth day. Spirit begins its 33rd sol on Mars at 2:43 a.m. Thursday, Pacific Standard Time. Opportunity begins its 13th sol on Mars at 3:04 p.m. Thursday, PST.

JPL, a division of the California Institute of Technology in Pasadena, manages the Mars Exploration Rover project for NASA's Office of Space Science, Washington, D.C. Images and additional information about the project are available from JPL at http://marsrovers.jpl.nasa.gov and from Cornell University, Ithaca, N.Y., at http://athena.cornell.edu .

Guy Webster (818) 354-5011 Jet Propulsion Laboratory, Pasadena, California
Donald Savage (202) 358-1547 NASA Headquarters, Washington, D.C.
NEWS RELEASE: 2004-051

Poised for Discovery

This image taken by the front hazard-identification camera onboard the Mars Exploration Rover Opportunity shows the rover's arm in its extended position. The arm, or instrument deployment device, was deployed on the ninth martian day, or sol, of the mission. The rover, now sitting 1 meter (3 feet) away from the lander, can be seen in the foreground.

Image credit: NASA/JPL

Mars in a Grain of Sand

This image highlights the patch of soil examined by the rover's microscopic imager on the 10th day, or sol, of its mission. The outer image was taken by the rover's navigation camera, the middle image by the panoramic camera and the inner image by the microscopic imager. Opportunity is currently sitting 1 meter (3 feet) away from its now-empty lander in a shallow crater at Meridiani Planum, Mars.

Image credit: NASA/JPL/Cornell

Mars Under the Microscope

This magnified look at the martian soil near the Mars Exploration Rover Opportunity's landing site, Meridiani Planum, shows coarse grains sprinkled over a fine layer of sand. The image was captured by the rover's microscopic imager on the 10th day, or sol, of its mission. Scientists are intrigued by the spherical rocks, which can be formed by a variety of geologic processes, including cooling of molten lava droplets and accretion of concentric layers of material around a particle or "seed."

The examined patch of soil is 3 centimeters (1.2 inches) across. The circular grain in the lower left corner is approximately 3 millimeters (.12 inches) across, or about the size of a sunflower seed.
Image credit: NASA/JPL/US Geological Survey

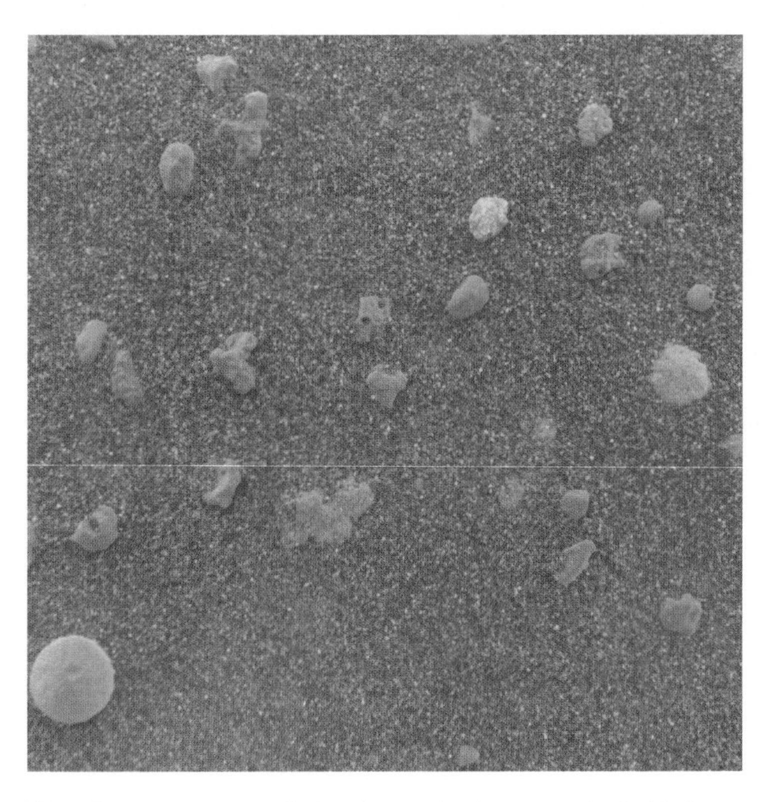

Mars Under the Microscope (color)

This magnified look at the martian soil near the Mars Exploration Rover Opportunity's landing site, Meridiani Planum, shows coarse grains sprinkled over a fine layer of sand. The image was captured by the rover's microscopic imager on the 10th day, or sol, of its mission and roughly approximates the color a human eye would see. Scientists are intrigued by the spherical rocks, which can be formed by a variety of geologic processes, including cooling of molten lava droplets and accretion of concentric layers of material around a particle or "seed."

The examined patch of soil is 3 centimeters (1.2 inches) across. The circular grain in the lower left corner is approximately 3 millimeters (.12 inches) across, or about the size of a sunflower seed. This color composite was obtained by merging images acquired with the orange-tinted dust cover in both its open and closed positions. The blue tint at the lower right corner is a tag used by scientists to indicate that the dust cover is closed.

Image credit: NASA/JPL/US Geological Survey

Mars Under the Microscope (stretched)

This magnified look at the martian soil near the Mars Exploration Rover Opportunity's landing site, Meridiani Planum, shows coarse grains sprinkled over a fine layer of sand. The image was captured on the 10th day, or sol, of the rover's mission by its microscopic imager, located on the instrument deployment device, or "arm." Scientists are intrigued by the spherical rocks, which can be formed by a variety of geologic processes, including cooling of molten lava droplets and accretion of concentric layers of material around a particle or "seed."

The examined patch of soil is 3 centimeters (1.2 inches) across. The circular grain in the lower left corner is approximately 3 millimeters (.12 inches) across, or about the size of a sunflower seed.

This stretched color composite was obtained by merging images acquired with the orange-tinted dust cover in both its open and closed positions. The varying hints of orange suggest differences in mineral composition. The blue tint at the lower right corner is a tag used by scientists to indicate that the dust cover is closed.

Image credit: NASA/JPL/US Geological Survey

Thar be Hematite!

This map of a portion of the small crater currently encircling the Mars Exploration Rover Opportunity shows where crystalline hematite resides. Red and orange patches indicate high levels of the iron-bearing mineral, while blue and green denote low levels. The northeastern rock outcropping lining the rim of the crater does not appear to contain much hematite. Also lacking hematite are the rover's airbag bounce marks. This image consists of data from Opportunity's miniature thermal emission spectrometer superimposed on an image taken by the rover's panoramic camera.

Image credit: NASA/JPL/Arizona State University/Cornell

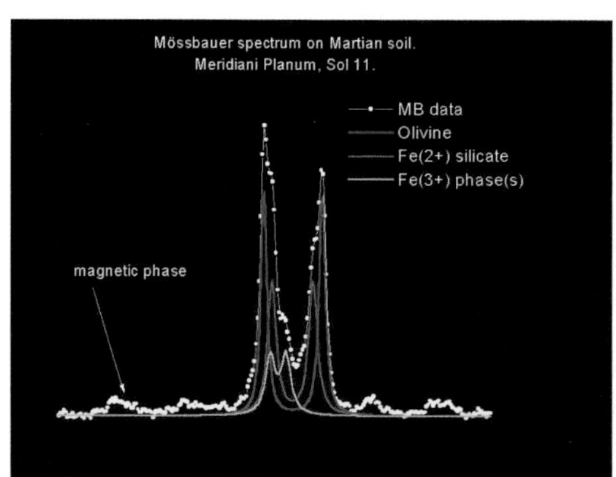

Meridiani's Autograph

This spectrum of the soil at the Mars Exploration Rover Opportunity's landing site, Meridiani Planum, shows the presence of the shiny green mineral called olivine also seen at the Mars Exploration Rover Spirit's landing site, Gusev Crater. Based on this data, scientists believe the soil at Meridiani is made-up of in part of finely grained basalt, a type of volcanic rock. The spectrum was captured by Opportunity's Moessbauer spectrometer.

Image credit: NASA/JPL/University of Mainz

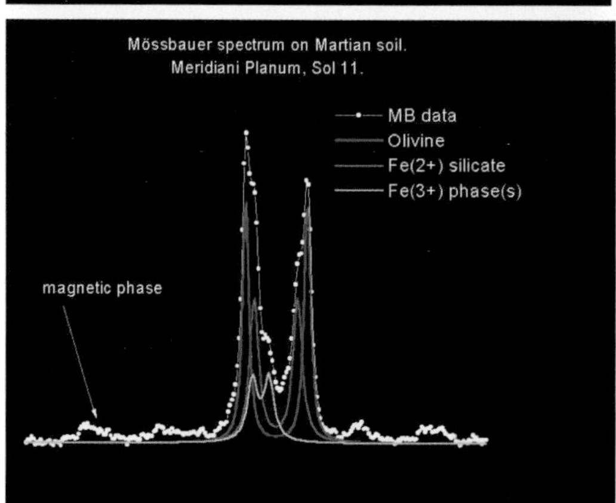

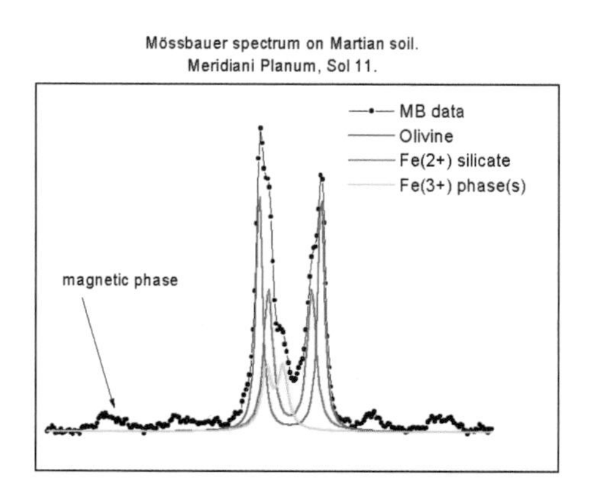

Mars Exploration Rover Mission Status

February 05, 2004

Opportunity's looks through its Rear Hazard Camera after taking a drive.

NASA's Opportunity rover drove about 3.5 meters (11 feet) early Thursday toward a rock outcrop in the wall of a small crater on Mars, and mission controllers plan to send it the rest of the way to the outcrop late Thursday.

Opportunity's twin, Spirit, successfully reformatted its flash memory on Wednesday. Flash is a type of rewritable memory used in many electronic devices, such as digital cameras, to retain information even while power is off. Problems with the flash memory interfered with Spirit's operations from Jan. 22 until this week. Engineers prescribed the reformatting to prevent recurrence of the problem.

On Thursday, Spirit's main assignment is to brush off an area on the rock nicknamed "Adirondack" to prepare for a dust-free examination of its surface. On Friday, controllers at NASA's Jet Propulsion Laboratory, Pasadena, Calif., plan to have Spirit grind off a small patch of Adirondack's outer surface and inspect the rock's interior. Spirit may start driving over the weekend toward a crater about 250 meters (about 270 yards) to the northeast.

For Opportunity, halfway around Mars from Spirit, controllers changed plans Thursday morning. They postponed a trenching operation until the rover gets to an area of its landing-site crater where the soil has a higher concentration of large-grain hematite. That mineral holds high interest because it usually forms under wet conditions. The main science goal for both rovers is to find geological clues about past environmental conditions at the landing sites, especially about whether conditions were ever watery and possibly suitable for sustaining life.

Instead of trenching, Opportunity will be commanded after it next wakes up to drive about 1.5 meters (about 5 feet) farther, possibly to within arm's reach of one of the rocks in the exposed outcrop.

Before it began driving on Wednesday, Opportunity finished using its alpha particle X-ray spectrometer for the first time. This spectrometer, which assesses what chemical elements are present, took readings on an area of soil that the rover had previously examined with its microscope.

Each martian day, or "sol," lasts about 40 minutes longer than an Earth day. Spirit begins its 34rd sol on Mars at 3:22 a.m. Thursday, Pacific Standard Time. Opportunity begins its 14th sol on Mars at 3:43 p.m. Friday, PST.

JPL, a division of the California Institute of Technology in Pasadena, manages the Mars Exploration Rover project for NASA's Office of Space Science, Washington, D.C. Images and additional information about the project are available from JPL at http://marsrovers.jpl.nasa.gov and from Cornell University, Ithaca, N.Y., at http://athena.cornell.edu .
Guy Webster (818) 354-5011 Jet Propulsion Laboratory, Pasadena, Calif.

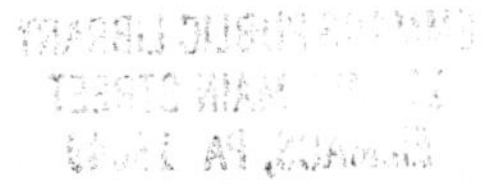

Donald Savage (202) 358-1547 NASA Headquarters, Washington, D.C.
NEWS RELEASE: 2004-052

Healthy Spirit Cleans a Mars Rock; Opportunity Rolls

February 06, 2004

"Our patient is healed, and we're very excited about that," said Jennifer Trosper of NASA's Jet Propulsion Laboratory, Pasadena, Calif., mission manager for Spirit.

Spirit temporarily stopped communicating Jan. 22; the problem was later diagnosed as a memory-management issue. Engineers regained partial control of the spacecraft within days and reformatted Spirit's flash memory Wednesday to prevent recurrence of the problem.

JPL's Glenn Reeves, flight software architect for the Mars Exploration Rovers, said Friday, "We're confident we know what the problem is, and we have a procedure in place we believe can work around this problem indefinitely."

Spirit's first day of science operations after the memory reformatting featured the first brushing of a rock on a foreign planet to remove dust and allow inspection of the rock's cleaned surface. Steel bristles on the rover's rock abrasion tool cleaned a circular patch on the rock unofficially named Adirondack. The tool's main function is to grind off the weathered surface of rocks with diamond teeth, but the brush for removing the grinder's cuttings can also be used to sweep dust off the intact surface.

The brushing on Thursday was the first use of a rock abrasion tool by either Spirit or its twin rover, Opportunity. The brush swirled for five minutes, said Stephen Gorevan of Honeybee Robotics, New York, lead scientist for the rock abrasion tools on both rovers.

"I didn't expect much of a difference. This is a big surprise," Gorevan said about a picture showing the brushed area is much darker than the rest of the rock's surface. "Ladies and gentlemen, I present you the greatest interplanetary brushing of all time."

One reason scientists first selected Adirondack for close inspection is because it appeared relatively dust free compared to some other rocks nearby. "To our surprise, there was quite a bit of dust on the surface," said Dr. Ken Herkenhoff of the U.S. Geological Survey's Astrogeology Team, Flagstaff, Ariz., lead scientist for the rovers' microscopic imagers.

Spirit was instructed Friday afternoon to grind the surface of Adirondack with the rock abrasion tool. After the grinding, the turret of tools at the end of the rover's robotic arm will be rotated to inspect the freshly exposed interior of the rock. Controllers plan to tell Spirit tomorrow to begin driving again.

Meanwhile, halfway around Mars, NASA's Opportunity drove early Friday for the second day in a row. It arrived within about a half a meter (20 inches) of the northeastern end of a rock outcrop scientists are eager for the rover to examine. "We expect to complete that approach tomorrow," said JPL's Matt Wallace, mission manager for Opportunity.

During Friday's drive, Opportunity did not travel as far as planned. The rover is climbing a slope of about 13 degrees, and the shortage in distance traveled is probably due to slippage in the soil, Wallace said.

The main task for both rovers is to explore the areas around their landing sites for evidence in rocks and soils about whether those areas ever had environments that were watery and possibly suitable for sustaining life.

Each martian day, or "sol," lasts about 40 minutes longer than an Earth day. Spirit begins its 35th sol on Mars at 4:02 a.m. Saturday, Pacific Standard Time. Opportunity begins its 15th sol on Mars at 4:23 p.m. Saturday, PST. JPL, a division of the California Institute of Technology in Pasadena, manages the Mars Exploration Rover project for NASA's Office of Space Science, Washington, D.C. Images and additional information about the project are available from JPL at http://marsrovers.jpl.nasa.gov and from Cornell University at http://athena.cornell.edu .

Guy Webster (818) 354-5011 Jet Propulsion Laboratory, Pasadena, Calif.
Donald Savage (202) 358-1547 NASA Headquarters, Washington, D.C.
NEWS RELEASE: 2004-053

Back in Action

This image shows the Mars Exploration Rover Spirit's "hand," or the tip of the instrument deployment device, poised in front of the rock nicknamed Adirondack, the rover's first science target since developing communication problems over two weeks ago. In preparation for grinding into Adirondack, Spirit cleaned off a portion of the rock's surface with a stainless steel brush located on its rock abrasion tool, seen here at the tip of its hand.The image was taken by the rover's panoramic camera.

Image credit: NASA/JPL/Cornell

Back in Action

This image shows the Mars Exploration Rover Spirit's "hand," or the tip of the instrument deployment device, poised in front of the rock nicknamed Adirondack, the rover's first science target since developing communication problems over two weeks ago. In preparation for grinding into Adirondack, Spirit cleaned off a portion of the rock's surface with a stainless steel brush located on its rock abrasion tool and seen here at the end of the yellow arrow. The image was taken by the rover's panoramic camera.

Image credit: NASA/JPL/Cornell

Back in Action-2

This image shows the Mars Exploration Rover Spirit's "hand," or the tip of the instrument deployment device, poised in front of the rock nicknamed Adirondack. In preparation for grinding into Adirondack, Spirit cleaned off a portion of the rock's surface with a stainless steel brush located on its rock abrasion tool, seen here at the top of its hand.The image was taken by the rover's panoramic camera.

Image credit: NASA/JPL/Cornell

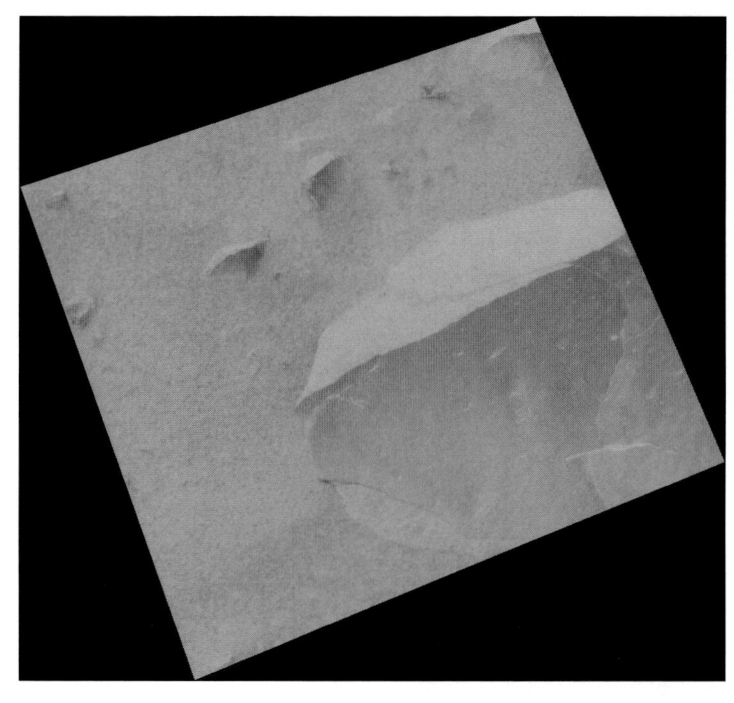

Dirty Adirondack

This image taken by the panoramic camera onboard the Mars Exploration Rover Spirit shows the rock dubbed Adirondack before the rover wiped off a portion of the rock's dust coating with a stainless steel brush located on its rock abrasion tool. Spirit cleaned off the rock in preparation for grinding into it to expose fresh rock underneath.

Image credit: NASA/JPL/Cornell

Adirondack in Need of Cleaning

This image taken by the microscopic imager onboard the Mars Exploration Rover Spirit shows the rock dubbed Adirondack before the rover wiped off a portion of the rock's dust coating with a stainless steel brush located on its rock abrasion tool. Spirit cleaned off the rock in preparation for grinding into it to expose fresh rock underneath.

Image credit:
NASA/JPL/Cornell/USGS

Adirondack's True Self

This image taken by the panoramic camera onboard the Mars Exploration Rover Spirit shows a cleaned off portion of the rock dubbed Adirondack. In preparation for grinding into the rock, Spirit wiped off a fine coat of dust with a brush located on its rock abrasion tool. Scientists plan to analyze the newly-exposed patch of rock with the rover's suite of science instruments, both before and after the top layer is removed.

Image credit: NASA/JPL/Cornell

Shiny and New

This microscopic image shows a cleaned off portion of the rock dubbed Adirondack. In preparation for grinding into the rock, Spirit wiped off a fine coat of dust with a stainless steel brush located on its rock abrasion tool. Some of this dust coating can be seen to the left of the image. Scientists plan to analyze the newly-exposed patch of rock with the rover's suite of science instruments, both before and after the top layer is removed. The image was taken by the rover's microscopic imager. The observed area is 3 centimeters (1.2 inches) across.

Image credit: NASA/JPL/Cornell/USGS

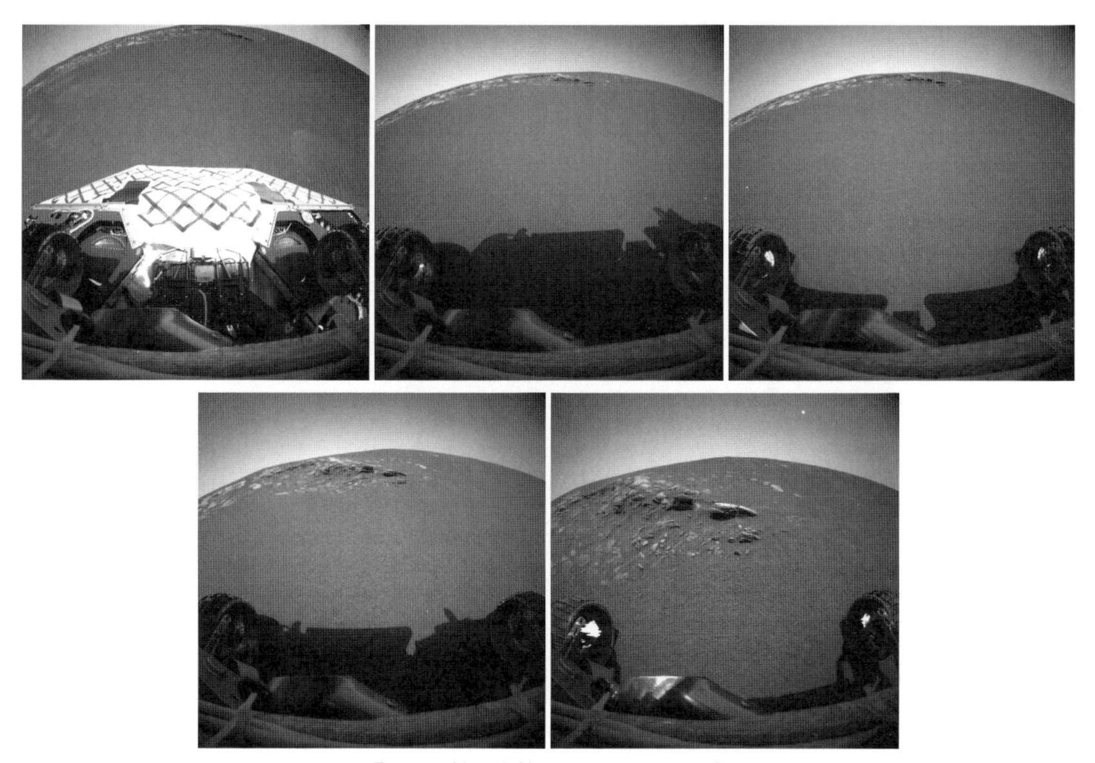

Forging Ahead (Animation sequence)

This animation shows the front view from the Mars Exploration Rover Opportunity as it drives north towards the eastern edge of the rock outcropping near its landing site at Meridiani Planum, Mars. The movie strings together images taken over the past six martian days, or sols, of its journey, beginning with a 1 meter (3 feet) stroll away from the lander on sol 7. On the 12th sol, Opportunity drove another 3 1/2 meters (11 feet), and then, one sol later, another 1 1/2 meters (5 feet). On its way, the rover twisted and turned in a test of its driving capabilities. This movie is made up of fish-eye images taken by the rover's front hazard-identification camera. Image credit: NASA/JPL

Rear View of Opportunity's Drive (Animation sequence)

This animation shows the rear view from the Mars Exploration Rover Opportunity as it drives north away from the lander and towards the eastern edge of the rock outcropping near its landing site at Meridiani Planum, Mars. The movie strings together images taken over the past six martian days, or sols, of its journey, beginning with a 1 meter (3 feet) stroll away from the lander on sol 7. On the 12th sol, Opportunity drove another 3 1/2 meters (11 feet), and then, one sol later, another 1 1/2 meters (5 feet). On its way, the rover twisted and turned in a test of its driving capabilities. This movie is made-up of fish-eye images taken by the rover's rear hazard-identification camera. Image credit: NASA/JPL

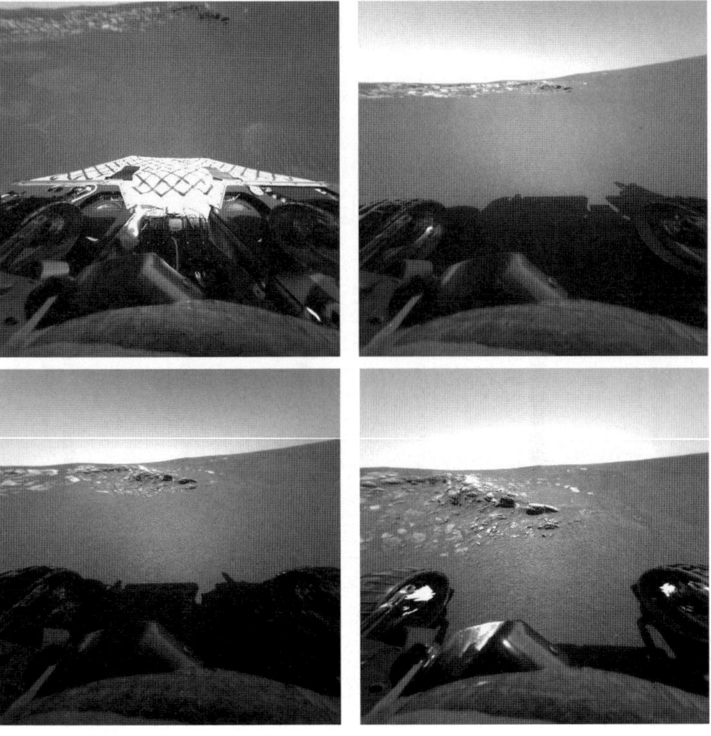

Forging Ahead (linearized) (Animation sequence)

This animation shows the front view from the Mars Exploration Rover Opportunity as it drives north towards the eastern edge of the rock outcropping near its landing site at Meridiani Planum, Mars. The movie strings together images taken over the past six martian days, or sols, of its journey, beginning with a 1 meter (3 feet) stroll away from the lander on sol 7. On the 12th sol, Opportunity drove another 3 1/2 meters (11 feet), and then, one sol later, another 1 1/2 meters (5 feet). On its way, the rover twisted and turned in a test of its driving capabilities. This movie is made up of images taken by the rover's front hazard-identification camera, which were corrected for fish-eye distortion.

Image credit: NASA/JPL

Traversing Martian Terrain

This 3-D view from behind the Mars Exploration Rover Opportunity shows the path the rover has traveled since rolling 1 meter (3 feet) away from its now-empty lander on the seventh martian day, or sol, of its mission. On the 12th sol, Opportunity drove another 3 1/2 meters (11 feet), and then, one sol later, another 1 1/2 meters (5 feet). On its way, the rover twisted and turned in a test of its driving capabilities. Opportunity is headed toward the eastern edge of the rock outcropping along the inner wall of the crater where it landed. This image was taken by the rover's rear hazard-identification camera.

Image credit: NASA/JPL

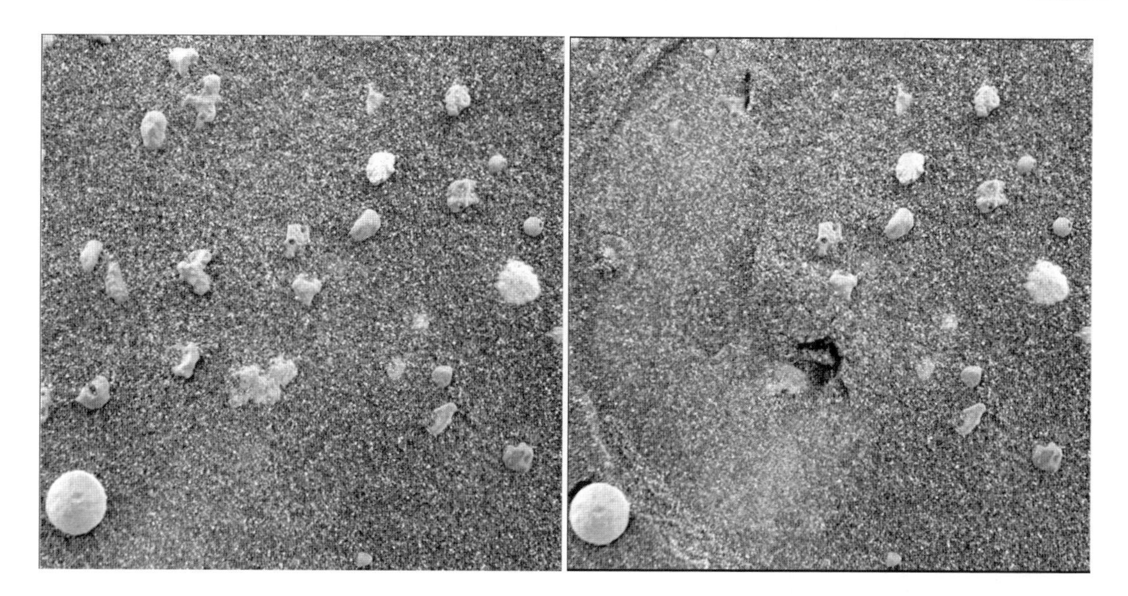

Now You See Them, Now You Don't (Animation sequence)

This animation flips back and forth between microscopic images taken of the soil at Meridiani Planum, Mars, before and after the Mars Exploration Rover Opportunity's Moessbauer spectrometer was pressed down to take measurements. The disappearing rocks indicate that the sandy soil is loosely packed.
Image credit: NASA/JPL/Cornell/USGS

SPIRIT UPDATE: Rock Interior Inspected - sol 35, Feb 08, 2004

NASA's Spirit examined the interior of a rock during Spirit's 35th sol on Mars, which ended at 4:41 a.m. Sunday, PST. Beginning late in the previous sol, Spirit took turns placing its Mössbauer spectrometer, alpha particle X-ray spectrometer and microscopic imager over the portion of the rock called Adirondack where Spirit's rock abrasion tool had cut away the rock's surface.

Spirit did not begin driving on sol 35, because a precautionary software setting to prevent driving was still in effect from the beginning of the anomaly two weeks ago. The rover is being commanded during sol 36, which ends at 5:21 a.m. Monday, PST, to back away from Adirondack, drive past the south side of the now-empty lander, and begin a trek northeast toward a crater nicknamed "Bonneville."

OPPORTUNITY UPDATE: Opportunity Succeeds with First "Touch and Go" - sol 14, Feb 07, 2004

Opportunity performed her first "touch and go" maneuver on the rover's 14th sol on Mars, which ended at 4:23 p.m. Saturday, PST. The activity included deploying the arm, taking microscopic images of the soil in front of the rover, re-stowing the arm and finishing its drive to Stone Mountain.

The panoramic camera and miniature thermal emission spectrometer instruments were used to make observations both before and after the "touch and go" sequence.

OPPORTUNITY UPDATE: Opportunity Gets a Closer Look at the Outcrop - sol 15, Feb 08, 2004

On Opportunity's 15th sol on Mars which ends at 5:02 p.m. Sunday, PST, the rover took microscopic images of a rock in the outcrop and nearby soil. The rock is called Stone Mountain (formerly called "Snout") and the target area for the microscope is called Robert E. The day's activities also include examination of Robert E with the alpha particle X-ray spectrometer and the Mössbauer spectrometer. Opportunity's panoramic camera and navigation camera were used to get pictures of the outcrop from the rover's current position.

In the coming sols, the plan is to move along the outcrop to examine other points along it.

Mars Rover Pictures Raise 'Blueberry Muffin' Questions

February 09, 2004

NASA's Spirit rover has begun making some of its own driving decisions while its twin, Opportunity, is presenting scientists with decisions to make about studying small spheres embedded in bedrock, like berries in a muffin.

Both rovers are on the move. Late Sunday, Spirit drove about 6.4 meters (21 feet), passing right over the rock called "Adirondack," where it had finished examining the rock's interior revealed by successfully grinding away the surface. The drive tested the rover's autonomous navigation ability for the first time on Mars.

"We've entered a new phase of the mission," said Dr. Mark Maimone, rover mobility software engineer at NASA's Jet Propulsion Laboratory, Pasadena, Calif. When the rover is navigating itself, it gets a command telling it where to end up, and it evaluates the terrain with stereo imaging to choose the best way to get there. It must avoid any obstacles it identifies. This capability is expected to enable longer daily drives than depending on step-by-step navigation commands from Earth. Tonight, Spirit will be commanded to drive farther on a northeastward course toward a crater nicknamed "Bonneville."

Over the weekend, Spirit drilled the first artificial hole in a rock on Mars. Its rock abrasion tool ground the surface off Adirondack in a patch 45.5 millimeters (1.8 inches) in diameter and 2.65 millimeters (0.1 inch) deep. Examination of the freshly exposed interior with the rover's microscopic imager and other instruments confirmed that the rock is volcanic basalt.

Opportunity drove about 4 meters (13 feet) today. It moved to a second point in a counterclockwise survey of a rock outcrop called "Opportunity Ledge" along the inner wall of the rover's landing-site crater. Pictures taken at the first point in that survey reveal gray spherules, or small spheres, within the layered rocks and also loose on the ground nearby.

NASA now knows the location of Opportunity's landing site crater, which is 22 meters (72 feet) in diameter. Radio signals gave a preliminary location less than an hour after landing, and additional information from communications with NASA's Mars Odyssey orbiter soon narrowed the estimate, said JPL's Tim McElrath, deputy chief of the navigation team.

As Opportunity neared the ground, winds changed its course from eastbound to northbound, according to analysis of data recorded during the landing. "It's as if the crater were attracting us somehow," said JPL's Dr. Andrew Johnson, engineer for a system that estimated the spacecraft's horizontal motion during the landing. The spacecraft bounced 26 times and rolled about 200 meters (about 220 yards) before coming to rest inside the crater, whose outcrop represents a bonanza for geologists on the mission.

JPL geologist Dr. Tim Parker was able to correlate a few features on the horizon above the crater rim with features identified by Mars orbiters, and JPL imaging scientist Dr. Justin Maki identified the spacecraft's jettisoned backshell and parachute in another Opportunity image showing the outlying plains.

As a clincher, a new image from Mars Global Surveyor's camera shows the Opportunity lander as a bright feature in the crater. A dark feature near the lander may be the rover. "I won't know if it's really the rover until I take another picture after the rover moves," said Dr. Michael Malin of Malin Space Science Systems, San Diego. He is a member of the rovers' science team and principal investigator for the camera on Mars Global Surveyor.

Opportunity's crater is at 1.95 degrees south latitude and 354.47 degrees east longitude, the opposite side of the planet from Spirit's landing site at 14.57 degrees south latitude and 175.47 degrees east longitude.

The first outcrop rock Opportunity examined up close is finely-layered, buff-colored and in the process of being eroded by windblown sand. "Embedded in it like blueberries in a muffin are these little spherical grains," said Dr. Steve Squyres of Cornell University, Ithaca, N.Y., principal investigator for the rovers' scientific instruments. Microscopic images show the gray spheres in various stages of being released from the rock.

"This is wild looking stuff," Squyres said. "The rock is being eroded away and these spherical grains are dropping out." The spheres may have formed when molten rock was sprayed into the air by a volcano or a meteor impact. Or, they may be concretions, or accumulated material, formed by minerals coming out of solution as water diffused through rock, he said.

The main task for both rovers in coming weeks and months is to explore the areas around their landing sites for evidence in rocks and soils about whether those areas ever had environments that were watery and possibly suitable for sustaining life. JPL, a division of the California Institute of Technology in Pasadena, manages the Mars Exploration Rover project for NASA's Office of Space Science, Washington, D.C. Images and additional information about the project are available from JPL at http://marsrovers.jpl.nasa.gov and from Cornell University at http://athena.cornell.edu .

Guy Webster (818) 354-5011 Jet Propulsion Laboratory, Pasadena, Calif.
Donald Savage (202) 358-1547 NASA Headquarters, Washington, D.C.
NEWS RELEASE: 2004-054

Opportunity Spies Its Backshell

From its new location at the inner edge of the small crater surrounding it, the Mars Exploration Rover Opportunity was able to look out to the plains where its backshell (left) and parachute (right) landed. Opportunity is currently investigating a rock outcropping with its suite of robotic geologic tools. This approximate true-color image was created by combining data from the panoramic camera's red, green and blue filters.

Image credit: NASA/JPL/Cornell

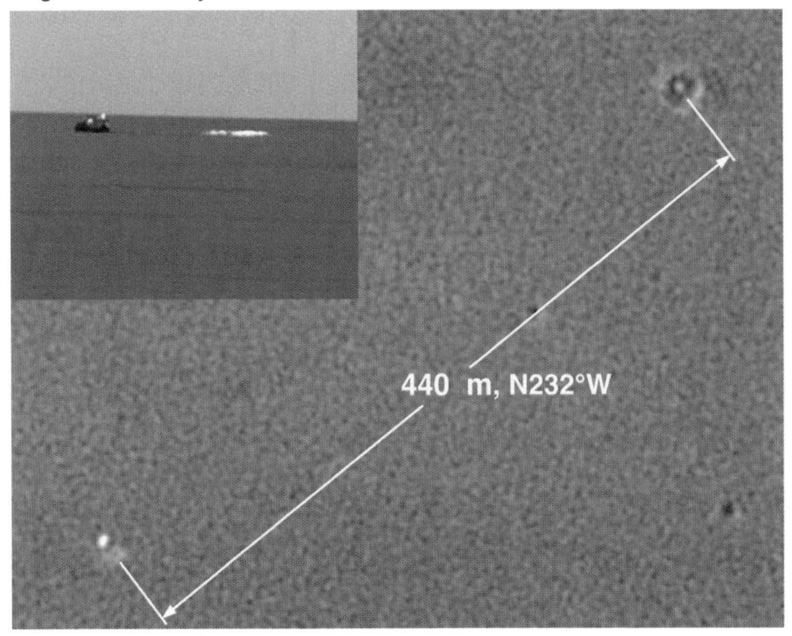

An Old Friend

This image shows two views of the backshell and parachute that helped deliver the Mars Exploration Rover Opportunity safely to the surface of Mars. The first, seen in the top left picture, is from the rover's perspective inside the small crater where it landed. The second, seen in the center, is from above and was taken by a camera onboard the Mars Global Surveyor orbiter. The white spot inside the crater in the upper right corner is the rover's lander, and the white mark in the lower left corner is the backshell.

Image credit: ASA/JPL/MSSS

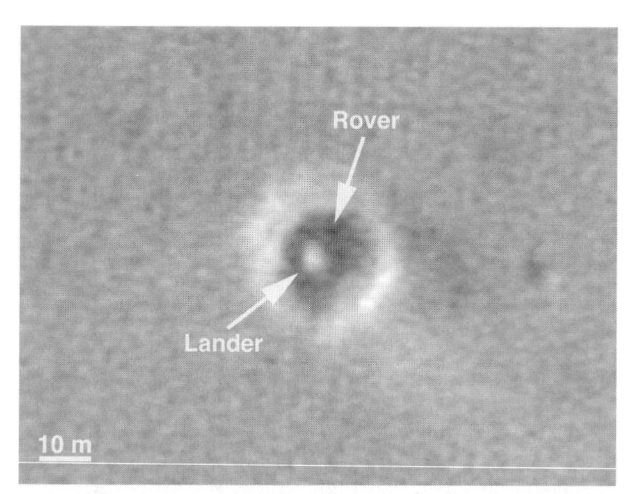

'You Are Here'

This map shows the Mars Exploration Rover Opportunity and its lander on the surface of Mars. The robotic geologist landed inside a small crater at Meridiani Planum on Jan. 24, 2004, PST. The image was taken by a camera onboard the Mars Global Surveyor orbiter.

Image credit: NASA/JPL/MSSS

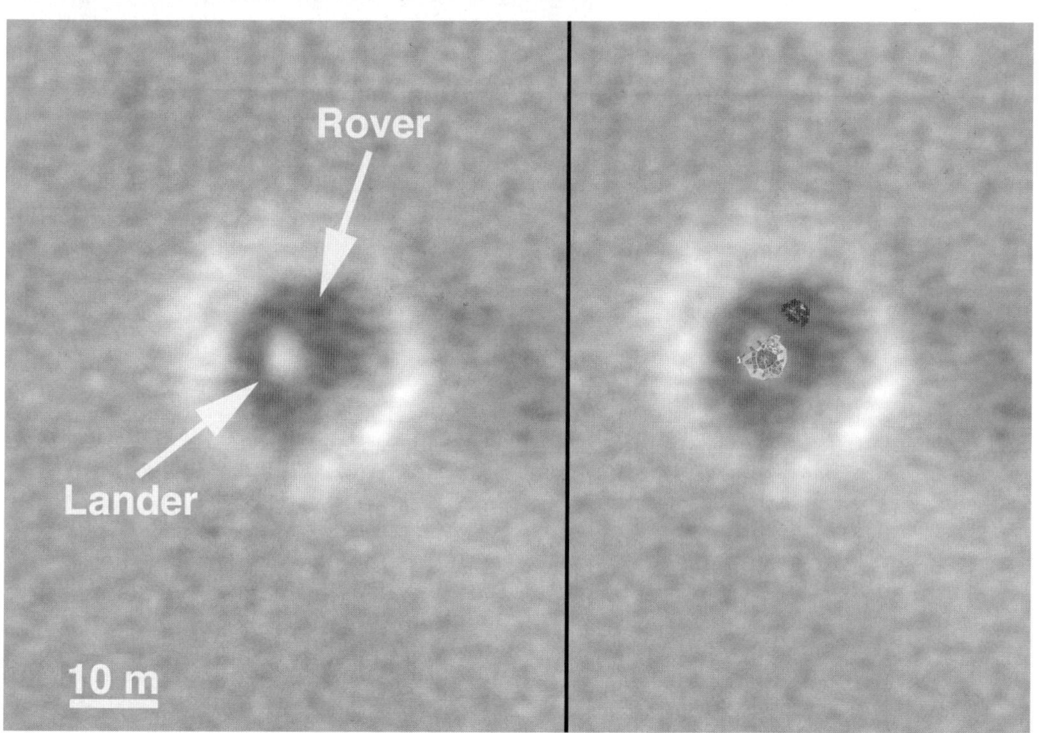

'You are here' (with models)

These maps show the Mars Exploration Rover Opportunity and its lander on the surface of Mars. The robotic geologist landed inside a small crater at Meridiani Planum on Jan. 24, 2004, PST. In the right image, computer-generated models of the rover and lander have been superimposed over the image of the actual rover and lander.

Image credit: NASA/JPL/MSSS/ASU

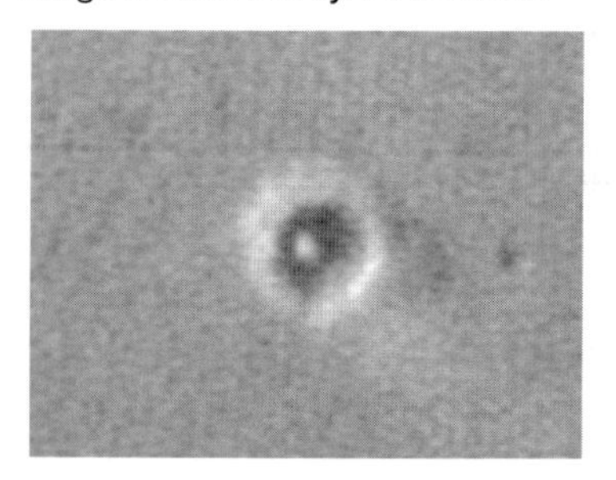

'You Are Here' (no labels)

This map shows the Mars Exploration Rover Opportunity and its lander on the surface of Mars. The robotic geologist landed inside a small crater at Meridiani Planum on Jan. 24, 2004, PST. The white spot is the lander, and the small black spot northeast of it is believed to be the rover. The image was taken by a camera onboard the Mars Global Surveyor orbiter.

Image credit: NASA/JPL/MSSS

Stone Mountain in Context

The colored square in this grayscale image taken by the panoramic camera onboard the Mars Exploration Rover Opportunity highlights the location of Stone Mountain, located within the rock outcrop at Meridiani Planum, Mars. Scientists are examining Stone Mountain with the instruments on the rover's instrument deployment device, or "arm," in search of clues about the composition of the rock outcrop.

Image credit: NASA/JPL/Cornell

Stone Mountain

This color image taken by the panoramic camera onboard the Mars Exploration Rover Opportunity shows the part of the rock outcrop dubbed Stone Mountain at Meridiani Planum, Mars. Scientists are examining Stone Mountain with the instruments on the rover's instrument deployment device, or "arm," in search of clues about the composition of the rock outcrop.

Image credit: NASA/JPL/Cornell

Speckled with Spherules

This false-color image taken by the panoramic camera onboard the Mars Exploration Rover Opportunity highlights the spherules that speckle the rock dubbed Stone Mountain. The colors in this picture were exaggerated or stretched to enhance the real difference in color between Stone Mountain and its collection of granular dots.

Image credit: NASA/JPL/Cornell

A Patch of Stone

The colorless square in this color image of the martian rock formation called Stone Mountain is one portion of the rock being analyzed with tools on the Mars Exploration Rover Opportunity's instrument deployment device, or "arm." The square area is approximately 3 centimeters (1.2 inches) across. Stone Mountain is located within the rock outcrop on Meridiani Planum, Mars. The image was taken by the rover's panoramic camera.

Image credit: NASA/JPL/Cornell/USGS

Mars Rock Formation Poses Mystery

This sharp, close-up image taken by the microscopic imager on the Mars Exploration Rover Opportunity's instrument deployment device, or "arm," shows a rock target dubbed "Robert E," located on the rock outcrop at Meridiani Planum, Mars. Scientists are studying this area for clues about the rock outcrop's composition. This image measures 3 centimeters (1.2 inches) across and was taken on the 15th day of Opportunity's journey (Feb. 8, 2004).

Image credit: NASA/JPL/Cornell/US Geological Survey

Mars Rock Formation Poses Mystery-2

This sharp, close-up image taken by the microscopic imager on the Mars Exploration Rover Opportunity's instrument deployment device, or "arm," shows a rock target dubbed "Robert E," located on the rock outcrop at Meridiani Planum, Mars. Scientists are studying the spherule, or small sphere, in the center of the image that appears to be protruding from the rock formation. This image measures 3 centimeters (1.2 inches) across and was taken on the 15th day of Opportunity's journey (Feb. 8, 2004).

Image credit: NASA/JPL/Cornell/US Geological Survey

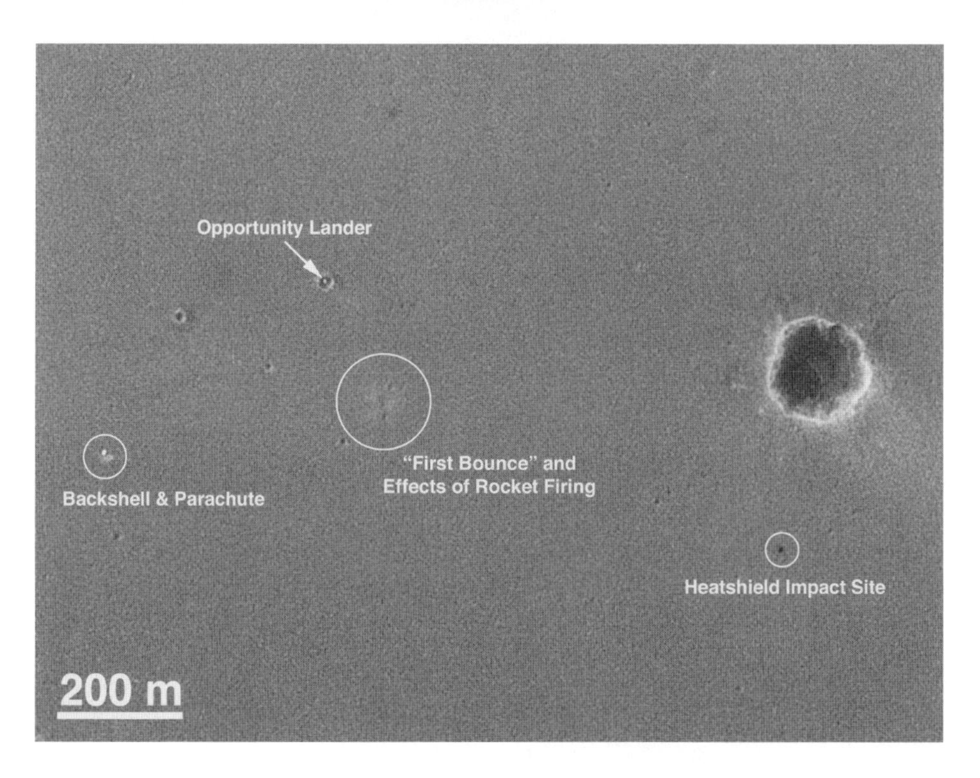

Reconstructing the Scene of Landing

This map of the Mars Exploration Rover Opportunity's new neighborhood at Meridiani Planum, Mars shows remnants of the rover's landing, including its lander; backshell and parachute; first bounce mark; and the site where its heat shield impacted the surface. The image was taken by a camera onboard the Mars Global Surveyor orbiter.

Image credit: NASA/JPL/MSSS

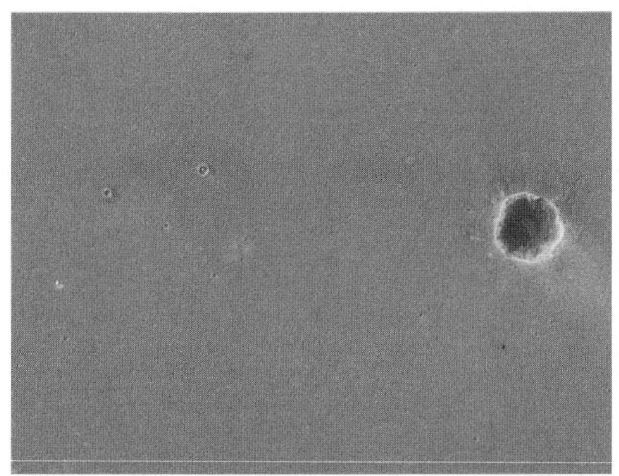

Reconstructing the Scene of Landing (no labels)

This map of the Mars Exploration Rover Opportunity's new neighborhood at Meridiani Planum, Mars shows remnants of the rover's landing, including its lander; backshell and parachute; first bounce mark; and the site where its heat shield hit the surface (see labeled version of image for exact locations). The image was taken by a camera onboard the Mars Global Surveyor orbiter.

Image credit: NASA/JPL/MSSS

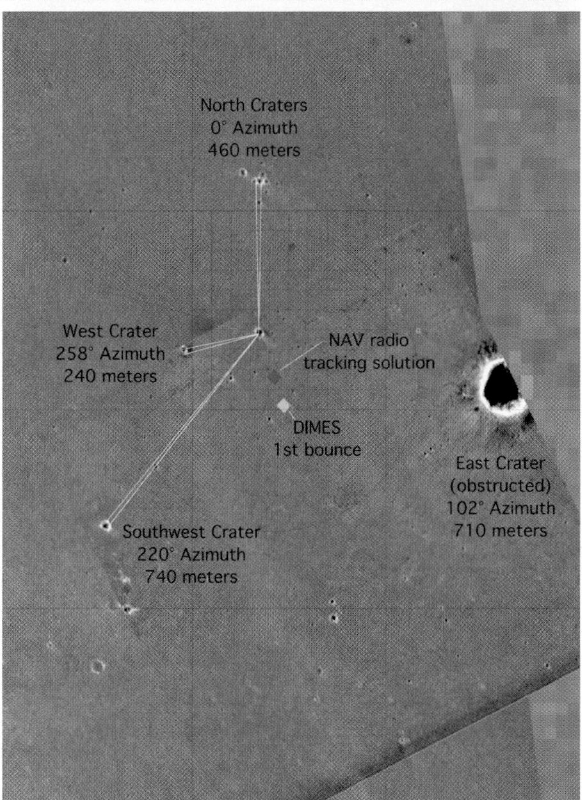

"X" Marks the Spot

This map of the Mars Exploration Rover Opportunity's new neighborhood at Meridiani Planum, Mars, shows the surface features used to locate the rover. By imaging these "bumps" on the horizon from the perspective of the rover, mission members were able to pin down the rover's precise location. The image consists of data from the Mars Global Surveyor orbiter, the Mars Odyssey orbiter and the descent image motion estimation system located on the bottom of the rover.

Image credit: NASA/JPL/MSSS/ASU

Opportunity's Hole in One (Side View)

In this side view of the path the Mars Exploration Rover Opportunity took when it landed at Meridiani Planum, Mars, a computer-generated red line shows the path of the spacecraft's descent and bouncing along the surface. The line is superimposed on a mosaic of the three images taken during descent by the descent image motion estimation system camera, located on the bottom of the lander. Initially, the Opportunity lander was traveling east, but near the end of its descent, it began moving north. When the lander was released from the parachute, the spacecraft bounced to the north into the crater shown at the top of the image. North is indicated by the red-tipped white arrow in the coordinate axes and east, by the green-tipped white arrow Image credit: NASA/JPL

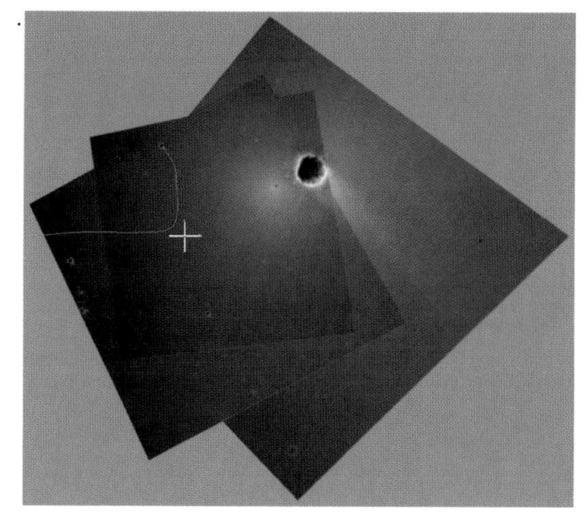

Opportunity's Hole in One

This computer-generated visualization depicts an overhead view of the path Opportunity took when it landed at Meridiani Planum, Mars. A red line shows the path of the spacecraft's descent and bouncing along the surface. The line is superimposed on a mosaic of the three images taken during descent by the descent image motion estimation system camera, located on the bottom of the lander. Initially, the Opportunity lander was traveling east, but near the end of its descent, it began moving north. When the lander was released from the parachute, the spacecraft bounced to the north into the crater shown at the top of the image. North is indicated by the red-tipped white arrow in the coordinate axes and east, by the green-tipped white arrow.

Image credit: NASA/JPL

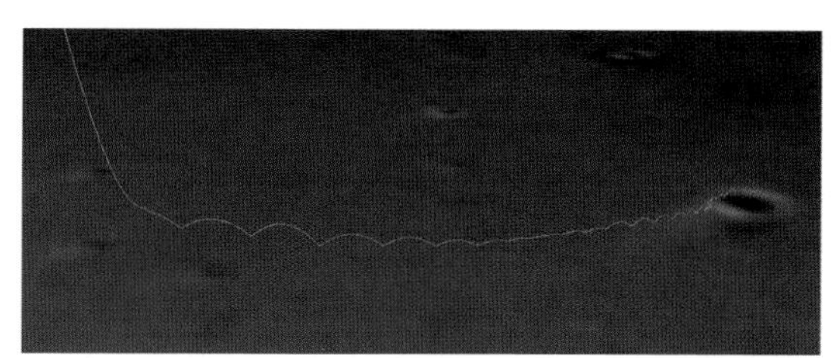

Opportunity Bounces to a Stop

In this close-up view of the path Opportunity took when it landed at Meridiani Planum, Mars, a computer-generated red line shows the spacecraft's bounce motions as it landed at Meridiani Planum, Mars. The spacecraft bounced north approximately 26 times while safely encased in airbags, until it came to a stop inside the crater to the right of the image. The red line is superimposed on a mosaic of the three images taken during descent by the descent image motion estimation system camera, located on the bottom of the lander.

Image credit: NASA/JPL

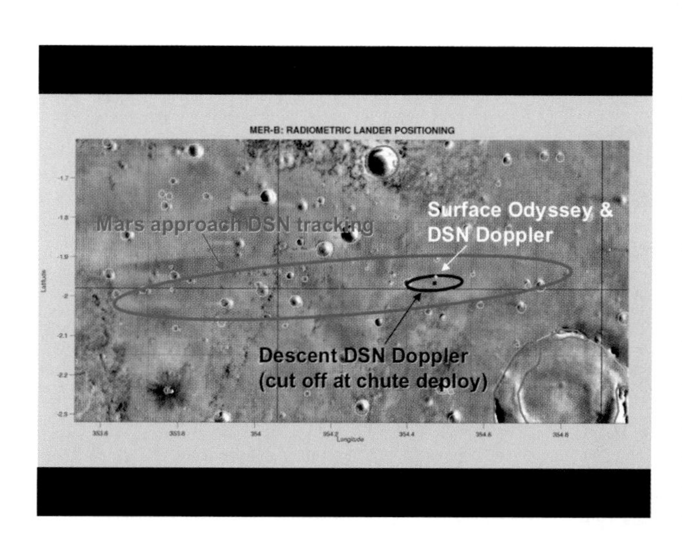

Found You!

This map of the Mars Exploration Rover Opportunity's new neighborhood at Meridiani Planum, Mars, demonstrates how engineers honed in on the location of the rover. The larger blue ellipse shows the projected landing area just before arriving at Mars. The black ellipse denotes the first approximation of the rover's location after landing based on radio signals received by NASA's Deep Space Network during entry, descent and landing. The white cross points to the rover's precise location based on radio signals sent from the surface both directly to Earth through the Deep Space Network and through NASA's Mars Odyssey orbiter.

Image credit: NASA/JPL/MSSS/ASU

Dusty Adirondack

This close-up image taken by the microscopic imager onboard the Mars Exploration Rover Spirit shows a portion of the rock dubbed Adirondack before dust was wiped from its surface by a brush on the rover's rock abrasion tool. Spirit cleaned the rock before grinding into it to expose fresh rock underneath. The observed area is 3 centimeters (1.2 inches) across.

Image credit: NASA/JPL/Cornell/USGS

Spirit Self-motivates

The Mars Exploration Rover Spirit drove itself 1 meter (3 feet) out of 6.4 meters (21 feet) at Gusev Crater, Mars, on Feb. 8, 2004, the 36th sol of its mission. This image shows the tracks it created in the martian soil as it drove straight ahead, then to the left. The rover also drove over Adirondack (seen in image bottom center), the bright rock that was targeted by Spirit's rock abrasion tool, on its way to a rock target called White Boat. This was the first test of the rover's autonomous system, which will be used many times in the days to come.

Image credit: NASA/JPL

Inner Adirondack

This close-up image taken by the microscopic imager onboard the Mars Exploration Rover Spirit shows the rock dubbed Adirondack after a portion of its surface was ground off by the rover's rock abrasion tool. The observed area is 3 centimeters (1.2 inches) across.

Image credit: NASA/JPL/Cornell/USGS

Spirit's First Grinding of a Rock on Mars

The round, shallow depression in this image resulted from history's first grinding of a rock on Mars. The rock abrasion tool on NASA's Spirit rover ground off the surface of a patch 45.5 millimeters (1.8 inches) in diameter on a rock called Adirondack. The hole is 2.65 millimeters (0.1 inch) deep, exposing fresh interior material of the rock for close inspection with the rover's microscopic imager and two spectrometers on the robotic arm. This image was taken by Spirit's panoramic camera, providing a quick visual check of the success of the grinding. The rock abrasion tools on both Mars Exploration Rovers were supplied by Honeybee Robotics, New York, N.Y.

Image credit: NASA/JPL/Cornell

SPIRIT UPDATE: Spirit is On the Move! - sol 36, Feb 09, 2004

On sol 36, which ended at 5:21 a.m. Monday, PST, Spirit drove 6.37 meters (20.9 feet), using the onboard navigation software and hazard avoidance system for the first time on Mars. The drive, intended to test the traverse commands, was extremely precise, taking Spirit to its intended goal - the rock called White Boat. Before leaving the rock Adirondack, Spirit took images and collected miniature thermal emission spectrometer data from the hole ground by the rock abrasion tool.

In the coming sols, Spirit will continue its drive toward Bonneville Crater.

OPPORTUNITY UPDATE: Shoot and Scoot - sol 15, Feb 09, 2004

Opportunity appears to have experienced slips during 50 percent of a drive on sol 15, so for sol 16, engineers played a lighthearted wake-up call: Paul Simon's "Slip Sliding Away." Regardless of the loose soil, Opportunity made it across

4 meters (12 feet) today and is positioned to continue observing parts of the outcrop up close tomorrow. In coming sols, Opportunity will "shoot and scoot," meaning the rover will shoot pictures of the terrain and acquire new scientific measurements of the rocks, then scoot up, down, and across the inside of the crater.

SPIRIT UPDATE: Record-breaking Sol - sol 37, Feb 10, 2004

On its 37th sol on Mars, which ends at 6 a.m. Tuesday, PST, Spirit broke the record for the farthest distance driven in one sol on Mars, traveling 21.2 meters (69.6 feet). Today's distance traveled shattered the Sojourner rover's previous record of 7 meters (23 feet) in one sol.

In the coming sols, Spirit will continue its drive towards the crater nicknamed "Bonneville."

11-Feb-2004

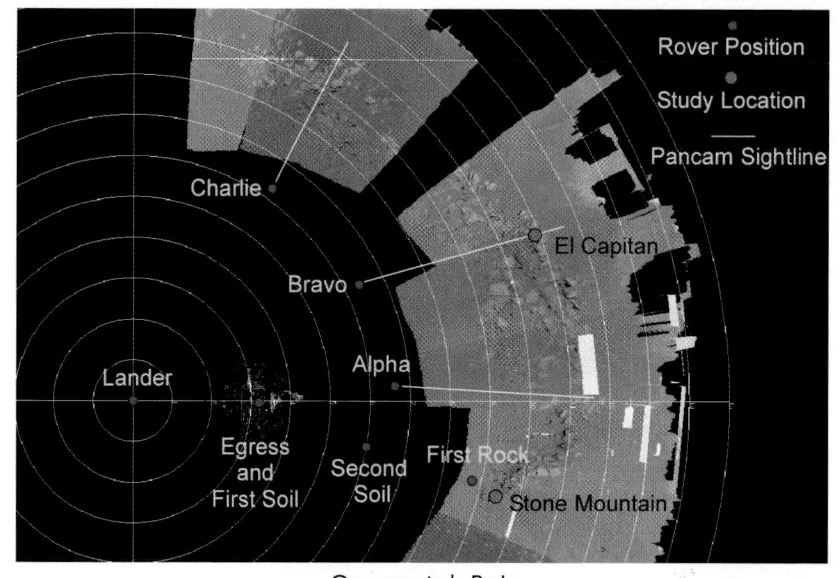

Opportunity's Path

This Long Term Planning graphic was created from a mosaic of navigation camera images overlain by a polar coordinate grid with the center point as Opportunity's original landing site. The blue dots represent the rover position at various locations.

The red dots represent the center points of the target areas for the instruments on the rover mast (the panoramic camera and miniature thermal emission spectrometer). Opportunity visited Stone Mountain on Feb. 5. Stone Mountain was named after the southernmost point of the Appalachian Mountains outside of Atlanta, Ga. On Earth, Stone Mountain is the last big mountain before the Piedmont flatlands, and on Mars, Stone Mountain is at one end of Opportunity Ledge. El Capitan is a target of interest on Mars named after the second highest peak in Texas in Guadaloupe National Park, which is one of the most visited outcrops in the United States by geologists. It has been a training ground for students and professional geologists to understand what the layering means in relation to the formation of Earth, and scientists will study this prominent point of Opportunity Ledge to understand what the layering means on Mars.

The yellow lines show the midpoint where the panoramic camera has swept and will sweep a 120-degree area from the three waypoints on the tour of the outcrop. Imagine a fan-shaped wedge from left to right of the yellow line.

The white contour lines are one meter apart, and each drive has been roughly about 2-3 meters in length over the last few sols. The large white blocks are dropouts in the navigation camera data.

Opportunity is driving along and taking a photographic panorama of the entire outcrop. Scientists will stitch together these images and use the new mosaic as a "base map" to decide on geology targets of interest for a more detailed study of the outcrop using the instruments on the robotic arm. Once scientists choose their targets of interest, they plan to study the outcrop for roughly five to fifteen sols. This will include El Capitan and probably one to two other areas.

Image created by John Grotzinger and the Long Term Planning Team.

Blue Dot Dates
Sol 7 / Jan 31 = Egress & first soil data collected by instruments on the arm
Sol 9 / Feb 2 = Second Soil Target
Sol 12 / Feb 5 = First Rock Target
Sol 16 / Feb 9 = Alpha Waypoint
Sol 17 / Feb 10 = Bravo Waypoint
Sol 19 or 20 / Feb 12 or 13 = Charlie Waypoint

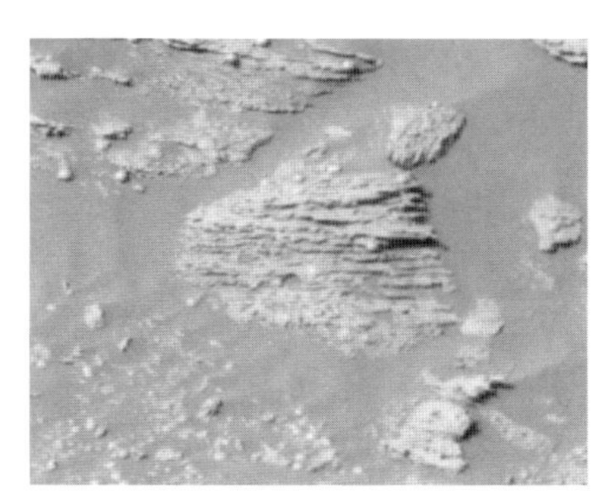

Unparallel Lines Give Unparalleled Clues

Scientists are excited to see new details of layered rocks in Opportunity Ledge. In previous panoramic camera images, geologists saw that some rocks in the outcrop had thin layers, and images sent to Earth on sol 17 (Feb. 10, 2004) now show that the thin layers are not always parallel to each other like lines on notebook paper. Instead, if you look closely at this image from an angle, you will notice that the lines converge and diverge at low angles. These unparallel lines give unparalleled clues that some "moving current" such as volcanic flow, wind, or water formed these rocks. These layers with converging and diverging lines are a significant discovery for scientists who are on route to rigorously test the water hypothesis. The main task for both rovers in coming weeks and months is to explore the areas around their landing sites for evidence in rocks and soils about whether those areas ever had environments that were watery and possibly suitable for sustaining life.

This is a cropped image taken by Opportunity's panoramic camera on sol 16 (Feb. 9, 2004). JPL, a division of the California Institute of Technology in Pasadena, manages the Mars Exploration Rover project for NASA's Office of Space Science, Washington, D.C.

Image credit: NASA/JPL/Cornell

OPPORTUNITY UPDATE: Extended Tour - sol 18, Feb 12, 2004

Opportunity had a couple of little hiccups on sol 18, February 11, which ends at 7:01 p.m. Wednesday, PST. The wrist on the real rover arm would not point as far vertically as the engineering rover's wrist did on Earth during a model test the night before. Because of this, the arm on Mars did not stow, and the rover did not move on to waypoint Charlie. The rover also automatically stopped use of the mast due to the fact that it believed a requested pointing position was in an area beyond its limits. Engineers solved both problems on sol 18. All systems are go for Opportunity to complete the tour of the outcrop by heading to outpost Charlie on sol 19, Thursday, February 12.

Student Programs Tap Into Mars Rover Adventures

February 12, 2004

NASA's Mars Exploration Rovers are not only providing scientists a flood of information about Mars -- including new insights today about winds -- they are also adding excitement to classrooms throughout the nation.

An assortment of programs giving students first-hand opportunities to work with information from NASA Mars missions help young people "see themselves as scientists in the future because they understand the process of science," said Sheri Klug of Arizona State University, Tempe, and NASA's Jet Propulsion Laboratory, Pasadena, Calif. She coordinates NASA Mars education programs for kindergarten through high school, part of the agency's goal to inspire the next generation of explorers. Silver Stage High School in Silver Springs, Nev., is one of 13 schools participating in one program that pairs selected students with researchers on the rover missions. "I actually get the opportunity to work with the scientists. It's really awesome!" said Shannon Theissen, 16, a Silver Stage junior.

Dr. Wendy Calvin, rover science team member from University of Nevada, Reno, and Shannon's mentor for a week at JPL, said, "This is the real stuff, not baby steps. The students are using the same tools we do."

Hundreds of other students from around the country participate in programs using pictures and other information from NASA Mars orbiters, and more than 1,000 have sent in rocks for a project to compare Earth rocks with Mars rocks.

Meanwhile, noted Art Thompson of JPL's rover flight team, "We have two very busy rovers on the surface of Mars." On Wednesday, Spirit broke its own record set earlier in the week for the longest one-day drive on Mars. The rover added 24.4 meters (80 feet) to its odometer, bringing the total to 57.4 meters (188 feet) and ending its day near a cluster of rocks dubbed "Stone Council."

In coming weeks, scientists and engineers plan for Spirit to drive up to the rim of a crater dubbed "Bonneville," still more than two football-field lengths away, in hopes of peering inside and seeing rock layers that could tell the geologic history and the potential role of water at the Gusev site.

Opportunity drove Friday morning to the fourth counterclockwise position in its survey of a rock outcrop along the inner slope of the crater in which it landed. Based on the survey, scientists will choose a small number of locations on the outcrop to come back to for more thorough examination later. The flight team has learned to compensate for wheel slippage in the soil on the slope. "When we attempt to drive up the slope we intentionally overdrive, and when we drive down a slope we intentionally underdrive," Thompson said.

Both rovers have used an infrared sensing instrument called the miniature thermal emission spectrometer to study the sky, as well as the ground. These atmospheric observations are revealing rapid temperature changes in the lower atmosphere. In mid-morning, the air temperature at about the height of an eight-story building swings up and down by several degrees within a minute.

"Warmer and colder blobs of air are intermittently passing over the rover," said Dr. Don Banfield, a rover science team collaborator from Cornell University, Ithaca, N.Y. "We're watching the overturning of the atmosphere as it's warming up in the morning." Rising warmer air carries heat to upper layers of the atmosphere. Observing the details of these changes helps scientists improve their models for understanding Mars' winds.

Better understanding of Mars' winds is important not only for the design of future landings on the planet, but also for interpreting some features on the surface. "We've been talking a lot about water on Mars in the past, but wind is currently the important agent of change on Mars," Banfield said.

Microscopic images indicate that windblown sand is eroding the outcrop that Opportunity is studying. Dr. Mark Lemmon, science team member from Texas A&M University, College Station, said that taking a series of images with that instrument at slightly different distances from the target allows creation of a three-dimensional view. "We're gathering as much information about the things we're looking at as we possibly can," he said.

The main task for both rovers in coming weeks and months is to explore for evidence in rocks and soils about whether the landing-site areas ever had environments that were watery and possibly suitable for sustaining life. JPL, a division of the California Institute of Technology in Pasadena, manages the Mars Exploration Rover project for NASA's Office of Space Science, Washington, D.C. Images and additional information about the project are available from JPL at http://marsrovers.jpl.nasa.gov and from Cornell University at http://athena.cornell.edu . Information about NASA school projects is available at http://education.nasa.gov .

Guy Webster (818) 354-5011 Jet Propulsion Laboratory, Pasadena, Calif.
Donald Savage (202) 358-1547 NASA Headquarters, Washington, D.C.
NEWS RELEASE: 2004-58

"Berries" on the Ground

This mosaic image shows an extreme close-up of round, blueberry-shaped formations in the martian soil near a part of the rock outcrop at Meridiani Planum called Stone Mountain. Scientists are studying these curious formations for clues about the area's past environmental conditions. The image, one of the highest resolution images ever taken by the microscopic imager, an instrument located on the Mars Exploration Rover Opportunity's instrument deployment device or "arm."

Image credit: NASA/JPL/Cornell/USGS

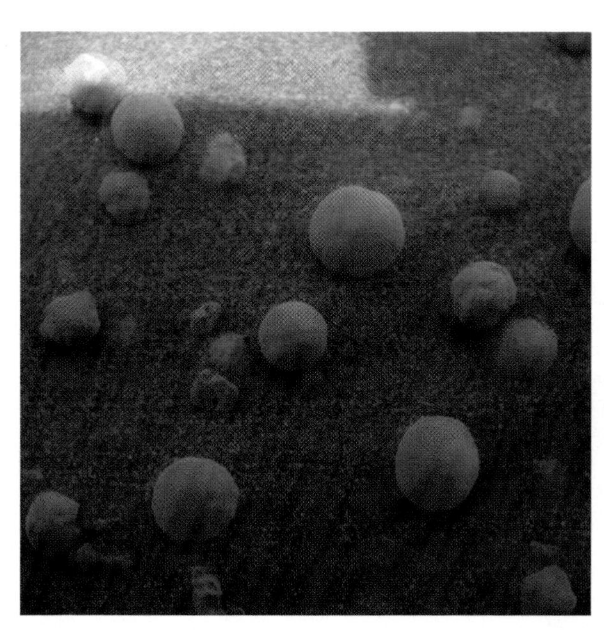

"Berries" on the Ground 2 (3-D)

This is the 3-D anaglyph showing a microscopic image taken of soil featuring round, blueberry-shaped rock formations on the crater floor at Meridiani Planum, Mars. This image was taken on the 13th day of the Mars Exploration Rover Opportunity's journey, before the Moessbauer spectrometer, an instrument located on the rover's instrument deployment device, or "arm," was pressed down to take measurements. The area in this image is approximately 3 centimeters (1.2 inches) across.

Image credit: NASA/JPL/Cornell/USGS/Texas A&M

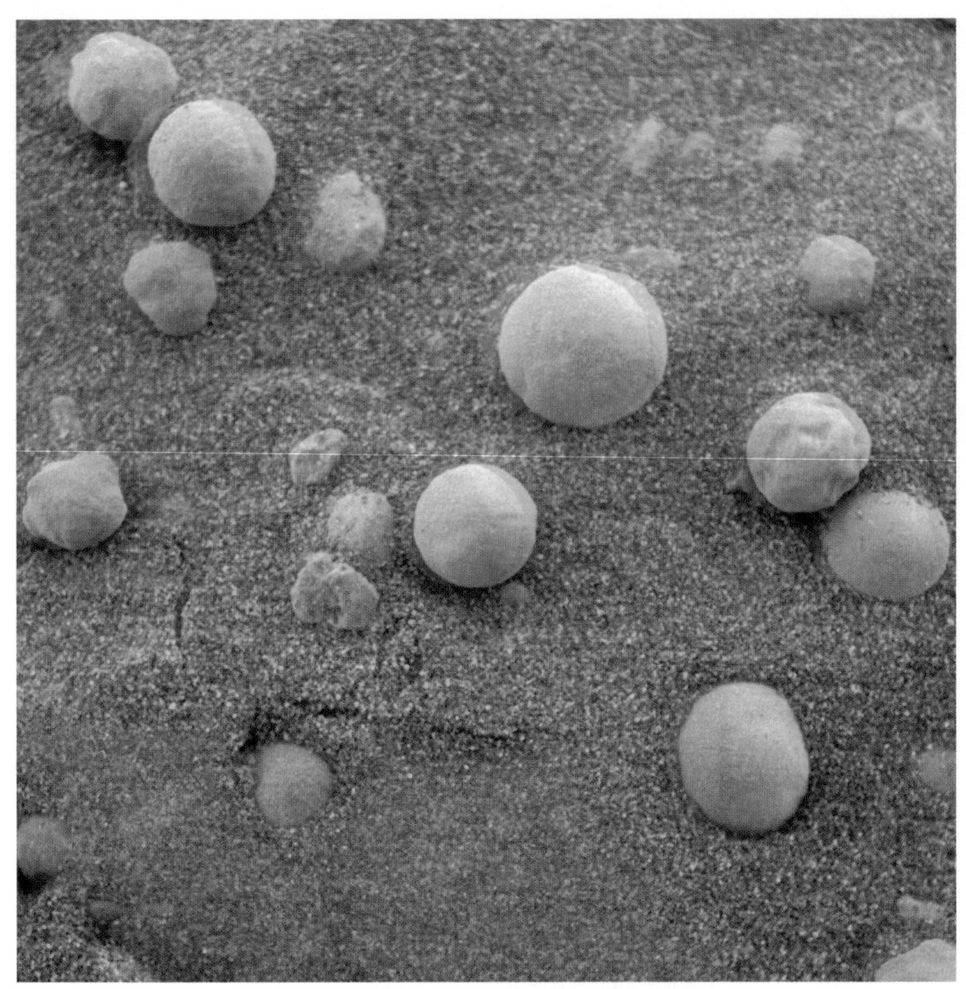

"Berries" on the Ground 2 (3-D)

This is the 3-D anaglyph showing a microscopic image taken of soil featuring round, blueberry-shaped rock formations on the crater floor at Meridiani Planum, Mars. This image was taken on the 13th day of the Mars Exploration Rover Opportunity's journey, after the Moessbauer spectrometer, an instrument located on the rover's instrument deployment device, or "arm," was pressed down to measure the soil's iron mineralogy. Note the donut-shaped imprint of the instrument in the lower part of the image. The area in this image is approximately 3 centimeters (1.2 inches) across.

Image credit: NASA/JPL/Cornell/USGS/Texas A&M

Inspector Gadget

This image is a still from a computer-generated animation showing the Mars Exploration Rover inspecting the rock dubbed Stone Mountain with its instrument deployment device, or arm.

Image credit: NASA/JPL

Martian Microscope

The microscopic imager (circular device in center) is in clear view above the surface at Meridiani Planum, Mars, in this approximate true-color image taken by the panoramic camera on the Mars Exploration Rover Opportunity. The image was taken on the 9th sol of the rover's journey. The microscopic imager is located on the rover's instrument deployment device, or arm. The arrow is pointing to the lens of the instrument. Note the dust cover, which flips out to the left of the lens, is open. This approximated color image was created using the camera's violet and infrared filters as blue and red.

Image credit: NASA/JPL/Cornell/Texas A&M

Strolling on Martian Ground

This animation is created from still images taken by the Mars Exploration Rover Spirit during its approximately 21.2-meter (69.6-foot) drive across the pebbly ground at Gusev Crater, Mars, on the 37th day, or sol, of its mission (Feb. 9, 2004). Two sols later, Spirit drove another 24 meters (78.7 feet) toward a rock target called White Boat. The images were captured by the rover's front hazard-avoidance camera after it began the autonomous portion of its drive.

Image credit: NASA/JPL

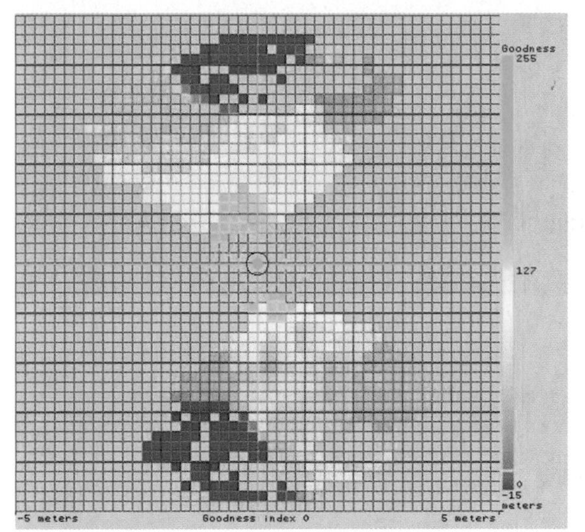

Mind of Its Own

This animation shows the path the Mars Exploration Rover Spirit traveled during its 24-meter (78.7-foot) autonomous drive across the bumpy terrain at Gusev Crater, Mars, on the 39th day, or sol, of its mission. The colored data are from the rover's hazard-avoidance camera and have been reconstructed to show the topography of the land. Red areas indicate extremely hazardous terrain, and green patches denote safe, smooth ground. At the end of its drive, Spirit decided it was safer to back up then go forward. The rover is now positioned directly in front of its target, a rock dubbed Stone Council.

Image credit: NASA/JPL

A Clean Adirondack (3-D)

This is a 3-D anaglyph showing a microscopic image taken of an area measuring 3 centimeters (1.2 inches) across on the rock called Adirondack. The image was taken at Gusev Crater on the 33rd day of the Mars Exploration Rover Spirit's journey (Feb. 5, 2004), after the rover used its rock abrasion tool brush to clean the surface of the rock. Dust, which was pushed off to the side during cleaning, can still be seen to the left and in low areas of the rock.

Image credit: NASA/JPL/Cornell/USGS/Texas A&M

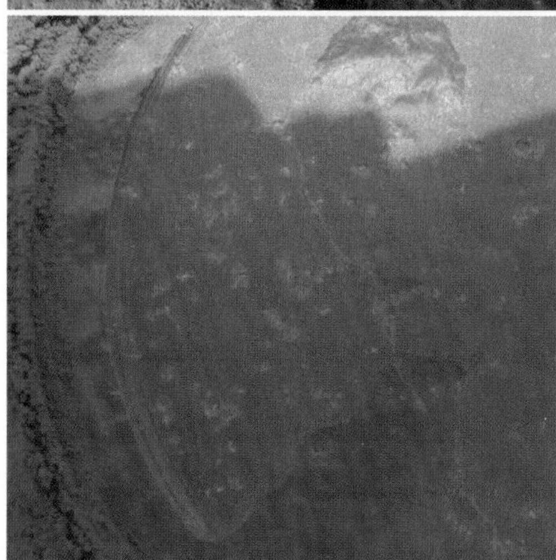

Adirondack Post-Drill (3-D)

This is a 3-D anaglyph showing a microscopic image taken of an area measuring 3 centimeters (1.2 inches) across on the rock called Adirondack. The image was taken at Gusev Crater on the 33rd day of the Mars Exploration Rover Spirit's journey (Feb. 5, 2004), after the rover used its rock abrasion tool to drill into the rock. Debris from the use of the tool is visible to the left of the hole.

Image credit: NASA/JPL/Cornell/USGS/Texas A&M

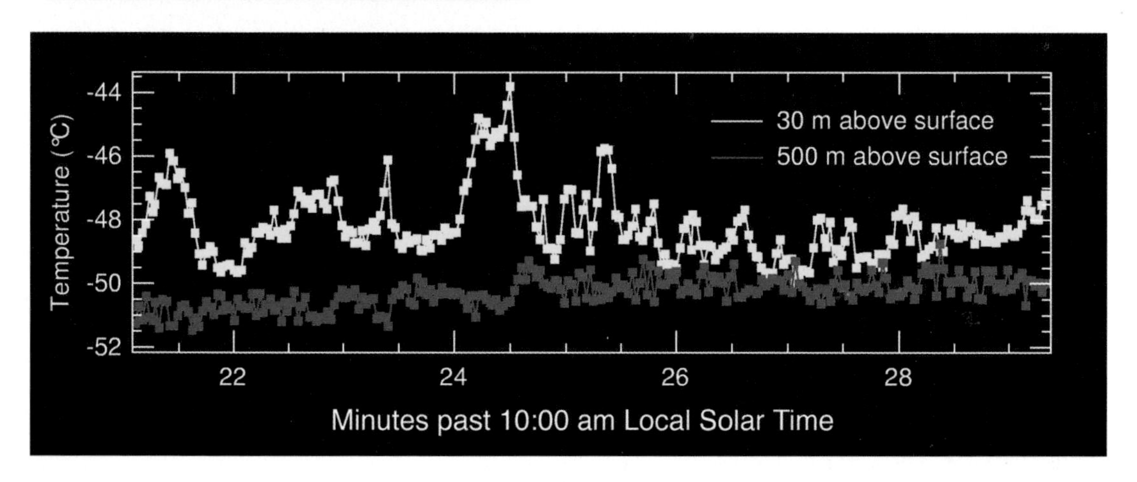

The Heat Below

This graph shows that the air 30 meters above the surface of Mars at Gusev Crater, Mars Exploration Rover Spirit's landing site, is hotter and fluctuates in temperature to a larger degree than the air higher up at 500 meters. These data, acquired by the rover's miniature thermal emission spectrometer, help scientists understand how the bottom layer of air closest to the surface behaves and interacts with global winds.

Image credit: NASA/JPL/Cornell/ASU

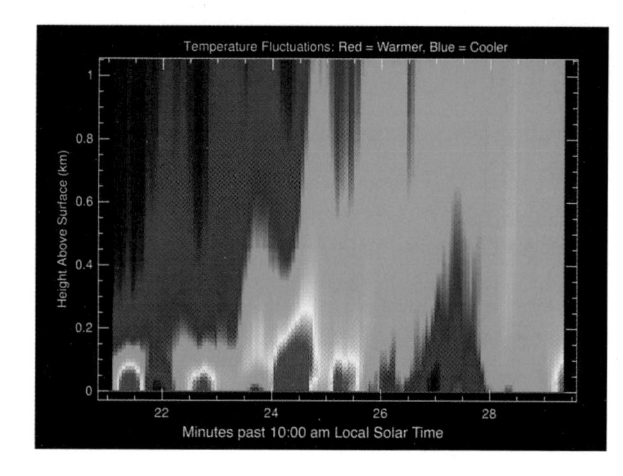

Martian Heat on the Rise

This graph shows that the atmospheric temperatures above the surface of Mars at Gusev Crater, Mars Exploration Rover Spirit's landing site, fluctuate to a significant degree. The color red denotes warmer temperatures, while blue is cooler. The red and yellow waves of color represent thermals, or pockets of heat, which rise and fall across the surface. These data, acquired by the rover's miniature thermal emission spectrometer, help scientists understand how the bottom layer of air closest to the surface behaves and interacts with global winds.

Image credit: NASA/JPL/Cornell/ASU

International Interplanetary Networking Succeeds

February 13, 2004

A pioneering demonstration of communications between NASA's Mars Exploration Rover Spirit and the European Space Agency Mars Express orbiter succeeded.

On February 6, while Mars Express was flying over the area Spirit was examining, the orbiter transferred commands from Earth to the rover and relayed data from the robotic explorer back to Earth.

"This is the first time we have had an in-orbit communication between European Space Agency and NASA spacecraft, and also the first working international communications network around another planet," said Rudolf Schmidt, the European Space Agency's project manager for Mars Express. "Both are significant achievements, two more 'firsts' for Mars Express and the Mars Exploration Rovers."

Jennifer Trosper, Spirit mission manager at NASA's Jet Propulsion Laboratory, Pasadena, Calif., said, "We have an international interplanetary communications network established at Mars."

The European Space Agency and NASA planned this demonstration as part of continuing efforts to cooperate in space and to enable plans to use joint communications assets to support future missions to the surface of Mars.

The commands for the rover were transferred from Spirit's operations team at JPL to the European Space Operations Centre in Darmstadt, Germany, where they were translated into commands for Mars Express.

The translated commands were transmitted to Mars Express, which used them to successfully command Spirit. Spirit used its ultra-high frequency antenna to transmit telemetry information to Mars Express. The orbiter relayed the data back to JPL, via the European Space Operations Centre.

"This is excellent news," said JPL's Richard Horttor, project manager for NASA's role in Mars Express. "The communication sessions between Mars Express and Spirit were pristine. Not a single bit of data was missing or added, and there were no duplications."

This exercise demonstrated the increased flexibility and capabilities of interagency cooperation and highlighted the spirit of close support essential in undertaking international space exploration.

Spirit and its twin Mars Exploration Rover, Opportunity, frequently use two NASA orbiters, Mars Odyssey and Mars Global Surveyor, for relaying communications. The rovers also can communicate directly with the Earth-based antennas of NASA's Deep Space Network in California, Spain and Australia, another layer of international cooperation.

JPL, a division of the California Institute of Technology in Pasadena, manages the Mars Exploration Rover Project and NASA participation in Mars Express for NASA's Office of Space Science, Washington, D.C.

For information about NASA and Mars programs on the Internet, visit: http://www.nasa.gov .

For images and information about the Mars Exploration Rover project on the Internet, visit: http://marsrovers.jpl.nasa.gov and http://athena.cornell.edu .

For images and information about Mars Express on the Internet, visit: http://www.esa.int/science/marsexpress and http://marsprogram.jpl.nasa.gov/express .

Guy Webster 818/354-5011 Jet Propulsion Laboratory, Pasadena, Calif.
Donald Savage (202) 358-1547 NASA Headquarters, Washington, D.C.
NEWS RELEASE: 2004-061

SPIRIT UPDATE: A Wayside Stop, Then Back to Driving - sol 42, Feb 15, 2004

Spirit used instruments on its robotic arm to examine an unusual-looking rock called "Mimi" during the rover's 42nd sol on Mars, which ended at 9:15 a.m. Sunday, PST. Scientists will be examining images and spectra to understand this rock's structure and composition and what those can tell about the environment in which the rock formed.

For sol 43, which will end at 9:58 a.m. Monday, PST, controllers have planned what they are calling a "mega drive": commanding a morning drive of about 25 meters (82 feet), then taking pictures of the scene ahead and letting the rover have a brief rest before using those mid-day pictures to guide an optional afternoon drive. Spirit is currently about 270 meters from the crater nicknamed "Bonneville," its mid-term destination.

OPPORTUNITY UPDATE: Dig this Place - sol 21, Feb 15, 2004

Opportunity completed its longest drive so far -- about 9 meters or 30 feet -- during its 21st sol on Mars, which ended at 9 p.m. Saturday, PST. The rover finished the drive with its first U-turn, arriving at a location selected for the mission's first trenching operation. Plans call for examining the hematite-rich surface of this location, called "Hematite Slope," during sol 22, then spinning one wheel to dig below the surface on sol 23.

Controllers at JPL chose "Send Me on My Way," by Rusted Root, and "Desert Drive," by Tangerine Dream, as Opportunity's wake-up music for sol 21. The rover worked a long day. It awoke earlier than usual for an early morning observation with its panoramic camera. It made additional observations from its new location just before finishing the drive, and again after finishing the last bit of the drive. Then it was woken after dark to make the mission's first nighttime observations with its infrared sensor, the miniature thermal emission spectrometer.

Opportunity Digs; Spirit Advances

February 17, 2004

NASA's Mars Exploration Rover Opportunity has scooped a trench with one of its wheels to reveal what is below the surface of a selected patch of soil.

"Yesterday we dug a nice big hole on Mars," said Jeffrey Biesiadecki, a rover planner at NASA's Jet Propulsion Laboratory, Pasadena, Calif.

The rover alternately pushed soil forward and backward out of the trench with its right front wheel while other wheels held the rover in place. The rover turned slightly between bouts of digging to widen the hole. "We took a patient, gentle approach to digging," Biesiadecki said. The process lasted 22 minutes.

The resulting trench -- the first dug by either Mars Exploration Rover -- is about 50 centimeters (20 inches) long and 10 centimeters (4 inches) deep. "It came out deeper than I expected," said Dr. Rob Sullivan of Cornell University, Ithaca, N.Y., a science-team member who worked closely with engineers to plan the digging.

Two features that caught scientists' attention were the clotty texture of soil in the upper wall of the trench and the brightness of soil on the trench floor, Sullivan said. Researchers look forward to getting more information from observations of the trench planned during the next two or three days using the rover's full set of science instruments.

Opportunity's twin rover, Spirit, drove 21.6 meters closer to its target destination of a crater nicknamed "Bonneville" overnight Monday to Tuesday. It has now rolled a total of 108 meters (354 feet) since leaving its lander

34 days ago, surpassing the total distance driven by the Mars Pathfinder mission's Sojourner rover in 1997.

Spirit has also begun using a transmission rate of 256 kilobits per second, double its previous best, said JPL's Richard Cook. Cook became project manager for the Mars Exploration Rover Project today when the former manager, Peter Theisinger, switched to manage NASA's Mars Science Laboratory Project, in development for a 2009 launch.

Spirit's drive toward "Bonneville" is based on expectations that the impact that created the crater "would have overturned the stratigraphy and exposed it for our viewing pleasure," said Dr. Ray Arvidson of Washington University in St. Louis, deputy principal investigator for the rovers' science instruments. That stratigraphy, or arrangement of rock layers, could hold clues to the mission's overriding question -- whether the past environment in the region of Mars where Spirit landed was ever persistently wet and possibly suitable for sustaining life.

Both rovers have returned striking new pictures in recent days. Microscope images of soil along Spirit's path reveal smoothly rounded pebbles. Views from both rovers' navigation cameras looking back toward their now-empty landers show the wheel tracks of the rovers' travels since leaving the landers.

Each martian day, or "sol" lasts about 40 minutes longer than an Earth day. Opportunity begins its 25th sol on Mars at 10:59 p.m. Tuesday, PST. Spirit begins its 46th sol on Mars at 11:17 a.m. Wednesday, Pacific Standard Time. The two rovers are halfway around Mars from each other. JPL, a division of the California Institute of Technology in Pasadena, manages the Mars Exploration Rover project for NASA's Office of Space Science, Washington, D.C. Images and additional information about the project are available from JPL at http://marsrovers.jpl.nasa.gov and from Cornell University at http://athena.cornell.edu .

Guy Webster (818) 354-5011 Jet Propulsion Laboratory, Pasadena, Calif.
Donald Savage (202) 358-1547 NASA Headquarters, Washington, D.C.
NEWS RELEASE: 2004-062

At Home in the Crater

The wheel tracks seen above and to the left of the lander trace the path the Mars Exploration Rover Opportunity has traveled since landing in a small crater at Meridiani Planum, Mars. After this picture was taken, the rover excavated a trench near the soil seen at the lower left corner of the image. This image mosaic was taken by the rover's navigation camera.

Image credit: NASA/JPL/Cornell

Opportunity Trenches Martian Soil

The Mars Exploration Rover Opportunity dragged one of its wheels back and forth across the sandy soil at Meridiani Planum to create a hole (bottom left corner) approximately 50 centimeters (19.7 inches) long by 20 centimeters (7.9 inches) wide by 9 centimeters (3.5 inches) deep. The rover's instrument deployment device, or arm, will begin studying the fresh soil at the bottom of this trench later today for clues to its mineral composition and history. Scientists chose this particular site for digging because previous data taken by the rover's miniature thermal emission spectrometer indicated that it contains crystalline hematite, a mineral that sometimes forms in the presence of water. The brightness of the newly-exposed soil is thought to be either intrinsic to the soil itself, or a reflection of the Sun. Opportunity's lander is in the center of the image, and to the left is the rock outcrop lining the inner edge of the small crater that encircles the rover and lander. This mosaic image is made up of data from the rover's navigation and hazard-avoidance cameras.

Image credit: NASA/JPL

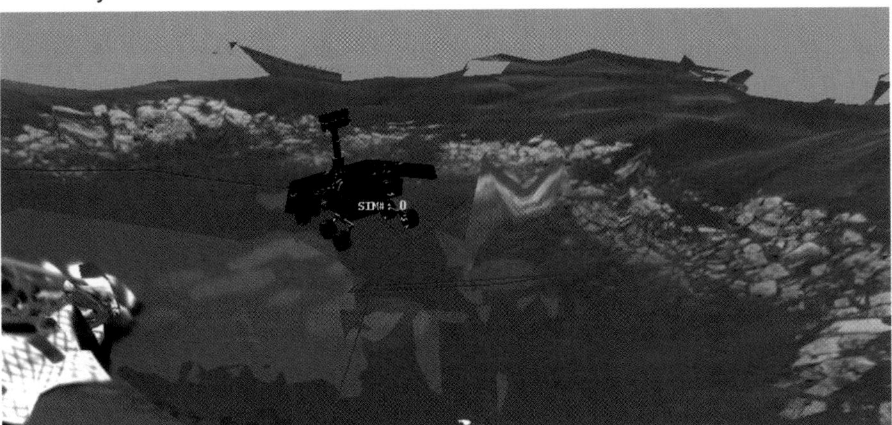

The Road Less Traveled

This image is a screenshot from a computer-generated animation showing the path the Mars Exploration Rover Opportunity traveled between the 16th and 21st days, or sols, of its mission. On sol 16, the rover followed a "V-shaped" route, driving backwards 2.5 meters (8.2 feet), turning in place 49 degrees counterclockwise, driving forward by 1 meter (3.3 feet), then driving forward again another 0.6 meters (2 feet). On sol 17, the rover traversed a "U-shaped" path, driving backwards 1.3 meters (4.3 feet), turning in place 90 degrees counterclockwise, driving forward 2.9 meters (9.5 feet), turning in place 90 degrees clockwise, driving forward 0.6 meters (2 feet), then finally driving forward again 0.5 meters (1.6 feet). On sol 19, the rover took another "V-shaped" journey, driving south backwards for approximately 1.9 meters (6.2 feet), turning in place to face the west-northwest rock target Zugspitze, then driving forward 2.5 meters (8.2 feet). On sol 21, the rover backed up 1.4 meters (4.6 meters), turned to the southwest, then via a series of arc turns, arrived at its trench location. Once there, the rover rotated 180 degrees to face 10 degrees to the west of north, then drove a short 0.5-meter-arc (1.6-foot-arc) forward. The drive ended with the rover facing approximately north, ready to dig a hole in the martian surface. Data from the rover's onboard sensors were used to make this animation.
Image credit: NASA/JPL

Opportunity Digs

This image is a screenshot from a computer-generated animation showing the Mars Exploration Rover Opportunity trenching a hole in the sandy soil at Meridiani Planum, Mars. Data taken during trenching by the rover's onboard sensors were used to create the movie.

Image credit: NASA/JPL

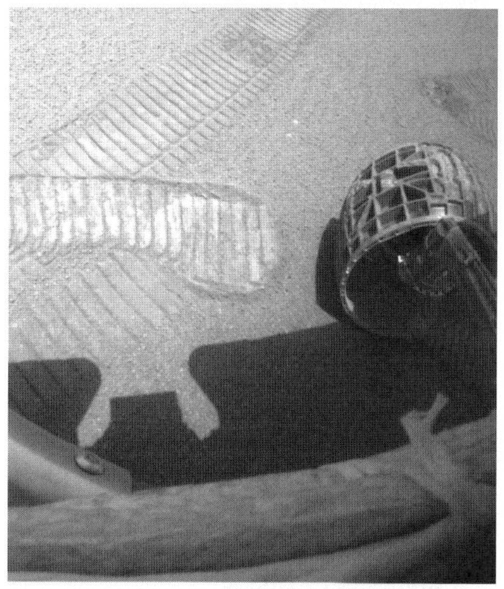

What Lies Beneath

The Mars Exploration Rover Opportunity dragged one of its wheels back and forth across the sandy soil at Meridiani Planum to create a hole (center) measuring approximately 50 centimeters (19.7 inches) long by 20 centimeters (7.9 inches) wide by 9 centimeters (3.5 inches) deep. The rover's instrument deployment device, or arm, will begin studying the fresh soil at the bottom of the trench later today for clues to its mineral composition and history. Scientists chose this particular site for trenching because previous data taken by the rover's miniature thermal emission spectrometer indicated that it contains crystalline hematite, a mineral that sometimes forms in the presence of water. The brightness of the newly-exposed soil is thought to be either intrinsic to the soil itself, or a reflection of the Sun. This image was taken by the rover's hazard-avoidance camera.

Image credit: NASA/JPL

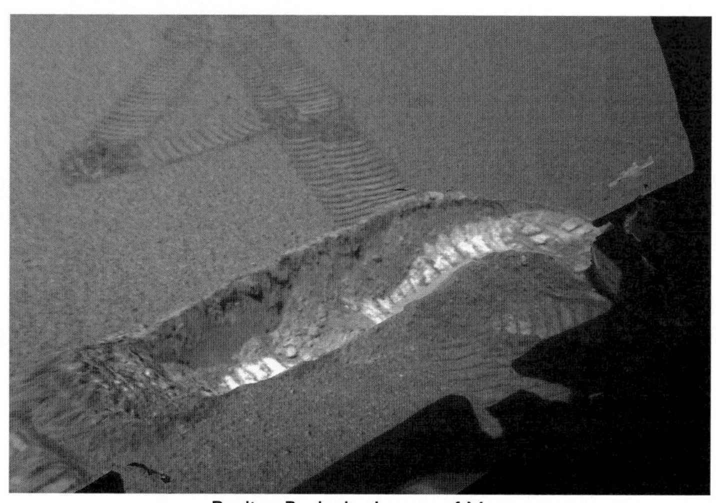

Peeling Back the Layers of Mars

This is a 3-D model of the trench excavated by the Mars Exploration Rover Opportunity on the 23rd day, or sol, of its mission. An oblique view of the trench from a bit above and to the right of the rover's right wheel is shown. The model was generated from images acquired by the rover's front hazard-avoidance cameras.

Image credit: NASA/JPL/Ames

Fresh Soil for Inspection

The Mars Exploration Rover Opportunity dragged one of its wheels back and forth across the sandy soil at Meridiani Planum to create a hole (bottom of image) measuring approximately 50 centimeters (19.7 inches) long by 20 centimeters (7.9 inches) wide by 9 centimeters (3.5 inches) deep. The rover's instrument deployment device, or arm, will begin studying the fresh soil at the bottom of this trench later today for clues to its mineral composition and history. Scientists chose this particular site for digging because previous data taken by the rover's miniature thermal emission spectrometer indicated that it contains crystalline hematite, a mineral that sometimes forms in the presence of water. The brightness of the newly-exposed soil is thought to be either intrinsic to the soil itself, or a reflection of the Sun. The rock outcrop lining the inner edge of the small crater encircling the rover and lander can be seen on the horizon. This fish-eye image was taken by the rover's hazard-avoidance camera.
Image credit: NASA/JPL

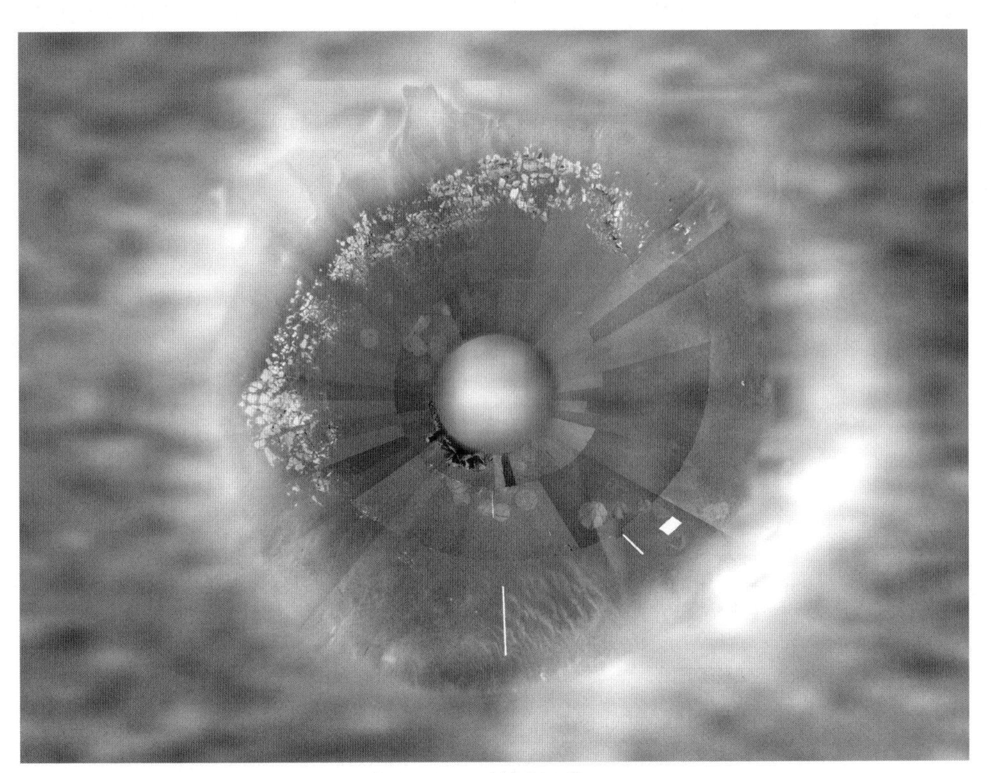

Opportunity Within Crater

This composite image shows a bird's-eye view of the crater occupied by the Mars Exploration Rover Opportunity at Meridiani Planum, Mars. A portion of the lander can be seen to the bottom left of the image's circular center. Bounce marks can be seen below and to the top left of the center. The rock outcrop containing many of the rover's rock targets runs from the top right of the image to the left of the image. The rover, which cannot be seen in the image, is located in the southwest quadrant, just left of the lander. Data depicting the inside of the crater wall is from the rover's panoramic camera. Data depicting the outside of the crater wall is from the camera on the orbiting Mars Global Surveyor. The top of the image faces north, and the image area measures approximately 22 meters (72 feet) wide.

Image credit: JPL/NASA/Cornell/MSSS

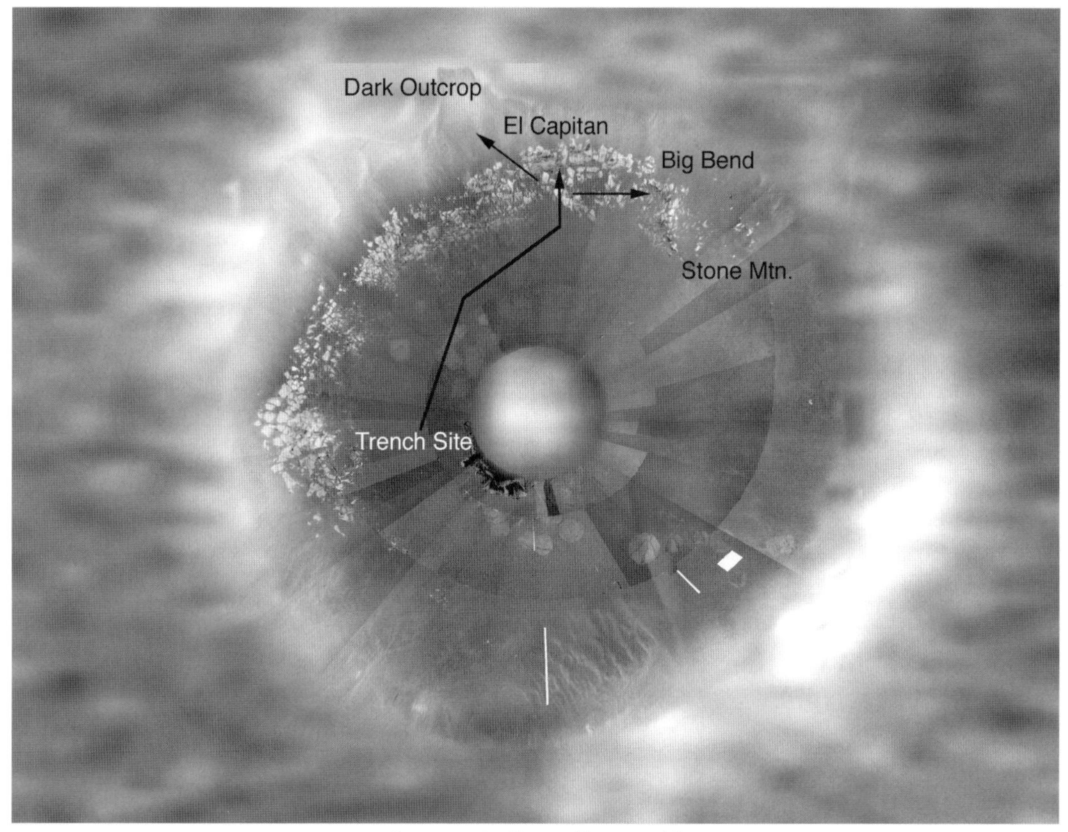

Opportunity Future Traverse Map

This composite image maps a future travel itinerary for the Mars Exploration Rover Opportunity at Meridiani Planum, Mars. The rover is currently located at the trench site toward the bottom left of the image. It has already been to the location named "Stone Mountain." Controllers plan to investigate the other labeled areas on the rock outcrop. Data depicting the inside of the crater wall is from the rover's panoramic camera. Data depicting the outside of the crater wall is from the camera on the orbiting Mars Global Surveyor. The top of the image faces north, and the image area measures approximately 22 meters (72 feet) wide.

Image credit: JPL/NASA/Cornell/MSSS

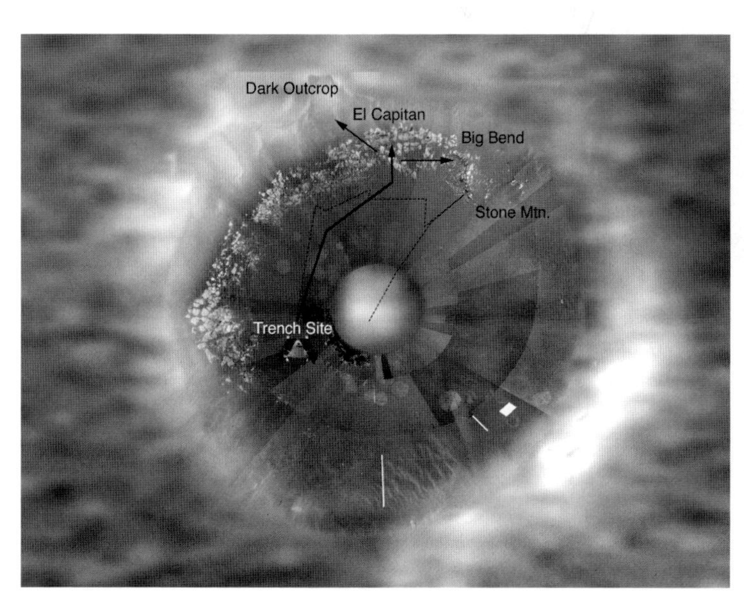

Opportunity Traverse Map

This composite image maps the areas traveled by the Mars Exploration Rover Opportunity at Meridiani Planum, Mars. Dotted lines represent areas the rover has already traveled. Solid lines represent areas still on the rover's travel agenda. Data depicting the inside of the crater wall is from the rover's panoramic camera. Data depicting the outside of the crater wall is from the camera on the orbiting Mars Global Surveyor. The top of the image faces north, and the image area measures approximately 22 meters (72 feet) wide.

Image credit: JPL/NASA/Cornell/MSSS

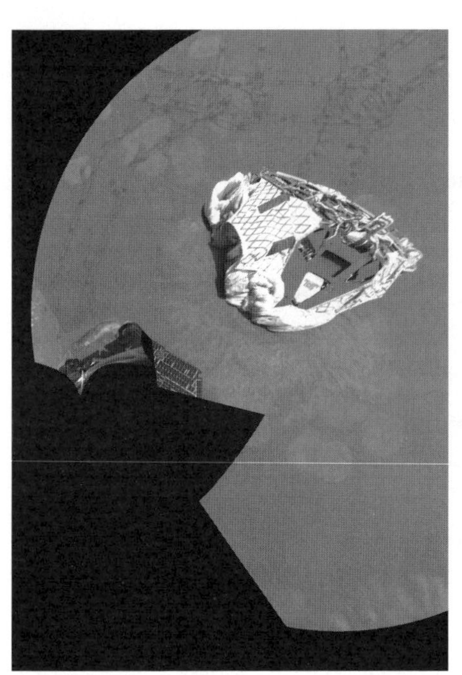

Track-and-Trench

This image shows the tracks and trench marks made by the Mars Exploration Rover Opportunity at Meridiani Planum, Mars. The rover can be seen to the lower left of the lander. The trench is visible to the upper left of the rover, which has traveled a total of 35.3 meters (116 feet) since leaving the lander on sol 7 (January 31, 2004). On sol 23 (February 16, 2004), the rover used one of its wheels to dig a trench measuring approximately 10 centimeters (4 inches) deep, 50 centimeters (20 inches) long, and 20 centimeters (8 inches) wide. This vertically projected image was created using a combination of images from the rover's navigation camera and hazard-avoidance cameras.

Image credit: NASA/JPL/Cornell

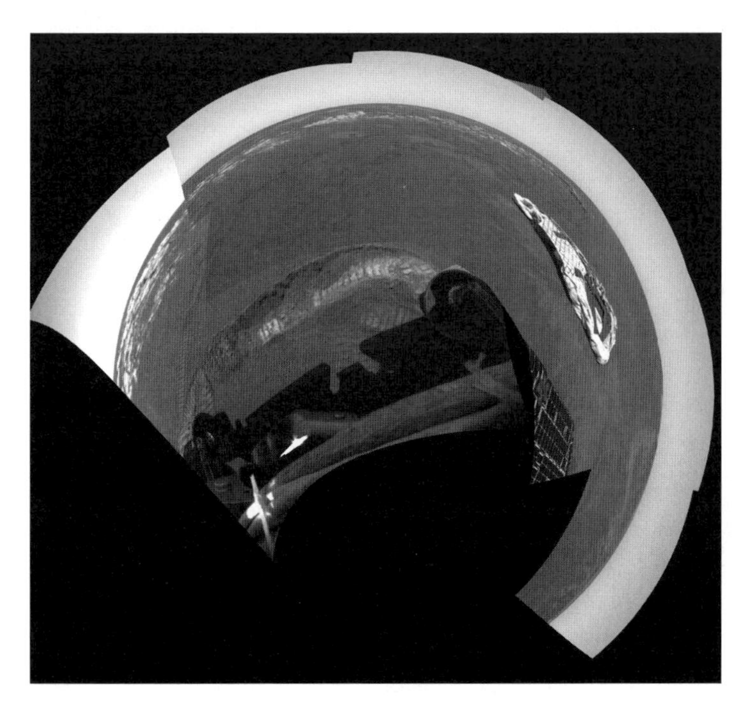

Track-and-Trench 2

This image shows the tracks and trench marks made by the Mars Exploration Rover Opportunity at Meridiani Planum, Mars. The rover can be seen to the lower left of the lander. The trench is visible to the upper left of the rover, which has traveled a total of 35.3 meters (116 feet) since leaving the lander on sol 7 (January 31, 2004). On sol 23 (February 16, 2004), the rover used one of its wheels to dig a trench measuring approximately 10 centimeters (4 inches) deep, 50 centimeters (20 inches) long, and 20 centimeters (8 inches) wide. This polar-projected image showing the horizon was created using a combination of images from the rover's navigation camera and hazard-avoidance cameras.

Image credit: NASA/JPL/Cornell

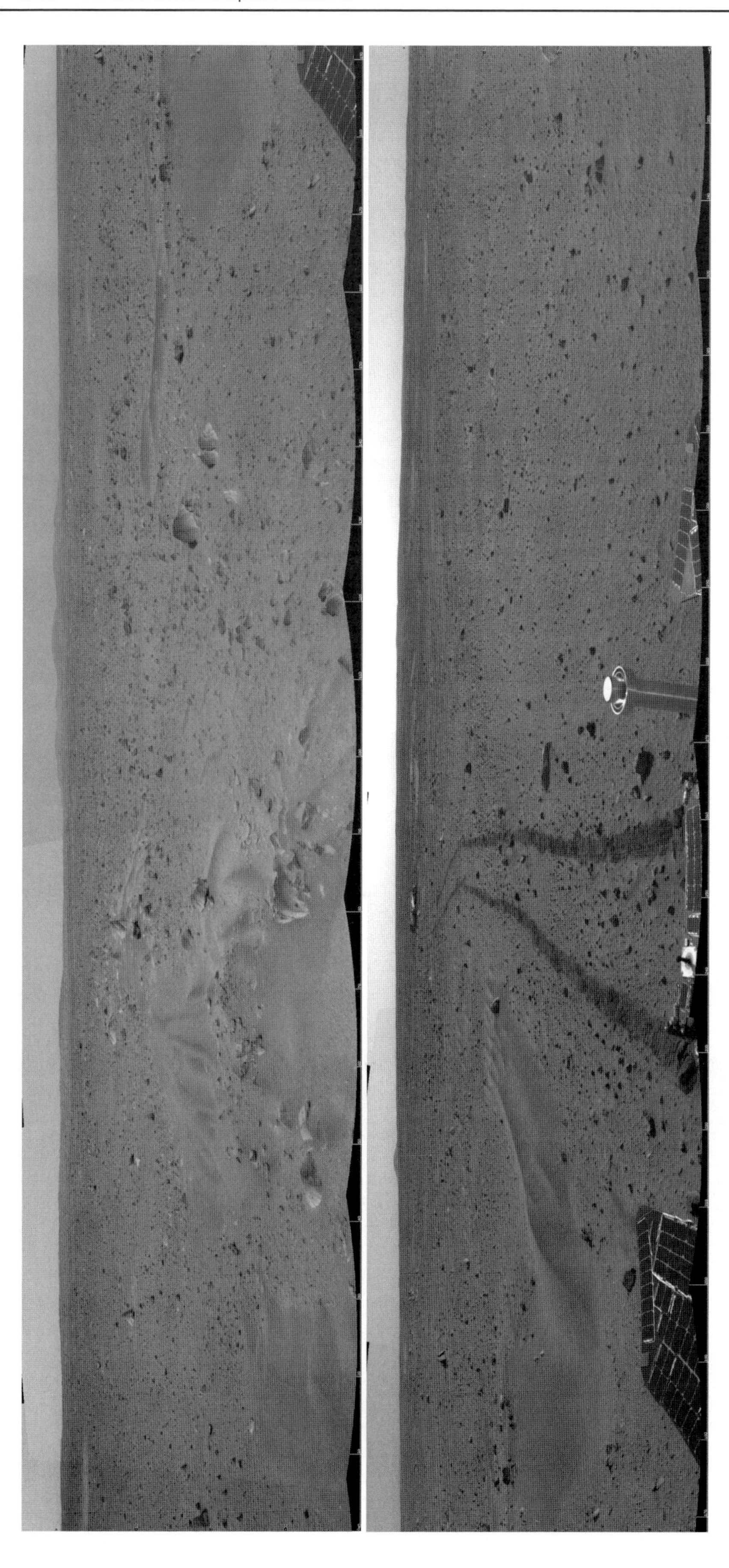

Spirit Keeps Rollin'

This 360-degree mosaic panorama image, taken by the navigation camera on the Mars Exploration Rover Spirit, includes a view of the lander. The lander is located to the south-southwest of the rover, which is moving toward a crater nicknamed "Bonneville. Sleepy Hollow can be seen to the right of the lander. As of Sol 44, which ended on February 17, 2004, the rover had moved a total of 106.6 meters (350 feet) since leaving the lander on January 15, 2004. This image was taken on Sol 39 (February 11, 2004).

Image credit: NASA/JPL

Spirit Spies "Bonneville"

This mosaic image from the panoramic camera on the Mars Exploration Rover Spirit shows the area in front of the rover after its record 27.5 meters (90.2 feet) drive on Sol 43, which ended February 16, 2004. Spirit is looking toward one of its future targets, the rim of a crater nicknamed "Bonneville."

Image credit: NASA/JPL/Cornell

Mark of the Moessbauer

This image, taken by an instrument called the microscopic imager on the Mars Exploration Rover Spirit, reveals an imprint left by another instrument, the Moessbauer spectrometer. The imprint is at a location within the rover wheel track named "Middle of Road." Both instruments are located on the rover's instrument deployment device, or "arm."

Not only was the Moessbauer spectrometer able to gain important mineralogical information about this site, it also aided in the placement of the microscopic imager. On hard rocks, the microscopic imager uses its tiny metal sensor to determine proper placement for best possible focus. However, on the soft martian soil this guide would sink, prohibiting proper placement of the microscopic imager. After the Moessbauer spectrometer's much larger, donut-shaped plate touches the surface, Spirit can correctly calculate where to position the microscopic imager.

Scientists find this image particularly interesting because of the compacted nature of the soil that was underneath the Moessbauer spectrometer plate. Also of interest are the embedded, round grains and the fractured appearance of the material disturbed within the hole. The material appears to be slightly cohesive. The field of view in this image, taken on Sol 43 (February 16, 2004), measures approximately 3 centimeters (1.2 inches) across.

Image credit: NASA/JPL/Cornell/USGS

Map of Moessbauer Placement

This elevation map of a soil target called "Peak" was created from images taken by the microscopic imager located on the Mars Exploration Rover Spirit's instrument deployment device or "arm." The image reveals the various high and low points of this spot of soil after the Moessbauer spectrometer, another instrument on the rover's arm, was gently placed down on it. The blue areas are farthest away from the instrument; the red areas are closest. The variation in distance between blue and red areas is only 2 millimeters, or .08 of an inch. The images were acquired on sol 39 (February 11, 2004). Image credit: NASA/JPL/USGS

Moessbauer Close-Up

This close-up image of the Mars Exploration Rover Spirit's instrument deployment device, or "arm," shows the donut-shaped plate on the Moessbauer spectrometer. This image makes it easy to recognize the imprint left by the instrument in the martian soil at a location called "Peak" on sol 43 (February 16, 2004). This image was taken by the rover's panoramic camera on sol 39 (February 11, 2004).

Image credit: NASA/JPL/Cornell

SPIRIT UPDATE: Spirit Does a "Wheel Wiggle" - sol 45, Feb 18, 2004

Spirit began sol 45, which ended at 11:17 a.m. February 18, 2004 PST, at its previous target, Halo, by conducting analysis with the alpha particle x-ray spectrometer, microscopic imager and Moessbauer spectrometer. Spirit also took panoramic camera images and miniature thermal emission spectrometer observations before its arm was stowed for the northeast drive toward a circular depression dubbed Laguna Hollow.

The first 19 meters of the drive toward Laguna Hollow was commanded using go-to waypoint commands with the hazard avoidance system turned off. This mode - which was used for the first time this sol - provides automatic heading correction during a blind drive. Some fine-tuning toward the target brought the total drive for this sol to 22.7 meters (74.5 feet).

After reaching Laguna Hollow, Spirit "wiggled" its wheels to disturb or scuff the fine dust-like soil at this location, which allows for more detailed observations with the instruments on the robotic arm. After adjusting position to put the disturbed soil in reach of the arm, Spirit backed up and completed a miniature thermal emission spectrometer scan of the new work area. Before the sol ended, Spirit made one more adjustment, putting it in perfect position to analyze the scuffed area beginning on sol 46

The plan for sol 46, which will end at 11:57 a.m., February 19, 2004 PST, is to conduct observations on Laguna Hollow with the instruments on the robotic arm, including some higher resolution analysis that will involve an overnight tool change.

OPPORTUNITY UPDATE: Peering into the Hole - sol 24, Feb 18, 2004

On sol 24, which ended at 10:59 p.m. Tuesday, PST, Opportunity used science instruments on its robotic arm to examine the hole it dug with its right front wheel on sol 23. The trench is about 50 centimeters (20 inches) long by 20 centimeters (8 inches) wide by 10 centimeters (4 inches) deep.

Sol 24's wake-up music was "Trench Town Rock" by Bob Marley.

The plan for sol 25, which will end at 11:38 p.m. Wednesday, PST, is to continue examining the walls and floor of the trench for clues about the history of Mars. Opportunity will also peek at its right front wheel with the panoramic camera to see what materials got stuck on the wheel from the trenching activity. Then, Opportunity will use the panoramic camera high on the rover's mast to check out a former piece of itself -- the heat shield, which is sitting off in the distance. The heat shield protected the rover during cruise and during descent through the atmosphere on Jan. 4, 2004, PST.

The Trench Throws a Dirt Clod at Scientists

This picture, obtained by the microscopic imager on NASA's Opportunity rover during sol 24, February 17 PST, shows soil clods exposed in the upper wall of the trench dug by Opportunity's right front wheel on sol 23.

The clods were not exposed until the trench was made. The presence of soil clods implies weak bonding between individual soil grains. The chemical agent or mineral that causes the dirt to bind together into a clod, which scientists call the "bonding agent," is currently unknown. Moessbauer and alpha particle X-ray spectrometer measurements of this spot, planned for sol 25, might help explain the bonding, which would ultimately help the rover team understand how geological processes vary across the red planet. In any case, the bonds between soil grains here cannot be very strong because the wheel dug down through this layer with little trouble.

Image credit: NASA/JPL/Cornell/USGS

Spirit Does a "Jig" at Laguna Hollow

This front hazard-avoidance image taken by the Mars Exploration Rover Spirit on sol 45 shows Spirit in its new location after a drive totaling about 20 meters (65.6 feet).

The circular depression that Spirit is in, dubbed "Laguna Hollow," was most likely formed by a small impact.

Scientists were interested in reaching Laguna Hollow because of the location's abundance of very fine, dust-like soil. The fine material could be atmospheric dust that has settled into the depression, or a salt-based material that causes crusts in the soils and coating on rocks. Either way, scientists hope to be able to characterize the material and broaden their understanding of this foreign world.

To help scientists get a better look at the variations in the fine-grained dust at different depths, controllers commanded Spirit to "jiggle" its wheels in the soil before backing away to a distance that allows the area to be reached with the robotic arm. Spirit will likely spend part of sol 46 analyzing this area with the instruments on its robotic arm.

Credit: NASA/JPL

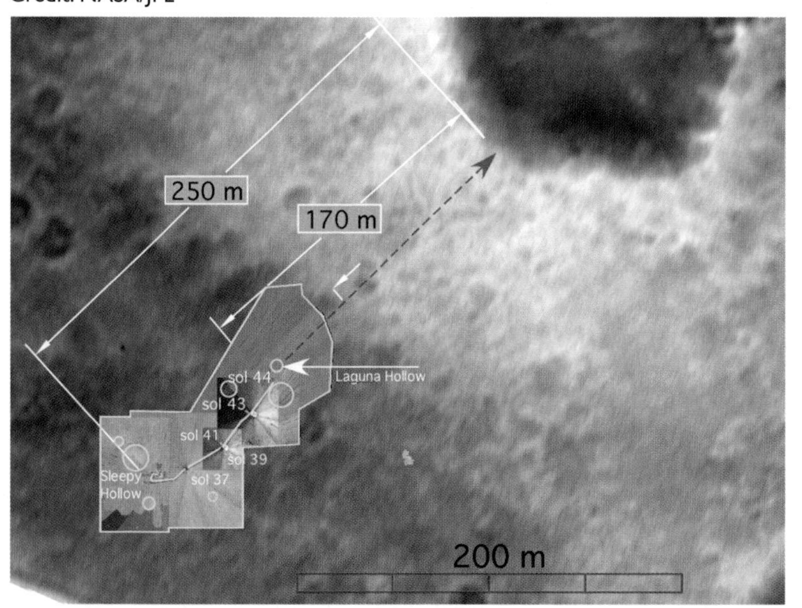

Spirit's Path to Bonneville

Scientists created this overlay map by laying navigation and panoramic camera images taken from the surface of Mars on top of one of Spirit's descent images taken as the spacecraft descended to the martian surface. The map was created to help track the path that Spirit has traveled through sol 44 and to put into perspective the distance left to travel before reaching the edge of the large crater nicknamed "Bonneville."

The area boxed in yellow contains the ground images that have been matched to and layered on top of the descent image. The yellow line shows the path that Spirit has traveled and the red dashed line shows the intended path for future sols. The blue circles highlight hollowed areas on the surface, such as Sleepy Hollow, near the lander, and Laguna Hollow, the sol 45 drive destination. Scientists use these hollowed areas - which can be seen in both the ground images and the descent image - to correctly match up the overlay.

Field geologists on Earth create maps like this to assist them in tracking their observations.
Credit: NASA/JPL/Cornell

Opportunity Examines Trench As Spirit Prepares To Dig One

February 19 2004

This image, taken by the microscopic imager, an instrument located on Opportunity 's instrument deployment device, or "arm," reveals shiny, spherical objects embedded within the trench wall at Meridiani Planum, Mars.

By inspecting the sides and floor of a hole it dug on Mars, NASA's Opportunity rover is finding some things it did not see beforehand, including round pebbles that are shiny and soil so fine-grained that the rover's microscope can't make out individual particles.

"What's underneath is different than what's at the immediate surface," said Dr. Albert Yen, rover science team member at NASA's Jet Propulsion Laboratory, Pasadena, Calif.

Meanwhile, NASA's other Mars Exploration Rover, Spirit, has reached a site with such interesting soil that scientists have decided to robotically dig a hole there, too. Spirit's trenching at a shallow depression dubbed "Laguna Hollow" could answer questions about whether traits on the soil surface resulted from repeated swelling and shrinking of an upper layer bearing concentrated brine, among other possibilities.

Opportunity has manipulated its robotic arm to use its microscope on five different locations within the trench the rover dug on Monday. It has also taken spectrometer readings of two sites. "We've given the arm a very strenuous workout," said JPL's Dr. Eric Baumgartner, lead engineer for the arm. The accuracy of the tool placements -- within 5 millimeters, or less than a quarter inch -- is remarkable for mobile robotics on Earth, much less on Mars.

Once data are analyzed from the alpha particle X-ray spectrometer and the Mössbauer spectrometer about what elements and what iron-bearing minerals are present, the differences between the subsurface and the surface will be easier to interpret, Yen said.

While Opportunity has been digging and examining its trench this week, it has also been catching up on transmission of pictures and information from its survey last week of a rock outcrop along the inner wall of the small crater in which the rover is working.

Both rovers can communicate directly with Earth, but JPL's Andrea Barbieri, telecommunication system engineer, reported that 66 percent of the 10 gigabits of data they have returned so far has come via relays by NASA's Mars Odyssey orbiter and another 16 percent via relays by NASA's Mars Global Surveyor.

Based on the outcrop survey, scientists have chosen a feature they have dubbed "El Capitan" as the next target for intensive investigation by Opportunity.

"We've planned our assault on the outcrop," said Dr. Steve Squyres of Cornell University, Ithaca, N.Y., principal investigator for the rovers' science instruments. "The whole stack of rocks seems to be well exposed here," he said of the chosen target. Upper and lower portions appear to differ in layering and weathering characteristics. Planners anticipate that Opportunity's arm will be able to reach both the upper and lower parts from a single parking spot in front of "El Capitan."

Halfway around the planet, Spirit will be told to use a front wheel to dig a trench during the martian day, or "sol," that will end at 12:36 p.m. Friday, PST.

Some soil in "Laguna Hollow" appeared to stick to Spirit's wheels. Possible explanations include very fine-grained dust or concentrated salt making the soil sticky, said Dr. Dave Des Marais, a rover science team member from NASA Ames Research Center, Moffett Field, Calif. Pictures of the surface there also show pebbles arranged in clusters or lines around lighter patches Des Marais described as "miniature hollows." This resembles patterned ground on Earth that can result from alternating expansion and shrinkage of the soil. Possible explanations for repeated expanding and contracting include cycles of freezing and thawing or temperature swings in salty soil.

After trenching to seek clues about those possibilities, Spirit will continue on its trek toward the rim of a crater nicknamed "Bonneville," now estimated to be about 135 meters (443 feet) away from the rover. Spirit has already driven 128 meters (420 feet). The rovers' main task is to explore their landing sites for evidence in the rocks and soil about whether the sites' past environments were ever watery and possibly suitable for sustaining life.

JPL, a division of the California Institute of Technology in Pasadena, manages the Mars Exploration Rover project for NASA's Office of Space Science, Washington, D.C. Images and additional information about the project are available from JPL at http://marsrovers.jpl.nasa.gov and from Cornell University at http://athena.cornell.edu .

Guy Webster (818) 354-5011 Jet Propulsion Laboratory, Pasadena, California
Donald Savage (202) 358-1547 NASA Headquarters, Washington, D.C.
NEWS RELEASE: 2004-066

Innovative Web Site Brings Mars Exploration to Desktops

February 20, 2004

NASA's Jet Propulsion Laboratory, Pasadena, Calif., is enhancing the availability of all Mars rover images for students and the public by distributing them via the Internet. The images can be viewed on the NASA Web site at http://marsrovers.jpl.nasa.gov as well as the educational Web site MarsQuest Online at http://www.marsquestonline.org/mer .

The sites allow anyone with an Internet connection to participate in the adventure of Mars exploration. MarsQuest Online is making the full set of images from Spirit and Opportunity available for public viewing, along with daily updates, in an integrated exploration and education environment. The site is a powerful example of inquiry-based learning and public engagement in the thrill of exploration and discovery.

Dr. Eric De Jong of JPL heads the Science Data Visualization and Modeling team that produces images of the rovers and the martian surface. He's also the co-principal investigator for MarsQuest Online. "This is NASA's vision of 21st century exploration through the Internet, a shared experience of scientists, students and the public. The rovers' eyes are our eyes, and MarsQuest Online puts these eyes on your desktop."

MarsQuest Online enables the public, educators and students to gain a sense of what it's like to explore another world. According to Principal Investigator Daniel Barstow, MarsQuest Online was created "to provide the public with a highly engaging and interactive experience while learning about Mars by directly exploring it, just as the scientists do."

Students are able to learn more about the red planet by examining the most recent panoramic views of the two landing sites and the images of rocks and soil investigated by the instruments on the rovers' robotic arms. They can also follow the progress of the twin robotic field geologists as they navigate around the martian surface. Students will learn information about Mars and the search for water through an array of learning activities, such as an interactive feature that allows users to control 3-D virtual flyovers of prominent martian landform features. These activities support key elements of the National Science Education Standards, including core concepts of Earth and space science.

Adults will appreciate the site, with its daily updates and its intuitive point-and-click interface. Each image is visually organized so novice and expert users can easily navigate across the martian landscape through the eyes of the rovers. The site also provides current news about important scientific findings.

With support from the National Science Foundation, Arlington, Va., MarsQuest Online was built in close collaboration with NASA's Mars visualization team. It extends the power of NASA's very popular Mars Web site http://marsrovers.jpl.nasa.gov to offer a more in-depth exploration environment for the images.

"The design of MarsQuest Online is based on extensive research on the most effective methods of using images and visualizations as tools by students and the general public," said Barstow, who also serves as director of the Technical Education Research Center (Terc) for Science Teaching and Learning in Cambridge, Mass. Barstow emphasized that "research findings show that most people can grasp concepts more quickly and intuitively by interacting with visual imagery than by simply reading text. With MarsQuest Online, students experience authentic science, venturing into the unknown, asking questions and pursuing answers."

MarsQuest Online is funded by the National Science Foundation Informal Science Education Division and developed through collaboration between the Technical Education Research Center, the Space Science Institute in Boulder, Colo., and JPL.

The Technical Education Research Center is a non-profit educational research and development organization that specializes in inquiry-based science, math and technology education. The Space Science Institute is a science and education research and development organization that created the traveling MarsQuest museum exhibit associated with MarsQuest Online.

JPL manages the Mars Exploration Program for NASA and provides the technical expertise on Mars rovers and the rover imaging systems. Internet mirroring support for high-bandwidth use of this site is provided by the University Corporation for Atmospheric Research, Boulder, Colo., and the San Diego Supercomputing Center.

For more information on the Mars Exploration Rover mission visit: http://marsrovers.jpl.nasa.gov . To learn how to make your own 3-D images of Mars images, see http://www.jpl.nasa.gov/news/features/3d.cfm .

Natalie Godwin (818) 354-0850 Jet Propulsion Laboratory, Pasadena, Calif.
David Shepard (617) 547-0430 Technical Education Research Center (Terc), Cambridge, Mass.
James Harold 720-974-5858 Space Science Institute, Boulder, Colo.
NEWS RELEASE: 2004-067

Mars Sunset Clip from Opportunity Tells Dusty Tale

February 26, 2004

Sunset on Mars, imaged by Opportunity.

Dust gradually obscures the Sun during a blue-sky martian sunset seen in a sequence of newly processed frames from NASA's Mars Exploration Rover Opportunity.

"It's inspirational and beautiful, but there's good science in there, too," said Dr. Jim Bell of Cornell University, Ithaca, N.Y., lead scientist for the panoramic cameras on Opportunity and its twin, Spirit.

The amount of dust indicated by Opportunity's observations of the Sun is about twice as much as NASA's Mars Pathfinder lander saw in 1997 from another site on Mars.

The sunset clip uses several of the more than 11,000 raw images that have been received so far from the 18 cameras on the two Mars Exploration Rovers and publicly posted at http://marsrovers.jpl.nasa.gov . During a briefing today at NASA's Jet Propulsion Laboratory, Pasadena, Calif., Bell showed some pictures that combine information from multiple raw frames.

A patch of ground about half the area of a coffee table, imaged with the range of filters available on Opportunity's panoramic camera, has soil particles with a wide assortment of hues -- "more spectral color diversity than we've seen in almost any other data set on Mars," Bell said.

Opportunity is partway through several days of detailed observations and composition measurements at a portion of the rock outcrop in the crater where it landed last month. It used its rock abrasion tool this week for the first time, exposing a fresh rock surface for examination. That surface will be studied with its alpha particle X-ray spectrometer for identifying chemical elements and with its Mössbauer spectrometer for identifying iron-bearing minerals. With that rock-grinding session, all the tools have now been used on both rovers.

Dr. Ray Arvidson of Washington University, St. Louis, deputy principal investigator for the rovers' science work, predicted that in two weeks or so, Opportunity will finish observations in its landing-site crater and be ready to move out to the surrounding flatland. At about that same time, Spirit may reach the rim of a larger crater nicknamed "Bonneville" and send back pictures of what's inside. "We'll both be at the rims of craters," he said of the two rovers' science teams, "one thinking about going in and the other thinking about going out onto the plain."

Not counting occasional backup moves, Spirit has driven 171 meters (561 feet) from its lander. It has about half that distance still to go before reaching the crater rim. The terrain ahead looks different than what's behind, however. "It's rockier, but we're after rocks," Arvidson said.

Spirit can traverse the rockier type of ground in front of it, said Spirit Mission Manager Jennifer Harris of JPL. As it approached the edge of a small depression in the ground earlier this week, the rover identified the slope as a potential hazard, and "did the right thing" by stopping and seeking an alternate route, she said.

However, engineers are also planning to transmit new software to both rovers in a few weeks to improve onboard navigation capabilities. "We want to be more robust for the terrain we're seeing," Trosper said. The software revisions will also allow engineers to turn off a heater in Opportunity's arm, which has been wasting some power by going on during cold hours even when not needed.

As it heads toward "Bonneville" to look for older rocks from beneath the region's current surface layer, Spirit is stopping frequently to examine soil and rocks along the way. Observations with its microscope at one wavy patch of windblown soil allowed scientists to study how martian winds affect the landscape. Coarser grains are concentrated on the crests, with finer grains more dominant in the troughs, a characteristic of "ripples" rather than of dunes, which are shaped by stronger winds. "This gives us a better understanding of the current erosion process due to winds on Mars," said Shane Thompson, a science team collaborator from Arizona State University, Tempe.

The rovers' main task is to explore their landing sites for evidence in the rocks and soil about whether the sites' past environments were ever watery and possibly suitable for sustaining life. JPL, a division of the California Institute of Technology in Pasadena, manages the Mars Exploration Rover project for NASA's Office of Space Science, Washington, D.C. Images and additional information about the project are available from JPL at http://marsrovers.jpl.nasa.gov and from Cornell University at http://athena.cornell.edu .

Guy Webster (818) 354-5011 Jet Propulsion Laboratory, Pasadena, Calif.
Donald Savage (202) 358-1547 NASA Headquarters, Washington, D.C.
NEWS RELEASE: 2004-070

Opportunity Rover Finds Strong Evidence Meridiani Planum Was Wet

March 02, 2004

Scientists have concluded the part of Mars that NASA's Opportunity rover is exploring was soaking wet in the past.

Evidence the rover found in a rock outcrop led scientists to the conclusion. Clues from the rocks' composition, such as the presence of sulfates, and the rocks' physical appearance, such as niches where crystals grew, helped make the case for a watery history.

"Liquid water once flowed through these rocks. It changed their texture, and it changed their chemistry," said Dr. Steve Squyres of Cornell University, Ithaca, N.Y., principal investigator for the science instruments on Opportunity and its twin, Spirit. "We've been able to read the tell-tale clues the water left behind, giving us confidence in that conclusion."

Dr. James Garvin, lead scientist for Mars and lunar exploration at NASA Headquarters, Washington, said, "NASA launched the Mars Exploration Rover mission specifically to check whether at least one part of Mars ever had a persistently wet environment that could possibly have been hospitable to life. Today we have strong evidence for an exciting answer: Yes."

Opportunity has more work ahead. It will try to determine whether, besides being exposed to water after they formed, the rocks may have originally been laid down by minerals precipitating out of solution at the bottom of a salty lake or sea.

The first views Opportunity sent of its landing site in Mars' Meridiani Planum region five weeks ago delighted researchers at NASA's Jet Propulsion Laboratory, Pasadena, Calif., because of the good fortune to have the spacecraft arrive next to an exposed slice of bedrock on the inner slope of a small crater.

The robotic field geologist has spent most of the past three weeks surveying the whole outcrop, and then turning back for close-up inspection of selected portions. The rover found a very high concentration of sulfur in the outcrop with its alpha particle X-ray spectrometer, which identifies chemical elements in a sample.

"The chemical form of this sulfur appears to be in magnesium, iron or other sulfate salts," said Dr. Benton Clark of Lockheed Martin Space Systems, Denver. "Elements that can form chloride or even bromide salts have also been detected."

At the same location, the rover's Mössbauer spectrometer, which identifies iron-bearing minerals, detected a hydrated iron sulfate mineral called jarosite. Germany provided both the alpha particle X-ray spectrometer and the Mössbauer spectrometer. Opportunity's miniature thermal emission spectrometer has also provided evidence for sulfates.

On Earth, rocks with as much salt as this Mars rock either have formed in water or, after formation, have been highly altered by long exposures to water. Jarosite may point to the rock's wet history having been in an acidic lake or an acidic hot springs environment.

The water evidence from the rocks' physical appearance comes in at least three categories, said Dr. John Grotzinger, sedimentary geologist from the Massachusetts Institute of Technology, Cambridge: indentations called "vugs," spherules and crossbedding.

Pictures from the rover's panoramic camera and microscopic imager reveal the target rock, dubbed "El Capitan," is thoroughly pocked with indentations about a centimeter (0.4 inch) long and one-fourth or less that wide, with apparently random orientations. This distinctive texture is familiar to geologists as the sites where crystals of salt minerals form within rocks that sit in briny water. When the crystals later disappear, either by erosion or by dissolving in less-salty water, the voids left behind are called vugs, and in this case they conform to the geometry of possible former evaporite minerals.

Round particles the size of BBs are embedded in the outcrop. From shape alone, these spherules might be formed from volcanic eruptions, from lofting of molten droplets by a meteor impact, or from accumulation of minerals coming out of solution inside a porous, water-soaked rock. Opportunity's observations that the spherules are not concentrated at particular layers in the outcrop weigh against a volcanic or impact origin, but do not completely rule out those origins.

Layers in the rock that lie at an angle to the main layers, a pattern called crossbedding, can result from the action of wind or water. Preliminary views by Opportunity hint the crossbedding bears hallmarks of water action, such as the small scale of the crossbedding and possible concave patterns formed by sinuous crestlines of underwater ridges.

The images obtained to date are not adequate for a definitive answer. So scientists plan to maneuver Opportunity closer to the features for a better look. "We have tantalizing clues, and we're planning to evaluate this possibility in the near future," Grotzinger said.

JPL, a division of the California Institute of Technology in Pasadena, manages the Mars Exploration Rover project for NASA's Office of Space Science, Washington.

For information about NASA and the Mars mission on the Internet, visit http://www.nasa.gov .
Images and additional information about the project are also available at http://marsrovers.jpl.nasa.gov and http://athena.cornell.edu .
Guy Webster (818) 354-5011 Jet Propulsion Laboratory, Pasadena, California
Donald Savage (202) 358-1547 NASA Headquarters, Washington
NEWS RELEASE: 2004-074

Opportunity Landing Spot Panorama (3-D Model)

The rocky outcrop traversed by the Mars Exploration Rover Opportunity is visible in this three-dimensional model of the rover's landing site. Opportunity has acquired close-up images along the way, and scientists are using the rover's instruments to closely examine portions of interest.The white fragments that look crumpled near the center of the image are portions of the airbags. Distant scenery is displayed on a spherical backdrop or "billboard" for context.Artifacts near the top rim of the crater are a result of the transition between the three-dimensional model and the billboard. Portions of the terrain model lacking sufficient data appear as blank spaces or gaps, colored reddish-brown for better viewing.This image was generated using special software from NASA's Ames Research Center and a mosaic of images taken by the rover's panoramic camera.

Image Credit: NASA/JPL/Ames/Cornell

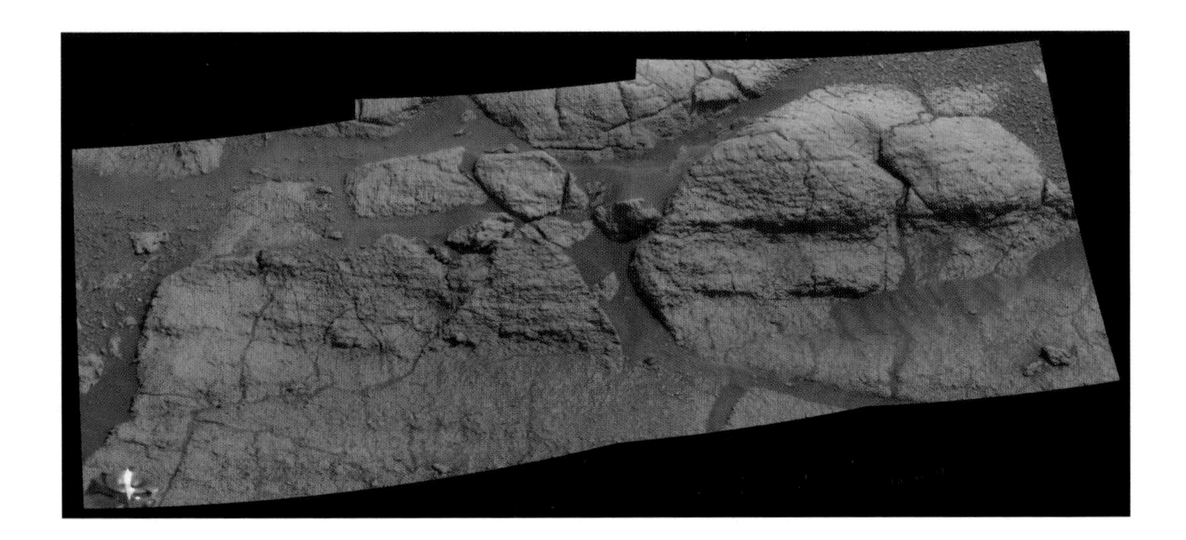

"El Capitan's" Scientific Gems

This mosaic of images taken by the panoramic camera onboard the Mars Exploration Rover Opportunity shows the rock region dubbed "El Capitan," which lies within the larger outcrop near the rover's landing site. "El Capitan" is being studied in great detail using the scientific instruments on the rover's arm; images from the panoramic camera help scientists choose the locations for this compositional work.The millimeter-scale detail of the lamination covering these rocks can be seen.The face of the rock to the right of the mosaic may be a future target for grinding with the rover's rock abrasion tool.

Image credit: NASA/JPL/Cornell

History Leaves Salts Behind

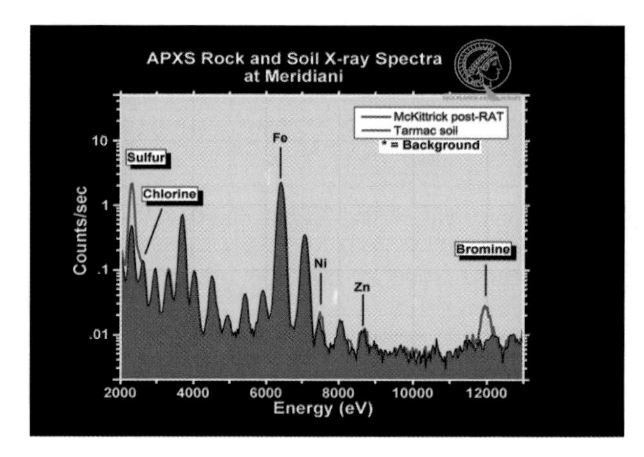

These plots, or spectra, show that a rock dubbed "McKittrick" near the Mars Exploration Rover Opportunity's landing site at Meridiani Planum, Mars, has higher concentrations of sulfur and bromine than a nearby patch of soil nicknamed "Tarmac." These data were taken by Opportunity's alpha particle X-ray spectrometer, which uses curium-244 to assess the elemental composition of rocks and soil. Only portions of the targets' full spectra are shown to highlight the significant differences in elemental concentrations between "McKittrick" and "Tarmac." Intensities are plotted on a logarithmic scale. A nearby rock named Guadalupe similarly has extremely high concentrations of sulfur, but very little bromine. This "element fractionation" typically occurs when a watery brine slowly evaporates and various salt compounds are precipitated in sequence.

Image credit: NASA/JPL/Cornell/Max Planck Institute

"McKittrick" Rich in Sulfur

These plots, or spectra, show that a rock dubbed "McKittrick" near the Mars Exploration Rover Opportunity's landing site at Meridiani Planum, Mars, possesses the highest concentration of sulfur yet observed on Mars. These data were acquired with the rover's alpha particle X-ray spectrometer, which produces a spectrum, or fingerprint, of chemicals in martian rocks and soil. This instrument contains a radioisotope, curium-244, that bombards a designated area with alpha particles and X-rays, causing a cascade of reflective fluorescent X-rays. The energies of these fluorescent X-rays are unique to each atom in the periodic table, allowing scientists to determine a target's elemental composition. The spectra shown here are taken from "McKittrick" and a soil patch nicknamed "Tarmac," both of which are located within the small crater where Opportunity landed. "McKittrick" measurements were acquired after the rover drilled a hole in the rock with its rock abrasion tool. Only portions of the targets' full spectra are displayed. The data are expressed as X-ray intensity (linear scale) versus energy. The measured area is 28 millimeters (1 inch) in diameter. When comparing two spectra, the relative intensities at a given energy are proportional to the elemental concentrations, however these proportionality factors can be complex. To be precise, scientists extensively calibrate the instrument using well-analyzed geochemical standards. Both the alpha particle X-ray spectrometer and the rock abrasion tool are located on the rover's instrument deployment device, or arm.

Image credit: NASA/JPL/Cornell/Max Planck Institute

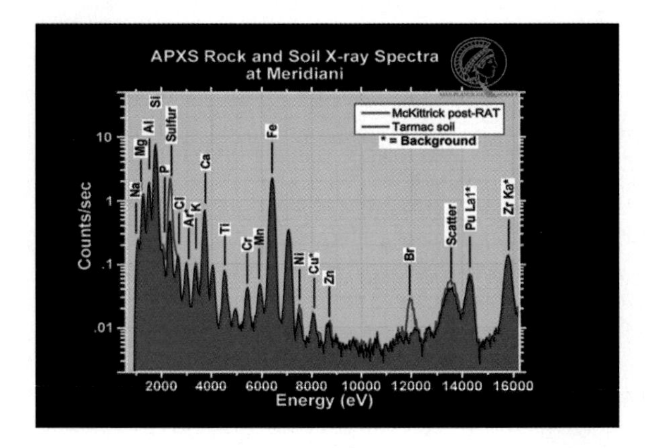

Salty Martian Rock

These plots, or spectra, show that a rock dubbed "McKittrick" near the Mars Exploration Rover Opportunity's landing site at Meridiani Planum, Mars, has higher concentrations of sulfur and bromine than a nearby patch of soil nicknamed "Tarmac." These data were taken by Opportunity's alpha particle X-ray spectrometer, which produces a spectrum, or fingerprint, of chemicals in martian rocks and soil. The instrument contains a radioisotope, curium-244, that bombards a designated area with alpha particles and X-rays, causing a cascade of reflective fluorescent X-rays. The energies of these fluorescent X-rays are unique to each atom in the periodic table, allowing scientists to determine a target's chemical composition.

Both "Tarmac" and "McKittrick" are located within the small crater where Opportunity landed. The full spectra are expressed as X-ray intensity (logarithmic scale) versus energy. When comparing two spectra, the relative intensities at a given energy are proportional to the elemental concentrations, however these proportionality factors can be complex. To be precise, scientists extensively calibrate the instrument using well-analyzed geochemical standards.

Both the alpha particle X-ray spectrometer and the rock abrasion tool are located on the rover's instrument deployment device, or arm.
Image credit: NASA/JPL/Cornell/Max Planck Institute

Mineral Tells Tale of Watery Past

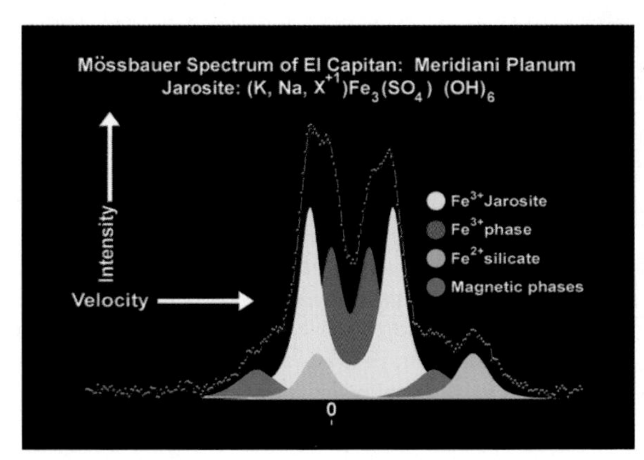

This spectrum, taken by the Mars Exploration Rover Opportunity's Moessbauer spectrometer, shows the presence of an iron-bearing mineral called jarosite in the collection of rocks dubbed "El Capitan." "El Capitan" is located within the rock outcrop that lines the inner edge of the small crater where Opportunity landed. The pair of yellow peaks specifically indicates a jarosite phase, which contains water in the form of hydroxyl as a part of its structure. These data suggest water-driven processes exist on Mars. Three other phases are also identified in this spectrum: a magnetic phase (blue), attributed to an iron-oxide mineral; a silicate phase (green), indicative of minerals containing double-ionized iron (Fe 2+); and a third phase (red) of minerals with triple-ionized iron (Fe 3+).

Image credit: NASA/JPL/University of Mainz

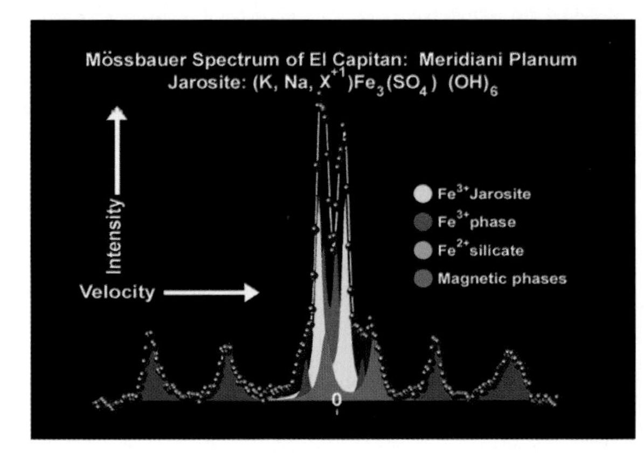

Mineral Tells Tale of Watery Past-2

This spectrum, taken by the Mars Exploration Rover Opportunity's Moessbauer spectrometer, shows the presence of an iron-bearing mineral called jarosite in the collection of rocks dubbed "El Capitan." "El Capitan" is located within the outcrop that lines the inner edge of the small crater where Opportunity landed. The pair of yellow peaks specifically indicates a jarosite phase, which contains water in the form of hydroxyl as a part of its structure. These data suggest water-driven processes exist on Mars. Three other phases are also identified in this spectrum: a magnetic phase (blue), attributed to an iron-oxide mineral; a silicate phase (green), indicative of minerals containing double-ionized iron (Fe 2+); and a third phase (red) of minerals with triple-ionized iron (Fe 3+).

Image credit: NASA/JPL/University of Mainz

"El Capitan" Exposed!

These plots, or spectra, show that the rock collection dubbed "El Capitan" near the Mars Exploration Rover Opportunity's landing site at Meridiani Planum, Mars, consists of three primary mineral groups. These data were taken by the rover's miniature thermal emission spectrometer, which uses infrared detectors to determine the mineral composition of rocks and soil.

The top curve in the graph is the spectrum of an average sulfate mineral. The two curves in the center are the spectrum of "El Capitan" (white) and the best modeled fit (green) to that spectrum. The bottom curve shows a rock composed primarily of silicates and oxides without a sulfate component.

Spectral features centered near the light wavelength of 24 micrometers signify the presence of iron oxides. The broad, bowl-shaped feature between 8 and 12 micrometers represents silicate minerals. The sharp slope from 8 to 9 micrometers shows that "El Capitan" contains a considerable amount of sulfate.

Image credit: NASA/JPL/Cornell/ASU

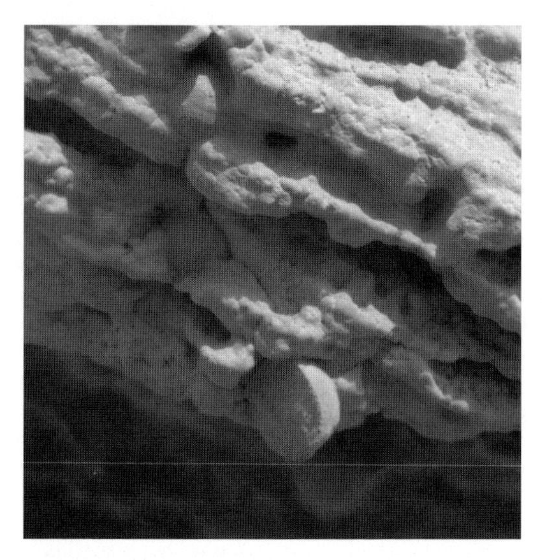

Focused on Robert E

This image, taken by the microscopic imager on the Mars Exploration Rover Opportunity, shows a geological feature dubbed "Robert E." Light from the top is illuminating the feature, which is located within the rock outcrop at Meridiani Planum, Mars. Several images, each showing a different part of "Robert E" in good focus, were merged to produce this view. The area in this image, taken on Sol 15 of the Opportunity mission, is 2.2 centimeters (0.8 inches) across.

Image Credit: NASA/JPL/US Geological Survey

Focus on El Capitan

This image, taken by the microscopic imager on the Mars Exploration Rover Opportunity, shows a geological region of the rock outcrop at Meridiani Planum, Mars dubbed "El Capitan." Light from the top is illuminating the region. Several images, each showing a different part of this region in good focus, were merged to produce this view. The area in this image, taken on Sol 28 of the Opportunity mission, is 1.3 centimeters (half an inch) across.

Image Credit: NASA/JPL/US Geological Survey

Focus on El Capitan-2

This image, taken by the microscopic imager on the Mars Exploration Rover Opportunity, shows a geological region of the rock outcrop at Meridiani Planum, Mars dubbed "El Capitan." Light from the top is illuminating the region. Several images, each showing a different part of this region in good focus, were merged to produce this view. The area in this image, taken on Sol 28 of the Opportunity mission, is 1.5 centimeters (0.6 inches) across.

Image Credit: NASA/JPL/US Geological Survey

Focus on El Capitan-3

This image, taken by the microscopic imager on the Mars Exploration Rover Opportunity, shows a geological region of the rock outcrop at Meridiani Planum, Mars dubbed "El Capitan." The region was in a shadow when the image was acquired. Several images, each showing a different part of this region in good focus, were merged to produce this view. The area in this image, taken on Sol 28 of the Opportunity mission, is approximately 3 centimeters (1.2 inches) across.

Image Credit: NASA/JPL/US Geological Survey

Over Here, Over There

This partial panoramic image from the navigation camera on the Mars Exploration Rover Opportunity shows the lander in the center of the crater at Meridiani Planum, Mars. The image, taken on sol 34 of Opportunity's journey, was not completely downlinked as of sol 35 of the rover's mission. Note the view of the plains outside the crater, the rover tracks in the center and right of the image, and the airbag bounce marks behind the lander.
Image Credit: NASA/JPL

The Texture of El Capitan

This image, taken by the panoramic camera on the Mars Exploration Rover Opportunity, shows a close up of the rock dubbed "El Capitan," located in the rock outcrop at Meridiani Planum, Mars. This image shows fine, parallel lamination in the upper area of the rock, which also contains scattered sphere-shaped objects ranging from 1 to 2 millimeters (.04 to .08 inches) in size. There are also more abundant, scattered vugs, or small cavities, that are shaped like discs. These are about 1 centimeter (0.4 inches) long.

Image Credit: NASA/JPL/Cornell

Focus on Guadalupe

This mosaic image, taken by the microscopic imager on the Mars Exploration Rover Opportunity, shows a portion of the rock outcrop at Meridiani Planum, Mars, dubbed "Guadalupe." Several images, each showing a different part of "Guadalupe" in good focus, were merged to produce this view.

Image Credit: NASA/JPL/US Geological Survey

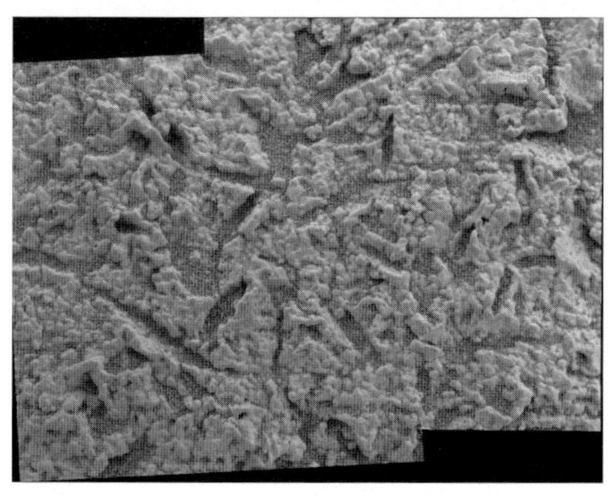

Vugs Provide Clues to Martian Past

This image, taken by the microscopic imager on the Mars Exploration Rover Opportunity, illustrates the shapes of the vugs, or small cavities, located on the region dubbed "El Capitan." The region is part of the rock outcrop at Meridiani Planum, Mars, which the rover is currently examining.

The image provides insight into the nature of the rock matrix -- the rock material surrounding the vugs. Several vugs have disk-like shapes with wide midpoints and tapered ends. This feature is consistent with sulfate minerals that crystallize within a rock matrix, either pushing the matrix grains aside or replacing them. These crystals are then either dissolved in water or eroded by wind activity to produce vugs. The rock matrix here exhibits a granular texture, delicately enhanced through wind abrasion. The primary sediment particles making up this granular layer are relatively uniform in size, ranging up to 1 millimeter (.04 inches). Note that some of these grains are well rounded, which could result from transport of rock fragments in air or water, or precipitation of mineral grains in water. Image credit: NASA/JPL/US Geological Survey

Larger Grains Suggest Presence of Fluid

The image, taken by the microscopic imager on the Mars Exploration Rover Opportunity, shows an extreme close-up of the "El Capitan" region, part of the rock outcrop at Meridiani Planum, Mars. As seen in panoramic images of "El Capitan," this region appears laminated, or composed of layers of firmly united material. The upper left portion of this image shows how the grains of the region might be arranged in planes to create such lamination.

At the upper right, in the zone surrounding two larger sphere-shaped particles, this image also shows another apparent characteristic at the scale of individual grains. The granularity of the matrix -- the rock in which the spherules are embedded -- is modified near the spherules compared with grains farther from the spherules. Around the upper spherule, the grain size is increased. This change in grain size might represent a "reaction rim," a feature produced by fluid interaction with the matrix material adjacent to the spherule during the growth of the spherule.

Image credit: NASA/JPL/US Geological Survey

Which Came First? Vug or Spherule?

This image, taken by the microscopic imager onboard the Mars Exploration Rover Opportunity, shows a close-up of the region dubbed "El Capitan," which lies within the rock outcrop at Meridiani Planum, Mars. In the lower left, a spherule, or sphere-shaped grain, can be seen penetrating the interior of a small cavity called a vug. This "cross-cutting" relationship allows the relative timing of separate events to be established. In this case, the spherule appears to "invade" the vug, and therefore likely post-dates the vug. This suggests that the spherules may have been one of the last features to form within the outcrop.

Image credit: NASA/JPL/Cornell/US Geological Survey

Ripples in Rocks Point to Water

This image taken by the Mars Exploration Rover Opportunity's panoramic camera shows the rock nicknamed "Last Chance," which lies within the outcrop near the rover's landing site at Meridiani Planum, Mars. The image provides evidence for a geologic feature known as ripple cross-stratification. At the base of the rock, layers can be seen dipping downward to the right. The bedding that contains these dipping layers is only one to two centimeters (.4 to .8 inches) thick. In the upper right corner of the rock, layers also dip to the right, but exhibit a weak "concave-up" geometry. These two features -- the thin, cross-stratified bedding combined with the possible concave geometry -- suggest small ripples with sinuous crest lines. Although wind can produce ripples, they rarely have sinuous crest lines and never form steep, dipping layers at this small scale. The most probable explanation for these ripples is that they were formed in the presence of moving water.
Image credit: NASA/JPL/Cornell

Sulfur and Chlorine at Meridiani
Cl
SO_3
Relative Amount
6.0
5.0
4.0
3.0
2.0
1.0
0.0
First Soil
McKittrick pre-RAT
McKittrick after RAT
Guadalupe after RAT

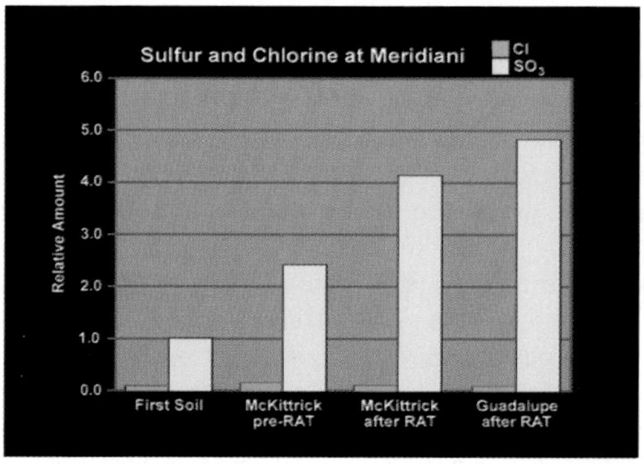

A Trail of Salts

This graph shows the relative abundances of sulfur (in the form of sulfur tri-oxide) and chlorine at three Meridiani Planum sites: soil measured in the small crater where Opportunity landed; the rock dubbed "McKittrick" in the outcrop lining the inner edge of the crater; and the rock nicknamed "Guadalupe," also in the outcrop. The "McKittrick" data shown here were taken both before and after the rover finished grinding the rock with its rock abrasion tool to expose fresh rock underneath. The "Guadalupe" data were taken after the rover grounded the rock. After grinding both rocks, the sulfur abundance rose to high levels, nearly five times higher than that of the soil. This very high sulfur concentration reflects the heavy presence of sulfate salts (approximately 30 percent by weight) in the rocks. Chloride and bromide salts are also indicated. Such high levels of salts strongly suggest the rocks contain evaporite deposits, which form when water evaporates or ice sublimes into the atmosphere.

Image credit: NASA/JPL/Cornell/Max Planck Institute

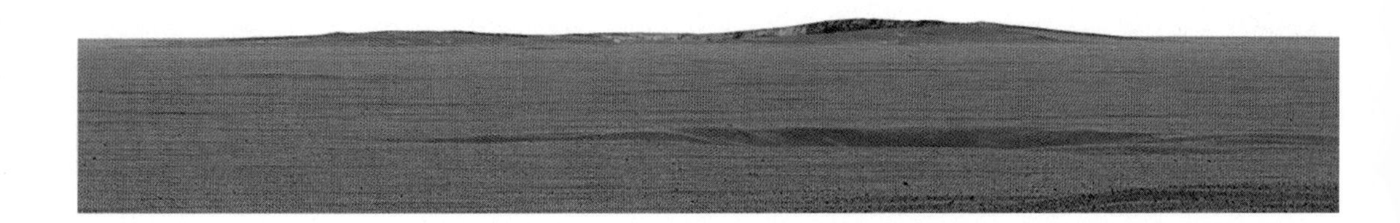

Opportunity Spies "Endurance" on the Horizon

This image taken by the Mars Exploration Rover Opportunity's panoramic camera shows the eastern plains that stretch beyond the small crater where the rover landed. In the distance, the rim of a larger crater dubbed "Endurance" can be seen.

This color mosaic was taken on the 32nd martian day, or sol, of the rover's mission and spans 20 degrees of the horizon. It was taken while Opportunity was parked at the north end of the outcrop, in front of the rock region dubbed "El Capitan" and facing east.

The features seen at the horizon are the near and far rims of "Endurance," the largest crater within about 6 kilometers (4 miles) of the lander. Using orbital data from the Mars Orbiter Camera on NASA's Mars Global Surveyor spacecraft, scientists estimated the crater to be 160 meters (175 yards) in diameter, and about 720 meters (half a mile) away from the lander.

The highest point visible on "Endurance" is the highest point on the far wall of the crater; the sun is illuminating the inside of the far wall.

Between the location where the image was taken at "El Capitan" and "Endurance" are the flat, smooth Meridiani plains, which scientists believe are blanketed in the iron-bearing mineral called hematite. The dark horizontal feature near the bottom of the picture is a small, five-meter (16-feet) crater, only 50 meters (164 feet) from Opportunity's present position.

When the rover leaves the crater some 2 to 3 weeks from now, "Endurance" is one of several potential destinations.

Image credit: NASA/JPL/Cornell/MSSS

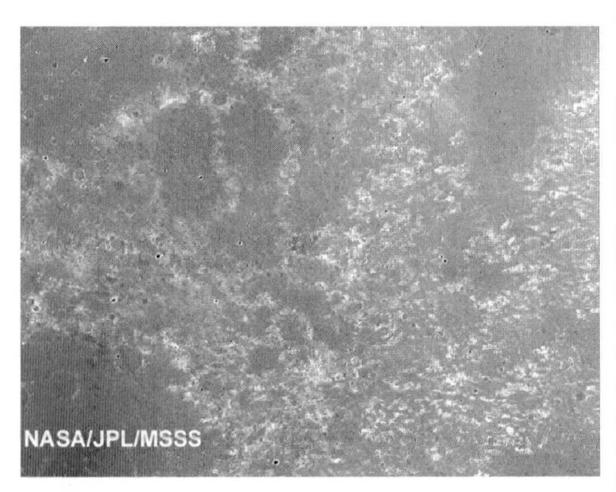

Meridiani Plains

This is a portion of a previously released image (PIA02397) taken by the Mars Orbiter Camera onboard NASA's Mars Global Surveyor, showing the dark, relatively smooth plains of Meridiani Planum, where the Mars Exploration Rover Opportunity landed. The larger circular features in the upper three-quarters of the image are thought to be the locations of buried craters formed by meteorite impacts. The cluster of smaller circular features in the bottom quarter of the scene represent a field of craters formed either by simultaneous impact of many meteorites, or impact of material thrown from a much, much larger nearby crater as it formed. The dark material covering these plains includes an abundance of the iron oxide mineral, hematite, that was detected by the Mars Global Surveyor thermal emission spectrometer. The scene is located near 2.2 degrees south, 3.7 degrees west and was acquired on August 19, 1999.

Image credit: NASA/JPL/MSSS

If You Thought That Was a Close View of Mars, Just Wait

September 23, 2003

Artist's concept of Mars Reconnaissance Orbiter.

As Earth pulls away from Mars after last month's close approach, NASA is developing a spacecraft that will take advantage of the next close encounter in 2005.

That spacecraft, Mars Reconnaissance Orbiter, will make a more comprehensive inspection of our planetary neighbor than any previous mission.

For starters, it will examine landscape details as small as a coffee table with the most powerful telescopic camera ever sent to orbit a foreign planet. Some of its other tools will scan underground layers for water and ice, identify small patches of surface minerals to determine their composition and origins, track changes in atmospheric water and dust, and check global weather every day.

"We're reaching an important stage in developing the spacecraft," said James Graf, project manager for Mars Reconnaissance Orbiter at NASA's Jet Propulsion Laboratory, Pasadena, Calif. "The primary structure will be completed next month." The structure weighs 220 kilograms (484 pounds) and stands 3 meters (10 feet) tall. At launch, after gear and fuel are added, it will support over 2 tons.

Also next month, the mission's avionics test bed will be assembled for the first time and put to use for testing of flight software.

Workers at Lockheed Martin Space Systems, Denver, have already assembled the spacecraft structure and will later add instruments being built for it at the University of Arizona, Tucson; at Johns Hopkins University Applied Physics Laboratory, Laurel, Md.; at the Italian Space Agency, Rome; at Malin Space Science Systems, San Diego, Calif.; and at JPL.

"In several ways, Mars Reconnaissance Orbiter will advance NASA's follow-the-water strategy for Mars exploration," said Dr. Richard Zurek, project scientist for the mission.

Current surveys of Mars' surface composition have found less evidence of water-related minerals than many scientists anticipated after earlier discoveries of plentiful channels that were apparently carved by water flows in the planet's past. A spectrometer on the Reconnaissance Orbiter is designed to identify some different types of water-related minerals and to see smaller-scale deposits. "Instead of looking for something as big as the Bonneville Salt Flats, we can look for something on the scale of a Yellowstone hot spring," Zurek said.

Probing below Mars' surface with penetrating radar, Reconnaissance Orbiter will check whether the frozen water that NASA's Mars Odyssey spacecraft detected in the top meter or two (yard or two) of soil extends deeper, perhaps as accessible reservoirs of melted water.

Above the surface, an atmosphere-scanning instrument will monitor changes in water vapor at different altitudes and might even locate plumes where water vapor is entering the atmosphere from underground vents, if that's happening on Mars.

Mars Reconnaissance Orbiter will stream home its pictures and other information using the widest dish antenna and highest power level ever operated at Mars. "The amount of data flowing back to Earth from Mars will be a giant leap over previous missions. It's like upgrading from a dial-up modem for your computer to a high-speed DSL connection," Graf said.

The Mars Reconnaissance Orbiter will lay the groundwork for later Mars surface missions in NASA's plans: a lander called Phoenix selected last month in a competition for a 2007 launch opportunity, and a highly capable rover called Mars Science Laboratory being developed for a 2009 launch opportunity. The orbiter's high-resolution instruments will help planners evaluate possible landing sites for these missions both in terms of science potential for further discoveries and in terms of landing risks. The orbiter's communications capabilities will provide a critical transmission relay for the surface missions.

Advantageous opportunities to launch Mars missions come in a rhythm of about every 26 months, shortly before each time Earth overtakes Mars in the two planets' concentric tracks around the Sun. NASA's two Mars Exploration Rovers and the European Space Agency's Mars Express mission were launched during the three months preceding Earth's most recent passing of Mars on Aug. 27. The Mars Reconnaissance Orbiter team has its work cut out for it to have the spacecraft ready for launch on Aug. 10, 2005, which is about 10 weeks before the next close approach.

JPL, a division of the California Institute of Technology in Pasadena, manages the Mars Reconnaissance Orbiter project for NASA's Office of Space Science, Washington, D.C.
Contact: Guy Webster (818) 354-6278 Jet Propulsion Laboratory, Pasadena, Calif.

This image taken by the Mars Exploration Rover Spirit's navigation camera shows the rocky path lying due east of the rover. Boulders as large as half a meter (1.6 feet) dot the landscape here near Bonneville Crater. The east hills, over two kilometers away (1.3 miles), can be seen to the far right. Spirit will most likely drive toward the rim of Bonneville crater along a safer route to the north of this area.

Image credit: NASA/JPL (March 8, 2004)

Deimos eclipse
March 4, 2004
Phobos eclipse
March 7, 2004
Pancam Images of the Sun and Martian
Moons from the Opportunity Rover

#	Title	ISBN	Bonus	US$	UK£	CN$
1	Apollo 8	1-896522-66-1	CDROM	$18.95	£13.95	$25.95
2	Apollo 9	1-896522-51-3	CDROM	$16.95	£12.95	$22.95
3	Friendship 7	1-896522-60-2	CDROM	$18.95	£13.95	$25.95
4	Apollo 10	1-896522-52-1	CDROM	$18.95	£13.95	$25.95
5	Apollo 11 Vol 1	1-896522-53-X	CDROM	$18.95	£13.95	$25.95
6	Apollo 11 Vol 2	1-896522-49-1	CDROM	$15.95	£10.95	$20.95
7	Apollo 12	1-896522-54-8	CDROM	$18.95	£13.95	$25.95
8	Gemini 6	1-896522-61-0	CDROM	$18.95	£13.95	$25.95
9	Apollo 13	1-896522-55-6	CDROM	$18.95	£13.95	$25.95
10	Mars	1-896522-62-9	CDROM	$23.95	£18.95	$31.95
11	Apollo 7	1-896522-64-5	CDROM	$18.95	£13.95	$25.95
12	High Frontier	1-896522-67-X	CDROM	$21.95	£17.95	$28.95
13	X-15	1-896522-65-3	CDROM	$23.95	£18.95	$31.95
14	Apollo 14	1-896522-56-4	CDROM	$18.95	£15.95	$25.95
15	Freedom 7	1-896522-80-7	CDROM	$18.95	£15.95	$25.95
16	Space Shuttle STS 1-5	1-896522-69-6	CDROM	$23.95	£18.95	$31.95
17	Rocket Corp. Energia	1-896522-81-5		$21.95	£16.95	$28.95
18	Apollo 15 - Vol 1	1-896522-57-2	CDROM	$19.95	£15.95	$27.95
19	Arrows To The Moon	1-896522-83-1		$21.95	£17.95	$28.95
20	The Unbroken Chain	1-896522-84-X	CDROM	$29.95	£24.95	$39.95
21	Gemini 7	1-896522-80-7	CDROM	$19.95	£15.95	$26.95
22	Apollo 11 Vol 3	1-896522-85-8	DVD*	$27.95	£19.95	$37.95
23	Apollo 16 Vol 1	1-896522-58-0	CDROM	$19.95	£15.95	$27.95
24	Creating Space	1-896522-86-6		$30.95	£24.95	$39.95
25	Women Astronauts	1-896522-87-4	CDROM	$23.95	£18.95	$31.95
26	On To Mars	1-896522-90-4	CDROM	$21.95	£16.95	$29.95
27	Conquest of Space	1-896522-92-0		$23.95	£19.95	$32.95
28	Lost Spacecraft	1-896522-88-2		$30.95	£24.95	$39.95
29	Apollo 17 Vol 1	1-896522-59-9	CDROM	$19.95	£15.95	$27.95
30	Virtual Apollo	1-896522-94-7		$19.95	£14.95	$26.95
31	Apollo EECOM	1-896522-96-3		$29.95	£23.95	$37.95
32	Visions of Future Space	1-896522-93-9	CDROM	$27.95	£21.95	$35.95
33	Space Trivia	1-896522-98-X		$19.95	£14.95	$26.95
34	Interstellar Spacecraft	1-896522-99-8		$24.95	£18.95	$30.95
35	Dyna-Soar	1-896522-95-5	DVD*	$32.95	£23.95	$42.95
36	The Rocket Team	1-894959-00-0	DVD*	$34.95	£24.95	$44.95
37	Sigma 7	1-894959-01-9	CDROM	$19.95	£15.95	$27.95
38	Women Of Space	1-894959-03-5	CDROM	$26.95	£17.95	$30.95
39	Columbia Accident Rpt	1-894959-06-X	CDROM	$25.95	£19.95	$33.95
40	Gemini 12	1-894959-04-3	CDROM	$19.95	£15.95	$27.95
41	The Simple Universe	1-894959-11-6		$17.95	£17.95	$29.95
42	New Moon Rising	1-894959-12-4		$TBA	£TBA	$TBA
43	Moonrush	1-894959-10-8		$TBA	£TBA	$TBA
44	Mars Volume 2	1-894959-05-1	DVD*	$TBA	£TBA	$TBA
45	Rocket Science	1-894959-09-4		$TBA	£TBA	$TBA
46	How NASA Learned	1-894959-07-8		$TBA	£TBA	$TBA
47	Virtual LM	1-894959-14-0		$TBA	£TBA	$TBA
48	Deep Space	1-894959-15-9	DVD*	$TBA	£TBA	$TBA

CG Publishing Inc home of **Apogee Books**
P.O Box 62034 Burlington,
Ontario L7R 4K2, Canada
TEL. 1 905 637 5737 FAX 1 905 637 2631
e-mail marketing@cgpublishing.com

* NTSC Region 0

Many more to come! Check our website for new titles.

www.apogeebooks.com

Collect them all!

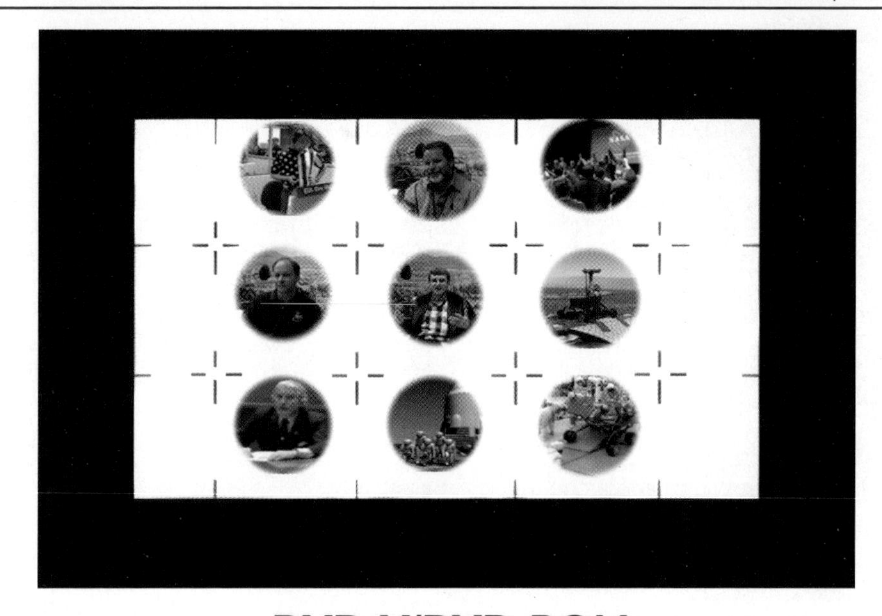

DVD-V/DVD-ROM
Side One of the attached disc will play on a home DVD-Video player*
To view Side Two requires a machine with DVD-ROM capability.

DVD video includes:

The Landing of Opportunity
JPL Opportunity Press Conference
MER Animation**
M2K4 Animation
MER Launch and Preflight Footage

Exclusive Interviews with:

MER Athena Science Payload Principal Investigator Steve Squyres,
MER EDL Manager Rob Manning
Opportunity Mission Manager Jim Erickson.

Also includes:

An extremely rare lecture by Dr Wernher von Braun from 1975 discussing missions to Mars.

DVD-ROM includes complete NASA special publications about Mars:

SP-179	The Book of Mars
SP2001-4521	Humans to Mars
SP-425	The Martian Landscape
CR-73159	Study of Life Support Systems for Space Missions Exceeding 1 year in Duration

PLUS!

Mars Exploration Rover Imagery up to Spirit Sol 47 & Opportunity Sol 29
Odyssey THEMIS Imagery
JPL Opportunity Meridiani Planum Water Press Conference March 2nd 2004
Canadian Space Agency Director Marc Garneau speech on Canadian Mars plans.

* NTSC Region 0
** Courtesy MAAS Digital/Cornell